Space Debris – Models and Risk Analysis

Heiner Klinkrad

Space Debris

Models and Risk Analysis

Springer

Published in association with
Praxis Publishing
Chichester, UK

PRAXIS

Dr. Heiner Klinkrad
European Space Agency
Darmstadt
Germany

Cover image shows a computer-generated snapshot of the observable population of space objects in 1997, based on orbit information provided by NASA.

SPRINGER–PRAXIS BOOKS IN ASTRONAUTICAL ENGINEERING
SUBJECT *ADVISORY EDITOR*: John Mason, M.Sc., B.Sc., Ph.D.

ISBN 3-642-42623-9 Springer-Verlag Berlin Heidelberg New York

Springer is part of Springer-Science + Business Media (springeronline.com)

Bibliographic information published by Die Deutsche Bibliothek

Die Deutsche Bibliothek lists this publication in the Deutsche Nationalbibliografie; detailed bibliographic data are available from the Internet at http://dnb.ddb.de

Softcover re-print-of the Hardcover 1st edition 2006

Cover design: Jim Wilkie
Completed in LaTeX: Heiner Klinkrad and Clare Martin
Cover image: © European Space Agency

Table of Contents

Prolog

Most colleagues who enter the solemn halls of space flight use the main entrance, following the footsteps of generations of famous predecessors in their endeavor to conceive and realize new, ground-breaking space missions. Since I am dealing with residues of space flight, sometimes disrespectfully denoted as space junk, I approached these solemn halls from the alley way, through the back door. This was in 1978, when I got involved in the re-entry assessment of the reactor-equipped Cosmos 954 satellite after it spread its radioactive debris across Canada. At that time almost 6,000 trackable objects were on orbit, with more than 90% thereof space debris. Space debris are useless space objects including spent orbital stages, de-commissioned satellites, items released during nominal missions, and fragments from deliberate and unintentional on-orbit explosions, but also exotic relicts of space activities, such as protective gloves, or a screwdriver lost by an astronaut. The space debris population has steadily increased since the early 1960s. By the year 2005 more than 14,000 space objects of diameters larger than 5 to 10 cm could be tracked, of which no more than 5% were operational spacecraft. With these numbers in mind, and with 20 years of involvement in the field, I felt that it was time to write a book on space debris, which today represent the overwhelming majority of man-made objects in Earth orbits.

Several years elapsed between the first ideas for this book and its completion. Not just the space debris environment itself, but also the related research activities undergo a highly dynamic evolution. As a consequence, a book on space debris runs the risk of becoming inconsistent during the write-up process, and obsolete by the time of its publication. My intention was to minimize this risk by providing a consistent snapshot of the space debris situation around the year 2002, and by developing and documenting a theoretical basis with a hopefully better durability and longevity that the environment itself.

This book was written by a single author. In many chapters, however, the research contributions of several colleagues deserved more than a citation or acknowledgment. In such cases, they have been listed as contributing authors in the chapter headings.

This is not the first book dealing with space debris, nor will it be the last one. There could have been many different ways of selecting the material, of presenting it, and of citing relevant work. What you see here is a very personal, probably unintentionally biased presentation and view of the subject, based on information which I collected during 25 years of work for the European Space Agency, with 20 years thereof at the European Space Operations Center in Darmstadt, Germany. It was this environment, and the professional and personal links and information exchange with ESA colleagues, and with colleagues of the European and international space debris community, which made this book possible. Above all, I would like to acknowledge the role of Walter Flury, who has led ESA's space debris activities since the mid 1980s. His dedication and enthusiasm helped to advance space debris research in ESA and within Europe, and it helped to develop a European space debris expertise which today has reached an internationally recognized level.

The content of the book is largely focused on European activities, in particular on those which were performed under the lead, or with the participation of ESA. The views expressed in this text, however, are those of the author, and they may not necessarily reflect those of ESA. Inconsistent or erroneous information which may be contained in this text are the sole responsibility of the author.

The topics which are addressed in this book follow the outline of a graduate course on *Space Debris* which I have given in a curriculum for aeronautical engineers since 2001. I have tried to keep the volume and technical detail of the presented material at a level which allows this text book to be used for teaching purposes at graduate level. Worked problems, which would have been helpful for teaching applications, are not yet contained in this text. They are, however, foreseen for a later release.

I would like to thank the publishing staff at Praxis and Springer for their endurance, and for their constructive comments and guidance during the preparation of the manuscript, and during the final typesetting of this book. The typesetting was done in LaTeX, by means of an expert-designed LaTeX class-file which was provided by Clare Martin, one of the contributing authors. Apart from the contributing authors I would also like to acknowledge support, contributions, or editorial comments from Ludger Leushacke, Thomas Schildknecht, Frank Schäfer, Michael Oswald, Sebastian Stabroth, Gerhard Drolshagen, Michel Lambert, Markus Landgraf, Cristina Hernández, Íñigo Mascaraque, Santiago Llorente, Dorothea Danesy, and Tom Mohr. There are many more individuals to whom I am grateful, and who may not recognize themselves in the previous acknowledgments. My apologies to all of you.

Finally, and most of all, I would like to thank my wife Angelika for her patience and for her continuous support.

Heiner Klinkrad

Seeheim-Jugenheim, July 2005

1

Introduction

H. Klinkrad

We seem to exist in a hazardous time, driftn' along here through space;
nobody knows just when we begun, or how far we've gone in the race.

Benjamin Franklin

1.1 HISTORY OF THE EVOLVING SPACE DEBRIS RESEARCH

On June 29, 1961, the US Transit-4A satellite was launched from Kennedy Space Center on a Thor-Ablestar rocket. The spacecraft was deployed into an orbit altitude between 881 km and 998 km, with an orbit inclination of 66.8°. Transit-4A was cataloged by the First Aerospace Control Squadron of the US Air Force as the 116th space object since the launch of Sputnik-1 on October 4, 1957. At 06:08:10 UTC on June 29, 77 minutes after the injection and separation of Transit-4A and two additional payloads, the Ablestar upper stage exploded, distributing its dry mass of 625 kg across at least 298 trackable fragments, of which nearly 200 were still on orbit 40 years later. This first on-orbit break-up event in space history increased the observable number of man-made space objects instantaneously by at least a factor 3.5 (Portree and Loftus, 1993). Since this date space debris has been the largest contributor to the observable space object population, with on-orbit explosions as its largest single source.

The terms *space debris* and *orbital debris* are often used as synonyms, with the following definition, as adopted by the Inter-Agency Space Debris Coordination Committee (IADC): "Space debris are all man made objects including fragments and elements thereof, in Earth orbit or re-entering the atmosphere, that are non functional" (anon., 2002a). According to this definition, 46.5% of the cataloged space objects prior the Ablestar explosion had already been space debris, for instance in the form of spent upper stages and mission-related release objects (not counting decommissioned satellites).

With increasing launch and deployment activities, the space debris environment also started to take shape. In August 1964 the first geostationary satellite, Syncom-3, was deployed. Since then, more than 800 objects have been placed in or close to the geostationary orbit (GEO). Fourteen years after the Syncom-3 launch, in June 1978, the first spacecraft explosion occurred in GEO. In 1979 Luboš Perek presented a paper on "Outer Space Activities *versus* Outer Space", which was the first to recommend space debris mitigation measures, including the re-orbiting of GEO spacecraft into a disposal orbit at the end-of-life (Portree and Loftus, 1993). All his recommended measures are still applicable today.

The low-Earth orbit regime (LEO), kept being the main proving ground for new space technologies. It experienced the first intentional explosion in October 1965, when the Cosmos-50 reconnaissance satellite was blown up after a mission-critical failure. In October 1968 Cosmos-249 was used as the first anti-satellite weapon (ASAT), destroying the Cosmos-248 target in a commanded explosion during a rendezvous operation. The growing space debris population in the 1970s was suspected by some analysts to have its origin in a series of such ASAT tests. However, novel analysis techniques devised by John Gabbard, and applied to a subset of the NORAD[1.1] catalog by Donald Kessler, revealed that explosions of nine Delta second stages between May 1975 and January 1981 were the main contributors. They alone accounted for 27% of the LEO catalog by 1981 (Portree and Loftus, 1993). Once the cause of these explosions had been identified by the stage manufacturer, remedial action was taken, and no further break-ups of Delta second stages were recorded thereafter. This can be regarded as one of the first effectively implemented space debris mitigation measures.

In 1977 Donald Kessler and Burton Cour-Palais predicted that man-made orbital debris would soon pose a higher collision risk in the LEO regime than natural meteoroids. One year later, in 1978, the same authors published a paper on "Collision Frequency of Artificial Satellites: The Creation of a Debris Belt". They postulated that within a few decades on-orbit collisions could become the primary source of new space debris (Portree and Loftus, 1993). In 1990 Donald Kessler investigated the problem further in a publication on "Collisional Cascading: The Limits of Population Growth in Low-Earth Orbits", which describes the consequences of a self-sustained growth of the space debris population, initially triggered by collisions between intact objects and ultimately sustained by collisions between collision fragments (this was later known as the "Kessler Syndrome"). Such a cascading process, which cannot be stopped in its advanced stage, could render certain altitudes shells in the LEO region unsafe for a long time. Seven years after this publication, in July 1996, the first accidental collision between two cataloged objects was recorded, when the Cerise satellite was damaged by a fragment of an Ariane orbital stage which exploded in November 1986.

Following the Salyut-1 precursor mission in 1971, which paved the way for a

[1.1]North American Aerospace Defense Command

permanent presence of humans in space, there was at least one space habitat on orbit after 1973, when Salyut-2 and Skylab were launched. These early space stations were manned for limited time periods of up to several months. After 1989, when Salyut-7 and Mir were operational, there was an uninterrupted presence of humans in space. After the launch of Columbia in 1981, the US Space Shuttle program also contributed to this record, although with limited residence times per mission of less than 17 days. In late 1998 the first elements of the International Space Station (ISS) were launched and integrated in orbit. Since the end of 2001 the ISS has been permanently manned. Space debris has played an important role both in the design and operation of the ISS, and of the Space Shuttle. The ISS manned modules are protected by debris and meteoroid protection shields which can defeat impactors of up to 1 cm size. The shields are based on a multi-layer design principle proposed by Fred Whipple in 1947. As of the early 1980s such shields were optimized in laboratory impact tests with high-performance accelerators which can propel objects of 1 cm size to orbital speeds. The energy release of such impactors is equivalent to an exploding hand grenade.

After the Challenger accident in January 1986 a flight rule was established for Space Shuttle operations, which outlines a procedure for collision avoidance of trackable space objects. A similar procedure is in place today for the ISS. On the average, the Shuttle performs one avoidance maneuver for every 3 months on orbit, while the ISS performs one avoidance maneuver per year.

The adopted definition of *space debris* also includes re-entry objects, which are captured by the Earth atmosphere. In fact about 66% of all cataloged objects in space history have decayed, with most of these burning up due to aerothermal heating. Some objects, however, can pose a risk either by surviving to ground impact, or by releasing dangerous substances into the atmosphere. The first risk object re-entry was of the latter kind. In April 1964, following a launcher failure, the Transit 5BN-3 satellite re-entered above the Indian Ocean, dispersing 1 kg of plutonium from its SNAP-9A radio-thermal generator into the atmosphere. In January 1978, three months after its launch, the reactor-equipped Cosmos-954 satellite re-entered uncontrolled over Canada. It spread its reactor fragments with 30 kg of radioactive uranium along a swath of $\sim$1,000 km length. This incident resulted in the first application of the 1972 UN Liability Convention, with Canada claiming six million dollars compensation. The event also triggered the preparation of "UN Principles Relevant to the Use of Nuclear Power Sources in Space" (anon., 2002b). In February 1983 Cosmos-1402, also equipped with a reactor, re-entered over the South Atlantic, with no evident environmental consequences. Since then, there has been no re-entry of a risk object with nuclear material on board. There were, however, re-entry events which posed a risk due to their large mass, and due to the resulting risk from surviving debris to the population on ground. This was for instance the case for Skylab, with a mass of 74 tons, which re-entered over the Indian Ocean and Australia in July 1979, and for Salyut-7, with a mass of 40 tons, which re-entered over South America in February 1991.

This short historic overview of the evolving field of space debris research has touched upon problems and identified solutions for a wide range of technical disciplines, and it has also highlighted fine examples of collecting scientific evidence for suspected debris sources, and their long-term effect on the debris environment. It was just a handful of dedicated scientists and engineers who realized in the 1970s that there may be a space debris problem. These few individuals propagated the topic against all odds to today's level of recognition in the scientific world. The author hopes that the spotlight on their debris-related work triggered the reader's interest to learn more about the theoretical background of space debris.

1.2 GUIDANCE FOR READING THIS BOOK

The following text is organized in such a way that individual chapters are largely self-standing, with chapter-wise bibliographies and numbering conventions for equations, figures, tables, and footnotes. Items which are common to the whole text or to several chapters are compiled in annexes, and in lists of symbols, abbreviations, tables, and figures. Finally, there is a list of bibliographic references covering the entire text, followed by an index.

A novice in the field of space debris is recommended to read Chapters 2 to 6 in sequence. They explain how the present space debris environment can be characterized, based on measurement data, how it can be modeled, based on different sources, how its future can be forecast, based on assumed traffic models, and how it can be positively influenced, based on recommended mitigation measures. Here and there it can be helpful to side-step to the Annexes A to C for some elementary orbit mechanics, and for some background information on the Earth atmosphere and on the geopotential. Chapters 7 to 9 are ordered with some logic, but they can also be studied individually, in arbitrary sequence. They describe special aspects of risk assessment and risk prevention from the viewpoint of on-orbit shielding, collision avoidance, and re-entry risk management. Annex D, with model data of the Earth population, should be referenced for the latter issue. Chapter 10 concludes the technical part of this text. It gives a short overview of the natural meteoroid and meteorite collision flux and its associated risks for spacecraft and for the population on ground. Finally, in Chapter 11 an overview is provided of the relevance of space debris research and associated policy and standardization issues in an international context.

1.3 REFERENCES

anon. (2002a). *IADC Space Debris Mitigation Guidelines.* issue 1, rev. 1.

anon. (2002b). *United Nations Treaties and Principles on Outer Space.* United Nations publications, Vienna, Austria.

Portree, D. and Loftus, J. (1993). Orbital Debris and Near-Earth Environmental Management: A Chronology. Technical Report 1320, NASA Reference Publication.

2

The Current Space Debris Environment and its Sources

H. Klinkrad

2.1 LAUNCH HISTORY AND THE RESULTING ORBITAL ENVIRONMENT

The space environment of man-made intact and debris objects is a mirror image of almost half a century of space activities following the launch of Sputnik-1 on October 4, 1957. The objects which have since been released into space are the result of launch activities, with the deployment of payloads, upper stages which injected them into orbit, and associated mission-related objects, such as launch adapters, lens covers, clamp-bands, and yo-yo de-spin devices. Further items were released unintentionally, such as screwdrivers or protective gloves during extra-vehicular activities of astronauts, slag particles produced during solid rocket motor burns, cooling liquids released from Russian reconnaissance satellites, or degradation products resulting from crack formations and small particle impacts on the coatings of satellites and upper stage surfaces. The most important single source of space objects, however, is on-orbit explosions of spacecraft and rocket stages, sometimes more than 20 years after their launch.

In order to understand the environment produced by man-made objects on Earth orbits, it is helpful to establish a proven starting point by first looking at consolidated, quasi-deterministic information on large-size objects, which can be observed and characterized in terms of orbit, origin (association with a launch event), and object properties. The most comprehensive data set for this category of objects is the NASA Satellite Situation Report (SSR), together with the so-called Two-Line Element (TLE) catalog of the US Strategic Command (USSTRATCOM, formerly US Space Command USSPACECOM). Both of these are based on observation data and orbit determinations from the US Space Surveillance Network (SSN). Due to limitations in the sensitivities of the SSN radars and telescopes, the lower size threshold of objects which can be observed and correlated with the catalog is on the order of 10 cm in the low-Earth orbit (LEO) regime, and on the order

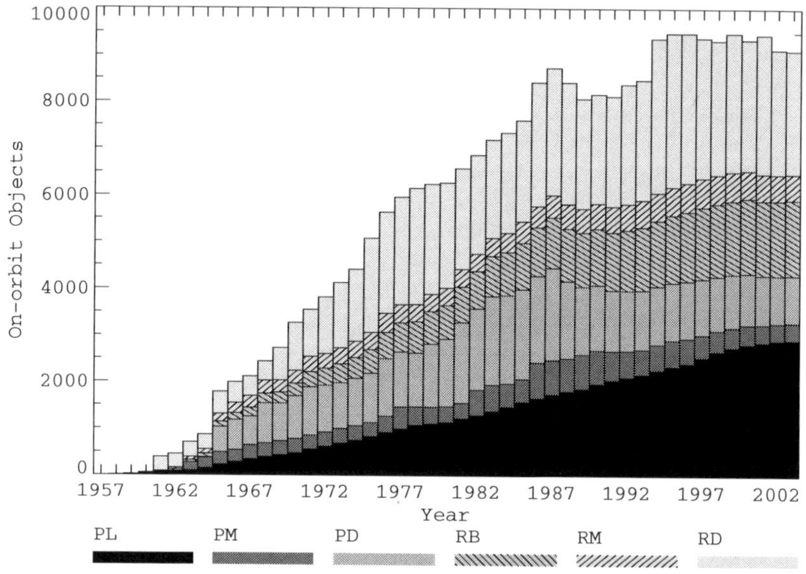

Fig. 2.1. Historic evolution of the number of trackable on-orbit catalog objects and their share between different source categories (PL = payloads, RB = rocket bodies, PM = PL mission-related objects, PD = PL debris, RM = RB mission-related objects, RD = RB debris).

of 1 m in the geostationary ring (GEO, see Fig. 6.16).

By January 2002 a total of 4,191 launches since 1957 had deployed 17,050 payloads, rocket bodies, and mission-related objects, which caused 27,044 detectable and trackable objects on Earth orbits. Of these 27,044 catalog objects, 18,051 had decayed into the atmosphere, leaving an on-orbit catalog population of 8,993. Fig. 2.1 shows the time evolution of the on-orbit catalog population according to source categories. As of 1962, a near-linear increase can be observed, at a rate of about 260 on-orbit objects per year. The total number of cataloged objects in the same time frame increased at a rate of about 710 per year. The linear rates are modulated with periods of 11 years, as a consequence of the solar cycle and its effect on air densities, which drive the orbit decay rates. With the end of the USSR around 1990/1991, the so far dominant launch nation, with peak contributions of about 80% to the overall launch rate, reduced its space activities significantly. The almost steady state annual launch rate of 110 $\pm$ 10 between 1965 and 1990 has dropped significantly since, to almost 50% of its original level. Since the years 2001/2002 it has seemed to settle at 60 launches per year. This decline in launch activities, a concurrent peak in solar activity, and a parallel reduction of on-orbit explosion rates due to post-mission passivation measures has resulted in an almost constant on-orbit catalog population close to 9,000 since 1994. Fig. 2.2 shows the historic evolution of the launch activities, and their share between major operators. Table 2.1 lists in more detail the originators of all 8,993 cataloged on-orbit objects for

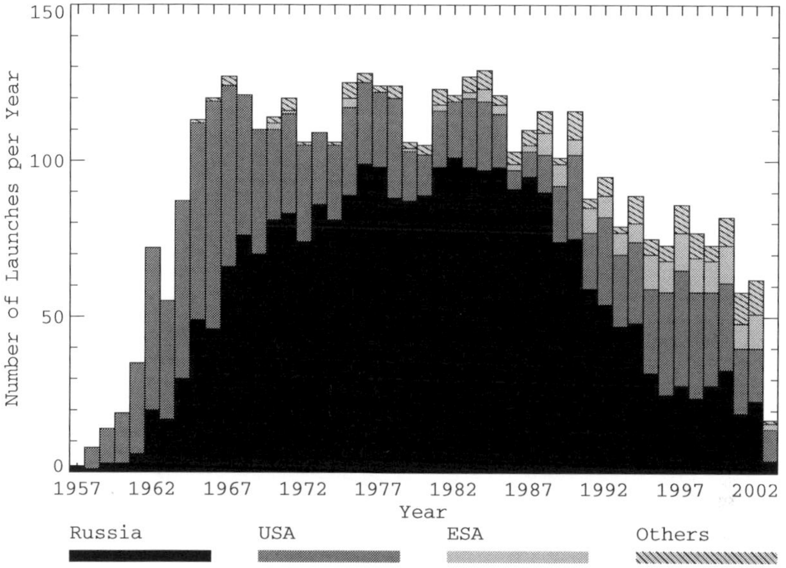

Fig. 2.2. Historic evolution of annual launch rates and their share between major operators (annual data for 2003 are incomplete).

January 1, 2002. When classified by object categories, 31.8% of these were payloads (~6% thereof active satellites), 17.6% were spent rocket upper stages and boost motors, 10.5% were mission-related objects, and the remainder of ~39.9% were debris, mainly from fragmentation events (28.4% caused by upper stages, and 11.5% caused by satellites). When classified according to orbit regime, 69.2% were in low-Earth orbits, at altitudes below 2,000 km, 9.3% were in the vicinity of the geostationary ring, 9.7% were on highly eccentric orbits (HEO), including the GEO transfer orbits (GTO), 3.9% were in medium Earth orbits (MEO), between LEO and GEO, and almost 7.8% were outside the GEO region. A small fraction, ~150 objects, were injected into Earth escape orbits.

The spatial distribution of catalog orbits is shown in Fig. 2.3 as a global view from outside the GEO region. This image is a computer-generated snapshot of the instantaneous positions of all objects contained in the USSPACECOM catalog of 1997. The geostationary orbit region (GEO), with its distinct distribution signature, is clearly noticeable in this image. Objects on these orbits, close to the equator plane, at inclinations $i \approx 0°$, eccentricities $e \approx 0.0$, and altitudes $H \approx 35,876$ km move synchronously with the rotating Earth, at orbital periods of one sidereal day (23h 56m 04.09s). They are hence ideally suited for communications, broadcasting, and meteorological applications, which have a high commercial value. Consequently, the annual launch rates into GEO have attained a mean level of about 30 (~25 of these payloads, and ~5 of these upper stages), as shown in Fig. 2.5, with no significant effects due to the political changes around 1990. In contrast to the LEO region, there are no energy-dissipating perturbations effective in GEO

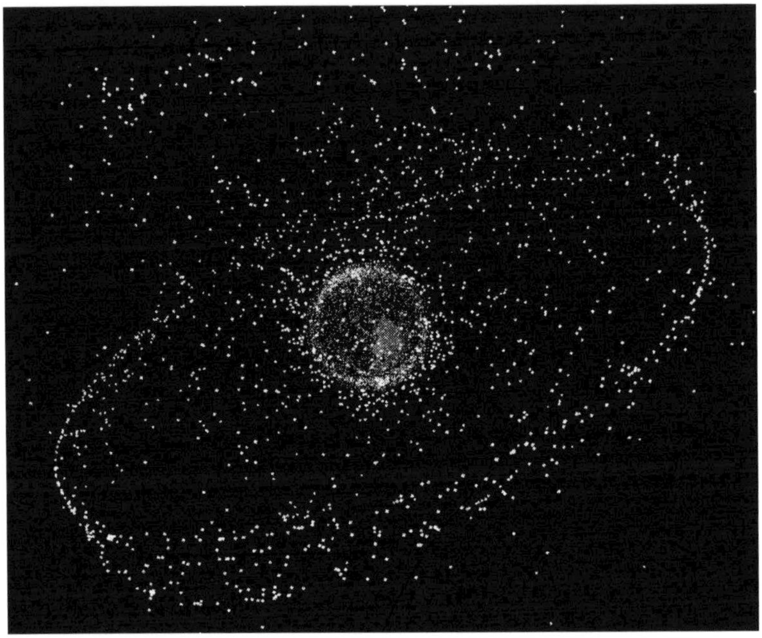

Fig. 2.3. Global snapshot of the catalog population in the year 1997 (source: ESA).

to remove objects from that altitude regime. Thus, the GEO catalog population is steadily increasing at a rate of about 30 per year. Fig. 2.4 indicates the growth history of trackable objects in GEO, and above the GEO ring (mainly in super-GEO "graveyard orbits"). In early 2002 the GEO ring was occupied by 485 payloads (active and non-active), and 108 rocket bodies or boost motors. 155 payloads and 60 rocket bodies were located on super-GEO orbits.

In later chapters, fragmentation processes (explosions and collisions) will be identified as dominating contributions to the evolution and stability of the future space debris environment. The determining parameters for the severity and frequency of fragmentation events are the masses and cross-sections of large orbital structures, predominantly of upper stages and spacecraft. Fig. 2.6 shows the time evolution of the on-orbit mass, which since the mid-1960s has increased by up to 1,500 tons per year, at a mean annual rate of about 110 tons, reaching a total of 5,100 tons by the year 2002. In the same time frame, the on-orbit cross-section, shown in Fig. 2.8 increased to 42,000 m^2, at a progressing rate over the last decade, with peak annual inputs of 3,600 m^2. This progression is due to the steady increase in the annually launched and deployed cross-section as indicated in Fig. 2.9, where, as for the annual launch masses in Fig. 2.7, the reduction of the USSR/Russian deployment activities was more than compensated by the USA.

The mass and cross-section deployment histories are strongly shaped by about 110 Shuttle launches, where each Shuttle had a dry mass of 78 tons, with a mean

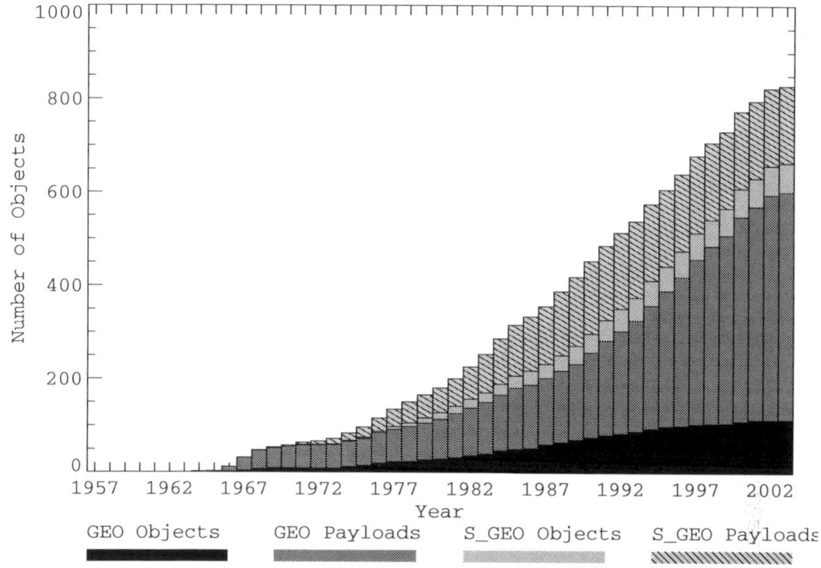

Fig. 2.4. Historic evolution of the number of trackable GEO and super-GEO (S_GEO) catalog objects and their share between payloads and upper stages.

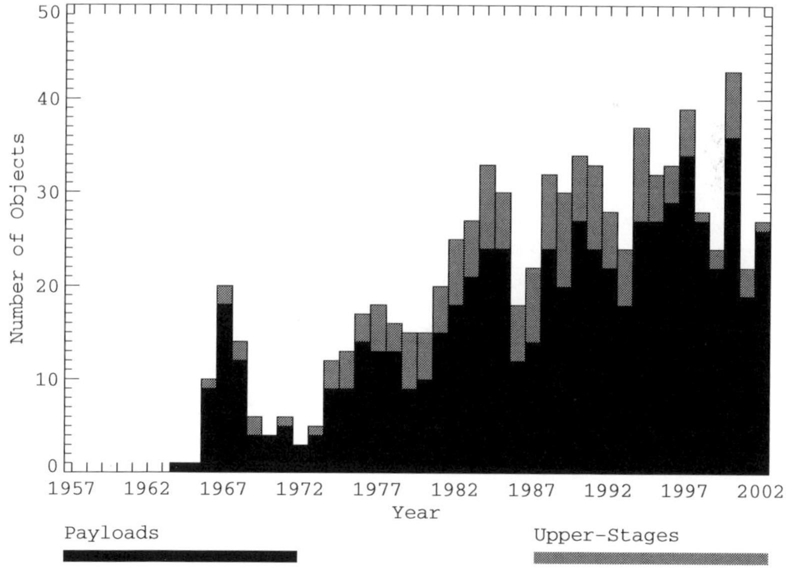

Fig. 2.5. Historic evolution of the annual launch rates into GEO and their share between payloads and upper stages (annual data for 2003 are incomplete).

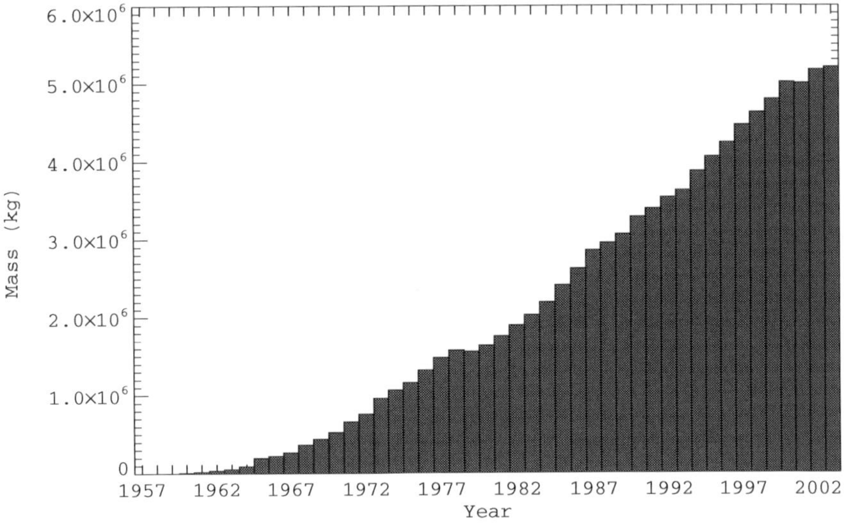

Fig. 2.6. Historic evolution of the on-orbit mass of catalog objects.

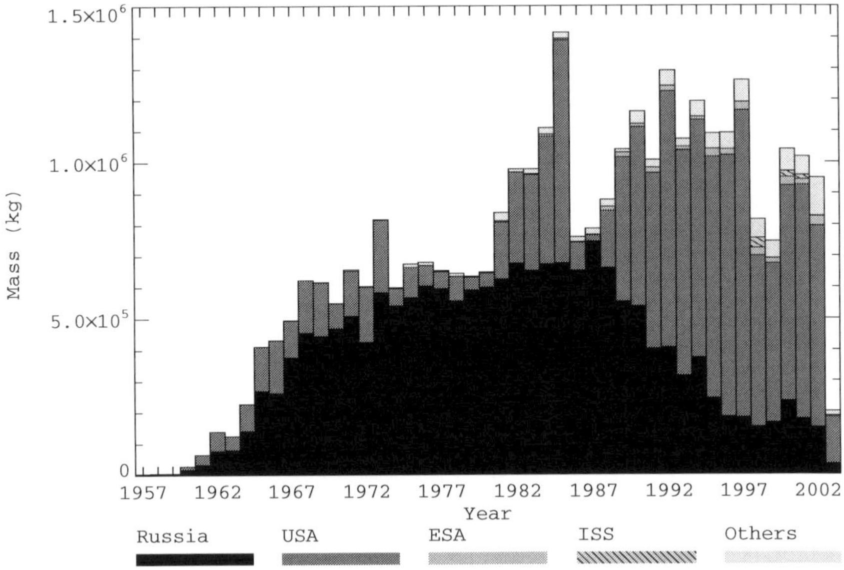

Fig. 2.7. Historic evolution of the annually launched mass of catalog objects and their share between major operators (annual data for 2003 are incomplete).

cross-section of 83 m^2, and a payload capacity to LEO of 24.4 tons. Moreover, several expandable launcher systems can deploy LEO payloads close to 20 tons

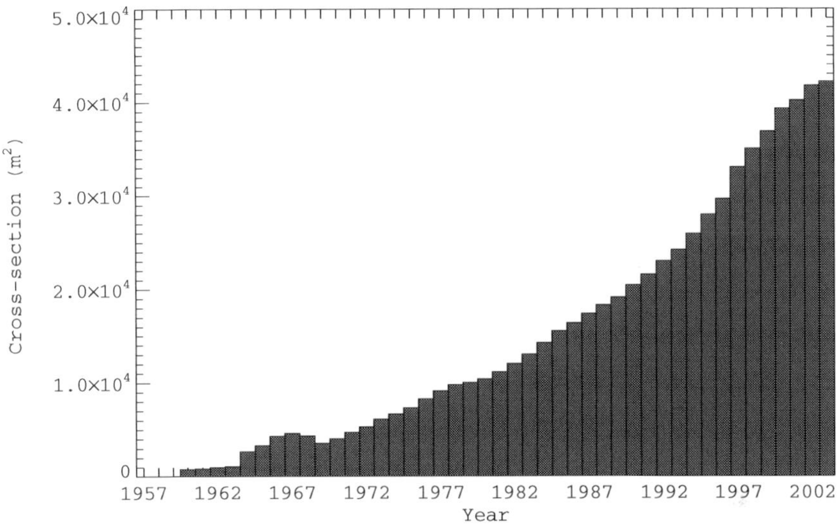

Fig. 2.8. Historic evolution of the on-orbit cross-section of catalog objects.

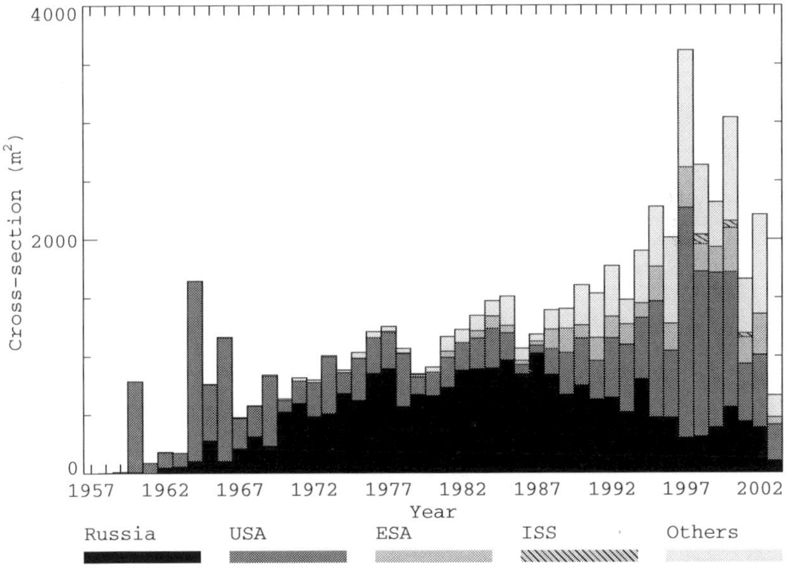

Fig. 2.9. Historic evolution of the annually launched cross-section of catalog objects and their share between major operators (annual data for 2003 are incomplete).

(Delta IV Heavy: 25.8 tons, Titan IV-B: 21.7 tons, Proton M: 21 tons, Atlas V 500: 20 tons, and Ariane 5: 18 tons). In this context, not just the launch capacity, but also the launch frequency of different transport systems plays an important role. By the year 2002, there had been 1,099 launches by Soyuz vehicles, 422 by Cosmos, 301 by Molniya, 286 by the Atlas series, 280 by Delta, 269 by Proton, 248 by Tsyklon, 199 by Titan, 149 by Vostok, and 147 by different Ariane versions. Orbital stages of such launchers have typical cross-sections between 15 and 30 m^2, and the empty masses, particularly for LEO missions, can reach 9.0 tons for Zenith, 5.5 tons for CZ-2F, 4.8 tons for Tsyklon, 4.1 tons for Proton, 2.8 tons for Titan and Delta, and 2.3 tons for the most frequently used Soyuz and Molniya launchers. The distribution of the on-orbit mass and cross-section in 2002 is concentrated in regions which are of major scientific, commercial, and defense interest. With 45.0% the main mass fraction is in the LEO region, 28.8% is in the GEO vicinity, 6.4% is in MEO, 8.7% is in GTO and HEO, an 11.1% is located outside the GEO ring. The ranking is inverted when looking at the distribution of cross-sections, where 40.9% is concentrated near GEO, and 34.8% is within the LEO region. The reason for this characteristic distribution lies in the large appendages (solar arrays and antennas) of GEO satellites, leading to a lower mass-to-area ratio, and in the more compact design of LEO spacecraft to overcome airdrag, which entails higher mass-to-area ratios.

Table 2.1. Space object statistics as of January 1, 2002. 'payloads' = operational + non-operational spacecraft, and 'debris' = rocket bodies + mission-related objects + fragments (continued on next page).

Source/organization	Objects in orbit			Objects decayed		
	payloads	debris	total	payloads	debris	total
Argentina	4	0	4	2	0	2
Australia	7	0	7	2	0	2
Brazil	10	0	10	0	0	0
Canada	17	1	18	1	2	3
Chile	1	0	1	0	0	0
China	33	317	350	28	227	255
Czechoslovakia	4	0	4	1	0	1
Denmark	1	0	1	0	0	0
Egypt	2	0	2	0	0	0
EUMETSAT	3	3	6	0	0	0
ESA	33	281	314	5	541	546
ESRO	0	0	0	7	0	7
EUTELSAT	16	0	16	0	0	0
France	34	18	52	8	59	67
France/Germany	2	0	2	0	0	0
Germany	20	1	21	10	0	10
Hong Kong	3	0	3	0	0	0
India	21	7	28	8	10	18
Indonesia	10	0	10	1	0	1
INMARSAT	9	0	9	0	0	0

Table 2.1. Space object statistics as of January 1, 2002. 'payloads' = operational + non-operational spacecraft, and 'debris' = rocket bodies + mission related objects + fragments (continued from previous page).

Source/organization	Objects in orbit			Objects decayed		
	payloads	debris	total	payloads	debris	total
Space Station (ISS)	4	0	4	0	10	10
INTELSAT	56	0	56	2	0	2
Israel	3	0	3	3	3	6
Italy	12	2	14	7	0	7
Japan	73	48	121	14	103	117
Korea, Rep. of	7	1	8	0	0	0
Luxembourg	12	0	12	0	0	0
Malaysia	3	0	3	0	0	0
Mexico	6	0	6	0	0	0
Morocco	1	0	1	0	0	0
NATO	8	0	8	0	0	0
Netherlands	0	0	0	1	0	1
Norway	3	0	3	0	0	0
Pakistan	1	0	1	1	0	1
Philippines	2	0	2	0	0	0
Portugal	1	0	1	0	0	0
Saudi Arabia	9	0	9	0	0	0
Singapore	1	0	1	0	0	0
South Africa	1	0	1	0	0	0
Spain	6	0	6	0	0	0
Sweden	10	0	10	0	0	0
Taiwan	1	0	1	0	0	0
Thailand	4	0	4	0	0	0
Turkey	3	0	3	0	0	0
United Emirates	1	0	1	0	0	0
United Kingdom	23	1	24	9	1	10
USA	1,007	2,905	3,912	718	3,940	4,658
USSR	1,119	2,036	3,155	1,690	9,750	11,440
Russian Fed.	246	519	765	161	726	887
Column totals	2,853	6,140	8,993	2,679	15,372	18,051
Sum total						**27,044**

As a result of airdrag perturbations on LEO orbits and luni-solar perturbations on highly eccentric orbits (mostly in concert with airdrag), a large fraction of the catalog objects which were released into orbits have in the meantime decayed into the atmosphere. A smaller fraction has been intentionally de-orbited, or recovered by controlled re-entries. By the year 2002, out of the 27,044 objects cataloged since 1957, 18,051 had re-entered (66.7%). In the same time frame, a total mass of about 27,050 tons (84%), and a cumulative cross-section of about 85,000 m^2 (67%)

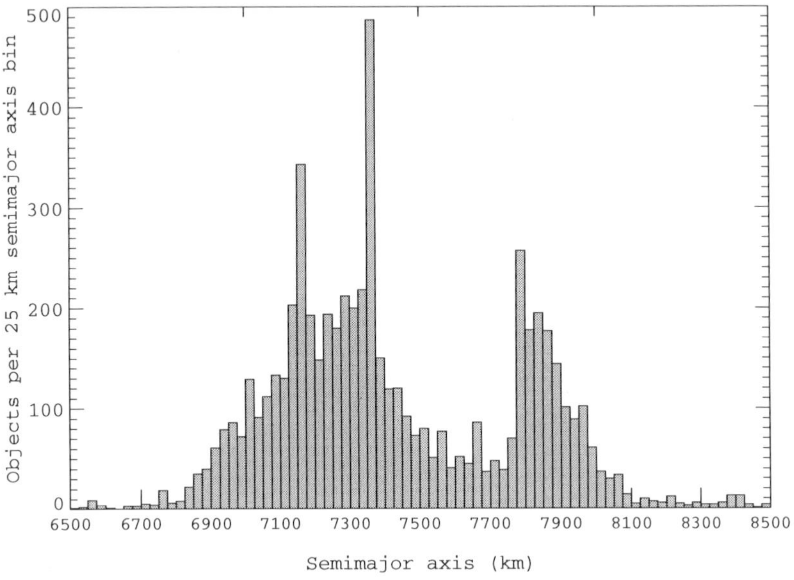

Fig. 2.10. Histogram of the distribution of LEO catalog objects with the semimajor axis of their orbit (class width: $\Delta a = 25$ km; status: June 2003).

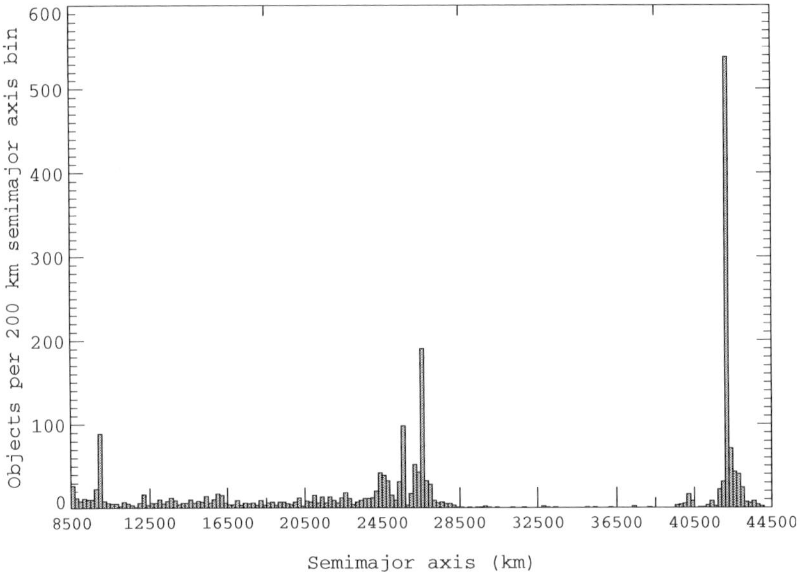

Fig. 2.11. Histogram of the distribution of super-LEO catalog objects with the semimajor axis of their orbit (class width: $\Delta a = 200$ km; status: June 2003).

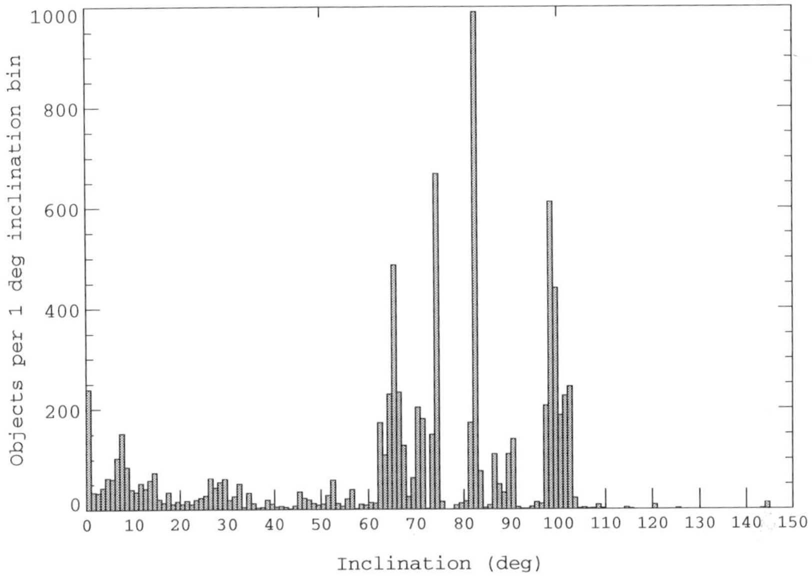

Fig. 2.12. Histogram of the distribution of catalog objects with the inclination of their orbit (class width: $\Delta i = 1°$; status: June 2003).

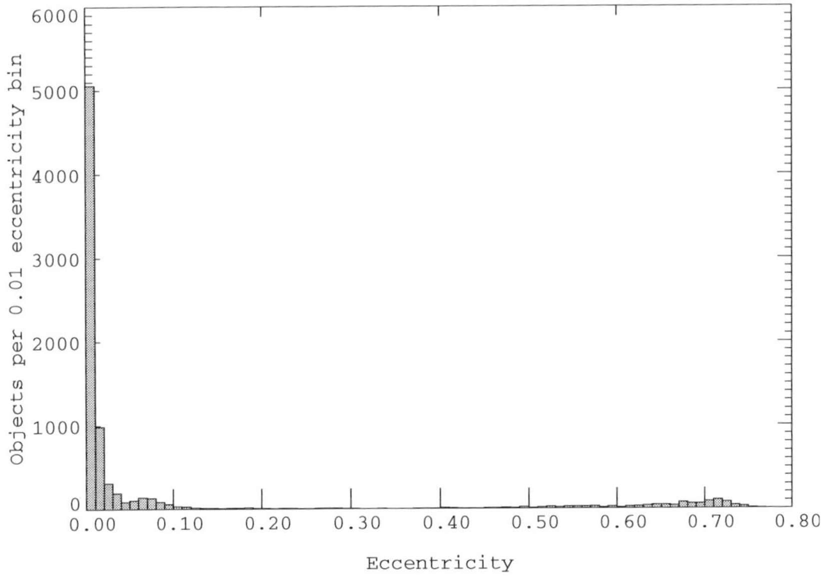

Fig. 2.13. Histogram of the distribution of catalog objects with the eccentricity of their orbit (class width: $\Delta e = 0.01$; status: June 2003).

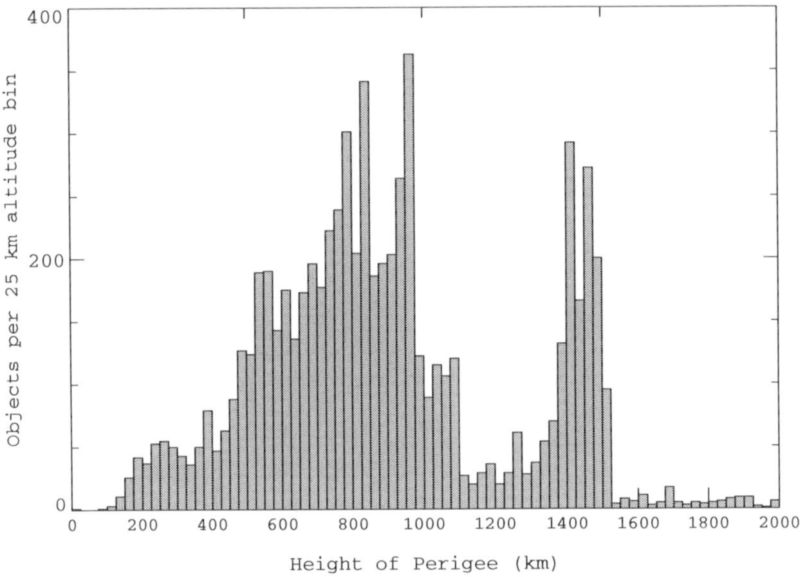

Fig. 2.14. Histogram of the perigee altitude distribution of LEO catalog objects (class width: $\Delta H_{pe} = 25$ km; status: June 2003).

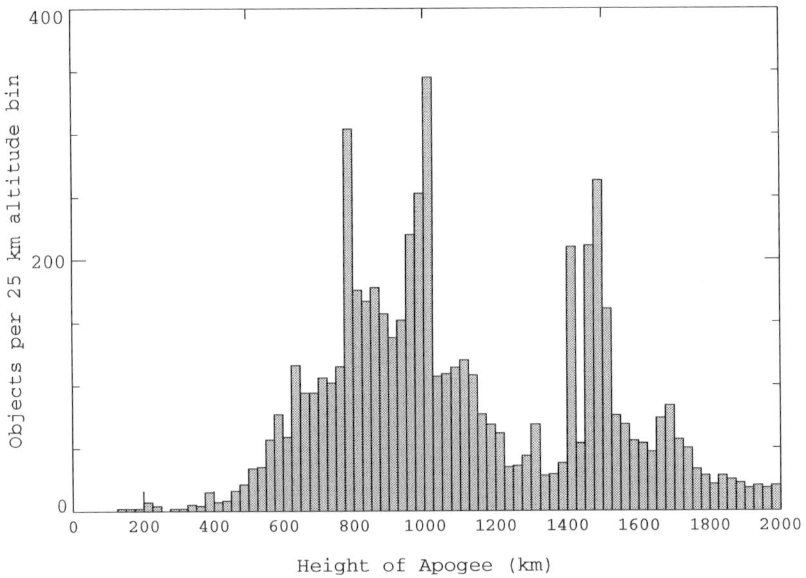

Fig. 2.15. Histogram of the apogee altitude distribution of LEO catalog objects (class width: $\Delta H_{ap} = 25$ km; status: June 2003).

decayed from orbit due to natural forces, or due to induced de-orbit maneuvers. This corresponds to a mean annual mass and cross-section loss-rate of on-orbit catalog objects of 800 tons and 2,100 m^2 during the past few decades.

So far, for the characterization of trackable catalog objects, their spatial distribution was only broadly classified according to orbital regimes (mainly LEO, MEO, GEO, GTO/HEO). A more detailed analysis must be based on the orbital element distributions within the catalog population. Such distributions identify preferred altitude shells, inclination bands, and orbital eccentricities of past space activities. Resulting clusters of preferred orbits form the origin of different classes of space debris which will be dealt with in the following chapters.

Fig. 2.13 shows the distribution of orbit eccentricities e for a catalog population in 2003. Clearly, the vast majority of the trackable objects reside on near circular orbits with $e \leq 0.01$ (54.5%), or moderately eccentric orbits of $0.01 < e \leq 0.1$ (32.0%). Another important contribution is from highly eccentric orbits of $0.6 < e \leq 0.8$ (8.0%), which include the GEO transfer orbits of typical eccentricities around 0.73, and Molniya orbits of semi-synchronous 12-hour periods, with eccentricities around 0.74. The large range of eccentricities $0.1 < e \leq 0.6$ contributes only 5.7%, and $0.8 < e \leq 1.0$ applies to only 0.1% of the catalog objects.

Fig. 2.10 (for the LEO regime) and Fig. 2.11 (for the super-LEO regime) show the distribution of the semimajor axes a of catalog orbits in 2003. Since the majority of orbits are near-circular, these histograms also give an indication of the distribution of mean altitudes $\overline{H} = a - a_e$, where $a_e = 6378.135$ km is the equatorial Earth radius. For the LEO regime, Fig. 2.14 and Fig. 2.15 confirm that the distribution of perigee altitudes $H_{pe} = a(1-e) - a_e$ and apogee altitudes $H_{ap} = a(1+e) - a_e$ closely resemble the profile in Fig. 2.10. Peak concentrations in LEO are at altitude shells near 800 km, 950 km, and 1,450 km due to preferred operational orbits of remote sensing, weather, and science missions. Local spikes in the distribution can to some extent be correlated with constellation deployments (Iridium at 780 km, OrbComm at 825 km, and GlobalStar at 1,415 km), but also with on-orbit fragmentation events. The asymmetries between the perigee and apogee distributions is mainly due to HEO and GTO orbits with perigees between 200 km and 600 km. In the super-LEO regime (Fig. 2.11) such eccentric orbits show peaks at $a \approx 26,560$ km ($\sim$12 h Molniya orbits) and 24,500 km ($\sim$10.5 h GTO orbits). Further contributions to medium Earth orbits are due to the navigation constellations GPS/Navstar ($\sim$12 h orbits at $a \approx 26,560$ km and $\overline{H} \approx 20,180$ km) and GLONASS ($\sim$11.25 h orbits at $a \approx 25,510$ km and $\overline{H} \approx 19,130$ km). The most densely populated altitude band above LEO is centered at the geostationary ring ($\sim$24 h orbits at $a \approx 42,164$ km and $\overline{H} \approx 35,786$ km), occupied by predominantly low eccentricity, low inclination trajectories, with close to 550 objects within a 200 km altitude shell. This compares with peak concentrations in LEO of almost 400 within a 25 km shell (with a corresponding mean distance between catalog objects of 2,500 km).

In contrast with the continuous distributions of semimajor axes and eccentricities (and hence of perigee and apogee altitudes), the histogram of inclinations, occupied by catalog objects in the year 2003 (Fig. 2.12), shows very distinct, narrow bands, with peak concentrations of up to 1,000 objects per 1° bin at $i \approx 82^\circ$.

Concentration maxima can be associated with special mission types, such as: Sun-synchronous orbits ($i = 100 \pm 5°$), polar orbits ($i \approx 90°$), navigation satellite orbits ($i \approx 55°$ and $65°$), orbits of critical inclination ($i \approx 63.4°$, see Annex A.9), and near-geostationary orbits ($i < 15°$). Further peaks are correlated with launch site latitudes, with injection azimuth constraints, or with the delivery of a maximum payload mass into orbit. Such maximum payload capacity is obtained by taking advantage of the Earth rotation via a due east launch. Launch sites which fulfill corresponding azimuth constraints are: Plesetsk (latitude of $\phi = 62.8°$, ΔV gain of $210\,\mathrm{m/s^2}$), Baikonur ($\phi = 45.6°$, ΔV gain of $328\,\mathrm{m/s^2}$), Tanegashima ($\phi = 30.4°$, ΔV gain of $400\,\mathrm{m/s^2}$), Kennedy Space Center ($\phi = 28.5°$, ΔV gain of $410\,\mathrm{m/s^2}$), Xichang ($\phi = 28.2°$, ΔV gain of $408\,\mathrm{m/s^2}$), and Kourou ($\phi = 5.2°$, ΔV gain of $463\,\mathrm{m/s^2}$). The latitude dependent gain in ΔV during the launch can increase the deliverable payload mass by up to 5%. Hence, launches optimized for payload mass criteria are eastward, into orbits with inclinations $i \approx \phi$, leading to concentration peaks at $i \approx 63°$, $52°$, $28.5°$, and $7°$. More often, however, mission objectives dominate injection conditions, and hence Plesetsk and Baikonur are frequently launching into north-easterly directions towards the 82° and 74° inclination band. High inclination orbits are also serviced by the Vandenberg launch site, which due to its launch azimuth constraints has a southward launch corridor.

A good survey of the cataloged space object population in terms of sources and distributions is provided by (anon., 1995), (anon., 2001), (Verger et al., 1997), and (Klinkrad et al., 2003). Most of the information compiled here to characterize the trackable space object environment was extracted from DISCOS, ESA's Database and Information System Characterizing Objects in Space (Hernández et al., 2001). This relational database, which covers the complete space history, contains mission data and orbit histories, mass and cross-section details, launch and decay information, and supporting data on launchers, launch sites, and associated launch statistics. Major portions of the DISCOS data originate from USSPACECOM (now USSTRATCOM). They are routinely provided through the official web interface of SpaceTrack.

2.2 HISTORIC ON-ORBIT BREAK-UP EVENTS

In Section 2.1 fragmentation debris was identified as the most important source of catalog objects, with a contribution of ~39.9% to the trackable population. By January 2002, in total 175 on-orbit fragmentation events were inferred from the detection of new objects and from the correlation of their determined orbits with a common source. In many instances, not just the source object, but also the fragmentation cause could be reconstructed with some confidence. Out of the 175 events, 48 are believed to have been deliberate explosions or collisions (causing 2,244 cataloged fragments), 52 may be attributed to propulsion system explosions (causing 3,558 cataloged fragments), 7 could be associated with electrical system failures (mainly battery explosions, causing 618 cataloged fragments), 10 may have been

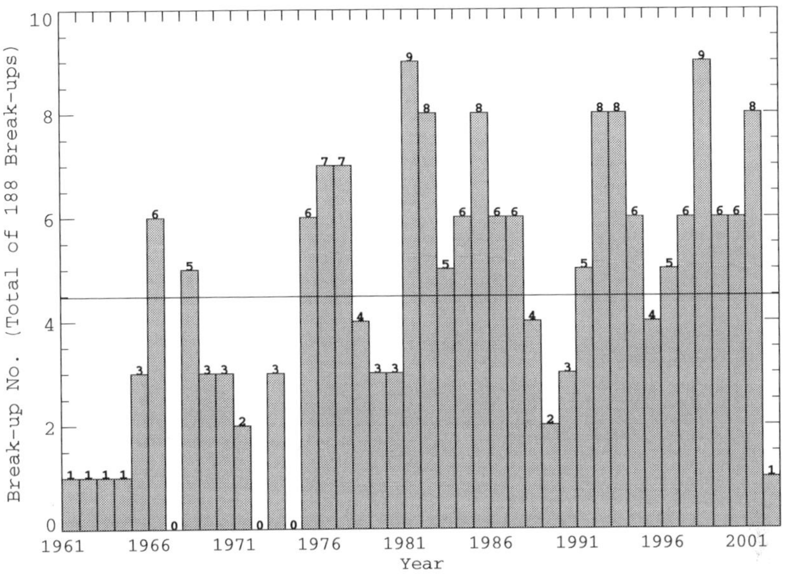

Fig. 2.16. Histogram of the annual count of on-orbit fragmentation events through January, 2002.

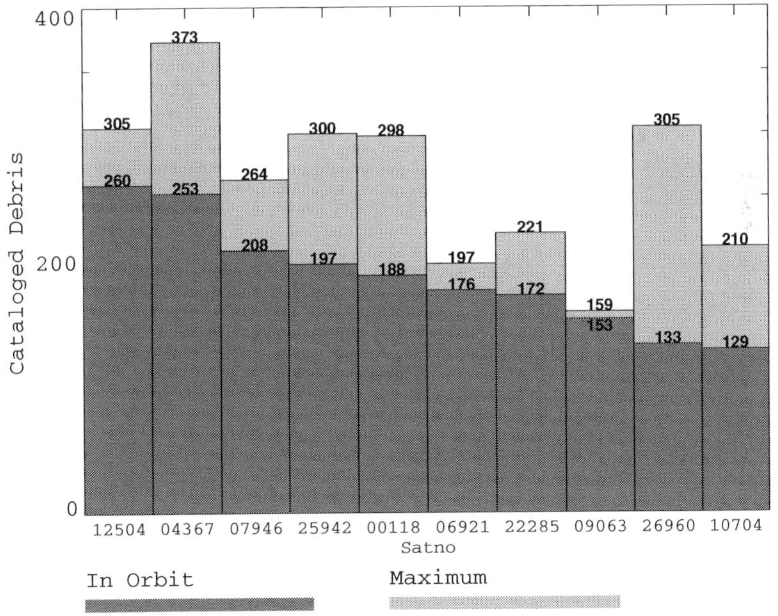

Fig. 2.17. Top ranking fragmentation events as of January 1, 2002, sorted by on-orbit catalog objects.

caused by aerodynamic forces, and at least one event was an accidental collision. The causes of 57 further break-up events could not be identified. With the exception of two GEO explosion events (an Ekran-2 satellite on June 22, 1978, and a Titan III-C Transtage on February 8, 1994), all known fragmentations occurred on orbits passing through altitudes below 2,000 km, with ~80% of the orbits within LEO, and ~17% on HEO and GTO.

The first unintentional collision in space history occurred between the French Cerise satellite (95-033B) and an Ariane-1 H-10 upper stage explosion fragment (86-019F), on July 24, 1996, at 09:48 UTC. The H-10 upper stage had previously delivered SPOT-1 into orbit on February 22, 1986, and exploded 9 months later, on November 13, 1986, causing one of the most severe fragment clouds in space history, with a total of 488 catalog entries (33 thereof still on orbit by January 2002, as shown in Table 2.2). Apart from this unintentional event, there were at least two deliberate collisions, both of them in context with SDI experiments (Strategic Defense Initiative). On September 13, 1985, Solwind P78-1 was destroyed by an anti-satellite (ASAT) missile, which was launched from a fighter aircraft, and intercepted the spacecraft during an ascending pass along the Californian coastline, on its 97.6° inclined orbit, at an altitude of about 530 km. The test scenario was selected such that the fresh fragment cloud could be observed by the SSN sensors in Alaska, and such that the orbital lifetime of the collision fragments was limited (Table 2.2 shows that only 2 out of 285 cataloged fragments were still on orbit in January 2002). Another intentional collision was brought about on September 5, 1986, between the USA-19 spacecraft and the Delta upper stage which previously injected it into orbit. The number of cataloged fragments in this case was only 13. Non-destructive collisions, for instance between a Progress cargo spacecraft and the Mir space station, are not included in the count of collision events. Fig. 2.16 indicates the annual break-up counts, which reached peak rates of 9 per year in 1981 and 1998, and a mean annual rate of about 4.5. Out of 175 events in total, there were 172 explosions or aerodynamic break-ups, and only 3 collisions.

Table 2.2 provides a summary of the top-ranking fragmentation events up to a cut-off date of January 1, 2002, sorted by maximum count of cataloged fragments on-orbit. Fig. 2.17 shows a corresponding histogram of the 10 most severe events, ranked by the number of on-orbit fragments for that date. The majority of the break-up events occurred within a few years after orbit injection, as indicated by the cumulation along a straight line in the plot of launch epoch versus fragmentation epoch in Fig. 2.18. In some cases, however, particularly upper stages exploded with a delay of more than a decade (the Titan III-C Transtage 1967-066G exploded 22 years after launch, and the Vostok stage 1964-006D exploded 33 years after launch). Fig. 2.18 also indicates than many fragmentations in recent years were caused by objects which were launched in the 1980s. Hence, end-of-life passivation measures (the release of latent on-board energies at mission termination), which have become common practice for many launch systems over the past decade, seem to prove their effectiveness.

Table 2.2. The most severe on-orbit fragmentation events as of reference epoch $t_{ref} = 01$ Jan 2002, sorted by the maximum number of cataloged debris. (∗) The number of correlated, on-orbit fragments of the PSLV upper stage fragmentation increased to 303 within one month after the reference epoch.

Catalog number	International designator	Source object name and type	Break-up date	Debris count max./t_{ref}	H_{pe}/H_{ap} [km]/[km]	i [°]	Assessed cause
23106	1994-029B	Pegasus R/B	03-Jun-1996	703/76	586/821	82.0	Unknown
16615	1986-019C	Ariane-1 R/B	13-Nov-1986	488/33	803/833	98.7	Unknown
1640	1965-082B	Titan III-C Transtage R/B	15-Oct-1965	470/41	658/761	32.0	Propulsion
4367	1970-025C	Thor Agena D R/B	17-Oct-1970	373/253	1063/1087	99.9	Unknown
12504	1981-053A	Cosmos 1275 S/C	24-Jul-1981	305/260	960/1014	83.0	Electrical
26960	2001-049D	PSLV R/B	19-Dec-2001	303/133 (∗)	549/674	97.9	Propulsion
25942	1999-057C	CZ-4B stage 3 R/B	11-Mar-2000	300/197	727/745	98.5	Propulsion
118	1961-015C	Thor Ablestar R/B	29-Jun-1961	298/188	882/997	66.8	Propulsion
11278	1979-017A	Solwind P-78 ASAT test	13-Sep-1985	285/2	515/546	97.6	Deliberate
7946	1975-052B	Delta-1 R/B	01-May-1991	264/208	1093/1102	99.6	Propulsion
4159	1969-82AB	Thor Agena D R/B	04-Oct-1969	259/75	917/1091	70.0	Unknown
9046	1976-072A	Cosmos 844 S/C	25-Jul-1976	248/0	172/353	67.0	Unknown
6127	1972-058B	Delta-1 R/B	22-May-1975	226/35	633/909	98.3	Propulsion
22285	1992-093B	Zenit-2 R/B	26-Dec-1992	221/172	847/854	71.0	Propulsion
10704	1978-026C	Delta-1 R/B	27-Jan-1981	210/129	879/912	98.9	Propulsion
7616	1975-004B	Delta-1 R/B	19-Jun-1976	206/33	734/918	97.8	Propulsion
26040	1999-072A	Cosmos 2367 S/C	21-Nov-2001	200/17	405/417	65.0	Accidental
6921	1973-086B	Delta-1 R/B	28-Dec-1973	197/176	1494/1516	102.1	Propulsion
6432	1973-021A	Cosmos 554 S/C	06-May-1973	195/0	147/349	72.8	Deliberate
17297	1987-004A	Cosmos 1813 S/C	29-Jan-1987	194/0	346/405	72.8	Deliberate
10144	1977-065B	Delta-1 R/B	14-Jul-1977	169/65	537/2036	29.1	Propulsion
1093	1965-012A	Cosmos 57 S/C	22-Feb-1965	167/0	169/419	64.8	Deliberate
9063	1976-077B	Delta-1 R/B	24-Dec-1977	159/153	1502/1522	102.0	Propulsion
14064	1983-044A	Cosmos 1461 S/C	11-Mar-1985	158/3	584/883	65.0	Unknown

The distribution of fragmentations is neither uniform in orbit altitude, nor in inclination. Since most of these events were unintentional, with ~3.7% of all launches entailing a break-up event, they were proportionally spread across the most frequented altitude and inclination bands. Hence, the altitude profile of the 2003 catalog population (Fig. 2.14 and 2.15), and its inclination frequency plot (Fig. 2.12) are strongly influenced by on-orbit break-ups, primarily due to explosions. This is particularly so in the inclination bands around $i = 100°$, $82°$, $72°$, and $65°$, and at altitudes in the vicinity of 800 km, 950 km, and 1,450 km (see also Table 2.2). Many historic break-ups are believed to originate from common causes, and they are often classified according to a progenitor event or common system component, such as "Cosmos self-destructs", "Delta second stage explosions", "Proton-K Block DM ullage motor explosions", or "Cosmos 699 class events". A detailed analysis of on-orbit fragmentation events and anomalous events through July 2001 is provided by (Johnson et al., 2001).

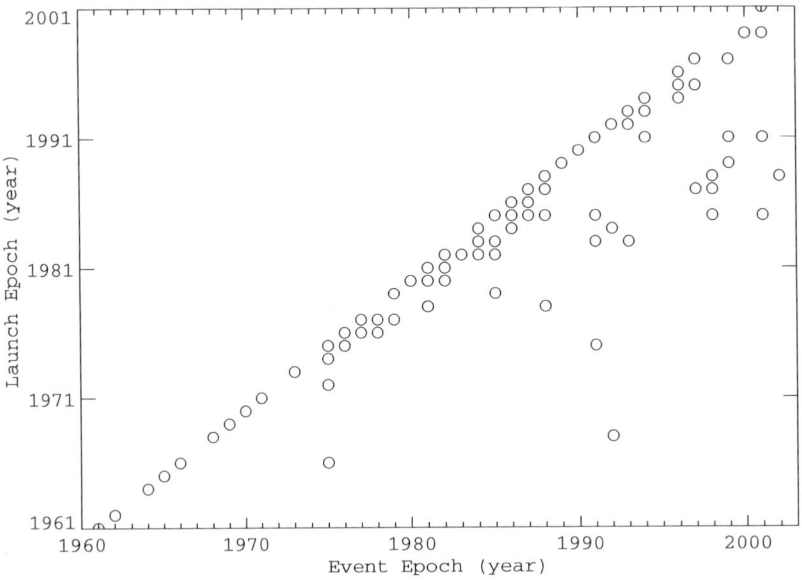

Fig. 2.18. Fragmentation event epoch versus launch epoch of all known on-orbit break-ups as of January 1, 2002.

Depending on the energy release in the generation processes, fragmentation debris is assumed to cover a size regime mainly within diameters of $0.1\,\text{mm} \leq d \leq 1\,\text{m}$. Only a small fraction thereof can be observed by routine space surveillance. Thus, additional experimental observation campaigns with high performance ground-based sensors, in-situ measurements in orbit, and explosion and hypervelocity impact (HVI) tests on ground and in orbit (Solwind P-78) can help to close the knowledge gap on the generated mass and size spectra, and on the imparted velocities, which are required for mathematical models of fragmentation processes.

2.3 NON-FRAGMENTATION DEBRIS SOURCES

The most important non-fragmentation source of space debris is solid rocket motor (SRM) firings and the resulting slag and dust particles, which are mainly composed of aluminum oxide (Al_2O_3). Aluminum powder is added to most solid fuels, typically with a mass fraction of ~18%, to stabilize the combustion process. It is assumed that about 99% thereof is continuously ejected with the exhaust stream during the main thrust phase in the form of Al_2O_3 dust of diameters largely within $1\,\mu m \leq d \leq 50\,\mu m$. Due to design constraints, several solid motors have nozzles penetrating into the burn chamber, causing cavities around the nozzle throats. During the burn phase, trapped Al_2O_3, melted Al droplets, and parts of released thermal insulation liner material can cumulate in this pool, and weld together to slag particles which can grow to sizes of typically $0.1\text{ mm} \leq d \leq 30\text{ mm}$ (particles as large as 50 mm were observed in ground tests). These slag particles are released at the end of the main thrust phase, as the internal motor pressure decreases. This release process is known to depend also on the spin rate of SRMs, which is used to stabilize the motor orientation during the payload injection maneuver.

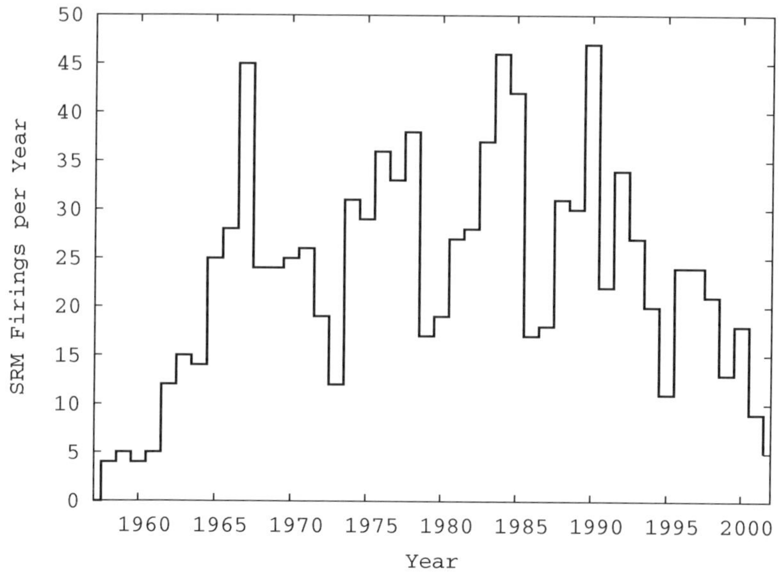

Fig. 2.19. Annual rate of solid rocket motor (SRM) firings through June, 2001.

The number of solid rocket motor firings between 1958 and June 2001 was 1,032, with peak rates of up to 47 events per year, and a mean annual rate of 23.5. Fig. 2.19 illustrates the historic evolution of SRM firings, to which missions of the United States alone contributed with almost 70%. Consequently, the injection orbits where SRMs were applied are mostly associated with US missions, such as deployments of GPS/Navstar constellation spacecraft in inclinations $i \approx 55°$ and

63° (first generation), GTO and GEO injections from Kennedy Space Center into orbits of $i \approx 28.5°$ (GTO) and $\approx 0°$ (GEO), or deployments into Sun-synchronous orbits and near-polar orbits of $80° \leq i \leq 110°$. Also the USSR used SRM boosters, for instance to deploy their Radar Ocean Reconnaissance Satellites at $i \approx 63°$, and several other launch operators used SRMs as GTO kick motors, and/or as GEO injection motors. The size of the solid motors, in terms of available propellant, covers a wide range. The most frequently used SRMs are the Star 37 motors, with a propellant mass of $m_p = 1,067$ kg, used for instance as final stage of Delta launchers to deploy GPS/Navstar payloads, the Payload Assist Module PAM-D, with $m_p = 2,011$ kg, also used as Delta final stage for instance for GTO injections, and the Inert Upper Stage (IUS), deployed from Titan IV or Space Shuttle, for instance to inject payloads into GTO with a first stage of $m_p = 9,709$ kg, and subsequently deliver the payload into a circular GEO by a second stage of $m_p = 2,722$ kg. Another powerful SRM engine, HS-601 with $m_p = 4,267$ kg, is used by Long March LM-2E launchers both for LEO and GTO payload injections.

It is assumed that during the 1,032 SRM firings up to June 2001 about 1,000 tons of propellant were released into space. Based on physical and mathematical model hypotheses, approximately 320 tons thereof were Al_2O_3 dust particles, and 4 tons were slag particles formed of Al_2O_3, metallic Al, and motor liner material. Due to orbital perturbations and their different action on μm-size dust and cm-size slag, merely 1 ton of Al_2O_3 dust and 3 tons of SRM slag particles are believed to be still on orbit. In some size regimes and altitude shells these contributions can dominate the debris environment.

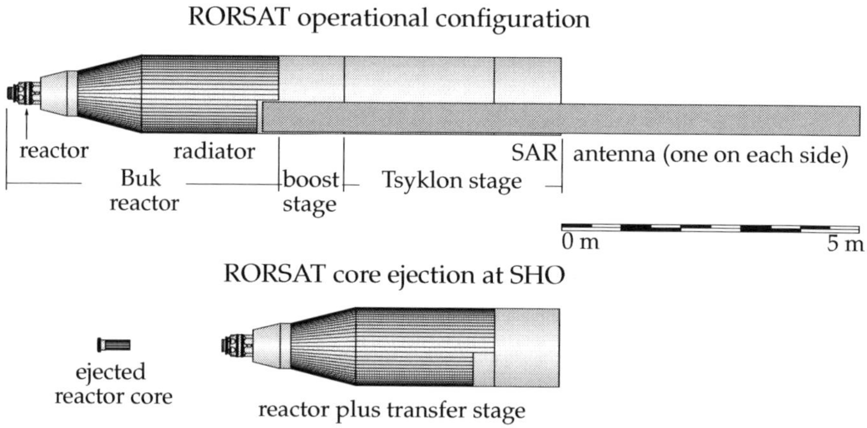

Fig. 2.20. Configuration of a RORSAT spacecraft during its operation at altitudes of about 250 km (top), and after its re-boost to a "Sufficiently High Orbit" (SHO) with subsequent reactor core ejection (bottom; source: TUBS/ILR).

Another source of cm-size objects also played an important role in LEO, in one particular inclination band. In the early 1990s, ground-based radar measurements, supported by optical observations and spot-check orbit determinations, provided

Table 2.3. List of core ejection events of RORSAT reactors. Assumption: core ejection date = transfer date to a "Sufficiently High Orbit" (SHO). Orbit inclination, perigee and apogee altitude refer to the achieved SHO. For events marked with "*" ejected cores could not be detected.

Object designator	RORSAT name (stage+reactor)	Launch date	Re-orbit date to SHO	H_{ap} [km]	H_{pe} [km]	i [°]
1980-034A	Cosmos 1176	29-Apr-1980	10-Sep-1980	923	888	64.8
1981-021A	Cosmos 1249	05-May-1981	19-Jun-1981	970	885	65.0
1981-037A	Cosmos 1266	21-Apr-1981	29-Apr-1981	944	892	64.8
1981-081A	Cosmos 1299	24-Aug-1981	05-Sep-1981	950	918	65.1
1982-043A	Cosmos 1365	14-May-1982	27-Sep-1982	965	872	65.1
1982-052A	Cosmos 1372	01-Jun-1982	11-Aug-1982	947	912	64.9
1982-099A	Cosmos 1412	02-Oct-1982	10-Nov-1982	944	913	64.8
1984-069A	Cosmos 1579	29-Jun-1984	27-Sep-1984	965	894	65.1
1984-112A	Cosmos 1607	31-Oct-1984	01-Feb-1985	947	929	65.0
1985-064A	Cosmos 1670*	01-Aug-1985	23-Oct-1985	992	912	64.9
1985-075A	Cosmos 1677*	23-Aug-1985	23-Oct-1985	997	887	64.7
1986-024A	Cosmos 1736	21-Mar-1986	21-Jun-1986	993	924	65.0
1986-062A	Cosmos 1771	20-Aug-1986	15-Oct-1986	968	916	65.0
1987-052A	Cosmos 1860	18-Jun-1987	28-Jul-1987	969	897	65.0
1987-101A	Cosmos 1900*	12-Dec-1987	30-Sep-1988	754	687	66.1
1988-019A	Cosmos 1932	14-Mar-1988	19-Mar-1988	961	940	65.0

evidence of a large population of spherical objects with metallic characteristics, densities similar to water, and sizes up to 5.6 cm. These were found to be concentrated within a narrow inclination band close to 65°, at altitudes between 900 and 950 km. All observed characteristics of this population (radar return, optical properties, size estimates, orbit decay rates) were consistent with droplets of a sodium–potassium alloy (NaK). Such a low-melting, eutectic alloy was used as coolant by Buk reactors on board of Russian RORSATs (Radar Ocean Reconnaissance Satellites). 31 RORSATs were launched between October 1970 (Cosmos 367) and March 1988 (Cosmos 1932). They consisted of an F-1m/SL-11 (Tsyklon 2) upper stage with two side-looking radars attached (Fig. 2.20), of a Buk reactor and a radiator system mounted in the in-flight direction of the compound of 14.6 m length and 1.3 m diameter, and of an orbital transfer stage in between the reactor and the payload. RORSATs were launched from Baikonur into operational, near-circular orbits of typically 250 to 270 km altitude, at 65° inclination. Due to the high-drag environment the RORSAT mission durations were generally shorter than 60 to 70 days. After the nominal end of payload operations, the F-1m/SL-11 segment with the antenna module was separated and left to decay into the Earth atmosphere. The transfer stage, with the switched-off reactor module attached was then raised to a near-circular, "sufficiently high orbit" (SHO) of about 950 km altitude, of unchanged inclination. In 16 cases, listed in Table 2.3, the reactor cores of 53 kg mass, 0.2 m diameter, and 0.6 m length were ejected in order to remove them from their beryllium enclosure, which serves as a reflector of slow neutrons during the reactor operation. During a later atmospheric re-entry, after hundreds of years, this protective beryllium mantle of high melting temperature (~1,285°C) would, however, prevent a burn-up of the reactor core. The ejection of the uranium-molybdenum core in the SHO deposit region is achieved at the cost of a rupture of the primary reactor coolant loop, which leads to a rapid release of NaK droplets of diameters in an estimated size regime of 0.1 mm $\leq d \leq$ 55 mm.

Table 2.3 lists the known core release events, assuming that the re-orbiting epoch is identical to the release epoch. For Cosmos 1900 the achieved orbit near 700 km altitude gives evidence of a non-nominal operation. It seems that in 1988 the Cosmos 1900 re-orbit maneuver could not be commanded from ground, and the spacecraft descended deep into the atmosphere, before an automatic back-up system, driven by atmospheric pressure gauges, induced the orbital transfer, which due to performance limitations could not reach the SHO deposit region. At two other occasions the re-orbit system completely failed. On January 24, 1978, Cosmos 954 with 30 kg of ^{235}U on-board, re-entered over Canada, dispersing fragments and radioactive debris along a ground swath extending from the Great Slave Lake to the Hudson Bay. This event triggered subsequent UN deliberations on nuclear power sources in space. In a second accident, Cosmos 1402 failed to perform its re-boost maneuver, and the reactor core re-entered over the South Atlantic, off the coast of Brazil, on February 7, 1983.

One can assume that a total of 208 kg NaK coolant was released in 16 events (~13 kg/event). Thermodynamic considerations suggest that only droplets of diameters $d >$ 0.1 mm could survive an extended period of time in a frozen state (~9.1 kg/event), while smaller objects tend to evaporate. For NaK droplets of $d >$ 3 mm, the time before evaporation exceeds the orbital lifetime. Hence, following the termination of Buk reactor deployments and core ejections between 1980 and 1988, airdrag most likely removed all NaK droplets smaller than 1 mm by the year 1993. The remaining population had an estimated mass of 50 to 60 kg. Ultimately, only droplets of $d >$ 1 cm maintain a durable population close to the initial release altitudes. Since RORSATs are no longer operated, NaK droplets can be regarded as a historic, non-regenerative source of debris.

Another historic and non-reproducing debris source are the so-called "Westford Needles". These are dipoles of thin copper wires, which were released in the early 1960s by the spacecraft Midas 4 (October 21, 1961) and Midas 6 (May 12, 1963), as part of a radio communication experiment. The intention was to deploy a layer of radio-frequency reflecting dipoles around the Earth, at altitudes of about 3,600 km, in near-polar inclinations. The release experiment consisted of a cylinder of 32 cm length and 12.8 cm diameter, containing millions of copper needles of 1.78 cm length, with diameters of 25.4 μm (Midas 4) and 17.8 μm (Midas 6). The needles were embedded in naphthalene, which was supposed to evaporate and release the needles from the rotating containers. The dispersion, however, was not executed successfully, and clusters of needles were forming (~40,000 in the first, and ~1,000 in the second experiment). About 150 of these were recorded in the USSPACECOM catalog. The total on-orbit mass of the remaining clusters of

Westford Needles by mid 2001 was estimated to be merely 60 g.

In contrast with the NaK droplets and the Westford Needles, release products from surface degradation and impact ejecta are a reproductive source. For reasons of thermal control most spacecraft and rocket upper stages are coated with paint, and/or with thermal blankets. Under the influence of the harsh space environment, in particular due to impinging atomic oxygen, due to aging and embrittlement by extreme ultra-violet (EUV) radiation, and due to micro-particle impacts, surfaces of space objects tend to erode. This combination of effects could be well observed on returned surfaces of LDEF (Long Duration Exposure Facility), which remained on orbit from April 1984 to January 1990 at altitudes of 470 km to 340 km, at an inclination of 28.5°. Atomic oxygen was found to cause a degradation of polyurethane binder material in paints, or damage of underlying substrate material through cracks in the paint surface. EUV radiation is known to cause defects in polymeric materials, such as Kapton, Teflon, or paints. This leads to an embrittlement of the material, with subsequent crack formations and possible delaminations of layered foils. The effect is enhanced through thermal cycling, mainly due to transits through the Earth shadow. Atomic oxygen corrosion and EUV radiation, often in combination with micro-particle impacts, leads to mass losses of surface coatings, and to the formation and detachment of paint flakes of µm to mm sizes. Traces of such materials could be found in impact craters, for instance on Space Shuttle windows (80 windows had to be replaced by 2001 due to micro-particle impacts).

All debris sources listed so far are estimated to result in a population of about 3×10^8 objects of diameters $d > 1$ mm, which increases to 3×10^{13} objects when looking at diameters of $d > 1$ µm. Together with the natural meteoroid population these objects are so abundant that they cause frequent impacts on spacecraft and spent upper stages, sometimes at high relative velocities. The impact craters or pits are correlated with the particle diameters d_p, leading to depths of typically $2 \times d_p$, and crater diameters of $5 \times d_p$ on ductile material like aluminum. On brittle material, like glass or ceramics, the diameter of the total damaged area can be $100 \times d_p$. During the impact, high velocity material is ejected from the crater, but also lower velocity products may be released from a concentric spall zone, within an area of about $5 \times D_c$ (where D_c is the crater diameter). Depending on the target material, an outer ring zone of approximately $15 \times D_c$ may also receive sufficient damage to cause a time-delayed separation of surface coatings, often in combination with EUV and atomic oxygen effects.

2.4 GROUND-BASED RADAR AND OPTICAL MEASUREMENTS

In Section 2.1 the trackable population of about 9,000 unclassified space objects was characterized, based on observational data and orbit determinations of the US Strategic Command and their Space Surveillance Network, who perform routine surveillance of orbital objects of sizes larger than about 10 cm in LEO, and larger than about 1 m in GEO. Surveillance can be described as the combined, routine tasks of operational detection, correlation with sources, characterization, and orbit determination of space objects. This task is mainly accomplished by ground-based radars (for LEO altitudes), and optical telescopes (for altitudes above LEO). Apart from the USA, only Russia has such an operational capability, supported by their Space Surveillance System (SSS). The corresponding catalog comprises orbit and characterization data of about 6,000 objects. Due to the lack of SSS sensors at lower latitudes, and probably due to a different detection threshold, this catalog is about 30% smaller than its US counterpart. Also France has an experimental surveillance system, the bi-static GRAVES installation (Grande Réseau Adapté à la Veille Spatial). The related catalog is limited to objects of typically 1 m size and larger in LEO, with a total count of about 2,500 entries.

Apart from sensors which are involved in routine space surveillance tasks (SSN and SSS), there are also sensors which track known catalog objects with higher accuracy (for instance the US Millstone and Goldstone radars, the German TIRA radar, and the French Armor radar). Alternatively, sensors may also acquire statistical information on sub-catalog, small-size objects (for instance the US Haystack and Goldstone radars, the German TIRA radar, the European EISCAT radar network, and the ESA space debris telescope). All these ground-based systems have to overcome a common problem: they need to use frequencies which pass the atmosphere with acceptable losses, and which are suitable to locate and characterize a space object with sufficient accuracy from a radar return echo, or from reflected light. The Earth atmosphere is transparent mainly in two wavelength regions, the so-called optical and radio windows.

A continuous optical window extends across wavelengths of $0.3\ \mu m \leq \lambda \leq 2\ \mu m$ (Fig. 2.24), with ultra-violet (UV) covering the short wavelengths ($0.1\ \mu m \leq \lambda_{UV} \leq 0.38\ \mu m$), near infra-red (IR) covering the long wavelengths ($0.78\ \mu m \leq \lambda_{IR} \leq 2\ \mu m$), and visible light (for which the human eye is optimized), carrying most of the emitted solar energy, located in the center of the window ($0.38\ \mu m \leq \lambda_{vis} \leq 0.78\ \mu m$). At short wavelengths the optical window is limited by resonances with O_2 and O_3 oxygen molecules, and at long wavelengths the limiting factors are resonances with O_3, carbon dioxide, and water vapor. Secondary absorption bands within the optical window can be significantly reduced by moving telescopes to high altitudes (at $H = 5$ km the air density drops to about 50% of its sea-level value).

Optical telescopes should be ideally placed at high altitudes, in regions with minimum light pollution, and with good meteorological and atmospheric conditions. Telescopes which are used for satellite tracking must be operated at night, with sufficient clearance from the terminator of the Sun's shadow (more than 18° below the horizon for "astronomical night" conditions). At the same time, the spacecraft to be tracked must still be illuminated by the Sun.

There are two basic types of telescopes: refractors, working with lens systems, and reflectors, using mirror surfaces to collect and focus the incoming light. Refractors are more complex, and technical constraints limit their aperture diameters to about $D \leq 1$ m. Reflectors are generally superior with respect to production costs, mass, image distortion, and resolution, which is also due to the achievable

Fig. 2.21. Ground-Based Electro-Optical Deep-Space Surveillance site (GEODSS) of the US Space Surveillance Network, at Socorro, New Mexico (source: SSN).

aperture diameters of $D \leq 10$ m. The reflector telescopes can be categorized into a few basic design types: (1) Newton telescopes, with a planar secondary mirror of 90° deflection angle, and a ratio of aperture to focal length in the range $1:5 \leq D/f \leq 1:3$, (2) Cassegrain telescopes, with a convex secondary mirror of 180° deflection angle, and $1:20 \leq D/f \leq 1:8$, (3) Coudé telescopes, with a 90° deflection by an extra secondary mirror, and large focal lengths of $1:50 \leq D/f \leq 1:30$, and (4) Ritchey–Chrétien telescopes, using the Cassegrain design with correcting optics.

Telescopes are pointed towards their observation targets by means of mounts which have two rotational degrees of freedom. If one of the rotation axes is aligned with the Earth's axis of rotation, the telescope can follow a celestial object by adjusting a single angle, the hour angle. Such an installation is called a parallactic mount. With the advent of automated telescope driver systems for pointing and tracking, also azimuth-elevation mounts have become viable alternatives, since they are more compact in their design.

A telescope collects light which is emitted or reflected by a space object. In the case of orbital objects, the light source is the Sun, and its photons are re-emitted by the target with an intensity which depends on the Sun-target-observer angle (Sun phase angle $\Phi_\odot$), and on the reflection efficiency of the target, which is denoted as geometric albedo α_t, with typical values near 0.1 for man-made space objects. If the target object is a simple plate of projected area A_t, then its re-emitted power as received by the telescope is

$$P = P_\odot \, \alpha_t \, \frac{A \cos \epsilon_{t,in} \cos \epsilon_{t,out}}{\rho_t^2} \tag{2.1}$$

Fig. 2.22. The ESA Space Debris Telescope at the Teide Observatory on Tenerife.

where $P_\odot = A_t\,F_\odot$, with the solar irradiance $F_\odot$ at one astronomical unit. $A = \pi D^2/4$ is the light collecting area of a telescope of aperture diameter D, $\epsilon_{t,in}$ and $\epsilon_{t,out}$ are the incident angles of the incoming and outgoing photons, $\Phi_\odot = \epsilon_{t,in} + \epsilon_{t,out}$ is the solar phase angle, and ρ_t^2 is the range from the telescope to the target. Hence, the received power, and the corresponding sensitivity of the telescope decreases with $1/\rho_t^2$ due to a spread of the re-emitted power over a certain solid angle $0 < \psi \leq 4\pi$, which translates into a spherical surface area of $\psi\,\rho_t^2$ at the range ρ_t. The relative brightness of an observed object is expressed in units of magnitude, abbreviated as "mag", and defined as

$$\text{mag} = -2.5\ \log_{10}(l/l_0) \tag{2.2}$$

where l is the brightness of the object, and l_0 is a reference brightness of a star of magnitude 0.0 (= 0^m0), such as Vega (α Lyrae). Some key magnitudes are: -27^m0 for the Sun, -12^m5 to -10^m3 for the Moon, -4^m5 for Venus, -1^m5 for Sirius, and $+6^m0$ for the detection limit of a human eye. High performance telescopes as used in space surveillance and space debris observations may have detection thresholds beyond $+20^m0$.

The light which is captured by the telescope aperture is manipulated by the optical assembly and translated into an image which can be viewed through an eyepiece, or which can be used to generate an exposure on a photographic plate. Nowadays, photographic plates are mostly replaced by Charge-Coupled Devices (CCDs), which are photosensitive, solid-state imaging sensors, based on metal oxide semiconductor technology (MOS). They convert the incoming photons into electric charges on an array of photodetectors. The charges of the array elements are then read out through one or several channels. The time-tagged array information can be used to reconstruct an image with a resolution determined by the granularity of the CCD (high-end systems use CCDs or CCD mosaics of up to

$4,096 \times 8,192$ pixels). CCDs are often cooled, for instance with liquid nitrogen, to reduce thermal noise in their read-out signals. The total energy received from a tracked object is proportional to the exposure time. In the tracking mode an increase in the integration time by a factor 3 may lead to an increase in sensitivity by $\Delta \text{mag} \approx 1^m0$. The theoretical image resolution, often characterized by an instrument-specific "point spread function", is governed by the aperture diameter and the wavelength of the incoming light via the ratio D/λ. The corresponding angular resolution of the human eye is ~60″ (arc seconds), while a small telescope with $D = 0.2$ m can reach ~0.65″. The resolution limit of ground-based telescopes (typically <1″) is determined by local seeing conditions. The instrument performance is further constrained by the light collection efficiency and signal-to-noise (S/N) ratio. Depending on the observation mode (inertially staring or tracking), the Earth orbiting objects can be noted as trails on a fixed star background, or as point sources on a background of streaks caused by stars on the celestial sphere. Optical observations should be performed during astronomical night (more than 18° from the Sun terminator), outside the two weeks centered on full Moon epochs.

The processed CCD or photographic plate exposures can provide time-tagged information on the optical detection statistics, on the magnitude of objects and its variability (light curves), and on angular directions either with respect to the telescope mount axes, or with respect to the star background. From a series of positions at corresponding epochs orbit determinations can be performed, if the observed orbit arc is large enough. Otherwise, for too short arcs, one often assumes a circular orbit for the initial orbit determination and later improves the eccentricity information with data from subsequent sightings. The sensitivity of optical telescopes decreases according to $1/\rho_t^2$ with the range ρ_t to the target (Eq. 2.1). This is considerably better than the performance of radars, which suffers a loss in sensitivity with $1/\rho_t^4$, due to the active illumination of the target (Eq. 2.8). As a consequence, optical systems are used to observe more distant objects, typically in MEO and GEO, while radars have advantages for LEO observations.

USSTRATCOM collect tracking data for the catalog maintenance of high altitude objects by means of their GEODSS (Ground-Based Electro-Optical Deep-Space Surveillance) and MOTIF installations (Maui Optical Tracking and Identification Facility), which are part of the Space Surveillance Network (SSN). The GEODSS locations are uniformly distributed around the globe at Socorro (New Mexico), Mount Haleakala (Maui/Hawaii), and Diego Garcia (Indian Ocean), to have a good coverage, particularly of the GEO ring. As shown in Fig. 2.21, each of these systems consists of 3 telescopes of 1 m aperture diameter each, equipped with electronic photon multipliers and CCD image processors. The Russian Space Surveillance System (SSS) also maintains optical observation sites to maintain their catalog of high altitude objects. Within Europe and Japan, several telescopes are used for space debris observations, but none of them yet for the operational maintenance of a catalog.

Apart from their operational SSN optical sensors, the USA collected debris data with a small aperture, wide field of view (1.5°) CCD debris telescope (CDT), and with a Liquid Mirror Telescope (LMT) of 3 m aperture, and a 0.34° field of view. The LMT, with 1,322 hours of operation from 1997 to 2001, was located at 32.97°N, at Cloudcroft (New Mexico). It consisted of a pool of liquid mercury, which rotates at 10 rpm, forming a low-cost, near-perfect parabolic reflector, which directs the light to an optical assembly with a video camera, or a $2,048 \times 2,048$ CCD array. Due to its special design the LMT could only perform statistical observations near the zenith, with a detection size threshold of ≤ 5 cm in LEO (object albedos of 0.1 shall hereafter be assumed for all comparisons). The mobile CDT telescope, with 1,954 hours of operation from 1997 to 2001, was mainly used for GEO observations. It had an aperture diameter of 32 cm, a field of view of $1.7^\circ \times 1.7^\circ$, and a CCD detector array of 384×576 pixels. For integration times of $\sim$20 s the limiting magnitude was about 17.5, equivalent to a size detection threshold of about 0.6 m in GEO.

The University of Michigan operates the Michigan Orbital Debris Survey Telescope (MODEST)[2.1], located in Chile at 30.2°S and 70.8°W. From this location 110° of the GEO ring can be observed (25°W to 135°W), covering most of the GEO slots assigned to the US. MODEST is an f/3.5 Schmidt telescope, with a 61 cm aperture and a field of view of $1.3^\circ \times 1.3^\circ$. The telescope is equipped with a backside illuminated CCD, mounted in the Newtonian focus. With a standard exposure time of 5 s a signal-to-noise ratio of 10 is obtained for objects of magnitude 18.

Since 1999 ESA has been routinely conducting orbital surveys of catalog objects and space debris in high-altitude orbits during $\sim$10 nights per year. The surveys are performed with a telescope of 1 m aperture and 0.7° field of view (FoV), which is located on Tenerife at 28.17°N (Fig. 2.22). The telescope is of the Cassegrain type, with an English mount, and with a Coudé focus for optical communication applications (e.g. with Artemis), and a Ritchey–Chrétien focus for space debris observations. The Ritchey–Chrétien system has an aperture diameter to focal length ratio of $D/f = 1 : 4.47$ (abbreviated as $f/4.47$). It leads the light through a hole in the primary mirror to a 4×4 array of CCD chips of $2,048 \times 2,048$ pixels each. The liquid nitrogen cooled CCD array cumulates the received energy of the photons during integration times of typically 2 s, before reading it out within $\sim$19 s. Up to 3 read-outs (images) per minute can be produced, with a detection threshold of $+19^m0$ to $+21^m0$ for a signal-to-noise ratio of $S/N \geq 5$. This makes it possible to detect and follow objects of diameters $d \geq 15$ cm at GEO altitudes. From its geodetic latitude of 28.17°N the ESA Space Debris Telescope (ESA SDT) covers a sector of $\sim$120° of the GEO ring. During a campaign in 1999, within 49 hours of observations in total, 206 unique objects could be identified, of which only 27% could be correlated with the USSTRATCOM catalog. From single observations, initial orbits can be derived which are generally adequate for re-acquisition of the object, and for subsequent successive orbit improvements. During the initial orbit determination it is often assumed that the orbit is near-circular, and emphasis is put on well-observable quantities such as the orbital period ($\rightarrow$ semimajor axis), and the orientation of the orbital plane ($\rightarrow$

[2.1]official designation: University of Michigan Curtis Schmidt telescope orbit inclination and right ascension of the ascending node). From a time history of observation data the eccentricity and perigee location can also be determined.

The ESA SDT surveys revealed a substantial population of small debris in and close to GEO, with a size distribution which is steadily rising up to the telescope sensitivity limit. There is strong evidence of several clouds of break-up fragments, which were monitored from 2002 to 2004. Their evolution in the orbit element space was consistent with theoretical results of propagated explosion fragments (see Section 3.10). Since 2002 about 50% of the ESA SDT observation time is devoted to the search for debris in low inclination GEO transfer orbits. For a subset of the detected objects real-time follow-up observations are performed, and reliable 6-parameter orbits are determined. Thereby, a significant population of objects could be detected with mean orbital motions of $n \approx 1\,\mathrm{d}^{-1}$, eccentricities of $0.05 \leq e \leq 0.6$, and large area-to-mass ratios (A/m up to three orders of magnitude larger than for a normal spacecraft). These objects are suspected to be pieces of lightweight insulation material, most likely stemming from orbital rocket stages. Due to large A/m ratios of the objects their orbits are strongly affected by solar radiation pressure, with periodic changes of considerable amplitudes in eccentricity and inclination (Klinkrad, 2002).

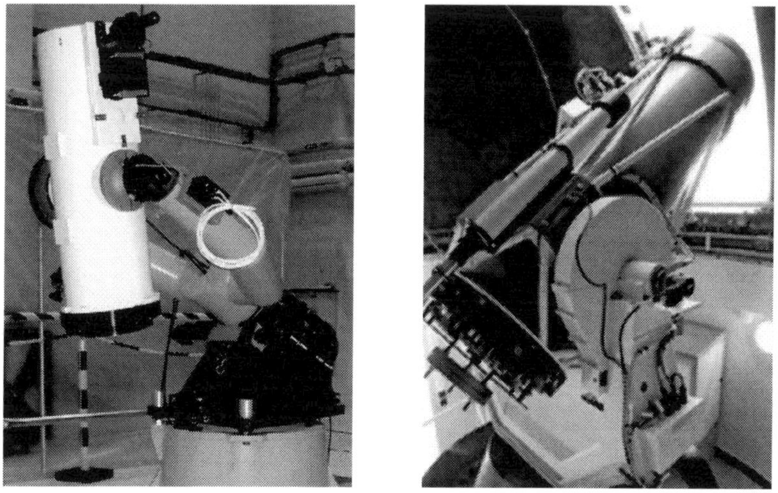

Fig. 2.23. The French telescopes TAROT at Calern (left image) and ROSACE at the Observatoire de Haute Provence (right image; source: CNRS).

France developed the optical SPOC system (Système Probatoire d'Observation du Ciel), a wide field-of-view sensor ($50^\circ \times 50^\circ$), as part of a network of tracking facilities of the French DGA (Délégation Générale pour l'Armement), with observation sites at Toulon and Odeillo. Each of the present 2 stations has 4 cameras, facing west, north, east (at elevations $h > 20^\circ$), and towards the zenith. The cameras are equipped with CCDs of 576×384 pixels, with threshold detection magnitudes of $\sim +6^m0$ to $+7^m0$. On clear nights, this allows the detection of 100 to

Table 2.4. Detection capabilities and modes of operation of some ground-based optical systems (source: (anon., 1999) and IADC). Capabilities of space surveillance networks (SSN and SSS) are not included in the list.

Country	Organization Detector type	D [m] FoV [°]	λ [°] ϕ [°]	Lim. magnitude [–] Operational status
Europe	ESA	1.0 m	16°W	19.5
	CCD	1.0°	29°N	operational
France	CNRS/CNES	0.5 m	2°E	19
	CCD	0.35°	48°N	experimental
France	CNRS/CNES	0.25 m	6°E	17
	CCD	2.0°	43°N	operational
Japan	JSP/NAL/NASDA	1.0 m	135°E	19.5
	CCD	3.0°	35°N	in development
Japan	JSP/NAL/NASDA	0.5 m	135°E	18.5
	CCD	2.0°	35°N	in development
Japan	Sendai	0.75 m	141°E	17
	CCD	0.04°	38°N	operational
Japan	CRL	1.5 m	137°E	18.7
	CCD	0.28°	35°N	operational
Russia	RSA	1.0 m	–	19
	CCD	0.2°	–	operational
Russia	RSA	0.6 m	–	18
	CCD	0.2°	–	operational
Switzerland	Univ. of Berne	1.0 m	7°E	19.5
	CCD	0.5°	47°N	operational
UK	RGO/MOD	0.4 m	–	18
	CCD	0.6°	–	operational (3 sites)
USA	NASA	0.3 m	105°W	17.1
	CCD	1.5°	33°N	operational
USA	NASA	3.0 m	105°W	21.5
	CCD	0.3°	33°N	operational
USA	USNO	1.3 m	112°W	21
	CCD	1.7°	35°S	operational
USA	Univ. of Michigan	0.6 m	70°W	18.5
	CCD	1.3°	30°S	operational
USA	AFRL	1.2 m	156°W	20
	CCD	1.2°	21°S	operational
USA	AFRL	0.5 m	107°W	18
	CCD	6.8°	34°N	in development
USA	AFRL	0.65 m	118°E	18.5
	CCD	0.5°	33°N	in development

400 objects in LEO. The SPOC system has also been used to deduce the intrinsic attitude motion of spacecraft (as for SPOT-3) from an analysis of photometric data (light curve histories). In the past, the French space agency CNES used the ROSACE Newton-type telescope for observations of slowly moving, near-GEO objects down to limiting magnitudes of $+19^m0$. ROSACE (Restitution d'Orbite par Système Autonome CCD d'Ecartométrie) has an aperture of 50 cm, and a CCD camera with an array of $1,024 \times 1,536$ pixels, corresponding to a field of view of $0.3° \times 0.4°$. Orbit determinations were performed on the basis of precise azimuth and elevation direction angles. More recently, the TAROT system (Télescope á Action Rapide pour les Objets Transitoires) was introduced for tracking operational GEO spacecraft and space debris on high orbits (see Fig. 2.23). Originally this instrument was designed for the robotic detection and observation of gamma-ray bursts. TAROT, located at Calern, is a f/3.5 Newton telescope of 25 cm aperture, with a large field of view of $2° \times 2°$, and with pointing speeds of up to 80°/s and angular accelerations of up to $120°/s^2$. It is equipped with a $2,048 \times 2,048$ CCD, with read-out speeds of 0.5 to 2 s and mean image processing cycles of 1 minute duration. The limiting magnitude is of the order of 17 for 10 s exposures. A second TAROT system, TAROT-S, will be installed at La Silla (Chile) (Klinkrad, 2002).

PIMS (Passive Imaging Metric Sensor) is an optical system for the surveillance of the GEO and MEO region, operated by the United Kingdom Ministry of Defence (MOD). PIMS telescopes are located at Herstmonceux (UK), Gibraltar, and Cyprus. The three sensors cover 165° of the GEO ring (65°W to 100°E). The azimuth-elevation mounted Cassegrain-type telescopes have an aperture of 40 cm, and a FoV of 40 arcmin $\times$ 40 arcmin. The images are captured on a CCD chip of $1,024 \times 1,024$ pixels, which can be read out within less than 5 s. The PIMS system can detect GEO objects down to diameters of 0.5 m.

A limited list of ground-based optical facilities for the observation of intact and debris objects in Earth orbits is provided in Table 2.4. This summary is mainly based on information compiled by the IADC (Inter-Agency Space Debris Coordination Committee). It excludes sensors of operational surveillance systems, but adds some sites which were not addressed in the foregoing discussion. Further details on in-orbit optical sensors are provided in Section 2.5.

Ground-based optical systems are strongly limited in their availability due to night-time observation constraints, the exclusion of nights with bright Moon light, and adverse meteorological conditions. These limitations do not play a role in the case of ground-based radar observations at frequencies within the radio window of the Earth atmosphere. That part of this window which is useful for debris observations extends across wavelengths of approximately 1 cm $\leq \lambda \leq$ 100 cm (see Fig. 2.24). This corresponds to an equivalent range of frequencies of about 30 GHz $\geq f \geq$ 0.03 GHz. The radio window is by convention partitioned into frequency bands, which are in ascending order of wavelengths: the Ku band (1.67 cm $\leq \lambda_{Ku} \leq$ 2.5 cm), the X band (2.5 cm $\leq \lambda_X \leq$ 3.75 cm), the C band (3.75 cm $\leq \lambda_C \leq$ 7.5 cm), the S band (7.5 cm $\leq \lambda_S \leq$ 15 cm), the L band (15 cm $\leq \lambda_L \leq$ 30 cm), the UHF band (30 cm $\leq \lambda_{UHF} \leq$ 100 cm), and the VHF band (1 m $\leq \lambda_{VHF} \leq$ 10 m). At longer wavelengths the window is closed due to

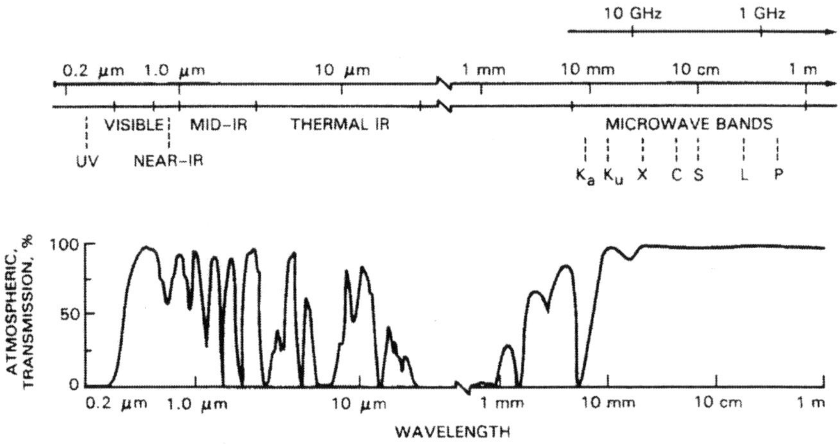

Fig. 2.24. Transmission windows of the Earth atmosphere for radio and optical wavelengths (Bleeker et al., 2001).

ionospheric attenuations and reflections, and at shorter wavelengths the window is limited by resonances with molecules of water vapor and oxygen.

There are two basic types of radar antenna systems: the traditional reflector antenna, and the phased array. A reflector or dish antenna is in general steerable within the local horizontal plane (around an azimuth angle A), and perpendicular to it (around an elevation angle h), to follow a single object during a station passage. The object is tracked by maintaining a maximum gain in the reception of a previously transmitted radar pulse. The ratio of pulse length to pulse period (normally $<10\%$) is denoted as duty cycle. To avoid ambiguities in the interpretation of returned radar echoes, a reception time gate is set, which is equivalent to a certain range gate. A typical reflector radar, such as the German Tracking and Imaging Radar TIRA (Fig. 2.30), consists of a 4-horn feed which illuminates a parabolic dish antenna via a hyperbolic sub-reflector. The emitted power is concentrated in a central beam with a beamwidth Θ_{3dB}, which is a function of the transmitting antenna diameter D_t and the signal wavelength λ_t.

$$\Theta_{3dB}[^\circ] \approx 70\ \lambda_t/D_t \tag{2.3}$$

Θ_{3dB} denotes the 3 dB (decibel) angular beamwidth, within which the power drops to 50% of its maximum. The relative magnitude of the power at a certain location in the antenna diagram can be described as

$$P[\text{dB}] = 10\ \log_{10}(P/P_{max}) \tag{2.4}$$

As with any antenna, radars also generate side lobes with power sub-maxima of typically $\leq 1/100\ P_{max}$ (corresponding to a power reduction by -20 dB). During the reception mode, the 4-horn feed serves as a controller of the antenna pointing direction by maintaining the 4 separate inputs at the same power level.

Fig. 2.25. Two-face phased array at Cape Cod, belonging to the US Space Surveillance Network (SSN) Phased Array Warning System (PAWS; source: SSN).

In the case of phased arrays (Fig. 2.25), a large number of elementary antennas (10,000 or more) is deployed in a planar array. Several of these small antennas can be combined into sub-arrays, each of which can be used to produce a synthetic beam from a superposition of phased outputs of the contributing elementary antennas. Due to the arrangement in a plane, the beam pointing capability is normally limited to angles within ~60° from the vertical of the plane. Hence, with a single planar array a coverage of ~120° can be obtained, and with three equally spaced arrays a total 360° coverage can be reached. Since no mechanical steering is involved, phased arrays can track a large number of objects simultaneously. Consequently, these systems are the dominating contributors to routine space surveillance by USSTRATCOM.

Most radars, both reflector antennas and phased arrays, can measure the 2-way signal travel time $\Delta t_{2w} = t_r - t_t$ (with the reception time t_r and transmission time t_t), the azimuth angle A and elevation angle h of the maximum gain pointing direction, the 2-way Doppler shift Δf_{2w} between received frequency f_r and transmitted frequency f_t, the received power P_r, and the polarization change in the radar pulse. From the 2-way signal travel time the range ρ to the target can be derived as

$$\rho \approx \frac{1}{2} c \, \Delta t_{2w} \tag{2.5}$$

where $c = 299,792.4562$ km/s is the speed of light. The 2-way Doppler shift determines the range-rate $\dot{\rho}$ to the target as

$$\dot{\rho} \approx -c \, \Delta f_{2w} / f_t \tag{2.6}$$

As indicated, these equations are only approximate relations. More accurate expressions require the implementation of relativistic effects, and of refraction corrections to account for neutral atmosphere and ionospheric effects on the path of the radar signal.

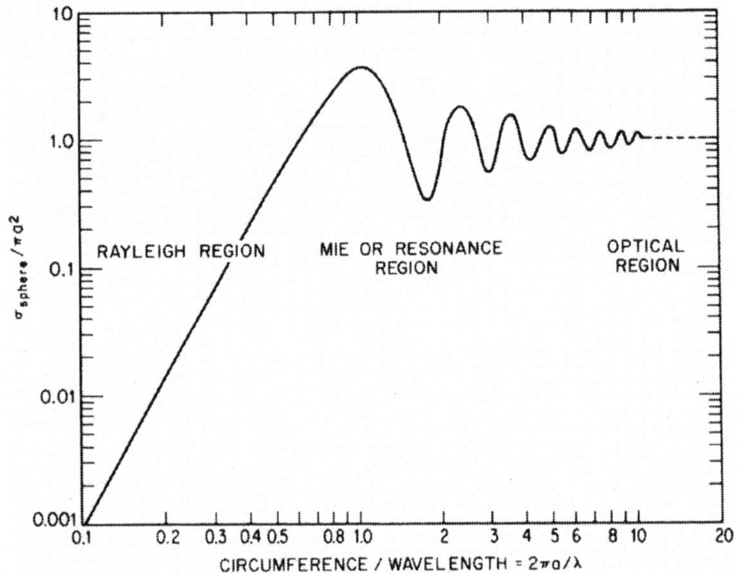

Fig. 2.26. Different domains for the radar cross-section interpretation as a function of the ratio $2\pi a/\lambda$ (object-circumference to wavelength ratio). The relative RCS is expressed as $\sigma_{sphere}/\pi a^2$, the interpreted radar cross-section for a sphere in relation to the geometric cross-section of a sphere of radius a (Skolnik, 1990).

In addition to data on range, range-rate, and direction angles, the radar echo gives an indication on the radar cross-section of the target, its history can reveal information on the attitude dynamics, and changes in its signal polarization can provide information on the object shape. The radar cross-section (RCS) of a target object depends on its material properties, shape, and orientation. It also depends on the ratio $r_\lambda = l_t/\lambda$ of a characteristic target object dimension to the radar wavelength (Fig. 2.26). The RCS, expressed in decibel square meters (dBsm), is a radar equivalent of the visual magnitude in optical observations (Eq. 2.2).

$$A_{rcs}[\mathrm{dBsm}] = 10 \log_{10}\left(A_{rcs}/[\mathrm{m}^2]\right) \tag{2.7}$$

For $r_\lambda \gg 1$ the radar signal is reflected like a light beam on a specular surface (optical region), for ratios of $0.5 < r_\lambda < 10$ the radar signal encounters resonances due to interferences of waves originating from scatter centers on the target which differ by up to few wavelengths (Mie region, or resonance region), and for $r_\lambda < 0.5$ the target is only recognized as a volume scatterer (Rayleigh region). In the optical region the target radar cross-section A_{rcs} and the geometric target cross-section A_t are of similar magnitude ($A_{rcs}/A_t \approx 1.0 = \text{const.}$), whereas in the Rayleigh region the radar cross-section as indicator of the detectability rapidly drops with decreasing target size according to $A_{rcs}/A_t \propto (l_t/\lambda)^4$. Due to ambiguities in the resonance region (Fig. 2.26), special care must be taken when converting measured radar cross-sections into equivalent geometric cross-sections.

In contrast to an optical telescope, radars actively illuminate the target and process the radar return signal. The power received P_r is a function of the power emitted P_e, the area of the emitter and receiver antenna A_e and A_r, the emitted wavelength λ_e, the range to the target ρ_t, and its radar cross-section A_{rcs}.

$$P_r = \frac{P_e A_e}{\lambda_e^2 \rho_t^2} \times \frac{A_{rcs}}{4\pi\rho_t^2} \times A_r \tag{2.8}$$

where A_e/λ_e^2 is a measure of the gain G_e of the emitting antenna according to $G_e = 4\pi\, A_e/\lambda_e^2$. A comparison with Eq. 2.3 shows that this gain is also related to the antenna beamwidth Θ_e via $G_e \propto 1/\Theta_e^2$. Eq. 2.8 demonstrates that for a given target range and target radar cross-section the received power of the radar echo increases with increasing emitter power, with increasing antenna aperture of the emitter and receiver (in terms of diameter or area), and with decreasing wavelength of the radar (increasing frequency).

In the case of a traditional tracking or mono-static radar, the emitting and receiving antenna aperture is identical ($A_e = A_r$). For bi-static observation scenarios, and for electronic fences, however, the receiving antenna is often considerably larger. An example of the bi-static mode is a cooperation between the German Effelsberg radio observatory with its 100 m receiving antenna, and the FGAN research institute with their 34 m emitting TIRA antenna at Wachtberg, at a ground distance of 21 km from Effelsberg. A well-known example of an electronic fence extending into space is the SPASUR installation (formerly Naval Space Surveillance system, operated since 2004 by the US Air Force). It consists of three triplets of one emitting antenna and two receiving antennas each, which are uniformly distributed across the continental United States along a small circle of latitude of 33°N. In both of these scenarios an emitting antenna illuminates a certain Earth-fixed volume in space, and orbiting objects passing through this volume reflect echos to a receiving antenna observing the same volume. In order to process the observations, the emitter and receiver must synchronize their data flows.

If the emitting radar antenna is also the receiving antenna ($A_e = A_r$), an alternative formulation of Eq. 2.8 is obtained by solving for the maximum target range $\rho_{t,max}$ which can be covered for a given radar system performance. This capability can be expressed in terms of $P_{r,min}$, the minimum received power required to reach a limiting signal-to-noise ratio $(S/N)_{min}$ for a given radar bandwidth B_{sys} and system noise temperature T_{sys}.

$$\rho_{t,max}^4 = \frac{P_e\, A_e^2}{4\pi\, \lambda_e^2\, P_{r,min}} \times A_{t,rcs} \quad \text{where} \quad P_{r,min} = kT_{sys}\, B_{sys} \left(\frac{S}{N}\right)_{min} \tag{2.9}$$

Here, $k = 1.380662 \times 10^{-23}$ W/(K Hz) is the Boltzmann constant, and $kT_{sys} B_{sys}$ is a system-specific threshold power, for a noise temperature T_{sys} [K] and system bandwidth B_{sys} [Hz], which needs to be exceeded by a signal-to-noise margin of at least $(S/N)_{min}$. Eq. 2.9 is the so-called radar equation, which for a given system determines the maximum detection range $\rho_{t,max}$ to a target of size $A_{t,rcs}$, or vice versa the minimum detectable radar cross-section at a given range.

Peaks in the spatial densities of LEO objects are noticeable at altitudes of 800 km to 1,000 km, and around 1,400 km. Peaks in the latitude distribution of objects are observed between 65° and 82°, were the dominant inclination bands have their highest resident probability. As a consequence, a zenith staring electronic fence deployed in Europe at 50°N will still be able to observe almost 80% of the entire catalog population. Due to the sparsely populated inclination bands at $i \leq 50°$, the coverage would only improve by about +5% when moving −20° to the south. On the other hand, it would deteriorate by −20% when moving +20° to the north. Hence, typical European latitudes are a good compromise between possible coverage of the orbit population, and frequent station passes (with pass frequency increasing with the station latitude ϕ according to $1/\cos\phi$ for an orbit of $i > \phi$). When pointing a radar located at $\phi = 50°$N in a southward direction, at an elevation of $h = 10°$, objects at the densely populated 900 km altitude band can still be observed on orbits of inclinations $i \leq 30°$, at a range of ~2,500 km.

For routine space observations the US Space Surveillance Network uses a core set of dedicated sensors with full availability, a set of collateral sensors with partial availability, and several contributing sensors which are not owned or operated by the US Government. Apart from the previously mentioned optical sites of the SSN (four GEODSS sites, plus two older Baker–Nunn cameras), the dedicated sensor network consist of a single-face phased array radar at Eglin (Florida), of the Globus II radar (Norway), and of the SPASUR electronic fence, which is distributed along the 33° northern parallel across the United States. Collateral SSN sensors belong to the Ballistic Missile Early Warning System (BMEWS), with locations at Thule (Greenland), Clear (Alaska), and Fylingdales (UK), or are part of the Phased Array Warning System (PAWS), with locations at Cape Cod (Massachusetts), Beale (California), and Robins (Georgia, decommissioned). Another powerful collateral sensor is the Cobra Dane radar on the Aleutian islands at Shemya (Alaska). All of these are 2-face phased array radars, with the exception of Fylingdales (three faces), and Cobra Dane (one face). Of several contributing SSN sensors the most important one, also in the light of debris research, is the Millstone/Haystack radar complex in Massachusetts, which consists of two dish radars. The SSN sensors are operated by USSTRATCOM from their Space Surveillance Center (SSC) in Cheyenne Mountain (Colorado), with a back-up control facility at Dahlgren (Virginia).

The single-face phased array radar at Eglin (Fig. 2.27) consists of two sub-arrays of 29 m diameter for transmission and 58 m diameter for reception. It operates in the UHF band, at 442 MHz, with a peak power of 32 MW. From its latitude at 30.57°N Eglin can track about 95% of all catalog objects within a southward oriented azimuth window of about 105°. With an average of 50,000 tracks per day it

Fig. 2.27. One-face phased array at Eglin, with separate transmitter and receiver sub-arrays, belonging to the US Space Surveillance Network (source: NASA).

is the major single contributor to SSN tracking data, particularly in LEO. Another main source of SSN tracking data is the electronic fence (SPASUR) of the Strategic Space Command, which extends along the 33° northern parallel across the continental USA and consists of three uniformly distributed transmitters, each of them centered between two receiver stations. The SPASUR fence covers a small circle of about 8,000 km length, with an extension into space up to MEO altitudes of 24,000 km (which includes circular 12 hour orbits). Its primary task is the detection and correlation of objects, based on position fixes from interferometric triangulation, with more than 30,000 detections every day. The SPASUR detection threshold is consistent with the tracking capability of the dedicated SSN sensors. Newly detected objects are forwarded to tracking radars or telescopes to determine more accurate orbits.

Of the collateral sensors used by the SSN, Cobra Dane, on the westernmost outskirts of the Aleutian islands, is a particularly powerful installation. Its one-face phased array radar of more than 15,000 elements, arranged in 96 sub-arrays within a circular area of 29 m diameter, operates in the L band (1.2 GHz) at peak powers of 15.5 MW. Also belonging to the class of collateral sensors are the PAVE PAWS radars at Beale, Cape Cod (Fig. 2.25), and Clear. They are 2-face phased arrays, with 1,792 elements distributed over array diameters of 22.1 m, leading to 240° azimuth coverage, and an elevation coverage from 3° to 85°. PAVE PAWS radars operate in the UHF band (420 to 450 MHz) at a peak power of 582 kW. The Fylingdales 3-face phased array radar in the UK is an example of the BMEWS sensor category. It operates in the UHF band (420 to 450 MHz) and is the only SSN associated sensor with a 360° azimuth coverage.

Fig. 2.28. Haystack Long-Range Imaging Radar (LRIR) and Haystack Auxiliary Radar (HAX) at Tyngsboro near Boston (source: NASA).

Of great interest for debris research activities are the SSN contributing sensors at Lexington (Massachusetts), at 42.62°N, operated by MIT. The Lexington site (Fig. 2.28) consists of the Haystack Long-Range Imaging Radar (LRIR, or just "Haystack"), the Haystack Auxiliary Radar (HAX), and the Millstone Hill radar. The LRIR system uses a 36 m parabolic antenna, operating in X band (10 GHz), with a 3 dB beamwidth of 0.058°, and a mean power of 0.4 MW. HAX is also an X band radar (16.6 GHz), with a 12.2 m parabolic antenna, and a 3 dB beamwidth of 0.1°. It has a detection threshold of about 2 cm. LRIR can detect objects down to 6 mm, and it was used to track objects smaller than 5 cm at an altitude of 900 km (orbit determination of NaK droplets released by RORSATs). Using synthetic aperture radar imaging techniques, LRIR can be used to characterize space objects, presumably with decimeter image resolution. In contrast to the monostatic Lexington radars, the Goldstone site (California), at 35.24°N, is a bi-static X band system (8.5 GHz) consisting of a 70 m transmitter antenna, with a 0.03° 3 dB beamwidth, and a 34 m receiver antenna at a ground distance of 497 m. Goldstone can detect LEO objects down 2 mm. Due to the mono-pulse capability of the Haystack radars they are, however, superior to the Goldstone system in deriving radar cross-section information.

Within Europe the initial development and deployment of tracking and surveillance sensors for space objects was mainly driven by national security requirements (Klinkrad, 2002). In some instances, sensors were also deployed under special agreements with USSPACECOM and later USSTRATCOM (e.g. the BMEWS phased array at Fylingdales in the UK, and the Globus II X band tracking radar at Vardø (Norway), at 70.35°N).

Fig. 2.29. The French electronic fence GRAVES (Grande Réseau Adapté à la Veille Spatial), with a transmitter, consisting of two phased arrays, at Dijon (top), and a receiver array at Apt (bottom; source: ONERA).

The French GRAVES system (Grande Réseau Adapté à la Veille Spatial) is presently the only European installation outside the US SSN and outside the Russian SSS which can perform space surveillance in its classical sense. GRAVES is owned by the French Department of Defense (DoD). Its concept is based on VHF transmitters with planar phased-array antennas of 15 m $\times$ 6 m each, which are located near Dijon (Fig. 2.29, top). Several of these tilted antennas can ultimately be arranged around a circle to deploy a conical detection fan up to altitudes of 1,000 km (at present there is one south-west facing and one south-east facing transmitter). Objects which pass through the detection volume (composed of individual detection fans) reflect the transmitted power, which is then received by a planar phased array of Yagi antennas located at Apt (Fig. 2.29, bottom), 380 km south of the transmitter. The array of receiver antennas is arranged in a circular area of 60 m diameter. The GRAVES system was developed during the 1990s, and started operational tests in 2001. It determines direction angles (azimuth and elevation), Doppler, and Doppler rates for a large number of simultaneous targets. From these data the processing software determines orbital element sets, of which initial estimates are sufficiently accurate to task other sensors, and to correlate subsequent detections of the same objects. Accordingly, GRAVES produces a "self-starting catalog" which can be autonomously built up and maintained. The detection size threshold up to 1,000 km altitude is on the order of 1 m, and the orbit coverage reaches down to inclinations of $\sim$28°. An experimental catalog build-up during

Fig. 2.30. FGAN Tracking and Imaging Radar (TIRA) at Wachtberg/Germany (artist's view, with partially removed radome; source: FGAN).

one month in 2001 produced more than 2,200 entries.

The German tracking and imaging radar (TIRA) belongs to the Research Establishment for Applied Science (FGAN) at Wachtberg, near Bonn. It is a monopulse radar, located at 50.6°N, with a parabolic dish antenna of 34 m diameter, housed in a 49 m diameter radome (Fig. 2.30). The radar uses the L band for tracking (1.333 GHz, 1 MW peak power, 0.45° 3 dB beamwidth), and the Ku band for ISAR imaging (16.7 GHz, 13 kW peak power, 0.031° 3 dB beamwidth, up to 2.1 GHz bandwidth). In its tracking mode, the TIRA system determines azimuth and elevation angles, range, and Doppler for a single target. A near real-time processing software can determine orbits from single station passes, which are compatible with the US SSN's Two-Line Element catalog format. The accuracy is generally sufficient to re-acquire the object a few days later. The detection threshold for TIRA is $\sim$2 cm at 1,000 km range. This sensitivity can be enhanced when using TIRA as transmitter, and the nearby (21 km away) Effelsberg 100 m radio telescope as receiver, in a bi-static beam-park mode. This was done during the COBEAM experiment in 1996, leading to detection statistics for objects as small as 0.9 mm, with orbit information limited to altitude and Doppler inclination. The TIRA L-band system has in the past provided time-critical information on orbits of re-entry risk objects (e.g. Skylab, Cosmos-1402, Salyut-7), and of potential collision partners of ESA satellites. The imaging capabilities of TIRA's Ku band radar have been called upon on several occasions for spacecraft emergencies, or for the post-launch verification of the deployment of spacecraft appendages. TIRA's ISAR imaging technique exploits the changing aspect angles due to the orbital motion and due to the spacecraft attitude change to produce images with range resolutions better than 7 cm (see Fig. 8.4).

Fig. 2.31. The French tracking ship *Monge* with its Armor and Gascogne radars (source: DGA).

DGA/DCE, the Systems Evaluation and Test Directorate of the French Ministry of Defense, is operating several radar and optical sensors throughout France. Their radar sensors are of four different types:

- Armor: C-band (∼5.5 GHz), monopulse system, with 1 MW peak power; 10 m dish antenna, with 0.4° 3 dB beamwidth; two units located on the *Monge*
- Béarn: C band, scanning system, with 1 MW peak power; 4 m dish antenna, with 0.96° 3 dB beamwidth; three units located at Toulon, two at Quimper, and four at Cazaux
- Provence: C band, monopulse system, with 1 MW peak power; 4 m dish antenna, with 0.96° 3 dB beamwidth; one unit located at Toulon
- Gascogne: C band, monopulse system, with 0.5 MW peak power; 4 m dish antenna, with 0.9° 3 dB beamwidth; one unit located on the *Monge*

The most powerful of these systems, Armor, is located on the tracking ship *Monge* (Fig. 2.31), which displaces 21,040 tons, has dimensions of 230 m × 25 m, and mainly supports French ballistic missile tests. Outside test and support campaigns the *Monge* is stationed at Brest. Each of the two Armor tracking radars can observe up to three objects simultaneously at up to 4,000 km range. They generate high-resolution azimuth and elevation measurements, and range data. The accuracy of Armor and TIRA tracking data has been cross-calibrated and verified against high-precision ephemerides of SPOT-4. Armor tracking data and orbit determinations have in the past been called upon for similar applications as TIRA (support for risk object re-entry and conjunction event predictions). As for TIRA, Armor data too can be processed to generate ISAR images of spacecraft.

Table 2.5. Detection capabilities and modes of operation of some ground-based radar systems (source: (anon., 1999) and IADC). Capabilities of space surveillance networks (SSN and SSS, with detection thresholds of $d_{min} \geq 10$ cm) are not included in the list.

Country	Facility Organization	Type Configuration	Θ_{3dB} [°] λ [m]	Sensitivity d_{min} [cm] Operational status
Europe	Tromsø	dish antenna	0.70°	1.9 cm at 500 km
	EISCAT	bi-static	0.320 m	experimental
Germany	TIRA	dish antenna	0.50°	2 cm at 1,000 km
	FGAN	mono-static	0.225 m	experimental
Germany	Effelsberg	dish antenna	0.16°	0.9 cm at 1,000 km
	MPIfR	bi-static with TIRA	0.225 m	experimental
Japan	MU Radar	phased array	3.70°	2 cm at 500 km
	Kyoto Univ.	mono-static	6.400 m	operational
Japan	Uchinoura	dish antenna	0.40°	2 cm at 500 km
	ISAS	bi-static	0.130 m	experimental
Japan	Usuda	dish antenna	0.13°	2 cm at 500 km
	ISAS	bi-static	0.130 m	experimental
USA	Arecibo	dish antenna	0.04°	0.4 cm at 575 km
	NASA/NSF	bi-static	0.130 m	one-time test
Ukraine	Evpatoria	dish antenna	0.10°	0.3 cm at 1,000 km
	NSAU	bi-static	0.056 m	developmental
USA	Haystack	dish antenna	0.058°	0.6 cm at 1,000 km
	NASA/DoD	mono-static	0.030 m	operational
USA	HAX	dish antenna	0.10°	2 cm at 1,000 km
	NASA/DoD	mono-static	0.018 m	operational
USA	Goldstone	dish antenna	0.02°	0.2 cm at 500 km
	NASA	bi-static	0.035 m	operational
USA	TRADEX	dish antenna	0.61°	3 cm at 500 km
	DoD	bi-static	0.23 m	operational

Apart from the Fylingdales SSN sensor, the UK operates another radar for experimental space observations at Chilbolton (Winchester). It is owned by the Rutherford Appleton Laboratory (RAL). This monopulse S band (3 GHz) system is used mainly for atmospheric/ionospheric and for radio communications research. It has a 25 m parabolic dish antenna with a 3 dB beamwidth of 0.28°. After an upgrade of the transmitter system the radar will be able to track LEO objects with a detection threshold diameter of ~10 cm at 600 km altitude.

Recently, the EISCAT system in northern Europe has demonstrated its usefulness for space debris research. EISCAT is a network of European Incoherent Scatter Radars, with sites at Tromsø/Norway (UHF transmitter and receiver at 69.58°N), Kiruna/Sweden (UHF receiver at 67.87°N), Sodankylä/Finland (UHF receiver at 67.37°N), and Longyearbyen/Svalbard (UHF transmitter and receiver at 78.15°N). The EISCAT system has been designed for high-latitude atmospheric and ionospheric research. Its data, however, also contain information on LEO objects. After an upgrade of their system, the Tromsø transmitter with its 32 m antenna, and the co-located receiver performed a proof-of-concept campaign in February 2002. Within 4.5 hours, 56 objects of sizes from 0.5 m down to 1.9 cm diameter could be detected at altitudes of $490\,\text{km} \leq H \leq 1{,}480\,\text{km}$. These EISCAT debris measurements generate range and Doppler data only, and they are not sufficient to deduce complete orbits. However, they can be generated as low-cost, secondary products during routine atmospheric research, and they provide a valuable means for the validation of catalogs and debris models.

Table 2.5 gives an overview of ground-based radar facilities used for the observation of intact and debris objects in orbit. As before (for the optical sites), this list excludes sensors which are primarily assigned to operational space surveillance systems. It also includes some sensors which were not discussed in the text.

2.5 IN-SITU MEASUREMENTS AND RETRIEVED SURFACES

So far, only ground-based radar and optical systems have been analyzed as contributors to measurement data on the terrestrial space environment of man-made objects. The resolution of these data is limited by the attenuating effects of the Earth atmosphere, both in the visible and radio frequencies, and by the fact that the energy of reflected signals is dispersed according to $1/\rho^4$ for a radar, and according to $1/\rho^2$ for a telescope, where ρ, the minimum slant range from ground to the object, must exceed 250 km for an orbit of sufficiently reduced atmospheric drag and sustained lifetime.

The main limitations of ground-based systems can be overcome by moving the sensors well outside the denser Earth atmosphere, into orbits within or from which object detections at higher rates, or with better resolution, can be expected. Many scientific payloads with optical instruments operating in the visible, infra-red (IR), ultra-violet (UV), X-ray, or γ-ray spectrum have been in operation on Earth orbits since the early 1960s. Some of these missions have also generated useful data for an analysis of the space debris environment, sometimes as unintentional by-products of their primary mission objectives.

The Infra-Red Astronomical Satellite (IRAS), for instance, a joint development of the USA, the UK, and the Netherlands, was launched in January 1983 into a Sun-synchronous orbit of 99.1° inclination and a mean altitude around 900 km. Its primary mission was a systematic scan of the celestial sphere in IR wavelengths during an operational period of 10 months. In the course of some "data mining", almost 10 years after mission completion, a total of ~200,000 tracks of non-celestial objects was detected in the archived images, all of which had focal plane crossing speeds clearly deviating from a nominal value of 3.85 arcmin/s for a celestial object. Assuming black spheres of a temperature of 300 K, the Space Research Organization of the Netherlands (SRON) extracted 2,735 observations of orbital debris of diameters $d \leq 10$ cm, with a lower detection threshold of $d \approx 0.5$ mm. Most of the detections, predominantly at small sizes, were recorded for the altitude regime of 920 km to 1,000 km. The sensitivity of the IRAS sensor and thus the observable object size decreased by a factor of 100 towards object altitudes of 2,000 km (at a range of $\rho \geq 1,100$ km).

Another spacecraft with valuable optical measurements for the detection and characterization of space debris is the MSX satellite (Mid-Course Space Experiment) which was launched in April 1996 into an orbit of ~900 km altitude at 99.2° inclination. MSX is part of a global space surveillance system initiated by the US DoD, with the objective of detecting, tracking, and characterizing space objects. The payload consists of radiometers and telescopes operating in the IR, UV, and visible wavelengths. An important contributor to MSX measurements is the Space Based Visible (SBV) optical assembly, with a thermo-electrically cooled CCD system, and hence with a sustained operational lifetime, in contrast to some cryogenically cooled IR sensors which lost their performance within the first year of operation. The SBV has an aperture diameter of 15 cm and is believed to generate high-resolution measurements of objects at LEO to GEO altitudes, with detection size thresholds considerably lower than for the operational, ground-based space surveillance system.

Fig. 2.32. The Hubble Space Telescope (HST) prior to its docking with STS-102 (Discovery) in March 2001 (source: NASA).

A well known optical sensor in orbit is the Hubble Space Telescope (HST). The HST was deployed by Space Shuttle *Columbia* on April 24, 1990, into an orbit of 610 km mean altitude, at an inclination of 28.5° (Fig. 2.32). This is close to the maximum reachable altitude for a Shuttle launch from Kennedy Space Center with an HST payload mass of 10.8 tons. Moreover, with a stowed dimension of 13.3 m length and 4.3 m diameter the HST fully occupied the Shuttle cargo bay. The supply of power to the payload, with its f/24 Cassegrain telescope of 2.3 m aperture diameter, is provided by two symmetric roll-out solar arrays of 2.3 m $\times$ 12.0 m each, which together generate close to 5,000 W. Due to nominal solar cell degradation in the harsh radiation environment the HST arrays have to be replaced periodically. This was done in December 1993 (*Endeavor*, STS-61), and in March 2002 (*Columbia*, STS-109). During the replacement in 1993, only one array could be stowed in the cargo bay and returned to Earth for analysis, while the second array, which could not be rolled up due to deformations, was cut off and left in orbit for a natural re-entry five years later, in October 1998. The retrieved array underwent an extensive survey, and a detailed analysis of impact features and chemical residues of impactors to determine the abundance of objects as a function of size, and to discriminate between man-made and natural impactors. Since the HST solar arrays up to 2002 were supplied by the European Space Agency, most of the post-flight analysis was performed in Europe. The total surface of the returned solar array was $\sim$55 m^2 (including front and back side), of which $\sim$20 m^2 were covered with solar cells. The HST solar arrays consisted of a flexible matrix structure, solar cells, and a cover glass layer with an overall thickness of 710 μm. Some 3,600 impacts could be detected on the brittle cover glass of the front side of the solar array at a 10-fold magnification, covering crater sizes down to $\sim$100 μm ($\sim$400 μm features could be identified on the more ductile back side of the array). The corresponding impact flux on the front side was on the order of 1.6×10^{-6} m^{-2} s^{-1}. In 148 cases pairs of impact features could be detected on the front and back side, indicating a complete penetration. The largest holes measured 2 to 3 mm across, and the largest impact features had diameters of 6 to 7 mm (Drolshagen, 1995). A chemical residue analysis of crater pits with scanning electron microscopes (SEM) concluded that in the size regime covered 69% of the features were caused by meteoroids, and 10% by man-made space debris (21% could not be correlated). Of the man-made impactors most were related to solid rocket motor firings (98%), and some were attributed to paint flakes (2%). In the course of the extra-vehicular maintenance activity for HST in 1993, an impact hole with a diameter close to 2 cm was noted on one of the two high-gain antennas. A similar impact and chemical residue analysis as in 1993 will be performed on the HST solar arrays which were recovered by STS-109 (*Columbia*) in March 2002.

A very important source of in-situ measurements of the terrestrial meteoroid and space debris environment was the Long Duration Exposure Facility (LDEF), which was deployed by STS-41C (*Challenger*) in April 1984, into an orbit of 28.5° inclination and 475 km altitude, and which was retrieved in January 1990 by STS-32 (*Columbia*) from an altitude of 334 km. LDEF had a mass of 9.7 tons, a length of 9.14 m and a diameter of 4.27 m, with a total exposed surface of $\sim$130 m^2, which was partitioned into twelve rows of six experiment trays each on the circumference, and seven trays on each of the two ends of the cylinder (Fig. 2.33). In total 57 experiments were included in the mission. LDEF was passively stabilized

Fig. 2.33. The Long Duration Exposure Facility (LDEF) during its retrieval by STS-32 (*Columbia*) in January 1990 (source: NASA).

in a gravity gradient mode (long axis towards the local vertical), and in a yaw-stabilized mode (row number 9 always in flight direction), exploiting the Earth gravitational and magnetic field, respectively. Hence, data could later be correlated with impact directions. In the post-flight analysis, after a space exposure of LDEF for 5.7 years, more than 30,000 impact features were visible to the naked eye, of which 5,000 were larger than 0.5 mm. The largest crater had an diameter of ~5 mm, and it was presumably caused by an impactor of about 1 mm size.

The European Retrievable Carrier (EURECA) was a technology and research platform of 4.5 tons mass which was deployed by STS-46 (*Atlantis*) in July 1992 into an orbit of 28.46° inclination, at an altitude of 505 km. The entire spacecraft was retrieved 11 months later by STS-57 (*Endeavor*), and returned to ground for a post-flight inspection, also regarding impact features. The total exposed surface of EURECA was ~140 m^2, of which ~96 m^2 belonged to the front and back sides of the two solar arrays. With analysis methods similar to those outlined for HST, about 2,000 impact features could be identified on the brittle front sides of the solar arrays. All impacts larger than 0.65 mm were analyzed at high resolution; 703 of them were on the solar array front sides, 144 on the back sides, and 85 on the main body. This main body was largely covered with beta-cloth (woven sheets of glass fiber), followed by 21 layers of aluminized Kapton as multi-layer thermal insulation (MLI). Of 60 visible impacts on the beta-cloth, several reached a penetration depth through more than 10 layers of the MLI (Drolshagen, 1995). On the cover glass of the solar cells zones of shattered material showed extensions of 10 to 20 times the deduced impactor diameter (Fig. 2.35), while some impacts even resulted in complete penetrations with spallation of brittle surface material

(Fig. 2.34). The largest impact crater, on an exposed, ductile metal surface, had a diameter of 6.4 mm. The EURECA impact flux causing crater pit sizes of 200 μm to 300 μm was about 2 to 8 times lower than for HST, indicating a significant decrease with lower altitudes.

There were several other occasions when returned space hardware was analyzed for impact features, such as for the Japanese Space Flyer Unit (SFU), which was released in March 1995 and retrieved in January 1996, or for the Russian Salyut flights 1 to 7 (1971 to 1991) and the Mir mission (1986 to 2001). During a servicing mission in January 1998 by STS-89 (*Endeavor*), 10 m^2 of solar array panels were retrieved from Mir, after 10 years of space exposure.

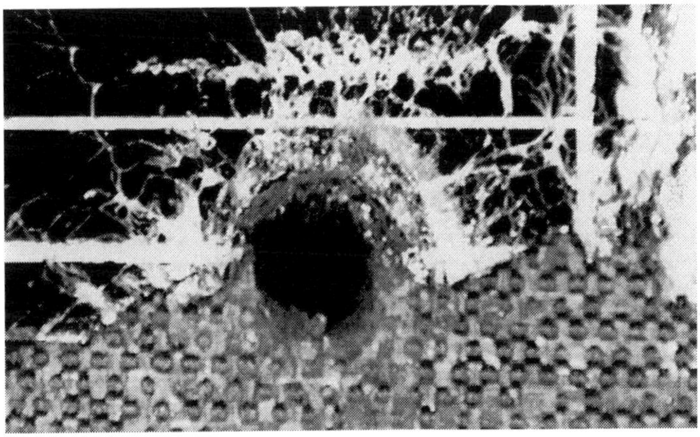

Fig. 2.34. Clear hole penetration of the EURECA solar array. The inner diameter of the hole is about 1.2 mm (grid finger spacing: 1.25 mm; source: ESA).

The analysis of returned hardware from space can give answers to a variety of questions, such as: origin of the impactor (man-made or meteoroid), deduced size of the impactor, impact velocity, and impact direction. Normally, there are only few chemical residues left of the impacting objects, which disintegrate and evaporate under pressures of more than 100 GPa and temperatures exceeding 10,000 K[2.2]. Such traces can, however, be sufficient to correlate the impactor with known chemical compositions of meteoroids, spacecraft surface materials, and man-made release products (e.g. SRM slag and dust, NaK droplets). A correlation of observed impact craters with the size of an impactor is more difficult, and it requires extensive test series of calibration shots with hypervelocity accelerators, which can reach orbital velocities for small particles under well defined laboratory conditions. From such tests damage equations can be derived, with coefficients which are calibrated for certain impactor materials and target configurations. A generic damage equation may look as follows,

$$d_c = c_0 \, d_p^{c_1} \, v_p^{c_2} \, (\cos \alpha_p)^{c_3} \, \rho_p^{c_4} / \rho_t^{c_5} \tag{2.10}$$

[2.2]at the Earth's center the pressure is ~365 GPa and the temperature is ~6,000 K with indices c, p, and t for crater, projectile, and target. d is a diameter, ρ is a density, v_p is the velocity, and α_p is the incident angle of the impact. The coefficients $c_0, ..., c_5$ are specific calibration constants for a certain impactor/target combination. A summary of representative sets of coefficients, and a more detailed excursus into the methodology of HVI damage equations is provided in Chapter 7, page 205ff, and in (McDonnell, 1992; Klinkrad et al., 2003).

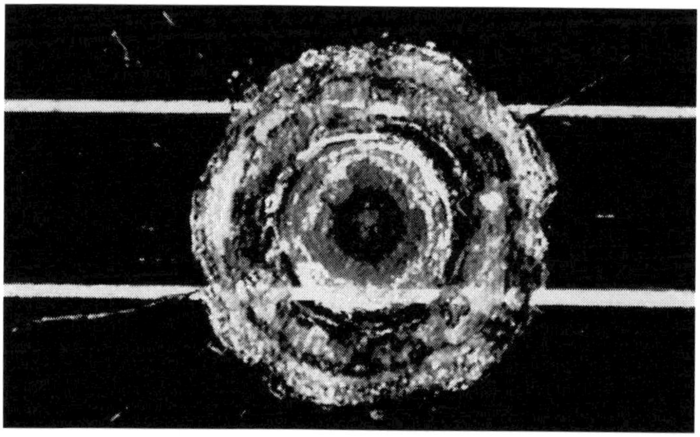

Fig. 2.35. Cratering impact on the EURECA solar array. The outer diameter of the crater is about 2 mm (grid finger spacing: 1.25 mm; source: ESA).

There is one quantity which cannot be recovered from the post-flight analysis of returned space surfaces, or which can only be obtained with very coarse time resolution: the epoch at which an impact occurred, and the corresponding position on the orbit. Since all impact features are convoluted, time-integrated data over the exposure duration, the mission length itself normally limits their time resolution. One way to improve the situation is by a controlled, time-tagged exposure of a surface, with prior and subsequent shielding inside a cassette (as in a camera). This idea was exploited in the Time-Band Capture Cell Experiment (TICCE) on-board EURECA, which recorded 126 visible impacts, some of which were clustered and time-correlated with certain orbit positions, probably due to passing a fresh cloud of objects from a fragmentation or SRM firing event.

A further step towards resolving impact time and location in space are on-orbit detectors, which time-tag impact measurements, characterize the events, and downlink corresponding data to ground for processing. In September 1996 ESA launched a Geostationary Orbit Impact Detector (GORID, Fig. 2.36) as secondary payload of the Russian Ekspress-2 satellite, which was later stationed at 80°E in the GEO ring. This detector is identical to those flown on the interplanetary Ulysses and Galileo missions. GORID has a circular capture area of 0.1 m^2 (with a diameter of 43 cm, covered by a lid during the launch phase), and an effective field-of-view of $\sim$140°. During its data collection in GEO, between April 1997 and June 2002, the axis of the instrument was aligned 5° earthward from the local

Fig. 2.36. The GORID impact plasma detector as used on the Ulysses, Galileo, and Ekspress-2 missions (sensor area: ~0.1 m^2; source: ESA).

horizontal plane, towards north, at an angle of 65° relative to the orbit velocity vector. GORID is an impact ionization detector, in which a particle impacting the gold-plated back-wall at hypervelocity speeds generates a plasma of electrons and ions. The two parts of the plasma are measured separately to deduce velocities from the signal rise times, and masses from the signal amplitudes, using empirical relationships derived from calibration experiments. The mass detection threshold at typical impact velocities is about 1×10^{-14} g. The mean number of impacts per day detected by GORID was on the order of 2.4, with peaks up to 50. There seems to be a strong correlation between the local solar time and the impact frequency, with a pronounced peak at midnight, and a minimum at noon. Also time-clusters of detections were noted during limited periods of a few days, perhaps indicating the pass of Ekspress-2 through the cloud of an SRM firing event.

A dedicated development of an in-situ particle flux detector, sponsored by ESA, is the Debris In-Orbit Evaluator (DEBIE, Fig. 2.37). This unit is very compact, with a mass of only 0.56 kg per sensor, and a data processor unit (DPU) with harness of 1.23 kg, which can service up to four sensors, each with a square capture area of 10 cm side length (0.01 m^2 area). DEBIE combines the physical principles of an impact plasma, momentum, and foil penetration detection device. Due to its small mass and low power consumption of 3 W it fits in as a secondary payload passenger on any terrestrial orbit. The first operational DEBIE was launched in October 2001, on-board the PROBA-1 satellite, into an eccentric orbit of 545 km × 663 km, at an inclination of 97.84°. Its data collection is still in progress. DEBIE and the development of other low-cost impact sensors, for instance based on metal oxide semiconductor (MOS) technology, tries to promote the idea of carrying in-situ impact detectors in larger numbers into all terrestrial

Fig. 2.37. The DEBIE impact plasma and momentum detector as used on the PROBA mission (sensor area: $\sim$0.01 m^2 per detector; source: ESA).

orbit regions, with near-complete spatial coverage, and continuous time coverage. Resulting measurements could greatly improve our knowledge particularly of the sub-millimeter size regime of debris and meteoroids.

In order to improve the knowledge on the share of debris and meteoroids in observed impact fluxes, also the idea of an intact capture of particulates has been studied. A particularly useful material for this purpose seems to be aero-gel (a transparent, flexible substance composed of air and silica-glass), which was for instance used on ESA's EURECA mission. It is expected than objects down to 1 μm can be identified within returned blocks of aero-gel. New technologies, applying the Raman effect of photon scattering at material-specific frequency shifts, may allow a non-destructive chemical analysis of the composition of the captured material.

During the previous discussion the NASA Space Shuttle fleet played in important role in retrieving exposed material and returning it to ground. However, as some of the largest structures shuttling in and out of space, the STS orbiters themselves are valuable impact flux detectors. Since 1981 the 2.5 m^2 window area on each returned Shuttle flight was inspected for impact damage. Through November 1995 a total count of 237 impact pits led to the replacement of 59 Shuttle windows. The largest impact craters were on *Endeavor* (STS-59, $d \approx 12$ mm, caused by a paint flake), on *Columbia* (STS-50, $d \approx 7$ mm, caused by metallic Titan), and on *Atlantis* (STS-7, $d \approx 4$ mm, caused by a paint flake). Depending on the Shuttle attitude, the window replacement rate for an average 10-day mission duration is between 0.2 (tail into velocity vector, cargo bay nadir pointing) and 3.6 (cargo bay into velocity vector, tail nadir pointing). Since 1992, about 120 m^2 of primary surfaces, including windows, radiators, and leading edges of the wings, are included in the post-flight survey (corresponding to $\sim$10% of the total exposed surface).

Table 2.6. Summary of retrieved surfaces and in-situ debris and meteoroid detectors (source: (anon., 1999) and IADC, sorted by time). The area refers to an exposed surface which was retrieved and analyzed, to the capture area of an in-situ detector, or to the aperture area of an on-orbit telescope. Sensors on extra-terrestrial orbits are not contained in the list.

Country	Organization Spacecraft	Timespan Orientation	H [km] i [°]	Area [m^2] Type
Russia	RSA Salyut 4 & 6	1974 – 1979 various	∼350 km 51.6°	7 m^2 retrieval
USA	NASA STS-7 window	Jun.1983 various	295 – 320 km 28.5°	2.5 m^2 retrieval
USA	NASA SMM	Feb.1980 – Apr.1984 Sun pointing	500 – 570 km 28.5°	2.3 m^2 retrieval
USA	JPL/RAL/SRON IRAS	Jan.1983 – Nov.1983 zenith pointing	∼905 km 99.1°	∼0.3 m^2 telescope
USA	NASA LDEF	Apr.1984 – Jan.1990 gravity gradient	340 – 470 km 28.5°	151 m^2 retrieval
Europe	NASA/ESA HST solar array	May 1990 – Dec.1993 Sun pointing	∼610 km 28.5°	62 m^2 retrieval
USA	NASA STS orbiters	1992 – present various	300 – 610 km 28.5° – 51.6°	> 100 m^2 retrieval
Europe	ESA EURECA	Jul.1992 – Jul.1993 Sun pointing	∼505 km 28.5°	131 m^2 retrieval
Canada	NASA/CSA STS-52	Oct.1992 various	350 km 28.4°	1 m^2 retrieval
Europe	NASA/ESA HST solar array	Dec.1993 – Mar.2002 Sun pointing	∼610 km 28.5°	62 m^2 retrieval
Japan	NASDA SFU	Mar.1995 – Jan.1996 Sun pointing	∼480 km 28.5°	50 m^2 retrieval
Europe	RSA/ESA Mir/EUROMIR	Oct.1995 – Dec.1993 gravity gradient	∼390 km 51.6°	0.06 m^2 retrieval
Russia	RSA Mir	1996 – 1998 various	∼390 km 51.6°	15 m^2 retrieval
USA	DoD/BMDO MSX	Apr.1996 – present various	∼900 km 99.2°	∼0.02 m^2 telescope
Europe	ESA Ekspress-2	Apr.1997 – Jun.2002 horizontal towards NE	∼35,800 km 0.2°	∼0.1 m^2 detector
Canada	CSA/NSAU Mir	Nov.1997 – Feb.1999 various	∼390 km 51.6°	1 m^2 retrieval
Europe	ESA PROBA	Oct.2001 – present various	550 – 680 km 97.0°	∼0.02 m^2 detector

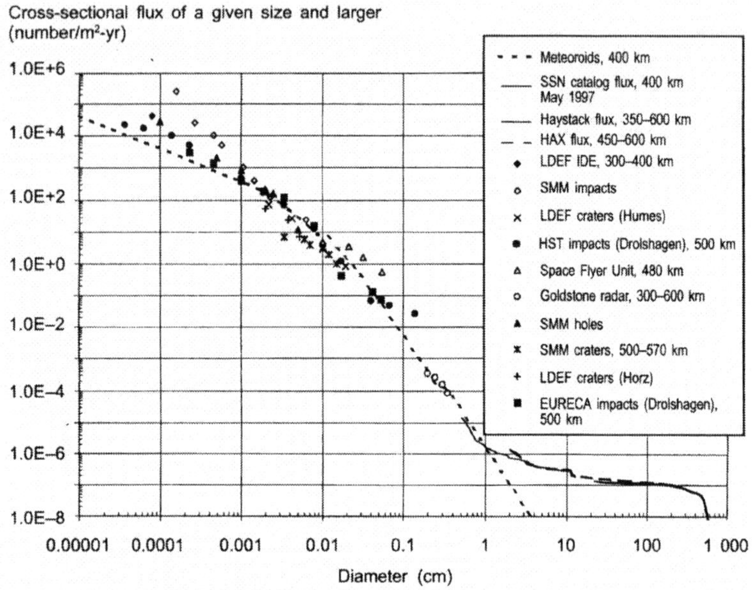

Fig. 2.38. Graphical overview of debris and meteoroid measurement data versus size. The data were acquired at inclinations near 28.5°, in various LEO altitudes, and at different epochs (anon., 1999).

A detailed investigation of 32 craters on STS-73 (*Columbia*) was conducted in October 1995. Of these larger-size impacts ($d > 0.35$ mm), 14 were on the payload door, 10 were on the windows, 6 were on the leading edge of the wing, and 2 were on elevon tiles. An analysis with a scanning electron microscope (SEM) and energy dispersive X-rays showed chemical residues in 15 of the craters. Of these, 9 were attributed to meteoritic origin and 6 to debris impacts. All top-ranking craters, with the largest one 17 mm across (on the payload door), were caused by man-made space debris, leaving chemical traces of metallic alloys (Al, Ti, Ni, Fe, Cr, Sn, Cu) and paint flakes (Si, Ca, Ti, Zn).

An overview of the history of in-situ sampling of the small-size terrestrial particulate environment is provided in Table 2.6. Fig. 2.38 shows how the in-situ measurements and retrieved surface data, of particles with typically $d < 1$ mm, can complete the picture of the collision flux environment, if they are combined with experimental ground-based observation data of objects with $d > 2$ mm (see Table 2.5), and with operational, ground-based surveillance data of objects with $d > 100$ mm. Such data of different origins must be carefully calibrated and translated into a common quantity, before they can be used as a combined dataset in the derivation of particulate environment models. Such calibration and conversion processes include a translation of photometric magnitude to size for optical observations, of radar cross-section to size for radar measurements, of crater diameter to size for in-situ impact measurements, and of voltages to size for in-situ flux detectors.

The measured space debris and meteoroid size spectrum in Fig. 2.38 (status of 1999) shows that the man-made environment exceeds the natural background for most sizes, except around 1 mm objects, where both sources are equal ranking. The chart also indicates that data are sparse in between diameters deduced from impact craters of returned surfaces (with $d \leq 1$ mm) and operationally tracked objects cataloged by USSTRATCOM (with $d \geq 10$ cm to 30 cm). This gap, which is at present only sampled by experimental ground-based radars (e.g. Haystack, Goldstone, TIRA/Effelsberg, and EISCAT), has recently been further reduced operationally, by extending the capabilities of the US Space Surveillance Network to track and correlate objects down to diameters of $d \geq 5$ cm to 10 cm. Between March 2002 and May 2003, almost 2,000 debris of this size class were tracked, and about 100 of these entered the catalog. This improvement is important also in view of ISS activities, since the Space Station is only capable of defeating objects of $d \leq 1$ cm with its given shield design, which constitutes a trade-off between mass and volume on one side, and maximum shieldable object diameter on the other side. If objects beyond the shield capacity can be reliably tracked, the ISS can perform evasive maneuvers to reduce the risk of a collision with objects which cannot be defeated by shielding.

2.6 REFERENCES

anon. (1995). *Orbital Debris – A Technical Assessment.* National Academy Press, Washington, DC.

anon. (1999). *Technical Report on Space Debris.* United Nations Publications, New York, USA.

anon. (2001). *Position Paper on Orbital Debris.* International Academy of Astronautics, Paris, France.

Bleeker, J., Geiss, J., and Huber, M. (2001). *The Century of Space Science – Volume I.* Kluwer Academic Publishers, Dordrecht, Boston, London.

Drolshagen, G. (1995). EURECA and HST Solar Array Post-Flight Impact Analysis. *Earth Space Review,* vol.4, no.2:15–18.

Hernández, C., Pina, F., Sánchez, N., Sdunnus, H., and Klinkrad, H. (2001). The DISCOS Database and Web Interface. In *Proceedings of the Third European Conference on Space Debris,* ESA SP-473, pages 803–807.

Johnson, N. L., Anz-Meador, P., Cisek, E., and Portman, S. (2001). History of On-Orbit Satellite Fragmentations, 12th edition. Technical Report JSC 29517, Orbital Debris Program Office, NASA/JSC, Houston, TX.

Klinkrad, H. (2002). Monitoring Space – Efforts Made by European Countries. In *Proc. of an International Colloquium on "Europe and Space Debris"*, ANAE Proceedings (on CD ROM).

Klinkrad, H. et al. (2003). *ESA Space Debris Mitigation Handbook.* ESA/ESOC, second edition. issue 1.0.

McDonnell, J. (1992). *Hypervelocity Impacts in Space.* University of Kent at Canterbury, Canterbury, UK.

Skolnik, M. I. (1990). *Radar Handbook.* McGraw-Hill, Boston, MA, second edition.

Verger, F., Ghiraradi, R., and Sourbès-Verger, I. (1997). *Atlas de la Géographie de l'Espace.* Éditions Belin, Paris, France.

3

Modeling of the Current Space Debris Environment

H. Klinkrad, P. Wegener, C. Wiedemann, J. Bendisch and H. Krag

The description of the space debris environment and its sources in Chapter 2 reflects a widely accepted, common understanding among space debris researchers. There are, however, different methods in existence for reproducing the observed environment by means of mathematical and physical models of release processes, for propagating orbits of release products, and for mapping the propagated environment onto spatial and temporal distributions of object densities, transient velocities, and impact fluxes. The subsequent chapters will focus on methods which have been developed at ESA, or under ESA contracts in context with ESA's MASTER-2001 model (Meteoroid and Space Debris Terrestrial Environment Reference, (Bendisch et al., 2002)). At the end of this chapter a general overview of some of the most prominent space debris environment models will be provided.

3.1 ORBIT PROPAGATION METHODS FOR LARGE POPULATIONS

A space debris environment as described in Chapter 2 is composed of individual source populations of a wide size range, which are generated at distinct times or during limited time spans, and which are released into source-specific orbital regimes. In order to analyze a combined population at a consistent 'snapshot' epoch, orbit prediction methods are required which make it possible to propagate individual objects, or clouds of debris over large timespans, with adequate accuracies, and with acceptable computation times. To meet such requirements, different orbit prediction methods are required for different propagation tasks.

The MASTER-2001 model assumes a population of more than 10^{+15} debris objects of diameters $d \geq 1\,\mu$m, $3.3 \times 10^{+8}$ of $d \geq 1$ mm, and $5.5 \times 10^{+5}$ of $d \geq 1$ cm. In order to propagate such large ensembles, the debris is sorted into classes of sufficient resolution, according to their size and orbit parameters. For each of these classes, only representative objects are propagated by means of a computer-efficient orbit theory. At small object sizes, the propagated objects may represent several thousand class members. At larger sizes, towards trackable objects, this representation factor converges to 1. Despite the reduction in CPU time achieved by class-wise propagation, the orbit prediction method needs to be fast, yet accurate enough to reflect the perturbation environment over long propagation time arcs.

The orbit propagator used for the MASTER development is known as FOCUS (Fast Orbit Computation Utility Software). FOCUS performs an integration in time of singly averaged Gauss perturbation equations (Eq. A.33–A.38), with mean time rates of change of the orbital elements determined from analytical solutions of Eq. A.46.

Assuming no cross-coupling between different perturbation types during one integration step, separate solutions of Eq. A.46 are determined for geopotential perturbations, airdrag, solar radiation pressure, and luni-solar perturbations. The adopted geopotential model considers the zonal harmonics J_2 through J_5 according to GEM-T1 (Seidelmann et al., 1992). These coefficients are applied to Liu's algorithms for singly averaged equations of motion, and to the first order (J_2) recovery of osculating elements (Liu and Alford, 1979). The airdrag model uses King-Hele's techniques (King-Hele, 1987) to derive rates of change of the mean elements for an object of a given ballistic parameter $B = c_D A/m$, and for an MSIS-77 based air density model which is dependent on epoch, orbit position, solar activities ($\overline{F}_{10.7}$, $F_{10.7}$), and geomagnetic indices (A_p). For the mean 10.7 cm solar radio flux $\overline{F}_{10.7}$ a moving average over 81 days (3 solar rotations) is adopted. The mean and daily solar activity proxies $\overline{F}_{10.7}$ and $F_{10.7}$, and the geomagnetic proxy A_p are read from a history file of observation data, or from a prediction file. The activity predictions are based on a McNish–Lincoln time-series analysis (Kerridge et al., 1989). The solar radiation pressure perturbations are computed with Aksnes' theory (Aksnes, 1976), for a given radiation pressure coefficient c_R and area-to-mass ratio A/m. The averaged perturbation equations consider an oblate, cylindrical Earth shadow, with closed-form solutions for the entry and exit positions. The third body perturbation analysis uses the Sun and Moon as point masses, with positions computed from analytical ephemerides (Seidelmann et al., 1992). The analytical averaging of the equations of motion is based on Roth's stroboscopic method (Roth, 1996).

The separately determined time rates of change of the mean elements due to the dominant perturbing forces are combined to form the right-hand sides of the 6 equations of motion $d\overline{\psi}_i/dt$ represented by Eq. A.46. Instead of mean Kepler elements $\underline{\overline{\psi}} = (\overline{a}, \overline{e}, \overline{i}, \overline{\Omega}, \overline{\omega}, \overline{M})$ equinoctial elements $\underline{\overline{\psi}} = (\overline{a}, \overline{e}\cos\overline{\omega}, \overline{e}\sin\overline{\omega}, \overline{i}, \overline{\Omega}, \overline{\omega} + \overline{M})$ are used, which do not have singularities for near-circular orbits. Since only long-periodic and secular variations are traced (curve [2] in Fig. A.5), the permissible propagation time steps can be on the order of a few days, instead of a few minutes when tracing the osculating states. The integration in time of the singly averaged perturbation equations is done with a 4th order Adams–Bashforth/Adams–

Moulton predictor/corrector method, which is initiated by a self-starting 4th order Runge–Kutta–Fehlberg method (see also (Montenbruck and Gill, 2000) and (Seidelmann et al., 1992)). In contrast to the Runge–Kutta–Fehlberg method the Adams–Bashforth predictor and Adams–Moulton corrector each evaluate the CPU demanding perturbation functions only once per integration step. The order of the integration algorithm is selected as 4, due to superior numerical stability as compared to higher orders.

Unless stated otherwise, all orbit predictions, and all analyses based on orbit state vectors will use an inertial, Earth-centered, mean equatorial coordinate system of 2000.0 (so-called "J2000"), which corrects for precession and nutation of the CIO pole (Conventional International Origin) at this date. In this system, $\underline{X}$ is towards the mean equinox, within the mean equatorial plane, $\underline{Z}$ is perpendicular to the mean equatorial plane, towards the CIO North pole, and $\underline{Y} = \underline{Z} \times \underline{X}$ completes the right-hand system (see Fig. A.2 and (Montenbruck and Gill, 2000)). All epochs hereafter will be given in coordinated universal time (UTC).

3.2 VOLUME DISCRETIZATION AND CELL-PASSAGE EVENTS

According to the laws of kinetic gas theory, the mean number of collisions c encountered by an object of collision cross-section A_c, moving through a stationary medium of uniform particle density D, at a constant velocity v, during a propagation time interval Δt is given by

$$c = v\,D\,A_c\,\Delta t \tag{3.1}$$

where $F = v\,D$ is the impact flux (in units of $\mathrm{m}^{-2}\,\mathrm{s}^{-1}$), and $\Phi = F\,\Delta t$ is the corresponding fluence (in units of m^{-2}). Since the collision process is following Poisson statistics, the probability $P_{i=n}$ of n impacts, and the probability $P_{i=0}$ of no impact is described by

$$P_{i=n} = \frac{c^n}{n!}\exp(-c) \quad \longrightarrow \quad P_{i=0} = \exp(-c) \tag{3.2}$$

The probability of one or more impacts is hence the complement of no impact.

$$P_{i\geq 1} = 1 - \exp(-c) \approx c \tag{3.3}$$

where c is given by Eq. 3.1. The approximation $P_{i\geq 1} \approx c$ yields results with less than 10% error for $c \leq 0.3$. Eq. 3.3 in context with Eq. 3.1 will provide the foundation for the subsequent impact probability and impact flux estimates. The focus will hence first be on the spatial and temporal distributions of the flux determining quantities v and D, and on their derivation for a large space debris population.

The MASTER model defines a 3-dimensional, time-varying space debris environment from altitudes below 200 km to well above the geostationary ring. The spatial resolution of the model is determined in inertial, spherical coordinates. A volume cell centered at geocentric distance r_i, declination δ_j, and right ascension

α_k, shall have the extensions Δr in radial direction, $\Delta\alpha$ along the small circle of latitude, and $\Delta\delta$ along the meridian of right ascension, as shown in Fig. 3.1.

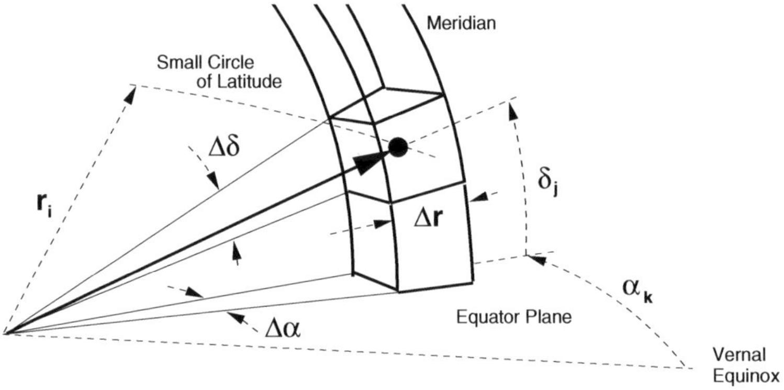

Fig. 3.1. Definition of control volume cells by means of spherical coordinates (r, α, δ) in an inertial, geocentric reference frame.

Let i, j, k be the indices in radial direction, declination, and right ascension, then the position vector $\underline{r}_{i,j,k}$ to the center of a given cell is defined as

$$\underline{r}_{i,j,k} = r_i \begin{pmatrix} \cos\alpha_k \cos\delta_j \\ \sin\alpha_k \cos\delta_j \\ \sin\delta_j \end{pmatrix} \tag{3.4}$$

The cell volume $V_{i,j,k}$ is determined from

$$V_{i,j,k} = \frac{2}{3}\left(3r_i^2 + \frac{1}{4}(\Delta r)^2\right) \cos\delta_j \sin\left(\frac{\Delta\delta}{2}\right) \Delta\alpha\,\Delta r \tag{3.5}$$

This equation is valid for all possible ranges of Δr, $\Delta\delta$, and $\Delta\alpha$, including spherical shells ($\Delta\delta = \pi$, $\Delta\alpha = 2\pi$) and latitude bands ($\Delta\alpha = 2\pi$). The latter option is used in MASTER for the medium altitude Earth orbit regime (MEO), where object densities are low. The full 3-dimensional discretization is used in the low-Earth orbit regime (LEO), and near the geostationary ring (GEO), where population densities are sufficiently high. The LEO, MEO, and GEO regimes are denoted as "control volumes". Each of these is discretized with individual resolutions in Δr, $\Delta\delta$, and $\Delta\alpha$, which are a trade-off between resource requirements (CPU time and mass memory) and requested accuracy. A control volume partition in uniform steps of $\Delta\delta$ and $\Delta\alpha$ goes along with a finer meridian grid spacing towards the poles. This will prove to be an advantage in the subsequent "cell passage event" (CPE) analysis, since flight azimuth angles in a horizontal system have the highest rates of change for high-inclination orbits at high latitudes.

After a set of adjacent control volumes with individual cell resolutions is defined, the intersections of a given space debris orbit with shells of constant radii

$\tilde{r}_i$, cones of constant declinations $\tilde{\delta}_j$, and meridian cross-section planes of constant right ascensions $\tilde{\alpha}_k$ can be determined.

$$\tilde{r}_i = r_{min} + i\,\Delta r \qquad \text{for} \quad i = 0, \ldots, I \tag{3.6}$$

$$\tilde{\delta}_j = \delta_{min} + j\,\Delta\delta \qquad \text{for} \quad j = 0, \ldots, J \tag{3.7}$$

$$\tilde{\alpha}_k = \alpha_{min} + k\,\Delta\alpha \qquad \text{for} \quad k = 0, \ldots, K \tag{3.8}$$

The corresponding cell centers are located at

$$r_i = \tilde{r}_i - \Delta r/2 \qquad \text{for} \quad i = 1, \ldots, I \tag{3.9}$$

$$\delta_j = \tilde{\delta}_j - \Delta\delta/2 \qquad \text{for} \quad j = 1, \ldots, J \tag{3.10}$$

$$\alpha_k = \tilde{\alpha}_k - \Delta\alpha/2 \qquad \text{for} \quad k = 1, \ldots, K \tag{3.11}$$

with $r_{min} = a_e + H_{min}$, $\Delta r = \Delta H$, and partitions as listed in Table 3.1. Each of the intersections with the cell borders $\tilde{r}_i$, $\tilde{\delta}_j$, and $\tilde{\alpha}_k$ is related to a position on the orbit defined by its true anomaly f.

For a given orbit the two possible intersections with a spherical shell of radius $\tilde{r}_i$ can be determined from Eq. A.6.

$$f_{\tilde{r},1} = \arccos\left(\frac{a\,(1-e^2) - \tilde{r}_i}{e\,\tilde{r}_i}\right) \tag{3.12}$$

$$f_{\tilde{r},2} = -f_{\tilde{r},1} \tag{3.13}$$

There is no intersection, if $\tilde{r}_i < r_{pe}$, or if $\tilde{r}_i > r_{ap}$.

The two possible crossings of a cone of constant declination $\tilde{\delta}_j$ are found from the spherical triangle between the ascending node, the orbit position, and the foot-point of the local meridian on the equatorial plane (see Fig. A.2).

$$f_{\tilde{\delta},1} = \arcsin\left(\frac{\sin\tilde{\delta}_j}{\sin i}\right) - \omega \tag{3.14}$$

$$f_{\tilde{\delta},2} = \pi - f_{\tilde{\delta},1} \tag{3.15}$$

No intersections are possible for inclinations $i < |\,\tilde{\delta}_j\,|$.

For a given meridian of right ascension $\tilde{\alpha}_k$, the same spherical triangle as before can be used to derive the single intersection point with an orbit according to Napier's rule.

$$f_{\tilde{\alpha}} = \arctan\left(\frac{\tan(\tilde{\alpha}_k - \Omega)}{\cos i}\right) - \omega + \hat{f}_{\tilde{\alpha}} \tag{3.16}$$

$$\hat{f}_{\tilde{\alpha}} = \begin{cases} 0 & \text{if} \quad \cos(\tilde{\alpha}_k - \Omega) \geq 0 \\ \pi & \text{if} \quad \cos(\tilde{\alpha}_k - \Omega) < 0 \end{cases} \tag{3.17}$$

For applications involving a stochastic debris population, the apparent singularities in Eq. 3.14 for $i \to 0$, and in Eq. 3.16 for $i \to \pi/2$ do not pose effective limitations.

For each true anomaly f of an intersection point according to Eq. 3.12–3.16 a corresponding elapsed time $t(f)$ since perigee passage can be computed.

$$t(f) = \sqrt{\frac{a^3}{\mu}} \left[2\,\mathrm{arctg}\left(\sqrt{\frac{1-e}{1+e}}\tan\left(\frac{f}{2}\right)\right) - \frac{e\sqrt{1-e^2}\sin f}{1+e\cos f}\right] \tag{3.18}$$

Let n be the counter for a debris object, let m be the cell passage event (CPE) counter along a single orbit of this debris object, and let ℓ be the counter for all orbit passes of a given debris population through one particular cell indexed by i, j, k. After sorting the events $m = 1, ..., M$ for one particular orbit n by ascending times, the centered CPE times $t_{\ell(i,j,k)}$, and the cell dwell times $\Delta t_{\ell(i,j,k)}$ for the ℓ-th cell transition are determined from

$$t_{\ell(i,j,k)} = \frac{1}{2}\left(t_{n,m+1} + t_{n,m}\right) \tag{3.19}$$

$$\Delta t_{\ell(i,j,k)} = t_{n,m+1} - t_{n,m} \tag{3.20}$$

For an analysis period of one single orbit with period T_n (larger time intervals can be shown to yield only minor improvements), the resident probability $P_{\ell(i,j,k)}$ in a given cell, and the resulting contribution $D_{\ell(i,j,k)}$ to the spatial object density in this cell are given by

$$P_{\ell(i,j,k)} = \frac{\Delta t_{\ell(i,j,k)}}{T_n} \tag{3.21}$$

$$D_{\ell(i,j,k)} = \frac{P_{\ell(i,j,k)}}{V_{i,j,k}} \tag{3.22}$$

If the total CPE count for a particular cell is $\ell(i,j,k) = 1, ..., L_{i,j,k}$, then the total object density is

$$D_{i,j,k} = \frac{1}{V_{i,j,k}} \sum_{\ell(i,j,k)=1}^{L_{i,j,k}} P_{\ell(i,j,k)} \tag{3.23}$$

In a collision flux analysis, the "debris medium" assumed in Eq. 3.1 is not at rest. Instead, each debris object n moves on a Kepler orbit, which at its m-th cell passage resides in a volume bin indexed by i, j, k. Let this be the ℓ-th CPE for this cell, then the debris velocity vector $(\underline{v})_{X,Y,Z}$ in the $\underline{X}, \underline{Y}, \underline{Z}$ coordinate system at the centered CPE time $t_{\ell(i,j,k)}$ is

$$\underline{v}_{\ell(i,j,k)} = \left(\underline{v}_{\ell(i,j,k)}\right)_{X,Y,Z} = \underline{v}_{n,m-1} + \frac{1}{2}\left(\underline{v}_{n,m} - \underline{v}_{n,m-1}\right) \tag{3.24}$$

One can show that the space debris population across all size regimes is fairly uniformly distributed in right ascension α. This characteristic can be exploited to perform averaging operations with respect to α, by considering CPE data from all cells along a common small circle of latitude. One can even go one step further and perform an additional averaging across a range of declination δ. These averaging operations are considerably simplified, if the transient CPE debris velocities are expressed in a local, quasi-horizontal coordinate system in radial, east, and north direction ($\underline{U}, \underline{E}, \underline{N}$). This conversion is accomplished by two successive rotations of $(\underline{v}_{\ell(i,j,k)})_{X,Y,Z}$ about $\underline{Z}$ by $+\alpha$, and about the new $\underline{Y}'$ by $-\delta$ to obtain $(\underline{v}_{\ell(i,j,k)})_{U,E,N}$.

$$(\underline{v}_{\ell(i,j,k)})_{U,E,N} = \mathrm{R}_{Y'}(-\delta)\,\mathrm{R}_Z(+\alpha)\;(\underline{v}_{\ell(i,j,k)})_{X,Y,Z} \tag{3.25}$$

$$\begin{pmatrix} v_{\ell(i,j,k),U} \\ v_{\ell(i,j,k),E} \\ v_{\ell(i,j,k),N} \end{pmatrix} = \begin{pmatrix} \cos\alpha\cos\delta & \sin\alpha\cos\delta & \sin\delta \\ -\sin\alpha & \cos\alpha & 0 \\ -\cos\alpha\sin\delta & -\sin\alpha\sin\delta & \cos\delta \end{pmatrix} \begin{pmatrix} v_{\ell(i,j,k),X} \\ v_{\ell(i,j,k),Y} \\ v_{\ell(i,j,k),Z} \end{pmatrix} \tag{3.26}$$

In the $\underline{U}, \underline{E}, \underline{N}$ coordinate system the transient velocity of the ℓ-th CPE in the cell indexed by i, j, k can be expressed by its magnitude $v_{\ell(i,j,k)}$, and its direction $A_{\ell(i,j,k)}$ and $h_{\ell(i,j,k)}$ (see Fig. A.3), where $A_{\ell(i,j,k)} \in [-\pi, +\pi]$ is the azimuth, measured from north over east in the quasi-horizontal plane, and $h_{\ell(i,j,k)} \in [-\pi/2, +\pi/2]$ is the elevation, measured from the quasi-horizontal plane, positive towards space.

$$v_{\ell(i,j,k)} = |\,\underline{v}_{\ell(i,j,k)}\,| = \sqrt{v^2_{\ell(i,j,k),U} + v^2_{\ell(i,j,k),E} + v^2_{\ell(i,j,k),N}} \tag{3.27}$$

$$A_{\ell(i,j,k)} = 2\arctan\left(\left(\sqrt{v^2_{\ell(i,j,k),E} + v^2_{\ell(i,j,k),N}} + v_{\ell(i,j,k),E}\right)/v_{\ell(i,j,k),N}\right) \tag{3.28}$$

$$h_{\ell(i,j,k)} = \arcsin\left(v_{\ell(i,j,k),U}/v_{\ell(i,j,k)}\right) \tag{3.29}$$

In computer programs the nominator and denominator in Eq. 3.28 are separately assessed for the arctan evaluation, avoiding singularities for $v_{\ell(i,j,k),N} \to 0$.

At a given epoch, the entire space debris population can be processed with the described procedure to obtain all cell passage events for every cell of each user-defined control volume. Apart from the spatial density contribution $D_{\ell(i,j,k)}$ and the transient velocity $v_{\ell(i,j,k)}$, with pass directions $A_{\ell(i,j,k)}$ and $h_{\ell(i,j,k)}$, any additional object-related information can be stored under the CPE index $\ell(i, j, k)$. This may include the debris mass, size, area-to-mass ratio, orbital elements, country of origin, launch date, type (e.g. intact or debris object, spacecraft- or rocket-related), and source (e.g. fragmentation, SRM firing, or NaK release event). Several such 3-dimensional population "snapshots" can be taken at successive epochs to arrive at a space debris environment model with a spatial and temporal resolution which is only limited by computer resources. The outlined CPE-based concept was first published by (Klinkrad, 1993).

Table 3.1 shows which discretization levels in H, δ, and α are applied in the MASTER-2001 model. The partitions vary between the different control volumes LEO, MEO, and GEO. Averaging in right ascension is applied only in MEO for macro source populations (TLE catalog objects, fragments, SRM slag, and NaK droplets), to yield acceptable spatial densities. All other sub-populations are discretized in 3-dimensional volume partitions. Different segmentations are also applied within single control volumes to account for characteristics of individual sub-populations and flux assessment methods. In spite of different partitions, results according to Eq. 3.21–3.29 can be combined in a global collision flux analysis.

Table 3.1. MASTER-2001 discretization of the control volumes LEO, MEO, and GEO, as a function of debris sub-populations. All ranges in right ascension are from $\alpha = -180°$ to $+180°$. Ranges in declination are from $\delta = -90°$ to $+90°$ ($-20°$ to $+20°$ in GEO). Micro sources comprise surface degradation products and SRM dust. Macro sources include TLE catalog objects, fragments, SRM slag, and NaK droplets. Altitudes refer to a spherical Earth of radius $a_e = 6378.137$ km.

	H_{min} [km]	H_{max} [km]	ΔH [km]	$\Delta\alpha$ [°]	$\Delta\delta$ [°]	Cells	Sources
LEO	186	2,286	10	10	2	680,400	Catalog, NaK droplets
	186	2,286	20	20	5	68,040	Fragments, SRM Slag
	186	2,286	100	10	4	34,020	Micro Sources
MEO	2,286	34,786	500	180	5	4,968	Macro Sources
	2,286	34,786	500	10	4	111,780	Micro Sources
GEO	34,786	36,786	20	10	2	324,000	Macro Sources
	34,786	36,786	100	10	4	32,400	Micro Sources

In Sections 3.3 to 3.7 the different space debris sources which contribute to the overall particulate environment will be analyzed with respect to physical interpretations and mathematical formulations of their release processes. A supporting general overview of the current space debris environment and its historic build-up is provided in Sections 2.1 to 2.3.

3.3 THE TRACKABLE SPACE OBJECT POPULATION

Sections 2.1 and 2.2 describe that part of the space object population (intact objects and debris) which is observable from ground-based sensors in an operational manner. As of January 2002, due to the detection size thresholds of operational space surveillance systems (predominantly between 10 cm in LEO and 1 m in GEO), this trackable population exclusively consisted of intact satellites, orbital stages, and mission-related objects (intact objects), and of fragments resulting from 172 explosions and 3 collisions. Together, these population members are denoted as "TLE catalog objects", since they are maintained by US Space Surveillance Network (SSN) in a list of about 8,500 unclassified, deterministically known orbits of correlated objects, each of which is described in terms of doubly averaged Kepler-like elements within two lines of data (known as "two-line elements", or TLE). Due to the detection limitations at the lower size threshold of the SSN catalog, the real number of objects at diameters of $d \geq 10$ cm is expected to be considerably larger (almost by a factor 3), primarily due to undetected fragments. This estimate is supported by experimental, optical observations in LEO, and by radar observations down to sizes of $d \geq 5$ cm, following the integration of the Cobra Dane phased array radar into the SSN in 2002. Though today the lower size regime of the upgraded catalog is still incomplete, the knowledge on the $d \geq 10$ cm population should be greatly improved. The reported number of objects of the extended SSN catalog (including so-called "analyst TLEs") of ~14,000 is supporting the trend shown by independent optical observations and experimental radar observations.

Of the two main populations of the trackable space object population (intact objects and debris) only the fragmentation debris is incompletely tracked from ground. Based on independent observation data, one can estimate its underrepresentation in the SSN catalog by an empirical calibration factor c_{TLE}.

$$c_{TLE} = 10^{\frac{1}{2}\exp\left(-2.464\left(\log_{10}\left(\frac{d_{SSN}}{[\mathrm{m}]}\right)+1.22\right)^2\right)} \tag{3.30}$$

c_{TLE} is a function of an altitude-dependent detection diameter threshold $d_{SSN}(H)$, which has been empirically fitted to published SSN performances.

$$d_{SSN}/[\mathrm{m}] = \begin{cases} 0.089 & \text{for } H \leq 620\,\mathrm{km} \\ 10^{-2.737+0.604\log_{10}\left(\frac{H}{[\mathrm{km}]}\right)} & \text{for } 620\,\mathrm{km} < H \leq 1,300\,\mathrm{km} \\ 10^{-6.517+1.819\log_{10}\left(\frac{H}{[\mathrm{km}]}\right)} & \text{for } 1,300\,\mathrm{km} < H \leq 3,800\,\mathrm{km} \\ 1.0 & \text{for } H \geq 3,800\,\mathrm{km} \end{cases} \tag{3.31}$$

Corrections by means of Eq. 3.30 should be applied to ground-based SSN observation data in the LEO and lower MEO region, in order to deduce a "true" space debris environment at small catalog and sub-catalog sizes. The correction shows a normal distribution in a logarithmic scale of d_{SSN}, with a peak $c_{TLE,max} \approx 3.2$ at diameters of $d_{SSN} \approx 6$ cm (i.e. below the operational threshold of SSN). The correction drops off to less than 5% for $d_{SSN} < 0.4$ cm and for $d_{SSN} > 100$ cm.

As a consequence of the calibration factor c_{TLE}, the modeled MASTER-2001 population with diameters $d > 10$ cm consist of 17,832 objects (as compared with a catalog size of ~8,500 unclassified objects), with 5,128 launch and mission-related objects, and with 12,704 pieces of fragmentation debris.

3.4 MODELING EXPLOSION AND COLLISION FRAGMENTS

Since the first recorded on-orbit explosion of a Thor Agena D orbital stage in 1961 (see Table 2.2) fragmentations have been the most dominant source of space debris in the size regime from a few millimeters to a few decimeters. This size range includes particularly hazardous objects with $1\,\mathrm{cm} \leq d \leq 10\,\mathrm{cm}$, which are neither shieldable by current on-orbit technology, nor operationally trackable by surveillance networks. The MASTER-2001 model estimates a total of ~370,000 of such fragments for the year 2001, with ~142,000 thereof in LEO altitudes.

Section 2.2 provides a summary of the time history, the assessed causes, and the top-ranking catalog object counts for 175 on-orbit fragmentations up to January 2002. Except for 3 collisions, all fragmentations were caused by explosions, of which only 2 could be detected in the GEO ring. The number of catalog objects in GEO at any epoch since 1970 has been on the order of 5% to 10% of the corresponding LEO population (compare Fig. 2.1 and 2.4). If one assumes that fragmentation rates are proportional to the number of satellites and upper stages, regardless of the orbit environment in which they reside, then one could expect between 9 and 18 GEO fragmentations in total. Indeed, observations of the GEO region with the ESA space debris telescope (SDT), with detection thresholds down to $d \geq 15$ cm, confirmed a large GEO population with 73% un-cataloged objects of magnitudes between 16 and 20, corresponding to diameters between 60 cm and 10 cm (Schildknecht et al., 2001). The deduced sizes, and the observed orbit clusters in a right ascension versus inclination diagram led to the conclusion that these objects originate from fragmentation events in the GEO region. From the position of the orbit inclination vectors $\underline{i} = (\sin i \cos\Omega, \sin i \sin\Omega)$ in the Ω, i-diagram the age of the debris clouds can be deduced, and corresponding fragmentation events can be generated at an inferred epoch. With this (iterative) procedure 11 simulated GEO explosion events were added to the 2 known GEO fragmentations in the MASTER-2001 model. The 168 historic LEO fragmentation events up to the model cut-off date in April 2001 were adopted with no amendments.

Fragmentation events in MASTER-2001 are simulated with a NASA break-up model (Johnson et al., 2001), which was produced for the long-term debris environment analysis software EVOLVE 4.0. This model, and its algorithms for the determination of cross-sections, masses, and imparted velocities, were validated for debris sizes of $d \geq 1$ mm against cataloged debris clouds (for explosions and collisions), and against ground-based impact and explosion test data.

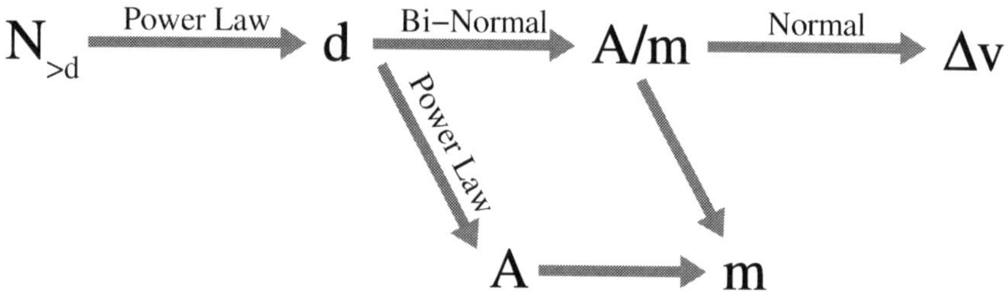

Fig. 3.2. Conceptual outline of the size-based derivation of cross-section A, mass m, and imparted velocity Δv in the NASA break-up model.

The principle of the NASA break-up model is sketched in Fig. 3.2. It starts off with the definition of a characteristic length $l_c = (l_x + l_y + l_z)/3$, which is the average of the 3 principal geometric axes of a fragment. For reasons of consistency, l_c will hereafter be treated as an "equivalent diameter" d. The cumulative number of explosion or collision fragments N_f of diameters $d \geq l_c$ can be determined from

$$N_f(d \geq l_c) = \begin{cases} 6\, c_S\, \hat{l}_c^{-1.6} & \text{for explosions} \\ 0.1\, \hat{m}^{0.75}\, \hat{l}_c^{-1.71} & \text{for collisions} \end{cases} \tag{3.32}$$

The following notations are used: $\hat{l}_c = l_c/[\text{m}]$ and

$$\hat{m} = \begin{cases} \dfrac{m_t + m_p}{[\text{kg}]} & \text{for } \tilde{E}_p \geq \tilde{E}_p^\star \\[2ex] \dfrac{m_p\, v_i}{1000\,[\text{kg m/s}]} & \text{for } \tilde{E}_p < \tilde{E}_p^\star \end{cases} \quad \text{with} \quad \tilde{E}_p = \frac{m_p\, v_i^2}{2\, m_t}\,,\ \tilde{E}_p^\star = 40\,[\text{kJ/kg}]$$

with ^ denoting normalized quantities, m_t the mass of a collision target object (mostly spacecraft or upper stage), m_p the mass of an impact projectile, v_i the impact velocity, $\tilde{E}_p$ the specific energy of the projectile, and $\tilde{E}_p^\star$ the specific energy threshold for a catastrophic collision (complete disintegration). The scaling parameter c_s in Eq. 3.32 is an event-specific calibration constant for historic events, and an empirical correction for certain classes of fragmentation events in the case of future projections (with $0.1 \leq c_s \leq 1.0$). For historic events the NASA model assumes a size limit of $l_c \geq 10$ cm for cataloged cloud objects of an on-orbit explosion. When inserting this threshold into Eq. 3.32 (top), then for a given number of trackable fragmentation objects $N_f(d \geq 10\,\text{cm})$ the calibration constant c_s is

$$c_s = c_{TLE} = \frac{N_f(d \geq 10\,\text{cm})}{238.864} \tag{3.33}$$

which is independent of the fragmentation orbit. Hence, for an improved approximation in the MASTER-2001 model, an altitude adjusted calibration factor c_{TLE} according to Eq. 3.30 is applied.

Following the outline of the NASA break-up model in Fig. 3.2, the area-to-mass assignments for generated fragments are performed by means of a bi-modal probability density function $p(\chi, \vartheta)$.

$$p(\chi, \vartheta) = \xi(\vartheta)\, p_1(\chi) + (1 - \xi(\vartheta))\, p_2(\chi) \tag{3.34}$$

with the normally distributed probability density functions

$$p_i(\chi) = \frac{1}{\sigma_i \sqrt{2\pi}} \exp\left(- \frac{(\chi - \mu_i)^2}{2\sigma_i^2} \right) \qquad \text{where} \quad i = 1, 2 \tag{3.35}$$

and with the area-to-mass parameter $\chi = \log_{10}(\{A/m\}/[\text{m}^2\,\text{kg}^{-1}])$, the size parameter $\vartheta = \log_{10}(d/[\text{m}])$, and the means μ_i and standard deviations σ_i of two overlaid and normalized Gaussian distributions. $\xi(\vartheta) \in [0, 1]$ are weighting functions. For objects of $d > 11$ cm the NASA model discriminates between fragments from spacecraft and rocket stages, with calibrated quantities $\xi(\vartheta)$, $\mu_{1,2}(\vartheta)$, and $\sigma_{1,2}(\vartheta)$. For spacecraft one gets $\xi \in [0.5, 1.0]$, $\mu_1 \in [-0.9, -0.45]$, $\mu_2 = -0.9$, $\sigma_1 = 0.55$, and $\sigma_2 \in [0.1, 0.28]$. For orbital stages one finds $\xi \in [0.0, 1.0]$, $\mu_1 \in [-0.95, -0.6]$, $\mu_2 \in [-2.0, -1.2]$, $\sigma_1 \in [0.1, 0.3]$, and $\sigma_2 \in [0.3, 0.5]$. For smaller fragments of $d < 8.0$ cm for spacecraft, and of $d < 1.7$ cm for upper stages, the second term in Eq. 3.34 is dropped ($\xi = 1.0$) to generate a single-mode, normal distribution with parameters $\mu_1 \in [-1.0, -0.3]$ and $\sigma_1 \in [0.2, 0.2 + 0.1333(\vartheta + 3.5)]$.

Fig. 3.3 illustrates the concept of bi-modal distributions at catalog sizes, and single-mode distributions at smaller sizes. In between there is a definition gap from $d = 8$ cm to 11 cm for spacecraft, and from $d = 1.7$ cm to 11 cm for rocket bodies. This gap can be closed by comparing a generated random number $\zeta_d \in [0.0, 1.0]$ with a value $\tilde{\zeta}(d)$ from Eq. 3.36.

$$\tilde{\zeta} = \begin{cases} 10\,\left(\log_{10}(d/[\mathrm{m}]) + 1.76\right) & \text{for rocket bodies} \\ 10\,\left(\log_{10}(d/[\mathrm{m}]) + 1.05\right) & \text{for satellites} \end{cases} \tag{3.36}$$

If $\zeta > \tilde{\zeta}$, then the appropriate bi-modal distribution for large size fragments is applied, else, the single-mode distribution for smaller objects is used.

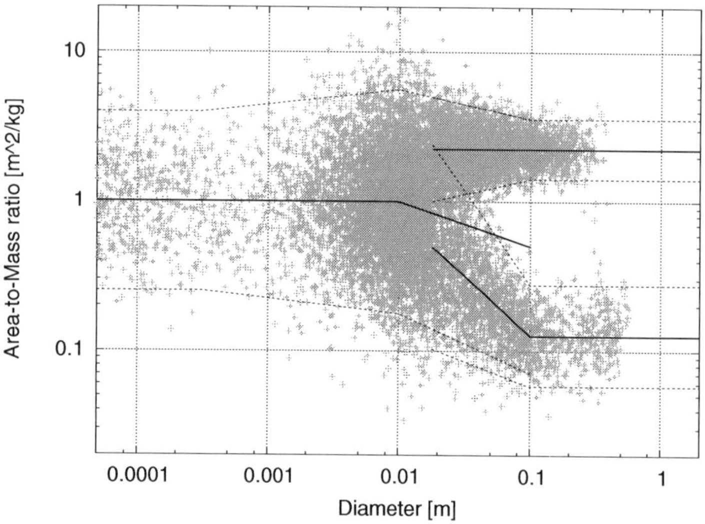

Fig. 3.3. Bi-modal area-to-mass distributions A/m for a Pegasus break-up event (see Table 2.2), with means and $\pm 2\,\sigma$ dispersions as a function of the characteristic length l_c, based on the NASA break-up model.

The effective cross-section A in the NASA break-up model is an explicit function of the fragment diameter d, according to

$$A/[\mathrm{m}^2] = \begin{cases} 0.540424\;(d/[\mathrm{m}])^2 & \text{for } d < 1.67\,\mathrm{mm} \\ 0.556945\;(d/[\mathrm{m}])^{2.0047077} & \text{for } d \geq 1.67\,\mathrm{mm} \end{cases} \tag{3.37}$$

The fragment mass is then simply determined from

$$m = \frac{A}{A/m} \tag{3.38}$$

with A from Eq. 3.37, and A/m statistically sampled via Eq. 3.34.

For implementations in computer programs a drawn pseudo-random number $\zeta_p \in [0.0, 1.0]$ has to be related to a probability $P(\chi, \vartheta)$ associated with the probability density function $p(\chi, \vartheta)$ of Eq. 3.35. This requires the integrals

$$P_i(\chi) = \int_{-\infty}^{\chi} p_i(\chi)\,\mathrm{d}\chi = \frac{1}{2}\,\mathrm{erf}\left(\frac{\chi - \mu_i}{\sigma_i \sqrt{2}}\right) + \frac{1}{2} \qquad \text{with} \quad i = 1, 2 \tag{3.39}$$

where $\mathrm{erf}(x)$ is the error function. This leads to the desired probability

$$P(\chi, \vartheta) = \xi(\vartheta)\, P_1(\chi) + (1 - \xi(\vartheta))\, P_2(\chi) \tag{3.40}$$

In the most general case of a bi-modal distribution (i.e. for large objects), the resulting A/m value needs to be determined iteratively.

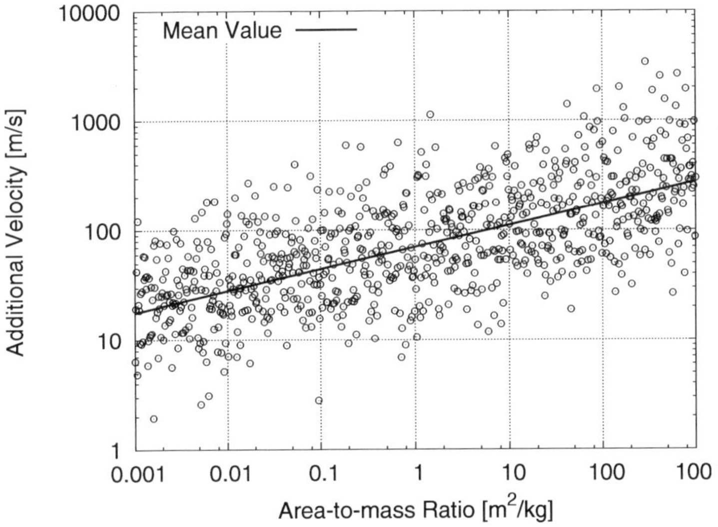

Fig. 3.4. Imparted velocities Δv as a function of the area-to-mass ratio A/m according to the NASA break-up model. A fictitious, uniform probability density distribution of A/m is assumed.

The probability density function of the imparted fragmentation velocities in the NASA model follows a normal distribution.

$$p(\nu) = \frac{1}{\sigma_\nu \sqrt{2\pi}} \exp\left(-\frac{(\nu - \mu_\nu)^2}{2\sigma_\nu^2}\right) \tag{3.41}$$

with $\nu = \log_{10}(\Delta v/[\mathrm{m\,s^{-1}}])$, $\chi = \log_{10}(\{A/m\}/[\mathrm{m^2\,kg^{-1}}])$, and the cause-dependent mean value and standard deviation given by

$$\begin{aligned} \mu_\nu &= 0.2\,\chi + 1.85 \qquad \sigma_\nu = 0.4 \qquad \text{for explosions} \\ \mu_\nu &= 0.9\,\chi + 2.90 \qquad \sigma_\nu = 0.4 \qquad \text{for collisions} \end{aligned} \tag{3.42}$$

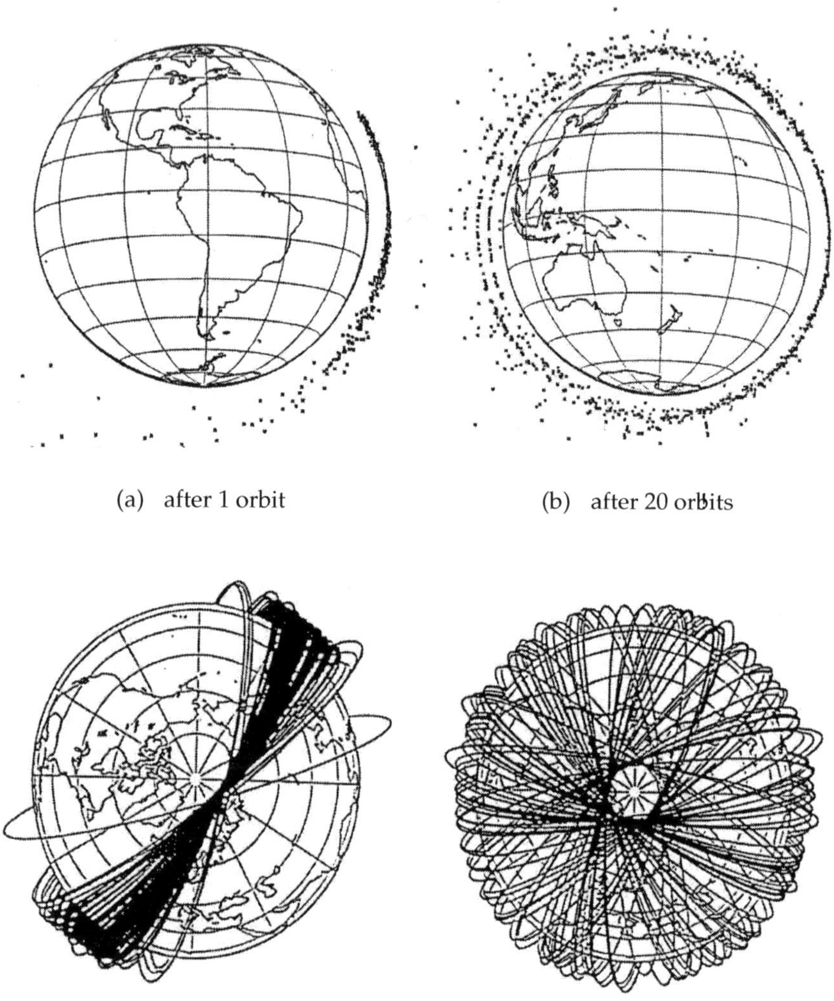

(a) after 1 orbit (b) after 20 orbits

(c) after 3 months (d) after 4 years

Fig. 3.5. Modeled time evolution of a fragmentation cloud generated by the explosion of an Ariane-1 H-10 upper stage on Nov. 13, 1986 (see also Table 2.2). The initial H-10 orbit was at $H_{pe} = 803$ km, $H_{ap} = 833$ km, and $i = 98.7°$. The explosion occurred during an ascending equator pass. Fig. (a, b) show along-track and radial dispersions, and Fig. (c, d) show cross-track dispersions of fragment orbits.

Due to its single-mode normal distribution, Eq. 3.41 can be directly solved with the aid of Eq. 3.39, yielding the probability

$$P_\nu(\chi) = \frac{1}{2}\,\mathrm{erf}\left(\frac{\nu - \mu_\nu}{\sigma_\nu\sqrt{2}}\right) + \frac{1}{2} \tag{3.43}$$

From this equation ν can be isolated and converted into Δv.

$$\Delta v/[\mathrm{m\,s}^{-1}] = 10^{\mu_\nu + \sigma_\nu \sqrt{2}\,\mathrm{erf}^{-1}(2P_\nu - 1)} \tag{3.44}$$

Fig. 3.4 shows a distribution of induced velocities as a function of the A/m ratio of fragments. A fictitious, uniform probability density in A/m is adopted to demonstrate the Δv spreading over the entire area-to-mass range.

Most of the historic fragmentation events occurred on near-circular orbits, and about 80% of all known events took place in LEO. The majority of these were associated with inclination bands between 65° and 100° (see Table 2.2). To understand the dynamics of a fragmentation cloud, the explosion of an Ariane-1 H-10 cryogenic upper stage (1986-019C) shall be investigated in more detail. This stage, which delivered the SPOT-1 satellite into orbit on February 22, 1986, exploded 9 months later, on November 13, 1986, near the equator, during an ascending pass on its near-circular, Sun-synchronous orbit of altitude 803 km × 833 km, and inclination 98.7°. Fig. 3.5 shows the resulting time evolution of a modeled fragmentation cloud (similar plots have been reproduced from catalog data).

For small imposed velocities the initial spreading of fragments is governed by the Gauss perturbation equations Eq. A.33–A.39. A closer inspection of these equations shows that for a given magnitude Δv, the most noticeable effects will be for $\Delta \underline{v}$ orientations tangential to the orbit in flight direction ($+\Delta v_t$), or opposite to the direction of motion ($-\Delta v_t$). These velocity changes have the largest effect on orbital energy, on the semimajor axis ($\Delta a \approx 2\,a\,\Delta v_t/v$), the orbit period ($\Delta T = \frac{3}{2}\,T\,\Delta a/a$), and the eccentricity ($\Delta e \approx \Delta a/a$). For a spherically uniform $\Delta \underline{v}$ distribution at the H-10 fragmentation epoch, Fig. 3.5 (a) and (b) show the cloud dispersion after 1 and after 20 orbits, with most of the fragments remaining close to the initial orbital plane. Those objects which are released opposite to the flight direction have a new perigee at $\Delta H_{pe} \approx -2\,\Delta a$. For $\Delta v_t \leq -200\,\mathrm{m/s}$, the perigees lead to an immediate re-entry, and for smaller magnitudes they lead to a reduced lifetime due to aero-capture in the Earth atmosphere. This explains an along-track cloud asymmetry, with the number of trailing cloud objects prevailing over the leading ones. Due to the laws of celestial mechanics, manifested by Eq. A.33–A.39, all members of a uniformly released fragmentation cloud, will (though at different times) revisit their point of origin, which is hence called the "pinch point". They will also pass through a common line, connecting the new apogees and perigees, opposite to the point of origin. This line is denoted as the "pinch line" (see Fig. 3.5 (a) and (b)).

During the early evolutionary phases of a fragmentation cloud the object concentrations near the "pinch point" and "pinch line" can greatly enhance local spatial densities and cause a temporal increase of collision risk in these areas, until orbit perturbations disperse the concentration peaks. Also for this reason several operators of space missions and launch services routinely check their planned flight trajectories for encounters with catalog objects, particularly after newly detected on-orbit fragmentation events.

The altitude and eccentricity evolution of a fragmentation cloud with time can

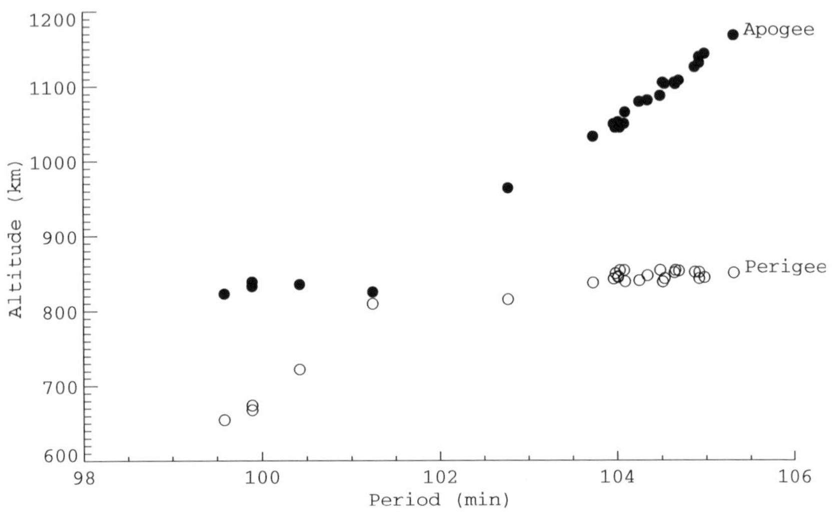

Fig. 3.6. Gabbard diagram of cataloged objects of a Zenith upper stage (1993-016B), 9 years after its explosion on March 27, 1993 (• → H_{ap} and ∘ → H_{pe} for each orbit).

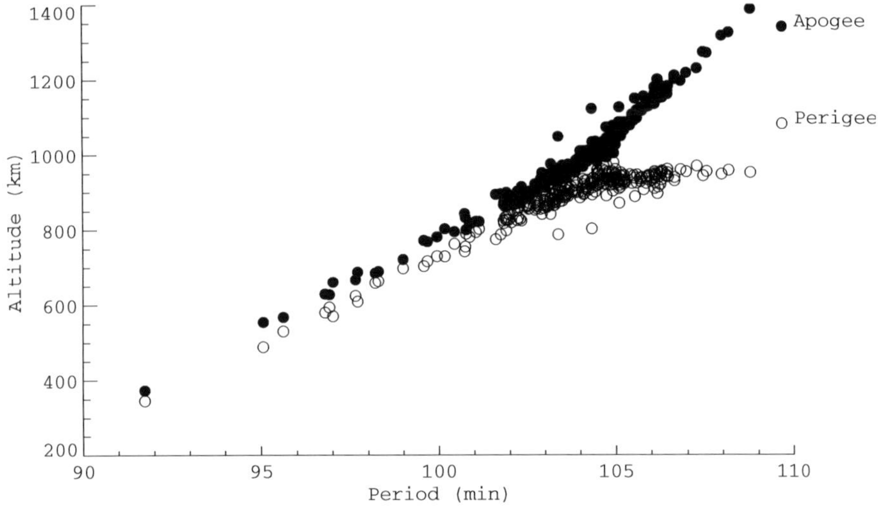

Fig. 3.7. Gabbard diagram of cataloged objects of the Cosmos 1275 satellite (1981-053A), 22 years after its explosion on July 23, 1981 (• → H_{ap} and ∘ → H_{pe} for each orbit).

be characterized by Gabbard diagrams, which show the perigee and apogee altitudes of all cloud members against their orbital period. Fig. 3.6 shows a Gabbard plot of the fragmentation cloud of a Zenith second stage 9 years after its explosion on March 28, 1993, on a near-circular orbit of ~845 km altitude and 71° inclination (in total 30 fragments of this event were cataloged by 2002). The Gabbard diagram shows the characteristic X-shape, with the top part representing the apogees, and the bottom part showing the perigees. Objects near the crossing point received no along-track Δv_t (hence no change in period). Objects left of the crossing point maintained their apogees and lowered their perigees (and orbital periods) due to $-\Delta v_t$, and objects towards the right maintained their perigees and raised their apogees (and orbital periods) due to $+\Delta v_t$. Over longer timespans, due to the exponential increase of atmospheric density with decreasing altitude, the perigees on the left-hand branch of the X-shape are affected by airdrag. The airdrag at perigee passes is acting like an impulsive braking maneuver and lowers the apogees until the orbits are circularized. At this point in time the leftmost objects in the Gabbard diagram start to re-enter. This phase is illustrated in Fig. 3.7 for the fragmentation cloud of the Cosmos 1275 satellite, 22 years after its explosion on 24 July, 1981, on an orbit of 940 km $\times$ 1,014 km at 83° inclination (for this event a total of 305 fragments were cataloged by 2002).

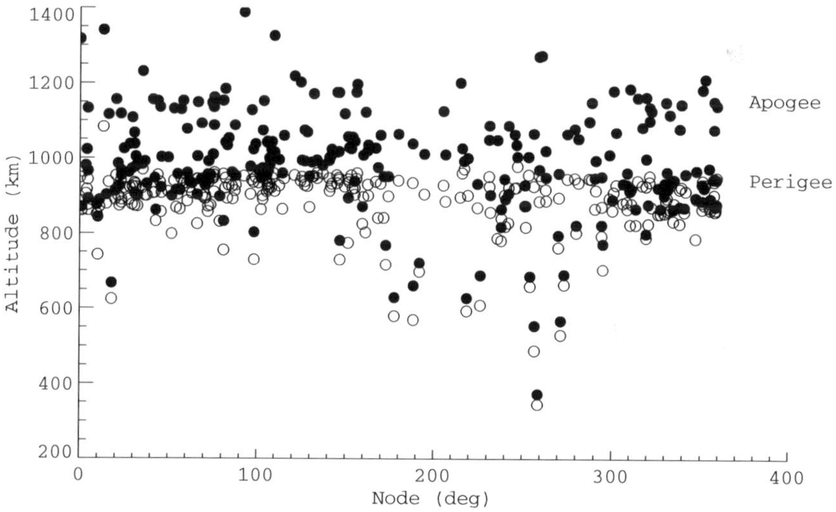

Fig. 3.8. Node dispersion of cataloged objects of the Cosmos 1275 satellite (1981-053A), 22 years after its explosion on July 23, 1981 ($\bullet \rightarrow H_{ap}$ and $\circ \rightarrow H_{pe}$ for each orbit).

Eq. A.35 and A.37 suggest that fragmentation-induced Δv components in an out-of-plane direction should affect the orientation of the orbit plane. This, however, requires a tilt of the orbit momentum vector, which is known to require a large Δv_w to become significant ($\Delta v_w \approx 140\,\mathrm{m/s}$ for $\Delta\Omega = 1°$). For area-to-mass ratios of catalog objects, this direct effect is overruled by a cumulative, secondary effect on the nominal node rotation $\dot{\Omega}(J_2)$ according to Eq. A.54. Due to changes in the semimajor axes a of the orbits (caused by the fragmentation process itself, or by airdrag), the nodal drift undergoes a differential drift $\Delta\dot{\Omega} = -\frac{7}{2}\,\dot{\Omega}\,\Delta a/a$. Fig. 3.5 (c) and (d) illustrate how this effect causes the fragment orbits to migrate

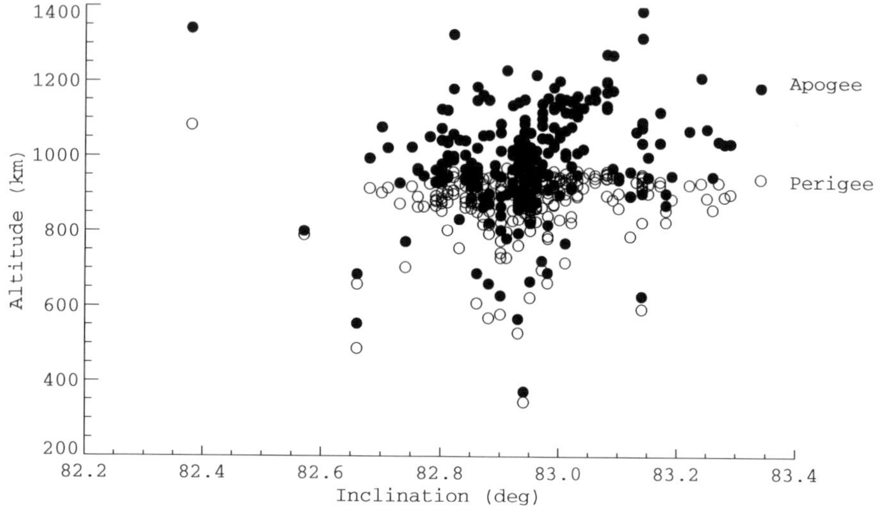

Fig. 3.9. Inclination dispersion of cataloged objects of the Cosmos 1275 satellite (1981-053A), 22 years after its explosion on July 23, 1981 (• $\rightarrow$ H_{ap} and ∘ $\rightarrow$ H_{pe} for each orbit).

from the initial orbital plane of the Ariane-1 H-10 upper stage. For the initial orbit altitude of 780 km and inclination of 98.52° the nominal node drift due to Earth oblateness is $\dot{\Omega} \approx -0.9856°/\text{day}$. It takes about 4 years until the orbits with prograde differential rotation catch up with those of retrograde differential rotation. Fig. 3.8 shows that even after 22 years the Cosmos 1275 fragmentation cloud has not reached a uniform nodal distribution.

In contrast with the orbit node, the inclination does not undergo a secular first-order perturbation (see Eq. A.53). Hence, as shown in Fig. 3.9, the initial changes during the fragmentation event are almost unaffected by time.

3.5 MODELING SOLID ROCKET MOTOR SLAG AND DUST

In Section 2.3 aluminum oxide residues (Al_2O_3) of more that 1,000 solid rocket motor (SRM) firings were identified as the most important non-fragmentation debris source. Of the estimated 1,000 tons of SRM propellant released into space, probably 320 tons were converted into Al_2O_3 dust, typically of sizes $0.01\,\mu m \leq d \leq 50\,\mu m$, and 4 tons may have been ejected as slag, of assumed sizes $0.1\text{ mm} \leq d \leq 30\text{ mm}$, composed of Al_2O_3 and motor liner material.

In order to reproduce the historic evolution of the SRM slag and dust population at orbital altitudes, it is essential to draw from a good database of known SRM firing events. For the MASTER-2001 development a database with 1,032 entries was compiled, based on a core set of data made available by NASA. The database contains information on the spacecraft or rocket body, the firing date, the spacecraft orbit before and after the firing, the firing position on the orbit, and the solid rocket motor (type, propellant mass m_p, ejection velocity w_e, motor case volume, and spin rate). The inertially fixed attitude of the motor axis during each firing event needs to be reconstructed from these core data.

Let δ_m and α_m describe the inertial pointing direction of the SRM thrust direction along the axis of symmetry of the motor. The velocity of an ejected SRM particle in the inertial frame is then determined as

$$\underline{v}_e(t) = \underline{v}(t) + w_e\, \underline{u}_e \tag{3.45}$$

where $\underline{u}_e$ is the thrust direction unit vector, aligned with the motor axis of symmetry.

$$\underline{u}_e = \begin{pmatrix} \cos\alpha_m \cos\delta_m \\ \sin\alpha_m \cos\delta_m \\ \sin\delta_m \end{pmatrix} \tag{3.46}$$

For some common maneuver types, $\underline{u}_e$ can be described in terms of the elements of the pre-burn orbit. For in-plane elliptic transfers with tangential $\Delta\underline{v}$ one gets

$$\underline{u}_{e,\parallel} = \begin{pmatrix} -\sin u \cos\Omega - \cos u \sin\Omega \cos i \\ -\sin u \sin\Omega + \cos u \cos\Omega \cos i \\ \cos u \sin i \end{pmatrix} \tag{3.47}$$

with $u = \omega + f$, and $\mp\underline{u}_{e,\parallel}$ in the case of orbit raising and orbit lowering maneuvers. For orbit plane changes one finds

$$\underline{u}_{e,\perp} = \begin{pmatrix} \sin\Omega \sin i \\ -\cos\Omega \sin i \\ \cos i \end{pmatrix} \tag{3.48}$$

where $\mp\underline{u}_{e,\perp}$ applies for inclination raise and reduction maneuvers, respectively.

During a firing event, the spacecraft velocity $\underline{v}(t)$ used in Eq. 3.45 changes both its magnitude and orientation according to

$$\underline{v}(t) = \underline{v}(t_0) - \Delta v(t)\, \underline{u}_e \tag{3.49}$$

The time-dependent $\Delta v(t)$ in Eq. 3.49 reflects the Ziolkowsky equation for the induced velocity change due to a constant mass ejection at velocity w_e with uniform propellant mass flow $\dot{m}_p$.

$$\Delta v(t) = w_e \ln\left(\frac{m_0}{m_0 - \dot{m}_p\, \Delta t}\right) \tag{3.50}$$

with $t_0 \le t < t_b$, for an ongoing burn of duration $\Delta t = t - t_0$. The complete system mass before ignition is m_0. For the total $\Delta v(t_b)$ obtained at the end of the firing $\dot{m}_p\, \Delta t_b = m_p$ is applied in Eq. 3.50.

In order to determine maneuver times and locations, one must try to detect an intersection of the pre-burn and post-burn orbits. If such a position is located with acceptable accuracy, such that $\underline{r}(t_0) = \underline{r}_{pre}(t_0) \approx \underline{r}_{post}(t_0)$ for an epoch t_0, then the required maneuver for the orbit change is given as $\Delta\underline{v}(t_0) = \underline{v}_{post}(t_0) - \underline{v}_{pre}(t_0)$. This quantity, derived from two sets of TLE catalog data, can be used to verify the foregoing model assumptions. In many cases, however, $\Delta\underline{v}(t_0)$ is necessary to fill knowledge gaps on the burn duration and mass flow.

Many solid rocket motors have nozzles which protrude into the burn chamber, creating a "dead water" zone during the burn process, where Al_2O_3 dust particles and torn-off motor liner material are trapped, and where they can cumulate to form SRM slag objects which can reach a few centimeters in size. The MASTER-2001 slag model is based on work at NASA (Ojakangas et al., 1996), and at MIT Lincoln Laboratory.

Results from ground-based firing tests suggest that 50% of the slag material is due to Al_2O_3 with density[3.1] $\rho_{s,Al} = 3.5\,\text{g/cm}^3$, and 50% results from liner material with density $\rho_{s,Li} = 1.8\,\text{g/cm}^3$. Based on radar observations of STAR-37 SRM plumes, the cumulative number of slag objects larger than a threshold size d_c can be expressed as

$$N_s(d > d_c) = N_s^\star(d > d^\star)\,\frac{m_p}{m_p^\star}\,\frac{\zeta + d^{\star 3}}{\zeta + d_c^3} \tag{3.51}$$

where ζ is an empirical fitting function introduced in (Bendisch et al., 2002), which allows one to extend the slag model beyond the lower bound of the originally specified range of validity of $5\,\text{mm} \le d \le 3\,\text{cm}$ (for which $\zeta = 0$).

$$\zeta = \frac{N_s^\star(d > \tilde{d}) \cdot \tilde{d}^3 - N_s^\star(d > d^\star) \cdot d^{\star 3}}{N_s^\star(d > d^\star) - N_s^\star(d > \tilde{d})} \tag{3.52}$$

In Eq. 3.51 and Eq. 3.52, $d^\star = 5\,\text{mm}$, $N_s^\star(d > d^\star) = 1{,}800$, and $m_p = 700\,\text{kg}$ are reference parameters for the STAR-37 experiment, m_p is the propellant mass of the SRM to be modeled, d_c is the cut-off diameter of the cumulative distribution, $\tilde{d} = 10\,\mu\text{m}$ is an extended lower size threshold, and $N_s^\star(d > \tilde{d}) = 3 \times 10^{+7}$ is the corresponding number of generated SRM slag particles with $d > \tilde{d}$.

During the burn process the slag particles are captured in the dead-water region between nozzle and motor case. As the chamber pressure drops towards the propellant depletion, the SRM slag can escape into space. The MASTER-2001 model assumes that 25% of all produced slag objects according to Eq. 3.51 are ejected in the last 25% of the burn duration, and 75% are released right after burn completion. Due to lack of detailed data, a mass-independent ejection velocity of $w_{e,s} = 75\,\text{m/s}$ is adopted, with ejection directions randomly distributed within a cone of 20° from the symmetry axis of the nozzle. All slag particles are considered to be solid spheres of a density of $\rho_s = \frac{1}{2}(\rho_{s,Al} + \rho_{s,Li}) = 2.65\,\text{g/cm}^3$. The

[3.1]Chemically pure Al_2O_3 has a density of $3.97\,\text{g/cm}^3$. The adopted value of $3.5\,\text{g/cm}^3$ takes into account gas inclusions.

corresponding, diameter-dependent SRM slag masses m_s and area-to-mass ratios $(A/m)_s$ are

$$\frac{m_s}{[\mathrm{kg}]} = 1.388 \times 10^{+3} \left(\frac{d}{[\mathrm{m}]}\right)^3 \qquad \frac{(A/m)_s}{[\mathrm{m}^2\,\mathrm{kg}^{-1}]} = 5.660 \times 10^{-4} \left(\frac{d}{[\mathrm{m}]}\right)^{-1} \tag{3.53}$$

A more in-depth theoretical analysis of SRM slag release dynamics is performed by (Meyer, 1992). Meyer takes into account spin-stabilization during firing, full gas and particle dynamics, and detailed information on the geometry and propellant distribution of the motor. He bases his analysis on a PAM-D (STAR-48) booster. Due to the poor availability of data on SRM spin and motor characteristics this model feature has not been activated in MASTER-2001.

Of the estimated 4 tons of released SRM slag particles in 1,032 historic firing events, an estimated 3 tons were still on orbit in 2001. This is due to the low release velocities, and due to the relatively low A/m ratios, which go along with reduced airdrag and radiation pressure perturbations. The mass of slag objects is mainly spread across a size spectrum of $10\,\mu\mathrm{m} \leq d_s \leq 3\,\mathrm{cm}$. The MASTER-2001 model assumes a population of $171,000$ of $d > 1\,\mathrm{cm}$ ($22,000$ thereof in LEO), $1.52 \times 10^{+8}$ of $d > 1\,\mathrm{mm}$ ($8.23 \times 10^{+6}$ thereof in LEO), and $5.05 \times 10^{+11}$ of $d > 10\,\mu\mathrm{m}$ ($8.33 \times 10^{+9}$ thereof in LEO).

A second, more abundant SRM-related population is called SRM dust. These small particles of typically $0.01\,\mu\mathrm{m} \leq d_d \leq 50\,\mu\mathrm{m}$ are Al_2O_3 residues of about 18% aluminum powder which is added to the solid propellant to improve the combustion process. In the form of Al_2O_3 grains in the exhaust stream they comprise about 34% of the original fuel mass. The SRM dust model which is used in MASTER-2001 is based on research by (Kessler, 1985) and (Akiba et al., 1990), who use data primarily obtained from ground-based SRM firing tests of a PAM-A and of a Mu-3B motor.

The number N_d of SRM dust particles larger than a cut-off diameter d_c can be computed from

$$N_d(d > d_c) = \begin{cases} \dfrac{m_p}{[\mathrm{kg}]}\,\hat{N}_c \exp\left(-2.0\,\dfrac{d}{[\mu\mathrm{m}]}\right) & \text{for} \quad 0.01\,\mu\mathrm{m} \leq d < 1.5\,\mu\mathrm{m} \\ \dfrac{m_p}{[\mathrm{kg}]}\,\hat{N}_c \exp\left(-0.66\,\dfrac{d}{[\mu\mathrm{m}]} - 2.0\right) & \text{for} \quad 1.5\,\mu\mathrm{m} \leq d \leq 50\,\mu\mathrm{m} \end{cases} \tag{3.54}$$

where m_p is the total mass of the solid propellant, and $\hat{N}_c$ is the number of dust particles with $d \geq 1.5\,\mu\mathrm{m}$, generated for 1 kg of fuel. $\hat{N}_c = 7.013 \times 10^{+14}$ can be determined by integrating mass over the applicable diameter range in the second function of Eq. 3.54 (for $\hat{N}_c = 1$ and $m_p = 1$ kg), according to

$$\frac{\hat{m}_d}{[\mathrm{kg}]} = 3.456\,\frac{\rho_d}{[\mathrm{kg}/\mu\mathrm{m}^3]} \int\limits_{d=1.5\,\mu\mathrm{m}}^{50\mu\mathrm{m}} \left(\frac{d}{[\mu\mathrm{m}]}\right)^3 \exp\left(-0.66\frac{d}{[\mu\mathrm{m}]} - 2.0\right) \mathrm{d}(d) \tag{3.55}$$

By comparison with the mass integral of the first function in Eq. 3.54 one finds that more than 90% of the mass is concentrated in diameters of $d > 1.5\,\mu\mathrm{m}$. The resulting $\hat{m}_d$ of Eq. 3.55 is in units of kilograms per $\hat{N}_c$. The inverse enters into Eq. 3.54 as a normalized number of objects for each kilogram of propellant mass.

The initial 18 mass-% of aluminum powder contained in the propellant mass m_p is converted into approximately 34 mass-% of Al_2O_3 during the combustion. Since slag particle mass m_s also contains 50 mass-% liner material, one can estimate the total mass available for SRM dust production from

$$m_d \approx 0.34\, m_p \;-\; 0.5\, m_s \tag{3.56}$$

From the analysis of test data it is known that the overall slag mass is small as compared to the ejected Al_2O_3 dust, with a ratio of typically $m_s/m_d \approx 0.01$. Hence, with good approximation the second term in Eq. 3.56 can be neglected (as was done in context with Eq. 3.54).

To obtain a steady thrust profile, solid rocket motors are generally designed such that the exposed propellant surface remains nearly constant while the fuel is consumed during the combustion process. Hence, SRM dust is produced at a steady rate throughout the motor firing. The magnitude of the dust ejection velocity v_d can be expressed as a function of the particle diameter d_d.

$$\frac{v_d}{[\mathrm{m/s}]} = 2{,}900\ \exp\left(-0.07\left(\frac{d_d}{[\mu\mathrm{m}]}\right)^{0.6}\right) \tag{3.57}$$

with velocities ranging from 2,900 m/s to 1,400 m/s between the lower and upper bound of the SRM dust size range ($0.01\,\mu\mathrm{m} \le d_d \le 50\,\mu\mathrm{m}$). According to (Kessler, 1985) the angular extension $\theta_{d,max}$ of the ejection cone, centered on the motor axis $\underline{u}_e$ (see Eq. 3.45–3.48), can be approximated by

$$\frac{\theta_{d,max}}{[^\circ]} = \begin{cases} 60.0\ \exp\left(-0.95\left(\frac{d}{[\mu\mathrm{m}]}\right)^{0.3}\right) & \text{for} \quad 0.01\,\mu\mathrm{m} \le d < 15\,\mu\mathrm{m} \\ 24.0 & \text{for} \quad 15\,\mu\mathrm{m} \le d \le 50\,\mu\mathrm{m} \end{cases} \tag{3.58}$$

The probability density function p_{θ_d} spanning the range of ejection cone angles $0 \le \theta_d \le \theta_{d,max}$ is assumed to be of elliptical shape, depending on the normalized cone angle $\hat{\theta}_d = \theta_d/\theta_{d,max}$.

$$p(\hat{\theta}_d) = \frac{4}{\pi}\sqrt{1-\hat{\theta}_d^2} \tag{3.59}$$

The resulting probability $P(\hat{\theta}_d)$ is hence

$$P(\hat{\theta}_d) = \frac{4}{\pi}\int_{\hat{\theta}_d}^{1}\sqrt{1-\hat{\theta}_d^2}\ \mathrm{d}\hat{\theta}_d \tag{3.60}$$

$$P(\hat{\theta}_d) = 1 - \frac{2}{\pi}\left(\hat{\theta}_d\sqrt{1-\hat{\theta}_d^2} + \arcsin\hat{\theta}_d\right) \tag{3.61}$$

For a randomly selected $P(\hat{\theta}_d) \in [0,1]$ the corresponding $\hat{\theta}_d$, and hence the angle θ_d, can be iteratively determined from Eq. 3.61. The distribution of the ejection

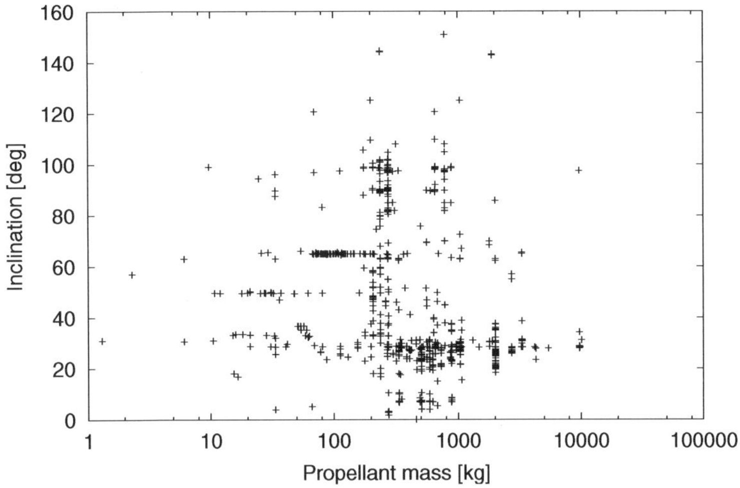

Fig. 3.10. Initial orbit inclination versus SRM propellant mass for SRM supported injection maneuvers.

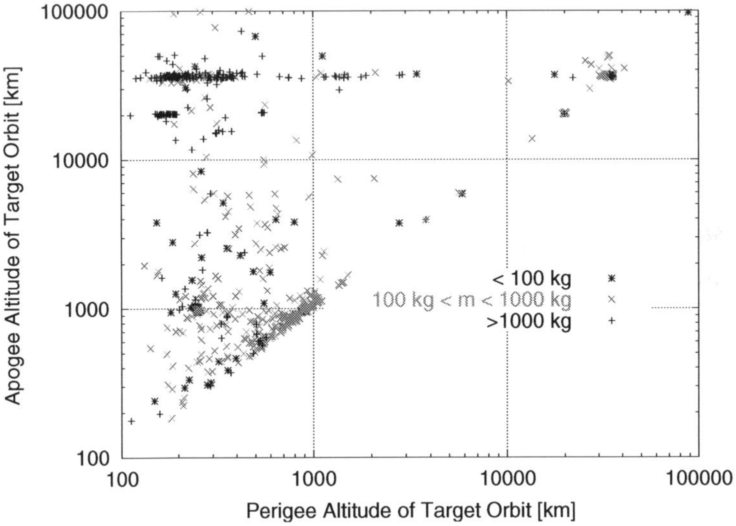

Fig. 3.11. Perigee and apogee altitudes of the delivery orbits for SRM supported injection maneuvers. The SRM propellant masses are indicated by color codes (see color plate on page 371).

azimuth angle A_d is assumed to be uniform, and rotationally symmetric around the motor axis, with random positions within $A_d \in [-\pi, +\pi]$.

Of the estimated 320 tons of released SRM dust particles up to the year 2001, only an estimated 2 tons were assumed to be still on orbit at that time. This is partly due to high release velocities (1,400 to 2,900 m/s), which for the frequent GEO insertion boosts exceed the initial GTO apogee velocity (~1,500 m/s), and may reach the final GEO velocity (~3,000 m/s), thus leading to an immediate re-entry of a large share (typically 50%) of the SRM dust. The remainder is often left in highly eccentric orbits, sometimes of retrograde inclinations. These orbits are strongly perturbed by solar radiation pressure, and (for low perigees) by airdrag, entailing short on-orbit resident times as a consequence of the high area-to-mass ratios of the small dust grains (since $(A/m)_d \propto 1/d_d$). The mass of dust objects is mainly concentrated in a size spectrum of $1\,\mu\text{m} \leq d_d \leq 50\,\mu\text{m}$. The MASTER-2001 model predicts a population of $1.38 \times 10^{+14}$ of $d > 10\,\mu\text{m}$ ($1.92 \times 10^{+13}$ thereof in LEO), and $7.22 \times 10^{+15}$ of $d > 1\,\mu\text{m}$ ($2.88 \times 10^{+13}$ thereof in LEO).

Fig. 3.10 shows the inclinations of initial orbits of 1,032 modeled, historic SRM firing events, along with the propellant mass of each motor. Since the early 1960s, with increasing payload masses the mean SRM propellant mass m_p also steadily increased. While around 1960 the propellant mass m_p per burn ranged from 20 kg to 200 kg, by the year 2000 it spanned masses from 800 kg up to 9,700 kg (first stage of the Inert Upper Stage, IUS; see Section 2.3 for more details). Fig. 3.10 indicates that SRM firings are concentrated in certain inclination bands, particularly at 28.5° to 30.5° (GEO or GTO injections for launches from Kennedy Space Center and Tanegashima), at 55° and 65° (injections of GPS satellites), and at 97° to 100° (injections of satellites into Sun-synchronous orbits). Fig. 3.11 compiles data on the final apogees and perigees of the destination orbits, with SRM propellant masses indicated by color codes. The most massive motors ($m_p \geq 1,000$ kg) are used for injections into GTO/GEO, and into the semi-synchronous orbits, or transfer orbits of navigation satellites. Motors of medium and small size ($m_p < 1,000$ kg) are mainly used for LEO insertions, and for GEO circularization burns.

SRM slag, due to its relatively high mass, low ejection velocity, and release at the end of the burn, remains close to the destination orbits. It shows distribution peaks at $i = 10^\circ \pm 5^\circ$ (with $\Delta m_s/\Delta i \approx 5\%/1^\circ$), at $i = 55^\circ \pm 1^\circ$ ($\Delta m_s/\Delta i \approx 3\%/1^\circ$), and at $i = 65^\circ \pm 1^\circ$ ($\Delta m_s/\Delta i \approx 1\%/1^\circ$). The eccentricities of SRM slag orbits are mostly moderate, with a flat peak for circularization burns at $e = 0.06$ (with $\Delta m_s/\Delta e \approx 5.5\%/0.01$), and secondary peaks for GPS and GEO transfer orbits at $e = 0.6$ and 0.73 (with $\Delta m_s/\Delta e \approx 1\%/0.01$). In contrast to slag, the SRM dust exhibits mass distributions relative to inclination and eccentricity which are smoother, with broader peaks and fewer structures. This is due to the high ejection velocities and small masses of the dust particles. The eccentricity distribution of surviving SRM dust orbits shows a single peak at $e \approx 0.45$ (with $\Delta m_d/\Delta e \approx 3\%/0.01$), which symmetrically falls off towards $e = 0$ and $e = 0.8$, with no dust particles of sustained lifetimes on orbits of $e > 0.8$. The majority of dust particles is left on orbits of inclinations between 40° and 150°, with two neighboring peaks at $i \approx 60^\circ$ and 85° (up to $\Delta m_d/\Delta i \approx 2\%/1^\circ$), and a secondary peak at retrograde inclinations of 140° (with $\Delta m_d/\Delta i \approx 1\%/1^\circ$).

In the MASTER-2001 model the SRM dust model is truncated at a lowest size threshold of $d_{d,min} = 1\,\mu\text{m}$, because of considerable uncertainties, and in order to reduce computational loads. The modeling approach for SRM slag and dust which has been described in this section does not exactly reflect the MASTER implementation, which uses additional internal scaling mechanisms, and an optional, more refined, alternative SRM dust model (Meyer, 1992).

3.6 MODELING SODIUM–POTASSIUM COOLANT DROPLETS

The second most important non-fragmentation source of space debris, with object sizes exceeding 1 cm, are sodium–potassium (NaK) coolant releases from 16 Russian RORSAT reactors between April 1980 and March 1988 (see Table 2.3 and Fig. 2.20). The release of the eutectic, liquid NaK alloy from the primary coolant loop of the Buk reactors was part of a post-mission disposal concept. A short description of the operational re-orbit of RORSAT reactors, and of their passivation procedure is provided in Section 2.3.

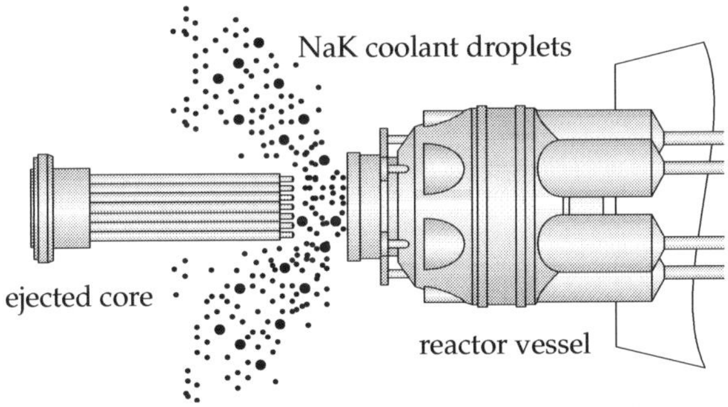

Fig. 3.12. Schematic illustration of the release of Buk reactor cores and of NaK coolant from RORSAT satellites in their disposal orbits.

RORSAT is a USSTRATCOM synonym for Russian Us-A satellites which performed radar ocean reconnaissance of NATO fleets between 1977 and 1988. The power-demanding payloads received their energy from fast neutron Buk reactors, which have an overall mass of 130 kg, with 53 kg reactor core material, 31.1 kg thereof reactor fuel. This fuel is contained in 37 rods which are composed of 90% enriched ^{235}U, with 3 mass-% of molybdenum added. The reactor produces a maximum thermal power of about 100 kW, from which a corresponding electrical power output of $\sim$3 kW is produced via electro-thermal converters. The stack of fuel rods is closed with two beryllium end reflectors, and the fission process is controlled by moving cylindrical beryllium side reflectors. The core is contained in a case of stainless steel, which also houses a 2-loop cooling system with a eutectic NaK alloy (77.8 mass-% of K) as cooling agent. 13 kg and 26 kg of NaK are assumed to be contained in the primary and secondary coolant loop. The maximum coolant temperature is estimated to be 370°C (Wiedemann et al., 2004).

After the RORSAT transfer stage, with the attached reactor, has reached its near-circular disposal orbit of mean altitude 900 km to 950 km at 65° inclination[3.2], the reactor core is ejected, presumably via gas pressure (see Fig. 3.12). During the ejection a seal of the primary cooling loop is ruptured, releasing most of the 13 kg coolant into space, in the form of a NaK-gas mixture. This release scenario for 16 out of 31 RORSAT missions left clouds of NaK objects within a common, tight inclination and altitude band. These clouds, with maximum observed object sizes of 5.54 cm were detected by the Goldstone and Haystack radars, where Goldstone alone observed 10 objects larger than 3.4 cm. Analysis of the polarization of the radar echos indicated that the objects were metallic spheres. Orbit determinations of the larger objects showed that the density of the spheres was similar to water ($\rho_{NaK} \approx 0.9\,\text{g}/\text{cm}^3$). Together with the derived orbital inclinations and altitudes the data pointed to RORSATs as potential source. This assumption was confirmed by Russian scientists in 1996.

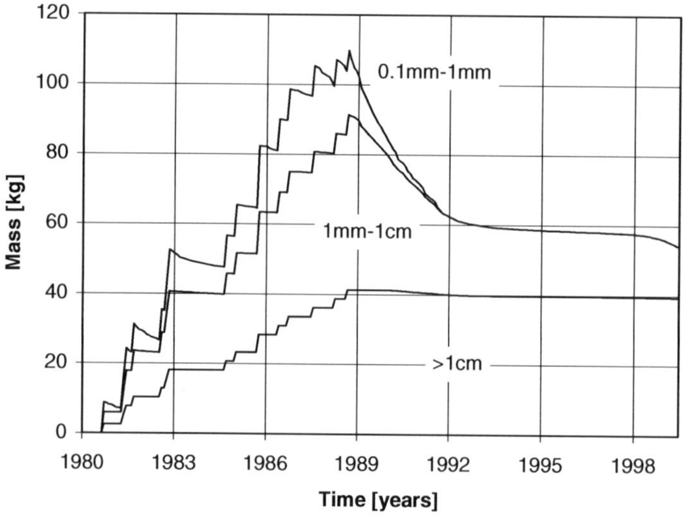

Fig. 3.13. Predicted time history of the size-dependent decay of NaK droplets from their release orbits (16 events modeled).

In the MASTER-2001 model the NaK population is composed from 16 individual release events listed in Table 2.3. The resulting clouds are then propagated to be compared with Haystack observation data between 1994 and 2002 (Stansbery et al., 1997), and with Goldstone observations between 1994 and 1996 (Goldstein et al., 1998). According to these sources, a total of 543,000 objects of diameters $d \geq 2.5\,\text{mm}$, 43,000 of $d \geq 6\,\text{mm}$, 15,000 of $d \geq 10\,\text{mm}$, and 2,000 of $d \geq 20\,\text{mm}$ were observed. Based on thermodynamic considerations, only NaK droplets larger than 0.1 mm can be expected to survive evaporation. The cumulative number N_n of released NaK droplets of $d > d_c$ shall hence be expressed by a

[3.2]due to non-nominal operations, Cosmos 1900 with its reactor was disposed in an altitude of 687 km × 754 km at 66.1° (see Table 2.3)

sequence of fitting functions for diameters $d \geq 0.1\,\text{mm}$ (Bendisch et al., 2002).

$$N_n(d > d_c) = \begin{cases} 6.67 \times 10^{+36} \exp\left(-1869.9 \dfrac{d_c}{[\text{m}]}\right) & \text{for } d_c > 44\,\text{mm} \\ 73.923 \exp\left(-41.336 \dfrac{d_c}{[\text{m}]}\right) & \text{for } 36\,\text{mm} \leq d_c < 44\,\text{mm} \\ \exp\left(9.825 \exp\left(-34.827 \dfrac{d_c}{[\text{m}]}\right)\right) & \text{for } 6\,\text{mm} \leq d_c < 36\,\text{mm} \\ 9.282 \times 10^{-6} \left(\dfrac{d_c}{[\text{m}]}\right)^{-3.823} & \text{for } 2.5\,\text{mm} \leq d_c < 6\,\text{mm} \\ 2.647 \times 10^{-3} \left(\dfrac{d_c}{[\text{m}]}\right)^{-2.880} & \text{for } 0.1\,\text{mm} \leq d_c < 2.5\,\text{mm} \end{cases} \tag{3.62}$$

For the ejection velocities a constant value of $v_n = 13\,\text{m/s}$ shall be adopted, independent of the droplet diameter d_n, with randomly distributed escape directions. The released NaK mass for each core ejection is assumed to be 13 kg (the whole content of the primary cooling loop), of which about 9.1 kg (70%) have a sustained lifetime as spherical droplets. The size-dependent droplet masses m_n and area-to-mass ratios $(A/m)_n$ (for $\rho_{NaK} \approx 0.9\,\text{g/cm}^3$) can be determined from

$$\frac{m_n}{[\text{kg}]} = 0.471 \times 10^{+3} \left(\frac{d}{[\text{m}]}\right)^3 \qquad \frac{(A/m)_n}{[\text{m}^2\,\text{kg}^{-1}]} = 1.668 \times 10^{-3} \left(\frac{d}{[\text{m}]}\right)^{-1} \tag{3.63}$$

The purely heuristic fitting functions in Eq. 3.62, which are suited to reproduce radar observation data collected in 1994 to 1996, will in future MASTER releases be replaced by a physically more justified approach, which is able to match observations up to 2002 in an open loop prediction, with no iterative tuning of parameters. The suggested new method, as documented in (Wiedemann et al., 2004), uses experimentally verified Rosin–Rammler equations to deduce an event-wise NaK number density distribution $\mathrm{d}N_n/\mathrm{d}d_c$ from the released NaK mass of $m_n \approx 13\,\text{kg}$, its mean temperature $T_n \approx 450\,\text{K}$, and the orifice diameters through which the NaK fluid escapes at the front end ($d_{o,f} \approx 3.0\,\text{cm}$) and at the rear end of the reactor vessel ($d_{o,r} \approx 0.5\,\text{cm}$). The total cumulative fraction $\hat{m}_n$ of the NaK release mass m_n with diameters $d > d_c$ is the weighted sum of individual contributions from the front and rear orifice ($\hat{m}_n \cdot m_n \in [0, m_n]$ is the resulting release mass).

$$\hat{m}_n = c_f\, \hat{m}_{n,f} + \left(1 - c_f\right) \hat{m}_{n,r} \qquad \text{with} \qquad 0 \leq \hat{m}_n \leq 1 \tag{3.64}$$

$c_f = 0.8356$ is the total mass fraction released from the front orifice of diameter $d_{o,f}$. $\hat{m}_{n,f}$ and $\hat{m}_{n,r}$ are cumulative mass fractions for $d > d_c$ released from the front and rear orifice.

$$\hat{m}_{n,f} = \exp\left(-\left(\frac{d_c/[\text{m}]}{0.03129}\right)^{2.6825}\right) \tag{3.65}$$

$$\hat{m}_{n,r} = \exp\left(-\left(\frac{d_c/[\text{m}]}{0.004583}\right)^{2.2954}\right) \tag{3.66}$$

The resulting number density distribution dN_n/dd_c has two peaks at diameters of 5 mm and 3 cm (corresponding to the diameters of the orifices). The maximum object size resulting from Eq. 3.64 is close to the observed $d_{c,max} \approx 5.54$ cm, and the total number of long-lived NaK droplets per release event is estimated to be $5.894 \times 10^{+5}$. This cumulative number of objects N_n with $d > 0.1$ mm must be determined by numerical quadrature of dN_n/dd_c. In a log-log scale the resulting curve of $N_n(d > d_c)$ shows a bi-modal shape, of which the part with $d > 2.5$ mm, where radar data are available, yields a very good fit of these measurements.

With decreasing sizes the NaK droplets encounter large perturbations due to radiation pressure and airdrag, leading to short orbital lifetimes. Fig. 3.13 shows that after termination of NaK release events in 1988 the small-size population was rapidly removed. As a consequence of increasing airdrag during a solar activity peak in 1990 the sub-millimeter droplets had decayed completely by 1992, and the sub-centimeter population was reduced by about 70%.

NaK release products can be regarded as a historic, non-reproducing source of space debris, which is only represented in the LEO region. For its reference epoch in May 2001 the MASTER-2001 model assesses the NaK population to consists of about 35,300 droplets of $d_c \geq 1$ mm, and about 15,000 droplets of $d_c \geq 1$ cm. All smaller sizes have decayed into the atmosphere.

3.7 MODELING WESTFORD NEEDLE CLUSTERS

Apart from NaK droplets, there is another historic, non-reproducing source of remnants of past space activities: the so-called "Westford Needles", which were deployed in two radio communication experiments in 1961 and 1963 (see Section 2.3). The first experiment (Midas 4) was launched on October 21, 1961. A cylinder filled with millions of copper needles of length 1.78 cm and diameter 25.4 μm was dispensed in an orbit of altitude 3,495.9 km × 3,756.1 km at 95.89° inclination. The cylindrical dispenser rotated at about 8 revolutions per second, inducing a release velocity of about 3.2 m/s. Probably due to cold welding most of the needles cumulated into clusters of which 150 entered the USSPACECOM catalog, while 250 more (down to an RCS of 27 mm^2) were detected by the Goldstone radar. A second experiment (Midas 6) was launched on May 9, 1963, into an orbit of 3,601.9 km × 3,682.1 km at 87.35° inclination. The needles in this test had the same length as for Midas 4, but a reduced diameter of 17.8 μm. In this case, due to improved surface coatings, fewer clusters formed, leading only to a few cataloged objects from this event. Single needles and small clusters are known to have short orbit resident times, due to their large area-to-mass ratios, and due to the corresponding level of radiation pressure and airdrag perturbations. Hence, for environment modeling purposes, only the clusters will be pursued.

The observed clusters of Westford Needles had radar cross-sections ranging from 27 mm^2 up to 466 mm^2. For the Goldstone radar wavelength of 3.5 cm the RCS can be converted into an "effective length", from which a mean diameter d of a mass-equivalent sphere can be deduced. From an analysis of available observation data the number of Westford clusters N_w with diameters $d > d_c$ can be approximated (Bendisch et al., 2002).

$$N_w(d > d_c) = N_{w,0} \left(c_0 + c_1 \left(\frac{d_c}{[\mathrm{m}]} \right)^{-1} \right) \quad \text{for} \quad 0.7\,\mathrm{mm} \leq d \leq 4\,\mathrm{mm} \tag{3.67}$$

For the Midas 4 experiment the coefficients are $N_{w,0} = 40,000$, $c_0 = -0.217$, and $c_1 = 8.475 \times 10^{-4}$. For the Midas 6 test, one finds $N_{w,0} = 1,000$, $c_0 = -0.201$, and $c_1 = 6.598 \times 10^{-4}$. The largest clusters probably comprised between 170 needles (for Midas 4) and 210 needles (for Midas 6). In total, the estimated number of needles contained in long-lived clusters is 739,000 for Midas 4 (with a total mass of 60 grams), and 20,700 for Midas 6 (with a mass of less than 1 gram). Due to the small overall mass involved, and due to the high orbit altitudes in which they reside, the effects of Westford Needle clusters on the space debris environment are of minor importance.

3.8 MODELING SURFACE DEGRADATION AND IMPACT EJECTA

In Sections 3.6 and 3.7 low-energy, deliberate debris release events are described. There is another class of unintentional releases of sub-millimeter space debris due to surface degradation, and due to surface ejecta from impacts of small-size debris and meteoroids on large-size, intact objects.

It is known from in-situ exposure experiments (e.g. on LDEF, or on STS-carried payloads) that surfaces of space hardware degrade in the harsh environmental conditions, particularly in the LEO regime. For mean atmospheric conditions atomic oxygen is the dominant species at orbital altitudes below ~750 km. It reacts with most surface materials. In the case of metal surfaces, it forms an oxide layer, which may crack and eventually flake. In the case of Kapton, it breaks chemical bonds, leading to the emission of CO and CO_2. Together with trapped atomic oxygen this can lead to the erosion of substrate underneath the atomic oxygen resistant coating, causing a growth of cavities, with possible undercutting, and an eventual delamination. Kapton, silver and some thermal control paint exhibit significant mass loss due to atomic oxygen impingement (up to $10^{-21}\,\mathrm{cm}^3$ per impacting oxygen atom). Most of the released particles are assumed to be smaller than 1 µm, with the exception of coating flakes released due to undercutting. This mechanism, however, only applies to susceptible substrate materials, and not to metallic surfaces. Since mostly paint is involved in the particle generation process, surface degradation products are often referred to as "paint flakes".

In 1992, during the EOIM-3 experiment on-board the Space Shuttle, several materials were exposed to atomic oxygen flux during a time span of 42 hours (Morton and Ferguson, 1993). When translated to LDEF conditions, the observed, average rates of mass loss of painted surfaces were up to 700 $\mathrm{g\,m^{-2}\,y^{-1}}$, with higher losses during the early exposure phase. This is about a factor 10 higher than for LDEF experiments.

Extreme ultra-violet (EUV) radiation is another contributor to surface degradation, often in combination with atomic oxygen. Exposure to EUV radiation causes embrittlement of some surface materials, mainly polymers, which are part of Kapton and paint coatings. Polymers have a wide absorption band in the UV and EUV spectrum. They dissipate the received energy by electron excitation, causing ruptures of inter-molecular bonds. This process increases the material brittleness, and enhances micro-cracking. There is no evidence, that EUV produces micro debris directly (e.g. by delamination), but surface cracks caused by EUV radiation make underlying laminates vulnerable to deterioration by thermal cycling and oxygen erosion.

According to (Maclay and McKnight, 1994) the number of surface degradation particles N_{sd} released into space during an on-orbit exposure of a surface A, over a timespan Δt, is a function of the atomic oxygen concentration n_O, of the thermal cycling frequency f_T and the associated temperature range ΔT, and of the surface material.

$$N_{sd} = \begin{cases} \left(1 + c_O \dfrac{n_O}{[\mathrm{m}^{-3}]} \dfrac{\Delta t}{[\mathrm{d}]} + c_T \dfrac{f_T}{[\mathrm{d}^{-1}]} \dfrac{\Delta T}{[\mathrm{K}]}\right) c_m \dfrac{A}{[\mathrm{m}^2]} \left(\dfrac{\Delta t}{[\mathrm{d}]}\right)^{\tau} & \text{for } \overline{H} < 2{,}000\,\mathrm{km} \\ c_m \dfrac{A}{[\mathrm{m}^2]} \left(\dfrac{\Delta t}{[\mathrm{d}]}\right)^{\tau} & \text{for } \overline{H} \geq 2{,}000\,\mathrm{km} \end{cases} \tag{3.68}$$

The coefficient c_m is material-dependent, c_O is a calibration constant for atomic oxygen effects, and c_T is a scaling coefficient for thermal cycling contributions to the release rate (Bendisch et al., 2002). $\tau \approx 1$ indicates the slope of the exposure time dependency. Above a mean altitude of 2,000 km the combined effect of atomic oxygen erosion and thermal cycling is neglected. For modeling the release process it is initially assumed that all surface degradation particles are generated with uniform probability over a size range $1\,\mu\mathrm{m} \leq d_{sd} \leq 200\,\mu\mathrm{m}$. Individual size bins within this range are subsequently adjusted (scaled) to match in-situ measurement data. The separation velocity is adopted to be $v_{sd} \approx 0$. The densities ρ_{sd} of released particles, and their area-to-mass ratios $(A/m)_{sd}$ for equivalent solid spheres are depending on the surface material. Two main classes of potential source objects can be distinguished: 3-axis stabilized and spin-stabilized spacecraft. For 3-axis stabilized satellites 33% of the surface shall be assigned to the body, covered by thermal blankets, and 67% shall be solar arrays, with their rear surface covered by Kapton, and their front side covered by solar cells with 95% boro-silicate glass and 5% silver conductors. For spinning satellites the surface shall be assumed to consist of 75% solar cells, and 25% thermal blankets. The number of spacecraft at a given epoch (see also Fig. 2.1 and 2.4), their orbital distribution, and their basic design characteristics can for instance be extracted from ESA's DISCOS database (Hernández et al., 2001).

When propagating the space object population through its entire evolution, up to 2001, one finds that the surviving surface degradation products ("paint flakes") concentrate in a broad inclination band around 20° (with up to $2.5 \times 10^{+12}$ per 1°), and between 60° and 100° (with up to $1.0 \times 10^{+12}$ per 1°). Concentration peaks with respect to altitude are in LEO, in semi-synchronous GPS/GLONASS orbits, and in GEO (each with up to $1.0 \times 10^{+10}$ per 50 km). The MASTER-2001 model predicts $1.51 \times 10^{+13}$ "paint flakes" of $d_{sd} \geq 10\,\mu m$ (8.4% thereof in LEO), and $9.63 \times 10^{+12}$ of $d_{sd} \geq 100\,\mu m$ (32.7% thereof in LEO).

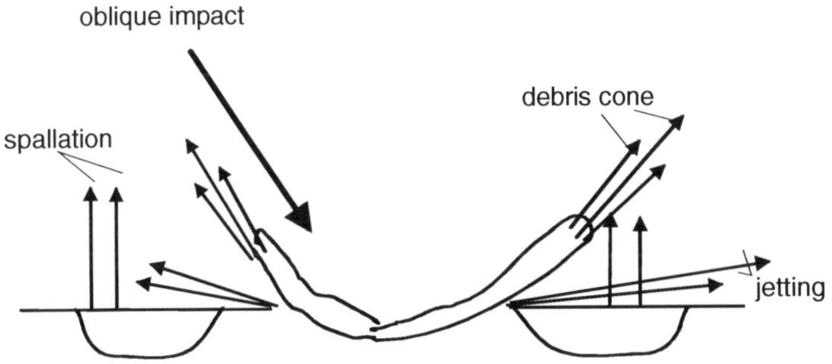

Fig. 3.14. Schematic illustration of impact ejecta and spall release due to small-particle impacts on surfaces of spacecraft or rocket bodies (McDonnell et al., 1999).

All debris sources addressed so far were assumed to be mutually independent, and associated with intact space objects only. However, the vast number of small-size debris, generated by processes described in Sections 3.3 to 3.8, leads to a considerable collision flux, particularly on large-size targets. Resulting impacts cause further small-size debris as ejecta and spall detachments. The principle of this feedback mechanism of space debris production (debris generated by impacting debris) is sketched in Fig. 3.14. The following procedures are based on work described in (McDonnell et al., 1999).

According to their emission velocities v, relative to the impact velocity v_i, impact ejecta can be discriminated as jet ejecta (with $v \leq 3 \times v_i$), cone ejecta (with $v \leq v_i$), and spallation particles (with $v \leq 0.8\,v_i$). In hypervelocity impact experiments under laboratory conditions one can demonstrate that about 60% of the released material is contained in cone ejecta, and about 40% in a few, relatively large spall ejecta. Jet ejecta account for less than 1%. They shall hence be neglected hereafter.

The size range of cone ejecta spans $0.1\,\mu m \leq d_{ce} < d_i$, where d_i is the diameter of the impactor. They are typically released within a cone width of 60° at velocities of approximately $10\,m/s \leq v_{ce} \leq v_i$. Let A_{ce} be the azimuth angle in the plane of the impacted surface, measured from the projection of the impact velocity vector $\underline{v}_i$ around the surface normal vector, and let θ_{ce} be the elevation co-angle, measured from the surface normal to the approach direction. The number density distribution of cone ejecta can then be described as

$$\frac{d^3 N_{sd}}{d d_{ce}\, d A_{ce}\, d\theta_{ce}} = c_{ce}\, f_1(d_{ce})\, f_2(A_{ce})\, f_3(\theta_{ce}) \tag{3.69}$$

where c_{ce} is a model calibration parameter. The distribution functions $f_1(d_{ce})$ of the ejecta size spectrum, and $f_2(A_{ce})$, $f_3(\theta_{ce})$ of the ejecta departure direction are defined hereafter.

The ejected mass m_t of a target material with density ρ_t, due to an object with mass m_i, density ρ_i, and velocity v_i, impacting at an incident angle θ_i, can be approximated by

$$\frac{m_t}{[\mathrm{kg}]} = 7.41 \times 10^{-6} \sqrt{\frac{\rho_i}{\rho_t}} \left(\frac{m_i}{[\mathrm{kg}]}\right)^{1.133} \left(\frac{v_i}{[\mathrm{m/s}]}\right)^{2.266} \cos^2\theta_i \tag{3.70}$$

The mass fraction of cone ejecta m_{ce} is

$$m_{ce} = \kappa\, m_t \tag{3.71}$$

where $\kappa = 0.6$ for ductile, metallic materials. For brittle surfaces, κ is a function of the material. Based on experimental results, the largest cone ejecta shall have a mass of $m_{ec,max} = 0.1\, m_{ce}$. The corresponding diameter $d_{ce,max}$ of a spherical droplet is

$$\frac{d_{ce,max}}{[\mathrm{m}]} = \left(\frac{6}{\pi}\frac{m_{ce}}{[\mathrm{kg}]}\right)^{1/3} \left(\frac{\rho_t}{[\mathrm{kg/m^3}]}\right)^{-1/3} \tag{3.72}$$

By definition, the smallest diameters shall be $d_{ce,min} = 0.1\,\mu\mathrm{m}$. The diameter dependent, normalized number density function $f_1(d_{ce})$ follows a power law.

$$f_1(d_{ce}) = \frac{1}{1.265 \times 10^{+17}} \left(\frac{d_{ce}}{[\mathrm{m}]}\right)^{-7/2} \tag{3.73}$$

Since the impact geometry cannot be modeled in detail for a large number of targets, it shall be assumed that all impacts are perpendicular ($\theta_i = 0°$). Consequently, the azimuth distribution of cone ejecta will be symmetric, with $f_2(A_{ce}) = 1$. For the density distribution of the co-elevation function $f_3(\theta_{ce})$ a normal distribution is assumed.

$$f_3(\theta_{ce}) = \frac{1}{\sqrt{2\pi}\,\sigma_\theta} \exp\left(-\frac{(\theta_{ce} - \theta_{ce,max})^2}{2\,\sigma_\theta^2}\right) \tag{3.74}$$

with $\theta_{ce,max} = 30°$ and $\sigma_\theta = 8°$ fitted to experimental data. For a normal impact ($\theta_i = 0°$), the total number of cone ejecta larger than a threshold diameter d_{ce} can be determined by numerical quadrature of Eq. 3.75.

$$N_{sd}(d > d_{ce}) = c_{ce} \int_{d_{ce}}^{d_{ce,max}} \int_0^{\theta_{ce,max}} f_1(d_{ce})\, f_3(\theta_{ce})\, d\theta_{ce}\, dd_{ce} \tag{3.75}$$

The size-dependent imparted velocities $v_{ce}(d_{ce})$ result from

$$v_{ce} = \frac{d_{ce,min}}{d_{ce,max} - d_{ce,min}} \left((v_{ce,max} - v_{ce,min})\frac{d_{ce,max}}{d_{ce}} + v_{ce,min}\frac{d_{ce,max}}{d_{ce,min}} - v_{ce,max}\right) \tag{3.76}$$

with $v_{ce,min} = 10\,\mathrm{m/s}$, $v_{ce,max} = 0.8\,v_i$, $d_{ce,min} = 10^{-7}\,\mathrm{m}$, and $d_{ce,max}$ from Eq. 3.72.

The modeling of spall ejecta follows the procedure for cone ejecta, with several simplifications. The spall mass $m_{se} = (1 - \kappa)\,m_t$ in most cases represents 40% of the entire mass ejection (depending on the surface material). It shall be assumed that this mass is equally distributed between exactly 10 spall ejecta (hence $N_{se} = 10$), of $m = 0.1\,m_{se}$ each. To determine the spall emission velocity $v_{se}(d_{se})$, Eq. 3.76 is applied with d_{se} instead of d_{ce}, where d_{se} is defined as

$$\frac{d_{se}}{[\mathrm{m}]} = \left(\frac{3}{5\,\pi}\,\frac{m_{se}}{[\mathrm{kg}]}\right)^{1/3} \left(\frac{\rho_t}{[\mathrm{kg/m^3}]}\right)^{-1/3} \tag{3.77}$$

Experiments show that particularly for oblique impacts the generated cloud of particles may impact neighboring surfaces, which are within the ejection cone. They may thus produce secondary ejecta (the so-called "ricochet phenomenon"). This second-order effect is not taken into account in the present assessment.

The MASTER-2001 model estimates that the population of impact ejecta in 2001 consist of $4.23 \times 10^{+10}$ cone and spall particulates of sizes $d \geq 1\,\mu\mathrm{m}$ (1.9% thereof in LEO), $1.44 \times 10^{+10}$ of $d \geq 10\,\mu\mathrm{m}$ (9.2% thereof in LEO), $2.52 \times 10^{+8}$ of $d \geq 100\,\mu\mathrm{m}$ (40.1% thereof in LEO), and $1.10 \times 10^{+6}$ of $d \geq 1\,\mathrm{mm}$ (72.2% thereof in LEO). Hence, larger ejecta can survive on LEO orbits of moderate eccentricities, while small-size ejecta only survive on high-altitude MEO and GEO orbits, or on highly eccentric orbits.

3.9 HISTORIC EVOLUTION AND SPATIAL DISTRIBUTION OF DEBRIS

The space debris environment is a highly dynamic system with sometimes strong temporal and spatial variations of distributions, particularly with respect to altitude and declination, and their related time histories. This variability, which can greatly differ for different size classes, is correlated with several space debris sources, with corresponding historic release events, with the size range of the objects produced during such events, and with the response of released objects to orbit perturbations.

Fig. 3.15, Fig. 3.17, and Fig. 3.19 show the size-dependent time evolution of LEO altitude profiles of spatial object densities since the beginning of the space age, for threshold diameters of $d > 10\,\mathrm{cm}$, 1 cm, and 1 mm. These size regimes were chosen, because $d > 10\,\mathrm{cm}$ marks the LEO tracking limit (collisions with most of these objects can be avoided), $d > 1\,\mathrm{cm}$ marks the limit of on-orbit shielding capabilities (most of these objects cannot be defeated by shielded satellites, and are likely to cause destruction of unshielded spacecraft), and $d > 1\,\mathrm{mm}$ covers objects which can damage satellites and degrade their missions.

In Fig. 3.15, Fig. 3.17, and Fig. 3.19 all debris sources are combined. A more detailed breakdown by sources of a recent[3.3], size-dependent space debris concentration profile is provided in Fig. 3.21 to 3.26. Fig. 3.15, for "catalog objects"

[3.3]the reference epoch is May 1, 2001

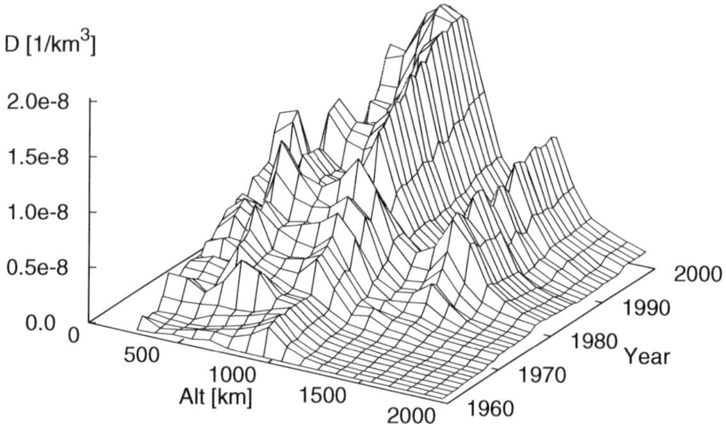

Fig. 3.15. Spatial object density versus altitude and year for objects of $d > 10\,\text{cm}$ according to the MASTER-2001 model.

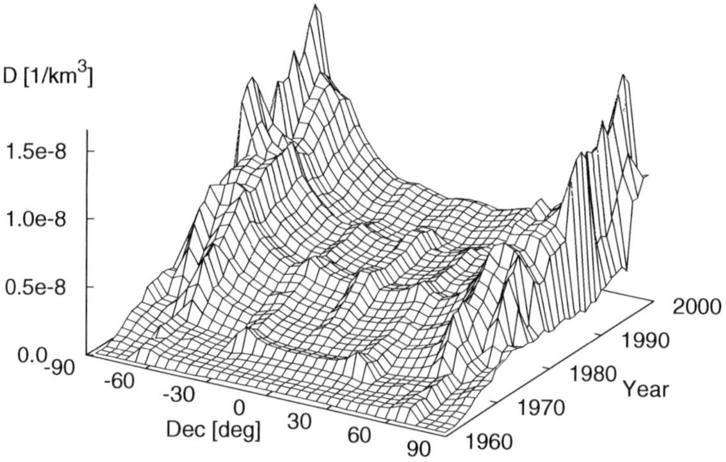

Fig. 3.16. Spatial object density versus declination and year for objects of $d > 10\,\text{cm}$ according to the MASTER-2001 model.

of $d > 10$ cm, shows for the evolution of the altitude distribution a build-up which is first limited to altitudes of 1,200 km and below, with a peak evolving near $\overline{H} \approx 1{,}000$ km. Only as of the early 1970s a secondary concentration peak develops at $\overline{H} \approx 1{,}400$ km. At certain times sharp increases indicate the release of catalog-size objects due to on-orbit fragmentations (see also Table 2.2). Some of the most dominant of such events shall be identified in ascending time order.

The first known fragmentation in space history, a Thor Ablestar stage 2 explosion on June 29, 1961, at $\overline{H} \approx 940$ km and $i = 66.8°$, released 298 trackable objects,

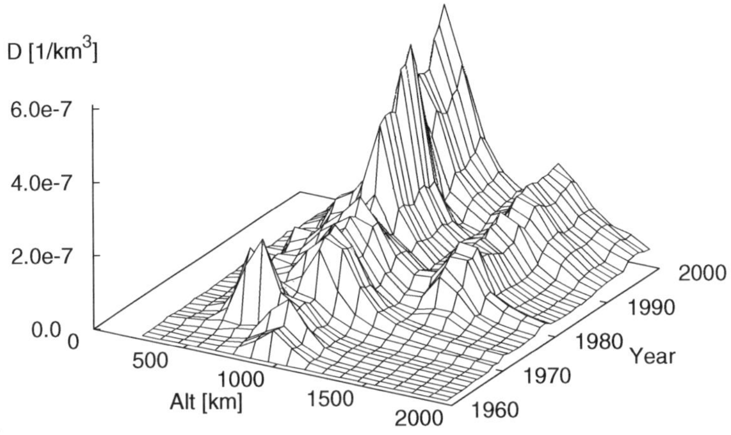

Fig. 3.17. Spatial object density versus altitude and year for objects of $d > 1$ cm according to the MASTER-2001 model.

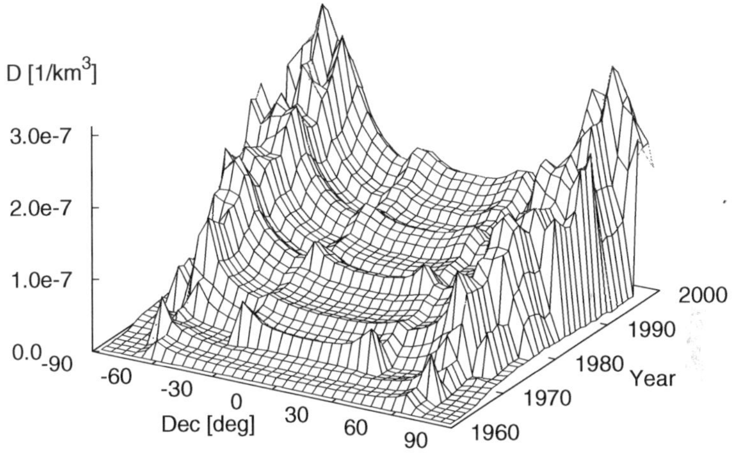

Fig. 3.18. Spatial object density versus declination and year for objects of $d > 1$ cm according to the MASTER-2001 model.

and more than tripled the catalog size at that time. On October 15, 1965, an OV2-1/LCS explosion released 470 trackable objects at $\overline{H} \approx 710$ km and $i = 32°$. Four years later two Thorad Agena D stages exploded within 12 months, on October 04, 1969, at $\overline{H} \approx 1,000$ km and $i = 70.0°$, and on October 17, 1970, at $\overline{H} \approx 1,070$ km and $i = 99.9°$, together causing a step increase by 631 cataloged fragments to the population at 1,000 km. This altitude class was further filled with 305 tracked fragments from the Cosmos 1275 explosion on July 24, 1981, at $\overline{H} \approx 980$ km and $i = 83.0°$. Another altitude band, near 1,500 km, started to evolve rapidly in the 1970s. Three explosions of Delta 300 and Delta 2310 second stages between De-

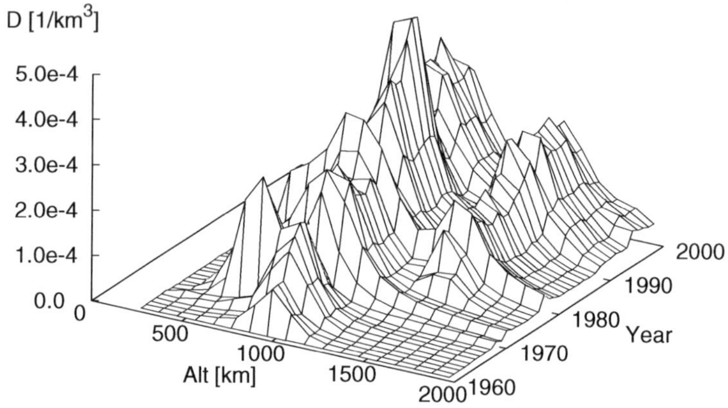

Fig. 3.19. Spatial object density versus altitude and year for objects of $d > 1$ mm according to the MASTER-2001 model.

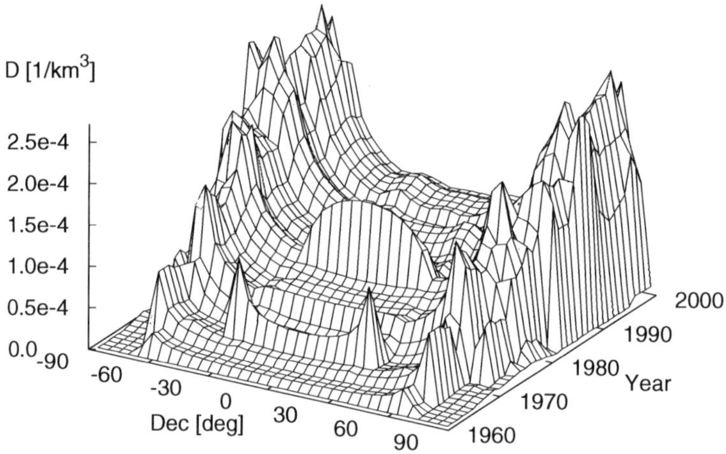

Fig. 3.20. Spatial object density versus declination and year for objects of $d > 1$ mm according to the MASTER-2001 model.

cember 28, 1973, and December 24, 1977, released a total of 502 tracked objects into almost identical orbits of $\overline{H} \approx 1,500$ km and $i \approx 102.0°$. Further explosion events replenished the 1,000 km and 1,500 km altitude regions in subsequent years.

Three distinct fragmentation events below 1,000 km after 1980 are clearly noticeable in Fig. 3.15. An Ariane-1 H-10 upper stage exploded on November 13, 1986, at $\overline{H} \approx 815$ km and $i = 98.7°$, leading to 488 cataloged fragments, a Pegasus orbital stage exploded on June 03, 1996, at $\overline{H} \approx 700$ km and $i = 82.0°$, leading to 703 cataloged fragments, and a deliberate on-orbit collision experiment, in which an aircraft-launched ASAT missile hit the Solwind satellite on September 13, 1985, caused 285 trackable objects in a short-lived orbit of $\overline{H} \approx 535$ km and $i = 97.6°$. Apart from a signature in the density distribution $D_{d>10\,\mathrm{cm}}(\overline{H}, t)$, each of the aforementioned fragmentation events caused an equally unique footprint in the orbital declination distribution $D_{d>10\,\mathrm{cm}}(\delta, t)$, as shown in Fig. 3.16. For each fragmentation on a moderately eccentric orbit of inclination i a U-shape declination profile develops, spanning the range $-i \leq \delta \leq +i$, with concentration maxima near $\delta \approx \pm i$, and minima at $\delta = 0$. The lower the inclination, the more pronounced the concentration maxima emerge, due to increased resident probabilities in the outermost declination bands. Most of the fragmentation events occurred at high inclinations, causing ridges of maximum spatial densities at $\delta \approx \pm 82°$ and $\pm 70°$. Some low inclination events are clearly discernible: an OV2-1/LCS explosion in 1965 at $i = 32°$ (see above), and a Delta 2914 stage 2 explosion on July 14, 1977, causing 170 trackable objects in an orbit of inclination $i = 29°$.

When lowering the size threshold of the debris environment model from $d >$ 10 cm to $d > 1$ cm, then solid rocket motor (SRM) slag and NaK droplets get added to fragmentation debris and intact objects, and can become significant in some scenarios. Fig. 3.17 shows the time history of the spatial object density $D_{d>1\,\mathrm{cm}}(\overline{H}, t)$ as a function of altitude. On top of the previously discussed fragmentation events, SRM slag starts to become relevant after the introduction of powerful motors with large amounts of propellants. On April 23, 1970, the first 1-ton class of SRM was fired. The CZ-1 stage 3 (GF-02), with a propellant mass of $m_p = 1{,}800$ kg, delivered a spacecraft into an orbit $H_{pe} = 440$ km and $H_{ap} = 2{,}388$ km at $i = 31.4°$ (8 more missions followed). On February 16, 1974, the first 3-ton SRM was employed. The Mu-3C stage 3 (M-3A), with $m_p = 3{,}280$ kg, injected its payload into an orbit of $H_{pe} = 281$ km and $H_{ap} = 3{,}264$ km at $i = 31.4°$ (13 more missions followed). The SRM kick-motors HS-381 ($m_p = 3{,}316$ kg, 6 missions) and HS-601 ($m_p = 4{,}267$ kg, 3 missions) belonged to the same, top performance class. This motor class is only rivaled by the most massive SRM compound, the Inert Upper Stage (IUS), with its stage 1 ($m_p = 9{,}709$ kg, mostly for GTO injection at $i \approx 28°$), and its stage 2 ($m_p = 2{,}722$ kg, mostly for GEO circularization). The IUS was first used on October 29, 1982, and has had 26 missions since. The most frequently used SRM in space history is the PAM-D motor ($m_p = 2{,}011$ kg) which was first introduced on November 20, 1980, and has completed 93 missions so far[3.4]. Since many solid motors are used for insertions into eccentric transfer trajectories, the slag they release in the destination orbits has a relatively short LEO residence time. The MASTER-2001 model predicts that less than 13% of the SRM slag objects larger than 1 cm reside within the LEO region at any time. IUS, the most massive SRM with a total propellant mass of almost 12.5 tons, is mainly used for GEO transfers, on low inclination orbits. As a result, the spatial density distribution $D_{d>1\,\mathrm{cm}}(\delta, t)$ in Fig. 3.18 clearly shows U-shape signatures of IUS stage 1 burns, with density peaks at $\delta \approx \pm 28°$ (for launches from KSC).

[3.4]all quoted SRM mission counts refer to a cut-off date on May 1, 2001

Table 3.2. Contributions by "macro objects" to the MASTER-2001 reference population for May 2001, discriminated by sources, size regimes, and orbital regions (sources not mentioned have no contributions to the indicated size regimes).

Source type	Orbit regime	>1 mm	>1 cm	>10 cm	>1 m
Launch/MROs	LEO	4,487	3,118	3,077	2,028
	LEO+MEO+GEO	30,561	5,171	5,128	3,808
Fragments	LEO	1.30e+08	149,266	6,602	228
	LEO+MEO+GEO	1.78e+08	382,716	12,704	462
NaK	LEO	35,280	15,012	0	0
	LEO+MEO+GEO	35,280	15,012	0	0
SRM slag	LEO	8.23e+06	22,065	0	0
	LEO+MEO+GEO	1.52e+08	171,046	0	0
Ejecta	LEO	795,870	0	0	0
	LEO+MEO+GEO	1.10e+06	0	0	0
Total count	LEO	1.39e+08	189,461	9,679	2,256
	LEO+MEO+GEO	3.31e+08	573,945	17,832	4,270

Table 3.3. Contributions by "micro objects" to the MASTER-2001 reference population for May 2001, discriminated by sources, size regimes, and orbital regions ("⟶" indicates that the object count is given by the number to the right, with no contributions from smaller size regimes).

Source type	Orbit regime	>1 μm	>10 μm	>0.1 mm	>1 mm
Launch/MROs	LEO	⟶	⟶	5,156	4,487
	LEO+MEO+GEO	⟶	⟶	45,342	30,561
Fragments	LEO	⟶	1.64e+11	2.37e+10	1.30e+08
	LEO+MEO+GEO	⟶	2.14e+11	3.20e+10	1.78e+08
NaK	LEO	⟶	⟶	⟶	35,280
	LEO+MEO+GEO	⟶	⟶	⟶	35,280
SRM slag	LEO	⟶	8.33e+09	7.66e+09	8.23e+06
	LEO+MEO+GEO	⟶	5.06e+11	4.13e+11	1.52e+08
SRM dust	LEO	2.88e+13	1.92e+13	0	0
	LEO+MEO+GEO	7.22e+15	1.39e+14	0	0
Paint flakes	LEO	5.03e+12	1.27e+12	3.15e+10	0
	LEO+MEO+GEO	4.23e+13	1.51e+13	9.63e+10	0
Ejecta	LEO	1.37e+09	1.33e+09	1.01e+08	795,870
	LEO+MEO+GEO	7.15e+10	1.44e+10	2.52e+08	1.10e+06
Total count	LEO	3.39e+13	2.06e+13	6.30e+10	1.39e+08
	LEO+MEO+GEO	7.26e+15	1.54e+14	5.41e+11	3.31e+08

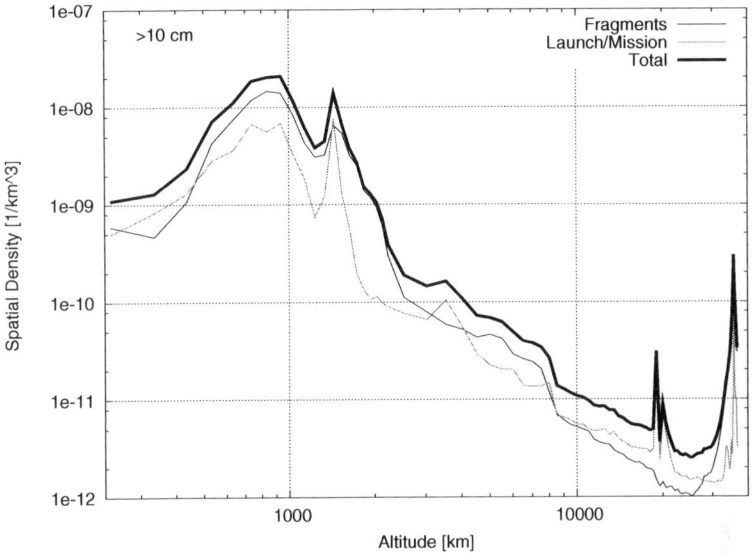

Fig. 3.21. Spatial density versus altitude for objects of diameters $d > 10\,\mathrm{cm}$ according to the MASTER-2001 model, for May 2001. Different sources are color coded. The thick line shows the envelope of total spatial density (see color plate on page 372).

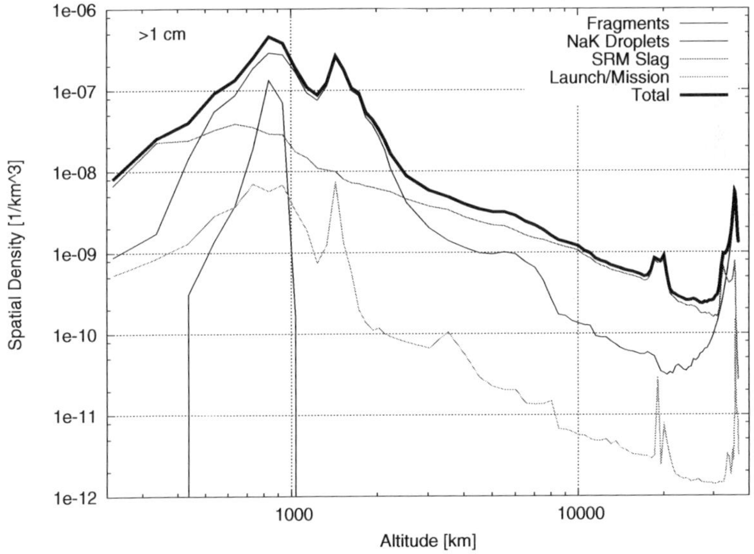

Fig. 3.22. Spatial density versus altitude for objects of diameters $d > 1\,\mathrm{cm}$ according to the MASTER-2001 model, for May 2001. Different sources are color coded. The thick line shows the envelope of total spatial density (see color plate on page 372).

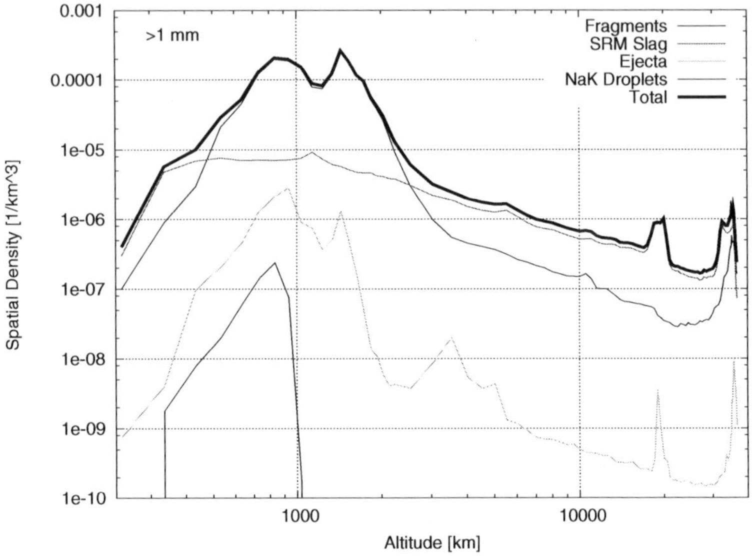

Fig. 3.23. Spatial density versus altitude for objects of diameters $d > 1$ mm according to the MASTER-2001 model, for May 2001. Different sources are color coded. The thick line shows the envelope of total spatial density (see color plate on page 373).

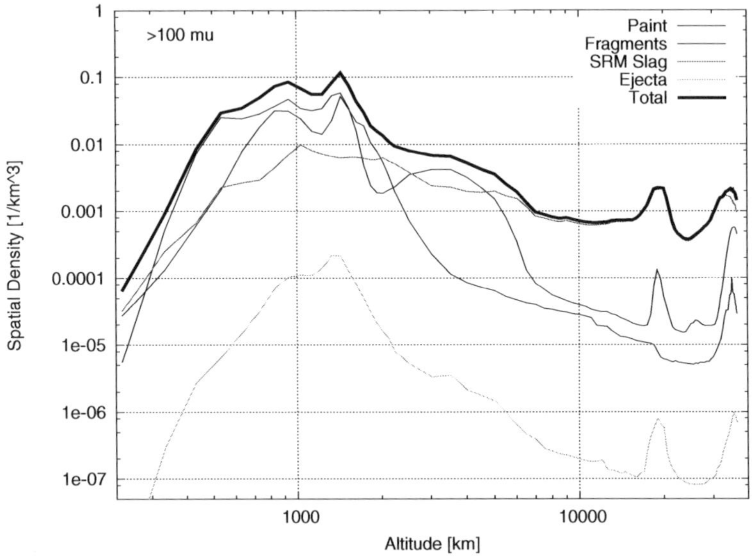

Fig. 3.24. Spatial density versus altitude for objects of diameters $d > 0.1$ mm according to the MASTER-2001 model, for May 2001. Different sources are color coded. The thick line shows the envelope of total spatial density (see color plate on page 373).

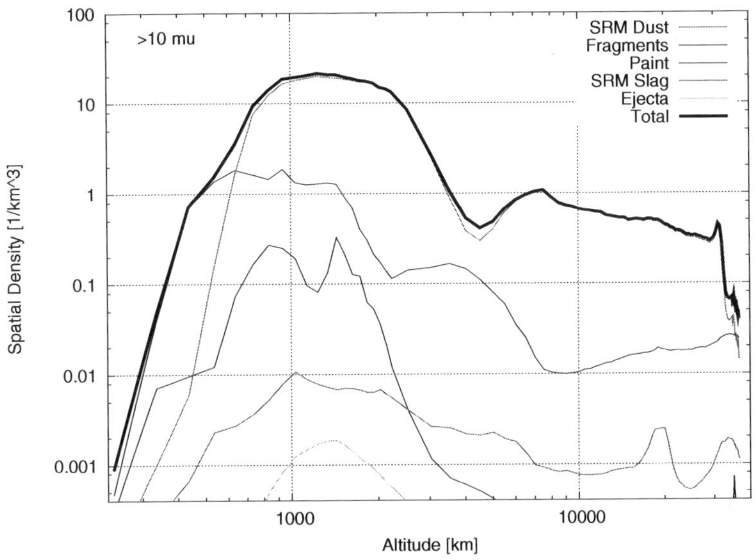

Fig. 3.25. Spatial density versus altitude for objects of diameters $d > 10\,\mu\mathrm{m}$ according to the MASTER-2001 model, for May 2001. Different sources are color coded. The thick line shows the envelope of total spatial density (see color plate on page 374).

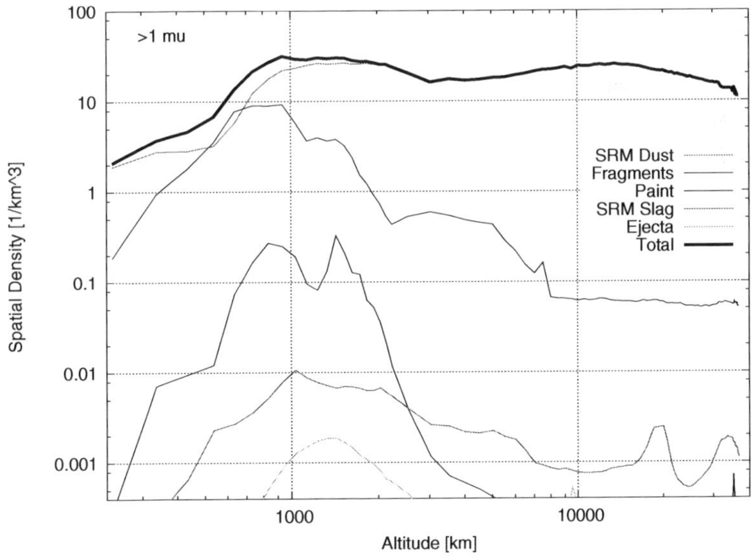

Fig. 3.26. Spatial density versus altitude for objects of diameters $d > 1\,\mu\mathrm{m}$ according to the MASTER-2001 model, for May 2001. Different sources are color coded. The thick line shows the envelope of total spatial density (see color plate on page 374).

At the beginning of the 1980s a new debris source entered the scene: sodium–potassium (NaK) droplets, released during 16 RORSAT reactor core ejections between April 29, 1980, and March 14, 1988 (see Table 2.3). These events ejected objects at sub-catalog sizes of 0.1 mm $\leq d \leq$ 4.54 cm into a tight, common orbit band at $\overline{H} \approx 950$ km and $i \approx 65°$. By 1988, an estimated 15,000 NaK droplets with $d > 1$ cm represented a share of ~10% of the LEO population in this size regime, which is apparent as a steep increase of object densities at 950 km within one decade (see Fig. 3.17). The effect is also noticeable in the declination distribution of Fig. 3.18, though with a smaller increment, due to the wide spreading across a declination band of ±65°.

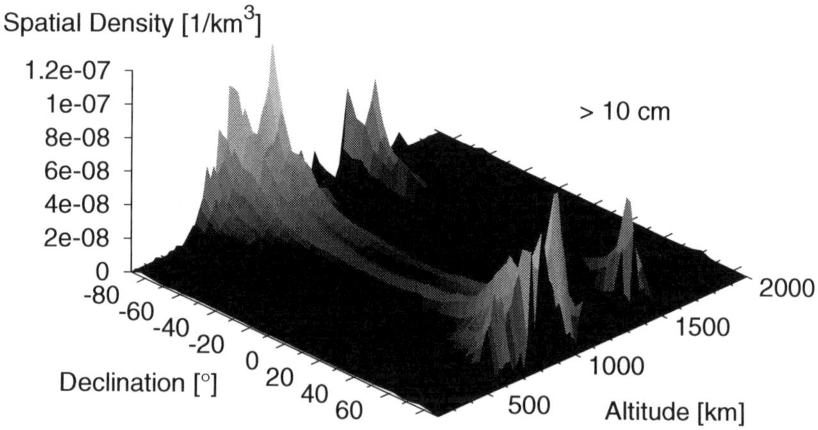

Fig. 3.27. Spatial density in LEO versus altitude and declination for objects of diameters $d > 10$ cm according to the MASTER-2001 model, for May 2001.

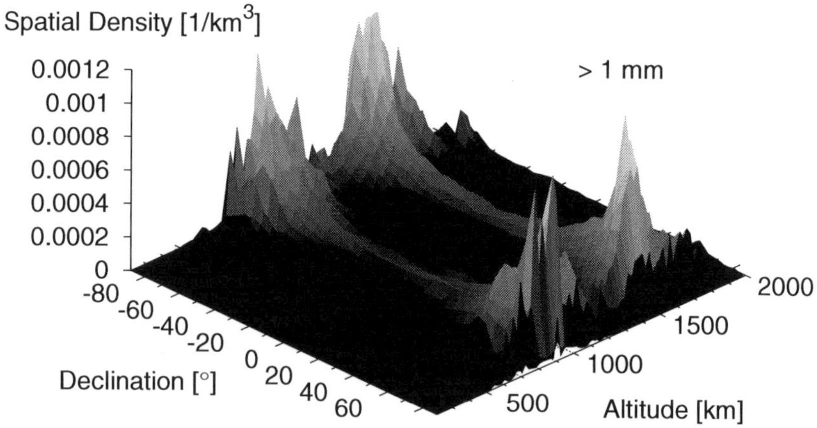

Fig. 3.28. Spatial density in LEO versus altitude and declination for objects of diameters $d > 1$ mm according to the MASTER-2001 model, for May 2001.

As compared to the catalog-size objects in Fig. 3.15, the centimeter- and millimeter-size populations in Fig. 3.17 and Fig. 3.19 show some airdrag cleansing at lower LEO altitudes (since drag is driven by the area-to-mass ratios $A/m \propto 1/d$). Since no extra source terms are added, the altitude profile of spatial densities $D_{d>1\,\mathrm{mm}}(\overline{H},t)$ looks similar to the counterpart for $d > 1$ cm, though with more noticeable population inputs during release events. Moreover, the NaK contribution at millimeter sizes in the 1980s is completely removed via airdrag by 1992. There is one striking difference in the declination distributions $D_{d>1\,\mathrm{mm}}(\delta,t)$ (Fig. 3.20) and $D_{d>1\,\mathrm{cm}}(\delta,t)$ (Fig. 3.18). Fig. 3.20 shows an inverse U-shape signature of the first SRM firing of the 3-ton class in February, 1974. The Mu-3C stage 3 injected its payload into an eccentric transfer orbit with a perigee at $H_{pe} = 281$ km and $\omega = 13°$, and an initial apogee at $H_{ap} = 3,264$ km, well outside the LEO region. Due to the low pericenter with high airdrag, the millimeter-size slag objects rapidly decayed, dragging the apogees into the LEO regime, while maintaining the pericenter altitudes. Due to the much longer residence times at the apocenter passes (which are close to the equator), the spatial density contribution increases accordingly, leading the the unusual, inverse U-shape of the profile $D_{d>1\,\mathrm{mm}}(\delta,t)$ at the "snapshot" epoch in mid 1974. The profile turns into a standard U-shape once eccentricities of the slag cloud orbits are sufficiently reduced.

For the MASTER-2001 reference epoch in May 2001 Table 3.2 shows the total count of modeled "macro objects" with $d \geq 1$ mm for 4 size thresholds. Separate counts are provided for each contributing space debris source, for LEO and for the combined LEO, MEO, and GEO environment. Table 3.3 provides equivalent data for the "micro objects" with $d < 1$ mm. These numbers correspond to integrals over the altitude profiles of spatial density as given by Fig. 3.21–3.26 for threshold sizes from $d \geq 10$ cm to $d \geq 1$ µm.

For the size class of $d \geq 10$ cm, Fig. 3.21 shows that fragments are dominating in all altitude regimes, except around 300 to 400 km, locally near 1,500 km, and in the vicinity of the GPS and GLONASS navigation satellite orbits, where launch and mission-related objects prevail.

In Fig. 3.22 (for $d \geq 1$ cm) the contributing sources are fragments, NaK droplets, SRM slag, and mission-related objects. Fragments dominate from 400 to 2,000 km and in the GEO ring. SRM slag is the largest contributor to all other altitudes, and especially dominates the 12 h navigation satellite orbits. In the vicinity of the RORSAT reactor deposit orbits near 900 km, 13 years after the last release event, NaK droplets still furnish a large share of the cm-size population. Launch and mission related objects only play a minor role (this source is omitted in the subsequent figures).

Fig. 3.23 depicts the spatial density distribution $D_{d>1\,\mathrm{mm}}(H)$ for the same source populations, plus impact ejecta. The altitude bands where fragments and SRM slag dominate are basically unchanged. Ejecta, however, start to become significant, especially in areas with high population densities of launch and mission-related objects. Due to atmospheric wash-out, the relative contribution of NaK droplets is considerably reduced, and no contributions can be expected for $d <$ 1 mm.

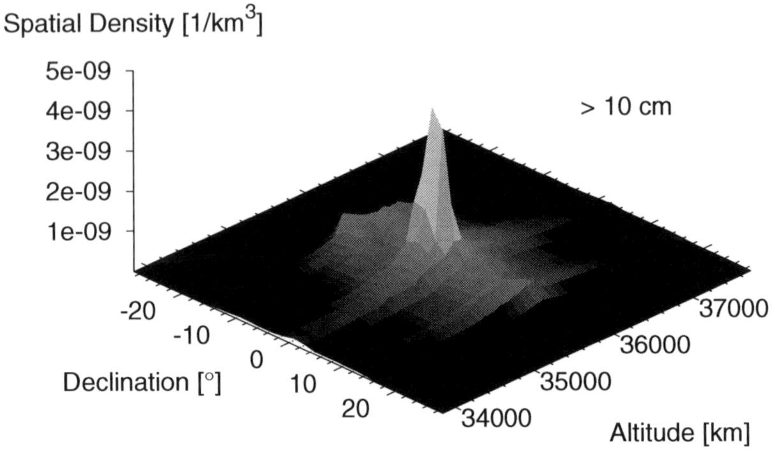

Fig. 3.29. Spatial density in GEO versus altitude and declination for objects of diameters $d > 10$ cm according to the MASTER-2001 model, for May 2001.

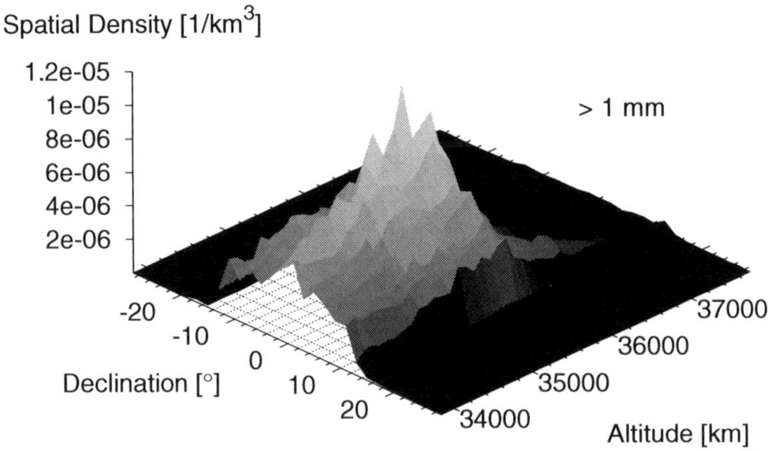

Fig. 3.30. Spatial density in GEO versus altitude and declination for objects of diameters $d > 1$ mm according to the MASTER-2001 model, for May 2001.

For sizes of $d > 0.1$ mm Fig. 3.24 shows that surface degradation products ("paint flakes") start to dominate at most altitudes below 5,000 km, only locally matched or exceeded by fragmentation debris. Above 5,000 km SRM slag contributions prevail by far. Since both "paint flakes" and ejecta have intact objects as their common source, their altitude profiles have a similar shape, though separated by 2 orders of magnitude.

When going to sizes below 10 μm, SRM dust is introduced as additional source. Fig. 3.25 shows the spatial density profile $D_{d>10\,\mu\mathrm{m}}(H)$, and Fig. 3.26 shows the equivalent profile for diameters of $d > 1\,\mu m$ (with the same scaling of axes). The peak concentrations in both charts are quite similar for all sources, with SRM dust dominating at all altitudes, except below 600 km, where "paint flakes" may prevail. The main difference between the 10 µm and 1 µm profiles lies in the altitude distribution of SRM dust. Dust particulates of $d > 1\,\mu m$ receive larger Δv increments (see Eq. 3.57), which generally take them to highly eccentric orbits, or to immediate re-entry. Due to dominating large orbit eccentricities, their distribution is quite uniform across all altitudes, in contrast with the 10 µm dust population. In general one can say that fine structures, and the ratios between extrema in the altitude and declination profiles of spatial densities are diminishing with decreasing object size. This is also reflected in the resident probabilities. From the population with $d > 10$ cm about 54.3% are contributing to the spatial densities in LEO, while this fraction reduces to 33.8% for $d > 1$ cm, to 11.6% for $d > 0.1$ mm, and to 0.5% for $d > 1\,\mu m$.

Fig. 3.27 through 3.30 show spatial density profiles in LEO and GEO, as a function of altitude and declination, for size thresholds of 10 cm and 1 mm. In Fig. 3.27 the spatial densities $D_{d>10\,\text{cm}}(H,\delta)$ have maxima in the U-shape declination profiles at $\delta = \pm 82°$ (with secondary peaks at $\delta = \pm 65°$) for altitudes of $\overline{H} \approx 900$ km, and $\approx 1,400$ km. The distribution $D_{d>1\,\text{mm}}(H,\delta)$ in Fig. 3.28 looks similar, but has equal ranking peaks at 900 km and 1,400 km. The decline of spatial densities at the upper range of LEO is much less pronounced in this case, while the elimination of these smaller objects due to airdrag below 500 km is considerably enhanced. The small band within $\delta = \pm 28.5°$, which is reaching further down to 300 km, is caused by SRM slag from GTO insertion burns of missions launched from KSC. At the locations of density peaks, the mean separation distance between two objects of the same size class is ~200 km for 10 cm objects, and ~10 km for 1 mm objects.

Fig. 3.29 illustrates the spatial density distribution $D_{d>10\,\text{cm}}(H,\delta)$ for the vicinity of the GEO region. The GEO ring is clearly discernible at $\delta \approx 0°$ and $\overline{H} \approx 35,786$ km. Objects which are no longer controlled undergo a cyclic variation of inclination with a period of 53 years and an amplitude of about 15° due to luni-solar perturbations in concert with stabilizing Earth oblateness effects. This is noted as a ridge across declinations of $\pm 15°$ at a near-constant GEO altitude. The object density decreases rapidly when moving away from the geosynchronous radius. This is quite different for $D_{d>1\,\text{mm}}(H,\delta)$ in Fig. 3.30. In this case, the density drop-off is only present for outward directions, while below the GEO ring SRM slag objects from GTO injections fill the space within ±15°of declination. Moreover, 2 known and 11 reconstructed explosions in GEO result in a wider spreading of small-size fragments, both in inclination and altitude. At concentration peaks, the spatial densities correspond to a mean separation distance of ~630 km for 10 cm objects, and of ~44 km for 1 mm objects.

3.10 COMPARISON OF MEASUREMENTS AND MODELED DATA

An important aspect in the development of a space debris environment model is its validation against measurement data of several independent sources, using different sensor techniques and performances to monitor the regions of interest, up to GEO altitudes, at different observation epochs. The MASTER-2001 model has been tuned and validated to achieve a good match with impact data on retrieved surfaces, with radar measurements, and with optical observations. Measurements by radars and telescopes are of particular interest, since they consolidate the knowledge on centimeter-size objects in LEO, and on decimeter-size objects in GEO, which constitute a risk to operational spacecraft and which have the potential to increase the growth of space debris in the future. Radar and optical data are very much dependent on the measurement system, its performance parameters, and its operational peculiarities. Hence, they cannot be readily interpreted and applied in the validation process of a space debris environment model. To bridge this deficiency a Program for Radar and Optical Observation Forecasts (PROOF-2001) was developed in support of the MASTER-2001 model (Krag et al., 2002). The PROOF-2001 tool has the task of determining detection rates for a given radar or optical sensor system (ground- or space-based), based on an input population provided by the MASTER-2001 model. The model population can then be adjusted until a consistent match is obtained with available measurement data. The basic principles and functionalities of PROOF-2001 shall be briefly outlined in the following.

Radars and telescopes have some common features: they have a sensor-specific field of view which can be pointed to a certain direction to detect objects which reflect energy in the sensitive electromagnetic spectrum of the instrument. One must distinguish between "detections" and "field-of-view crossings". For a given population model the latter events are straightforward to model. The challenge lies in the translation from the idealized crossing objects to the detected ones. This complex problem can be solved, if a detailed system model of the sensor is available.

For radars PROOF-2001 considers a pulsed system in which the received signals are processed into complex samples, which are forwarded to a matched-filter bank, and subsequently serve as input to a square-law detector. A predefined number of detections is then added (incoherently integrated) and compared against a detection threshold level to determine a real detection. It is assumed that the signal amplitude is known and constant in all samples (so-called "Swerling Case 0"), that the antenna pattern is rotationally symmetric, that the range is constant for each integration interval, and that the path offset is constant for each pass. Inputs required for the analysis are the pass geometry with range and range-rate, the object size, the desired false alarm probability, and the location and parameters of the radar system (e.g. antenna size, transmitted power, signal wavelength, pulse duration, pulse repetition frequency, range gate, number of integrated pulses, and noise-equivalent RCS for reference conditions).

The validation of the MASTER-2001 LEO population down to the centimeter level was performed on the basis of TIRA and Haystack beam-park experiments in 1999 and 2000. Deficiencies which were noted in the initial population model after simulating the campaigns were then corrected to obtain a better match of the data. The calibrated, final MASTER population generated simulated measurements which were in good agreement with real observation data (see Fig. 3.31, 3.32, and 3.34).

Between 1999 and 2004 the FGAN tracking and imaging radar (TIRA, see Section 2.4 and Fig. 2.30) performed seven beam-park experiments, most of these of approximately 24 hours duration, to scan a full small circle of latitude in inertial space. From its location at 7.1299° E and 50.6166° N the antenna beam was oriented due east (azimuth $A = 90°$), at an elevation of 75°, with a range gate set to cover LEO altitudes. With a noise-equivalent RCS of −47.8 dB at 1,000 km range, for a peak power of 1.5 MW, the expected detection threshold was on the order of $d \geq 2\,\text{cm}$. The observable parameters of the TIRA system for a fixed pointing direction are the power of the radar echo (leading to an RCS), the round-trip signal travel time (leading to a range), and the Doppler shift of the echo (leading to a range-rate). With an adopted size estimation model the RCS can be readily converted into an equivalent object diameter (see Fig. 2.26), the range can be translated into an orbit altitude, and the range-rate can be used to derive a "Doppler inclination", when assuming a near-circular orbit of the detected object. This range-rate interpretation is greatly simplified by the chosen pointing direction along a small circle of latitude. During two 24 hour campaigns in 1999 and one campaign in 2000, TIRA detected 378, 352, and 471 objects, of which 69, 83, and 94 could be correlated with the USSPACECOM catalog. When these experiments were reproduced by the PROOF program for a preliminary MASTER population, deficiencies were noted at $\overline{H} \approx 1,400\,\text{km}$ and $i \approx 55 \pm 3°$, and at $\overline{H} \approx 1,000 \pm 400\,\text{km}$ and $i \approx 90°$. These population gaps were closed, and the measurement data were adequately matched, by identifying and modeling five unrecorded fragmentation events, and by introducing two further fictitious explosions in LEO.

The revised, final LEO population of MASTER-2001 was used to simulate the TIRA beam-park experiments in 1999 and 2000, and to reproduce Haystack measurements of 1999. These comparisons provided satisfactory matches, as illustrated in Fig. 3.31 to 3.34 for the TIRA results. Fig. 3.31 shows the altitude distribution of detections rates for bin sizes of $\Delta H = 50\,\text{km}$. Peaks of 1.8 per hour at 950 km and 1.2 per hour at 1,450 km were observed and closely matched by simulations. For the same measurement data set Fig. 3.32 shows the detection rates versus range-rate-derived Doppler inclinations (with bin sizes of $\Delta i = 2°$). Again, the observed detection peaks of up to 1.3 per hour, and the fine structure of the distribution are well represented by the simulations. Fig. 3.33 illustrates that the real orbit inclinations can be quite different from the Doppler-derived ones, particularly in inclination bands with a large share of eccentric orbits. This is noticeable at $i \approx 65°$, where several Cosmos spacecraft exploded on HEO trajectories. In the same inclination band NaK droplets from RORSAT reactor ejections and SRM slag particles from GPS transfer orbits are contributing. Further population clusters are visible at 72°, 75°, 82°, and Sun-synchronous inclinations of 98° to 105°.

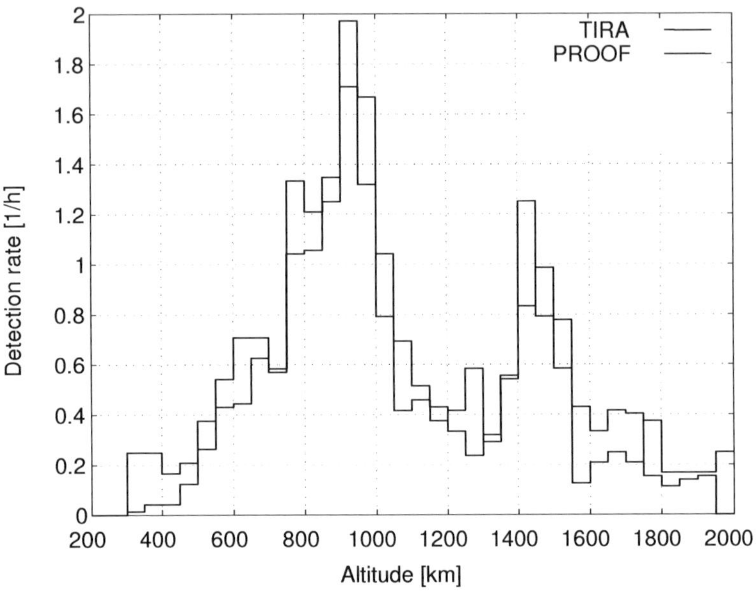

Fig. 3.31. Number of detections versus altitude for an FGAN/TIRA beam-park experiment in 2001, compared with PROOF simulations, using the MASTER space debris population (see color plate on page 376).

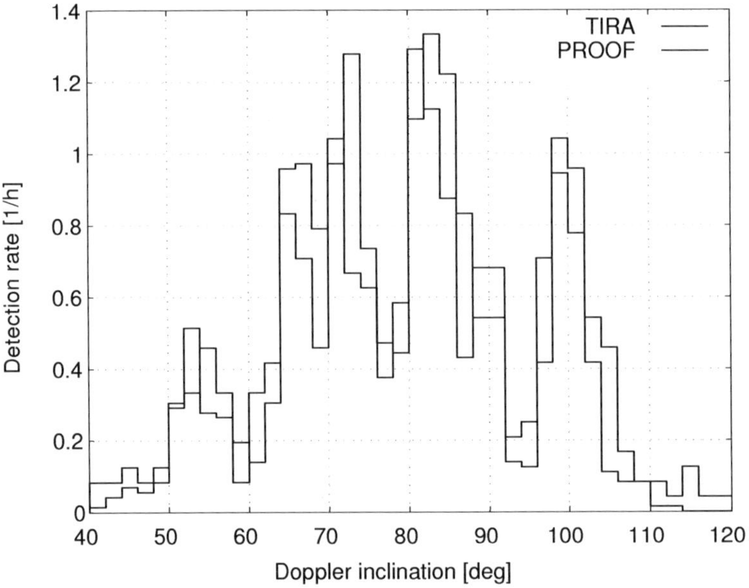

Fig. 3.32. Number of detections versus Doppler-inclination for an FGAN/TIRA beam-park experiment in 2001, compared with PROOF simulations, using the MASTER space debris population (see color plate on page 376).

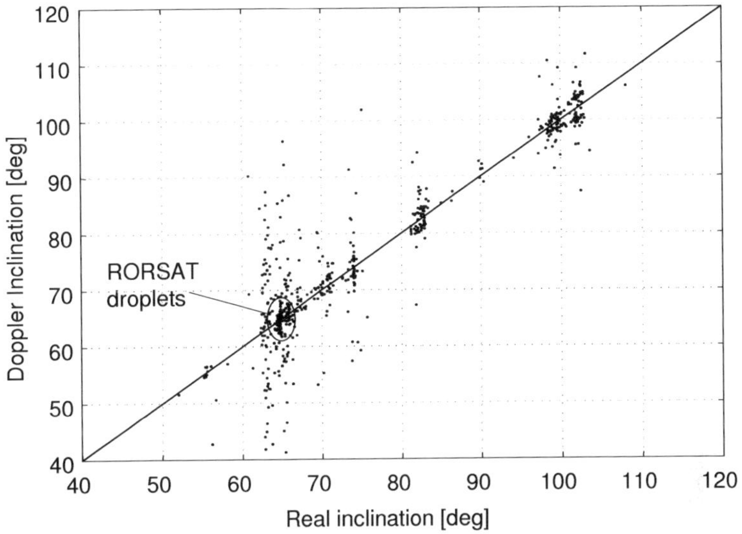

Fig. 3.33. Correlation between Doppler-derived inclination and the corresponding real inclination of a PROOF-simulated FGAN/TIRA campaign for the MASTER space debris population in 2001.

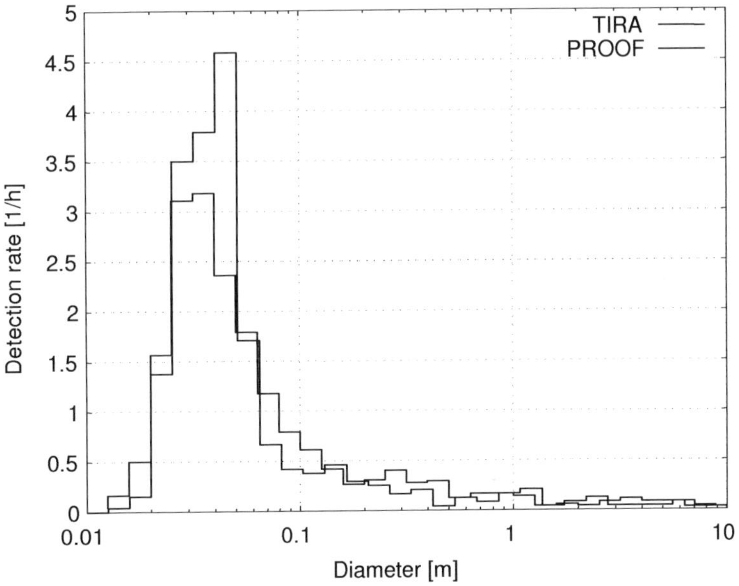

Fig. 3.34. Number of detections versus object diameter for an FGAN/TIRA beam-park experiment in 2001, compared with PROOF simulations, using the MASTER space debris population (see color plate on page 375).

Fig. 3.34 shows detection rates versus object diameters as interpreted by the TIRA system, using the NASA size estimation model (SEM), and as determined from the customized size-estimation model of PROOF. The sensitivity threshold of the radar at ~2 cm is well reproduced, as is the integral over all detections. The simulated distribution, however, is slightly broader than the reference profile, with a flatter peak. This is due to special assumptions introduced in PROOF for the object shapes as a function of their size. In contrast to the NASA size-estimation model, which considers spherical objects only, PROOF assumes solid, spherical objects for diameters $d \leq 5$ cm, oblate spheroids of different flattening for 5 cm $< d \leq 50$ cm, and flat plates of 1 cm thickness for $d > 50$ cm.

All radar data which supported the MASTER-2001 development were produced in mono-static observation scenarios of TIRA and Haystack for 24 hour beam-park experiments in 1999 and 2000. In all cases the pointing direction of the radar beam was at $A = 90°$ (due east) under elevation $h = 75°$. For the TIRA L-band system the 3 dB beamwidth was 0.5° (detection threshold ~2 cm), and for the Haystack LRIR radar, operating in X-band, the 3 dB beamwidth was 0.058° (detection threshold ~6 mm). During the second TIRA campaign in 1999 a total of 352 detections were made, of which 83 could be correlated with the US-SPACECOM catalog. For the same conditions, the PROOF simulations indicated 313 detections (−11.1%). The TIRA campaign in 2000 generated 471 detections (94 thereof catalog objects), as compared to 430 simulated ones for PROOF (−8.3%). The Haystack campaign in 1999 was synchronized with TIRA measurements, but lasted 28 hours. In this timespan 216 objects were detected (11 of those correlated with the catalog), as compared to 277 simulated detections predicted by PROOF (+28.2%). In view of the highly dynamic environment at sub-centimeter sizes, this is still a good match. Since PROOF is simulating detections for a fully characterized population model, the detections can also be correlated with individual space debris sources. For the TIRA campaign in 2000, with a detection threshold of $d \geq 2$ cm, 81.7% of the PROOF-simulated detections are due to fragments, 13.8% are due to catalog objects, 2.5% are NaK droplets, and 2.0% are SRM slag. In the case of Haystack, with a detection threshold of ~6 mm, 67.8% are fragments, 1.5% are catalog objects, 19.1% are NaK, and 11.6% are SRM slag objects. Hence, at lower sizes, NaK and SRM slag are more dominant.

Data which were generated by TIRA/Effelsberg and Goldstone/Goldstone in bi-static mode, and which reached down to sub-centimeter objects, cannot yet be processed by PROOF-2001, but will be incorporated in a successor model.

Radar systems are well suited to sample the LEO space debris environment. Due to their strongly decreasing sensitivity with increasing range (received power $P_r \propto 1/\rho^4$, see Eq. 2.8), they are less useful for MEO and GEO observations. In these orbital regions optical systems have advantages since their sensitivity limitation with range is less severe ($P_r \propto 1/\rho^2$, see Eq. 2.1).

For optical sensors PROOF-2001 considers an astronomical telescope, equipped with a CCD (Charge Coupled Device), operating in the visible light spectrum. An observation geometry filter first determines which objects of a given population are passing through the field-of-view, as determined by the optical properties (e.g.

aperture) and pointing direction of the telescope. The performance parameters of the CCD (e.g. sensitivity, quantum efficiency, read-out time, required signal-to-noise ratio), together with the pixel dwell times, integration times, and observation gap times govern the overall performance of the optical system for an object of known visual magnitude. Based on these performance characteristics the detectable objects are extracted from the super-set of crossing objects. For the simulated image processing chain discrete light sources (Sun, Moon, planets, bright stars), continuous light sources (faint stars, galaxies, airglow, scattered Sun and Moon light), and atmospheric effects are taken into account.

In order to verify the MEO space debris population of the MASTER-2001 model measurement data from 1999 of NASA's liquid mirror telescope (LMT) were processed (see Table 2.4). The LMT is located at 105°W and 33°N. Due to its liquid mirror, formed of a rotating pool of mercury, it only operates in zenith looking mode, with a field-of-view of 0.278°. When using a video camera equipment, with a frame rate of $30\,s^{-1}$, and a frame resolution of 1250×1250 pixels, its detection threshold of $\sim$17.5 mag allows a detection of objects with $d > 3\,cm$ at distances of up to several thousand kilometers. During an observation campaign in 1999/2000 the LMT made 544 detections (122 thereof catalog objects). Some of these (probably 14%) are suspected to be meteoroid detections. Based on the LMT system model, the PROOF program simulated 370 detections of un-cataloged objects (-12.7%), which is a very good match of the observations, considering the remaining uncertainties in the data interpretation. Most of the detections were at LEO altitudes of $H < 2{,}500\,km$, and at inclinations of 65° to 72°, 82°, 90°, and 100°. The most distant detections observed (and simulated) were at 9,000 km altitude.

Due its considerable distance from ground-based sensors the GEO vicinity is the most difficult region to be modeled and verified in the MASTER-2001 space debris environment. The tuning and verification of the MASTER-2001 GEO population was entirely based on observations made by the ESA Space Debris Telescope (SDT), which is located at 16°W and 19°N. The telescope has an aperture of 1 m, a field-of-view of 0.7°, and a CCD detector of $4{,}096 \times 4{,}096$ pixels (composed of a 2×2 mosaic). During its observation campaign in 2001 (as for most of its time) the SDT was operated in a tracking mode. In this mode the telescope viewing direction is kept Earth-fixed, in order to pseudo-track GEO satellites. After each exposure, the field-of-view is re-positioned to re-visit the same area of the sky again. The exposure repetition rate is 1 frame per minute (with 2 s integration time and 58 s gap time). It is chosen such that each GEO object is visible on 3 consecutive frames, allowing its orbit determination from 3 time-tagged position fixes. The ESA SDT observation campaign to be modeled was conducted in 2001, covering 266 search fields of about 0.5 hours length each. For a given MASTER-2001 population, all of these search arcs were modeled individually by the PROOF software, assuming a mean albedo of 0.1 for all objects in the field-of-view.

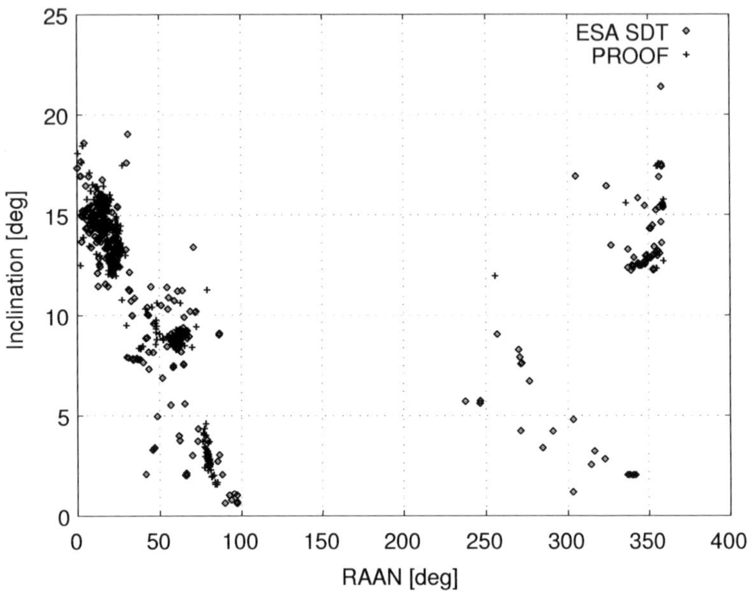

Fig. 3.35. Right ascension and inclination of detectable near-GEO objects, as observed by the ESA telescope, and as simulated by PROOF for the MASTER space debris population of 2001 (see color plate on page 377).

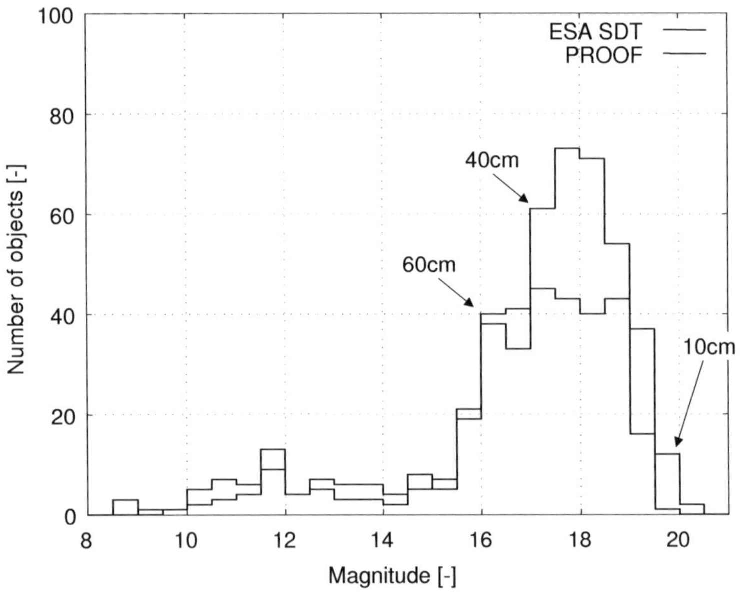

Fig. 3.36. Visual magnitudes of detectable near-GEO objects, as observed by the ESA telescope, and as simulated by PROOF for the MASTER space debris population of 2001 (equivalent detection threshold diameters are marked; see color plate on page 377).

Fig. 3.35 shows the distribution of SDT detections as a function of the ascending node position (RAAN) and inclination of the observed orbits. This diagram is a typical fingerprint of the GEO situation at a given epoch, providing a status of the inclination vector migration during the 53 years cyclic inclination drift within $0^\circ \leq i \leq 15^\circ$. Its highly dynamic evolution can be illustrated by different snapshots across the space history (the MASTER-2001 model provides such snapshots at 3-month intervals). Initial comparisons of SDT observations with simulations for a preliminary MASTER-2001 population indicated large deficits in the detection rates. Hence, the two known GEO fragmentations (explosion of the Ekran-2 satellite in 1978, and of a Titan III-C Transtage in 1992) were augmented by 11 fictitious explosion events with fragmentation epochs between 1986 and 1998. These epochs, and the corresponding inclinations and node positions of the source objects were chosen such that the detection clusters in Fig. 3.35 were well reproduced (the "banana"-shape clusters which are discernible in the observations can be shown to be an indication of fragmentation events). The adjusted MASTER-2001 population led to 442 detections simulated by PROOF, which is just 11.9% more than observed. In another open-loop comparison the final MASTER population was checked against results of an ESA SDT campaign in 1999. The 100 scans of 50 hours duration in total yielded 150 detections by the SDT, as compared to 152 simulated ones by PROOF.

The sensitivity of the ESA Space Debris Telescope is indicated in Fig. 3.36, where the number of detected objects (real and simulated) is plotted versus visual magnitude. For a mean albedo of 0.1 markers are also showing corresponding detection size thresholds. The overall match of the two distributions is acceptable, though PROOF is not able to reproduce the unusual flat peak in the observation data between magnitude 16 and 19 (this unusual feature was not present in observation data of 1999).

PROOF-2001, the Program for Radar and Optical Observation Forecasts, has been an essential tool in the calibration and validation process of the MASTER-2001 space debris population from sub-centimeter to catalog sizes. The validation of the population at sub-millimeter sizes was performed on the basis of predicted versus observed impact fluences for the LDEF and EURECA satellites, and for solar arrays of the Hubble Space Telescope which were put in orbit, and which were retrieved by the Space Shuttle (see Table 2.6 and (Bendisch et al., 2002)).

3.11 ALTERNATIVE SPACE DEBRIS ENVIRONMENT MODELS

The MASTER-2001 model and its associated tools are used as baseline throughout Chapters 3 to 6, because of the close involvement of some of the present authors in its development. Several other space debris environment models, however, do exist and are in use. For the following overview two of these models shall be focused on, which (together with MASTER-2001) took part in an IADC review of Debris Engineering Models in 2003: the ORDEM-2000 model of NASA (Liou et al., 2001), and the Russian SDPA-E model (Nazarenko and Menshikov, 2001).

SDPA-E (Space Debris Prediction and Analysis Engineering model) considers object sizes of $d > 1\,\mathrm{mm}$, and spans orbit altitudes up to super-GEO ($H \leq 36{,}200\,\mathrm{km}$. It is a semi-analytical stochastic model to assess the current and future space debris environment, and to determine associated collision fluxes on user-defined target orbits. The SDPA-E model does not separate source terms, but uses size-dependent probability distributions with altitude and declination to determine spatial object densities and resulting collision fluxes on spherical target objects. The flux contribution and the associated mean impact velocity varies with the impact azimuth in the horizontal plane. The reference epoch for the SDPA-E model is January 2000.

ORDEM-2000 (Orbital Debris Environment Model) was developed at NASA JSC to assess the debris environment and its resulting spatial densities and collision fluxes for object sizes of $d > 10\,\mu\mathrm{m}$, and for altitudes of $H \leq 2{,}000\,\mathrm{km}$. The model can be applied to epochs between 1991 and 2030. ORDEM-2000 is based on a 3D discretization of the LEO region in spherical sectors, storing for all transient orbits the spatial density contributions, velocities, and inclinations for different, discrete size classes. For a user-defined target orbit the collision flux, velocity, and azimuth direction is determined by interrogating all intersected volume elements. The underlying debris orbit distributions in terms of perigee, eccentricity, and inclination (for different size classes) are fitted to a large set of measurement data, including Haystack (LRIR and HAX) and Goldstone observations, and returned surfaces from LDEF, HST, and the Space Shuttle (STS). Intentionally, no attempt is made to discriminate between different space debris source terms. The reference epoch for the ORDEM-2000 model is January 1, 1999. The model is freely available from NASA, also as download from the homepage of the Orbital Debris Program Office at JSC.

A comparison between SDPA-E, ORDEM-2000, and MASTER-2001 for near-circular LEO orbits shows that there is a fair to good agreement in terms of spatial densities and collision fluxes for debris particles with $d > 1\,\mathrm{mm}$. The structures of azimuth and velocity dependent flux contributions can, however, show noticeable differences, predominantly at high inclinations. At smaller sizes of $d \leq 0.1\,\mathrm{mm}$, differences in flux levels between ORDEM-2000 and MASTER-2001 can reach one order of magnitude (SDPA-E does not cover this size range). Such discrepancies are particularly emerging when comparing results for eccentric orbits of $e \geq 0.1$.

3.12 REFERENCES

Akiba, R. et al. (1990). Behavior of Aluminum Particles Exhausted by Solid Rocket Motors. In *Orbital Debris Conference, Baltimore/MD*, AIAA/NASA/DoD.

Aksnes, K. (1976). Short-Period and Long-Period Perturbations of a Spherical Satellite Due to Direct Solar Radiation Pressure. *Celestial Mechanics*, vol.13:89–104.

Bendisch, J., Bunte, K., Sdunnus, H., Wegener, P., Walker, R., and Wiedemann, C.

(2002). Upgrade of ESA's MASTER Model. Technical Report (final report) ESA contract 14710/00/D/HK, ILR/TUBS, Braunschweig.

Goldstein, R., Goldstein, S., and Kessler, D. (1998). Radar Observations of Space Debris. *Planetary and Space Sciences*, vol.46, no.8:1007–1013.

Hernández, C., Pina, F., Sánchez, N., Sdunnus, H., and Klinkrad, H. (2001). The DISCOS Database and Web Interface. In *Proceedings of the Third European Conference on Space Debris*, ESA SP-473, pages 803–807.

Johnson, N., Krisko, P., Liou, J., and Anz-Meador, P. (2001). NASA's New Break-Up Model of EVOLVE 4.0. *Advances in Space Research*, vol.28, no.9:1377–1384.

Kerridge, D., Carlaw, V., and Beamish, D. (1989). Development and Testing of Computer Algorithms for Solar and Geomagnetic Activity Forecasting. Technical Report WM/89/22C [ESA CR(P) 3039], British Geological Survey.

Kessler, D. (1985). The Effects of Particulates from Solid Rocket Motors Fired in Space. *Advances in Space Research*, vol.5:77–86.

King-Hele, D. (1987). *Satellite Orbits in an Atmosphere: Theory and Applications*. Blackie, Glasgow, UK.

Klinkrad, H. (1993). Collision Risk Analysis for Low Earth Orbits. *Advances in Space Research*, vol.13:551–557.

Krag, H., Bendisch, J., Rosebrock, J., Schildknecht, T., and Sdunnus, H. (2002). Sensor Simulation for Debris Detection. Technical Report (final report) ESA contract 14708/00/D/HK, ILR/TUBS, Braunschweig.

Liou, J., Matney, M., Anz-Meador, P., Kessler, D., Jansen, M., and Theall, J. (2001). The New NASA Orbital Debris Engineering Model ORDEM2000. In *Proceedings of the Third European Conference on Space Debris*, ESA SP-473, pages 309–313.

Liu, J. and Alford, R. (1979). A Semi-Analytic Theory for the Motion of a Close-Earth Artificial Satellite with Drag. In *17th Aerospace Sciences Meeting, New Orleans*, AIAA Paper No. 79-0123.

Maclay, T. and McKnight, D. (1994). The Contribution of Debris Wakes from Resident Space Objects in the Orbital Debris Environment. *Safety & Rescue Science & Technology Series*, vol.88:215–228.

McDonnell, J. et al. (1999). Meteoroid and Debris Flux and Ejecta Models. Technical Report (final report) ESA contract 1887/96/NL/JG, Unispace Kent, Canterbury, UK.

Meyer, R. (1992). In-Flight Formation of Slag in Spinning Solid Propellant Rocket Motors. *Journal of Propulsion & Power*, vol.8, no.1:45–50.

Montenbruck, O. and Gill, E. (2000). *Satellite Orbits – Models, Methods, and Applications*. Springer, Berlin, Heidelberg, New York.

Morton, T. and Ferguson, D. (1993). Atomic Oxygen Exposure of Power System and Other Spacecraft Materials: Results of the EOIM-3 Experiment. Technical Report NASA TM 107427, NASA Lewis Research Center.

Nazarenko, A. and Menshikov, I. (2001). Engineering Model of the Space Debris Environment. In *Proceedings of the Third European Conference on Space Debris*, ESA SP-473, pages 293–298.

Ojakangas, G. et al. (1996). Solid-Rocket-Motor Contributions to the Large-Particle Orbital Debris Population. *Journal of Spacecraft & Rockets*, vol.33, no.4:513–518.

Roth, E. (1996). Construction of a Consistent Semi-analytic Theory of a Planetary or Moon Orbiter Perturbed by a Third Body. *Celestial Mechanics*, vol.28:155–169.

Schildknecht, T., Ploner, M., and Hugentobler, U. (2001). The Search for Debris in GEO. *Advances in Space Research*, vol.28, no.9:1291–1299.

Seidelmann, P. et al., editors (1992). *Explanatory Supplement to the Astronomical Almanac*. University Science Books, Mill Valley, CA.

Stansbery, G., Matney, M., Settecerri, T., and Bade, A. (1997). Debris Families Observed by the Haystack Orbital Debris Radar. *Acta Astronautica*, vol.41, no.1:53–56.

Wiedemann, C., Oswald, M., Stabroth, S., Vörsmann, P., and Klinkrad, H. (2004). NaK Droplet Size Distribution. Technical Report ILR-IB-2004-001, rel 0.9, Institute of Aerospace Systems, TU Braunschweig.

4

Modeling of Collision Flux for the Current Space Debris Environment

H. Klinkrad, P. Wegener, J. Bendisch and K. Bunte

The terrestrial space environment of intact objects and space debris was characterized in Chapters 2 and 3 for particle diameters ranging from 1 μm to catalog sizes. The collision flux resulting from this environment for an object of a certain size on a given target orbit will be determined in the present chapter.

4.1 DETERMINATION OF COLLISION FLUX

Eq. 3.3 in conjunction with Eq. 3.1 defined the collision or impact probability $P_c = v\,D\,A_c\,\Delta t$ as the product of impact flux $F = v\,D$, collision cross-section A_c, and resident time Δt. Δt refers to the time spent in a volume of uniform spatial density D (occupied by objects which are at rest), which is trespassed at a mean velocity v. The quantity of main interest here is the collision flux, both in terms of magnitude and direction in a target-centered, orbit-related coordinate system.

Based on the concept of volume discretization and cell-passage events, the spatial density $D_{i,j,k}$ in a given volume cell which is centered at a geocentric distance r_i, declination δ_j, and right ascension α_k is given by Eq. 3.23, for a total number $L_{i,j,k}$ of passes through this cell by different debris orbits. For each pass event $\ell(i,j,k)$, with $1 \leq \ell(i,j,k) \leq L_{i,j,k}$, the transient velocity $\underline{v}_{\ell(i,j,k)}$ is defined in a quasi-horizontal coordinate system (radial, east, north $= \underline{U}, \underline{E}, \underline{N}$), as indicated by Eq. 3.26. Eq. 3.27 to 3.29 translate the velocity vector components into an equivalent data set of velocity magnitude $v_{\ell(i,j,k)}$, pass azimuth angle $A_{\ell(i,j,k)}$ with respect to north, and pass elevation angle $h_{\ell(i,j,k)}$ with respect to the quasi-horizontal plane. In this format the MASTER-2001 model stores all cell-passage events (CPEs) by means of tailored compression algorithms (Bendisch et al., 2002).

In order to compute collision fluxes on a user-defined target orbit, all cell-passage events $1 \leq m \leq M$ for a single revolution of the target orbit are determined by applying the same control volume partitions which were previously used for a CPE snapshot of the space debris environment at the same epoch. Let the m-th CPE of the target object take place in the volume cell indexed by i, j, k (at radial shell r_i, small circle of declination δ_j, and great circle of right ascension α_k), then the flux contribution from this ℓ-th space debris orbit which passed this cell can be expressed as

$$\underline{F}_{m,\ell} = P_m \, D_\ell \, \Delta\underline{v}_{m,\ell} \tag{4.1}$$

where $D_\ell = D_{\ell(i,j,k)}$ is the spatial density contribution to this volume cell from the ℓ-th debris object according to Eq. 3.22, and $P_m = P_{m(i,j,k)}$ is the resident probability of the target object in the volume cell according to

$$P_m = \frac{\Delta t_m}{T} \tag{4.2}$$

with Δt_m the resident time, and T the orbital period of the target object (see Eq. 3.20–3.21). The remaining quantity to determine in Eq. 4.1 is the relative velocity $\Delta\underline{v}_{m,\ell}$ between the target and debris population objects in the presently interrogated volume cell. The CPE data of the entire debris population are stored in a quasi-horizontal $\underline{U}$, $\underline{E}$, $\underline{N}$ coordinate system, in radial, east, and north direction (see Fig. A.3). In this system, cell-passage velocities of debris objects are defined in terms of the pass azimuth A_ℓ with respect to north, pass elevation h_ℓ with respect to the quasi-horizontal plane, and transient velocity magnitude v_ℓ (see Eq. 3.27–3.29). In a first step the debris velocity data, which are provided in the $\underline{U}$, $\underline{E}$, $\underline{N}$ system, must be converted into a velocity vector $\underline{v}_\ell = (\underline{v}_\ell)_{X,Y,Z}$ in the geocentric equator system $\underline{X}$, $\underline{Y}$, $\underline{Z}$, in which the computations shall be performed.

$$\underline{v}_\ell = v_\ell \begin{pmatrix} \cos\alpha_k \left(-\sin\delta_j \cos h_\ell \cos A_\ell + \cos\delta_j \sin h_\ell\right) - \sin\alpha_k \cos h_\ell \sin A_\ell \\ \sin\alpha_k \left(-\sin\delta_j \cos h_\ell \cos A_\ell + \cos\delta_j \sin h_\ell\right) + \cos\alpha_k \cos h_\ell \sin A_\ell \\ \cos\delta_j \cos h_\ell \cos A_\ell + \sin\delta_j \sin h_\ell \end{pmatrix} \tag{4.3}$$

The approach velocity vector $\Delta\underline{v}_{m,\ell} = \left(\Delta\underline{v}_{m,\ell}\right)_{X,Y,Z}$ of a piece of debris relative to the target object is then given as

$$\Delta\underline{v}_{m,\ell} = \underline{v}_\ell - \underline{v}_m \tag{4.4}$$

where the target velocity $\underline{v}_m$ at central passage time of the m-th volume cell is generally provided by an orbit propagator in the $\underline{X}$, $\underline{Y}$, $\underline{Z}$ coordinate system, and $\underline{v}_\ell$ is computed from Eq. 4.3.

For the interpretation of debris approach directions relative to a moving target object, it is advantageous to map $\Delta\underline{v}_{m,\ell} = \left(\Delta\underline{v}_{m,\ell}\right)_{X,Y,Z}$ into an orbit-related coordinate system $\underline{U}$, $\underline{V}$, $\underline{W}$, centered at the target object, with $\underline{U}$ in radial direction, $\underline{V}$ in transversal direction, and $\underline{W}$ in out-of-plane direction of the target orbit.

$$\left(\Delta\underline{v}_{m,\ell}\right)_{U,V,W} = \begin{pmatrix} U_X & U_Y & U_Z \\ V_X & V_Y & V_Z \\ W_X & W_Y & W_Z \end{pmatrix} \left(\Delta\underline{v}_{m,\ell}\right)_{X,Y,Z} \tag{4.5}$$

The X,Y,Z components of the unit direction vectors $\underline{U}, \underline{V}, \underline{W}$ are given by Eq. A.11 to A.13 in Annex A.1. For an unperturbed Kepler motion they are a function of the argument of true latitude $u = \omega + f$ of the target position on its orbit.

In the $\underline{U}, \underline{V}, \underline{W}$ coordinate system the impact geometry and velocity of the ℓ-th debris object with the target object on its m-th cell pass through a volume element indexed by i, j, k can be expressed in terms of collision velocity magnitude $\Delta v_{m,\ell}$, and the approach direction angles $A_{m,\ell}$ and $h_{m,\ell}$. In this case, $A_{m,\ell} \in [-\pi, +\pi]$ is the debris approach azimuth, measured from the target flight direction to $-\Delta \underline{v}_{m,\ell}$ in the quasi-horizontal plane (positive around $-\underline{U}$), and $h_{m,\ell} \in [-\pi/2, +\pi/2]$ is the approach elevation, measured from the quasi-horizontal plane to $-\Delta \underline{v}_{m,\ell}$, positive towards space (see Fig. A.3).

$$\Delta v_{m,\ell} = |\Delta \underline{v}_{m,\ell}| = \sqrt{(\Delta v_{m,\ell,U})^2 + (\Delta v_{m,\ell,V})^2 + (\Delta v_{m,\ell,W})^2} \tag{4.6}$$

$$A_{m,\ell} = -2 \arctan\left(\left(\sqrt{(\Delta v_{m,\ell,V})^2 + (\Delta v_{m,\ell,W})^2} + \Delta v_{m,\ell,W}\right) / \Delta v_{m,\ell,V}\right) \tag{4.7}$$

$$h_{m,\ell} = -\arcsin\left(\Delta v_{m,\ell,U} / \Delta v_{m,\ell}\right) \tag{4.8}$$

In computer programs the nominator and denominator in Eq. 4.7 are separately evaluated during the arctan computation, avoiding singularities for $\Delta v_{m,\ell,V} \to 0$.

The total debris flux which a target object receives from M cell passes along one orbital revolution is computed from

$$F = \sum_{m=1}^{M} \sum_{\ell=1}^{L_{i,j,k}} P_m \, D_l \, |\Delta \underline{v}_{m,\ell}| \tag{4.9}$$

where the m-th cell pass of the target orbit points to a volume segment indexed by i, j, k (for the radial, declination, and right ascension partition), which has a total count of $L_{i,j,k}$ cell-passage events of debris orbits. As explained in Section 3.2 (see also Eq. 3.21), an analysis time interval of one orbital revolution is a good trade-off between an adequate sampling of the debris environment and required computing resources. The outlined collision flux determination concept was developed and first published by the author (Klinkrad, 1993).

Eq. 4.9 describes the impact flux on a spherical target object, with no sensitivity to the orientation. In many applications, however, directionalities of the flux with respect to an oriented, planar surface are of importance (e.g. for impact sensors, or for impact shields). Such oriented surfaces are mostly mounted to satellites which are 3-axis controlled in yaw (around $\underline{U}$), in roll (around $\underline{V}$), and in pitch (around $-\underline{W}$). Very often the satellite axes are kept aligned with or close to the $\underline{U}, \underline{V}, \underline{W}$ coordinate system of the orbit. In this case, the orientation of the normal unit vector $\underline{n}$ of the target surface can be expressed by the azimuth A_n and elevation h_n of the pointing direction.

$$(\underline{n})_{U,V,W} = \begin{pmatrix} \sin h_n \\ \cos A_n \cos h_n \\ -\sin A_n \cos h_n \end{pmatrix}_{U,V,W} \tag{4.10}$$

The total flux on an oriented surface can then be determined from Eq. 4.11, with a switch function $\delta_{n,v}$ suppressing all flux contributions which do not impact on the front side of the surface.

$$F = \sum_{m=1}^{M} \sum_{\ell=1}^{L_{i,j,k}} P_m \, D_l \, \delta_{n,v} \left| (\underline{n})_{U,V,W} \cdot \left(\Delta \underline{v}_{m,\ell}\right)_{U,V,W} \right| \tag{4.11}$$

$$\delta_{n,v} = \begin{cases} 1 & \text{if} \quad (\underline{n})_{U,V,W} \cdot \left(\Delta \underline{v}_{m,\ell}\right)_{U,V,W} \leq 0 \\ 0 & \text{if} \quad (\underline{n})_{U,V,W} \cdot \left(\Delta \underline{v}_{m,\ell}\right)_{U,V,W} > 0 \end{cases} \tag{4.12}$$

For large-scale space structures not only the orientation of a surface element, but also its view factor is important. The view factor determines which part of the visible hemisphere is obstructed by other components of the spacecraft, casting a "flux shadow" onto the target surface for certain approach directions. As an example, Fig. 4.1 shows how certain modules of the International Space Station ISS are particularly exposed to the debris impact flux (which mainly comes from the ram direction), and how these modules serve as passive protection for the downstream ISS elements.

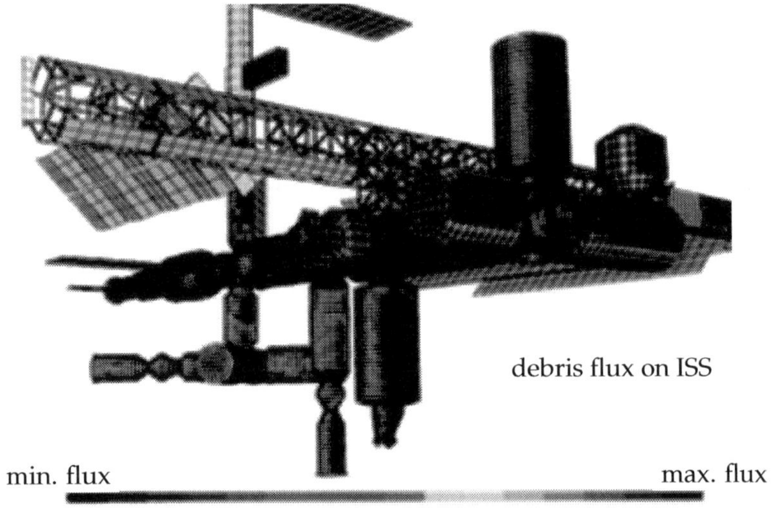

Fig. 4.1. Impact flux distribution on the ISS geometry for space debris objects of $d >$ 1 cm (view from approach azimuth $A \approx +45^\circ$ and elevation $h \approx -10^\circ$; source: NASA; see color plate on page 375).

The color-coded flux distribution in Fig. 4.1 for debris objects of $d > 1\,\text{cm}$ indicates which modules of the ISS need special shield deployments to defeat such high-risk projectiles for a nominal ISS orientation. A size threshold of $d \leq 1\,\text{cm}$ marks the current protection capability for on-orbit shielding technology based on stuffed Whipple shields.

4.2 ANALYSIS OF COLLISION GEOMETRIES

The vast majority of the catalog objects of $d > 10\,\mathrm{cm}$ are residing on near-circular orbits (see Fig. 2.13). This also applies to the majority of space debris objects at risk-relevant sizes of $d > 1\,\mathrm{mm}$. Moreover, even in the case of eccentric orbits, a collision between a target object and a member of the space debris population is in general most likely to occur at locations of highest resident probability. This would be at the apocenter, where the orbit velocity vector is parallel to the local horizontal plane. Hence, collision geometries within the quasi-horizontal plane (which is perpendicular to the radial unit vector $\underline{U}$) are of particular importance.

Fig. 4.2 illustrates the case of a collision within the horizontal plane, where the velocity $v_t = |\underline{v}_t|$ of the target object is smaller than the velocity v_d of the debris object. In this figure, [t] marks the target location. If the "debris medium" were at rest, it would impact on the target with an approach velocity of $\Delta v = |-\underline{v}_t|$ under an azimuth angle of $A = 0°$. Since the debris objects are also in motion, with an absolute velocity of $v_d > v_t$, two extreme cases can be identified: a head-on collision under $A = 0°$ at $\Delta v = v_d + v_t$ (case [a]), and a tail-on collision under $A = 180°$ at $\Delta v = v_d - v_t$ (case [b]). In the more general case [c] the impact velocity Δv and the approach azimuth A are defined as (see Fig. A.3)

$$\Delta \underline{v} = \underline{v}_d - \underline{v}_t \tag{4.13}$$

$$\sin A = -\frac{|\,\Delta \underline{v} \times \underline{v}_t\,|}{\Delta v\, v_t} \, \frac{\underline{U} \cdot (\Delta \underline{v} \times \underline{v}_t)}{|\,\underline{U} \cdot (\Delta \underline{v} \times \underline{v}_t)\,|} \tag{4.14}$$

$$\cos A = -\frac{\Delta \underline{v} \cdot \underline{v}_t}{\Delta v\, v_t} \tag{4.15}$$

$$A = 2 \arctan\left(\frac{1 - \cos A}{\sin A}\right) \tag{4.16}$$

For a given collision azimuth A, target velocity v_t, and debris velocity v_d the impact velocity Δv in the quasi-horizontal plane can be expressed as

$$\Delta v = |\,\Delta \underline{v}\,| = v_d \, \frac{\sin\left(A + \arcsin\left(\frac{v_t}{v_d} \sin A\right)\right)}{\sin A} \tag{4.17}$$

Fig. 4.3 shows another collision scenario, if the target velocity v_t is larger than the debris velocity v_d. In this case, the maximum and minimum impact velocities are again aligned with the target velocity vector, but now the lowest approach speed is produced in a head-on collision. Since the minimum and maximum impact velocities are both occurring for $A = 0°$, there must be intermediate, non co-linear cases with $A \neq 0$, and also a maximum possible azimuth angle A_{max}. This angle $A_{max} = \angle\,(\underline{v}_t, -\Delta \underline{v}_c)$ in Fig. 4.3 is given by Eq. 4.18.

$$A_{max} = \arcsin\left(\frac{v_d}{v_t}\right) \qquad \text{for} \qquad v_d \leq v_t \tag{4.18}$$

If both the target and the debris object are traveling on circular orbits at the same altitude, with $v_t = v_d$, then Eq. 4.17 can be rewritten in a simplified way.

$$\Delta v = 2\, v_t \cos A \tag{4.19}$$

$$\cos A = -\frac{\Delta \underline{v} \cdot \underline{v}_t}{\Delta v\, v_t} \tag{4.20}$$

In this case Eq. 4.18 yields $A_{max} = 90°$ for an approach velocity of $\Delta v = 0$. Hence, two objects on circular orbits can only collide under angles of $-90° < A < +90°$.

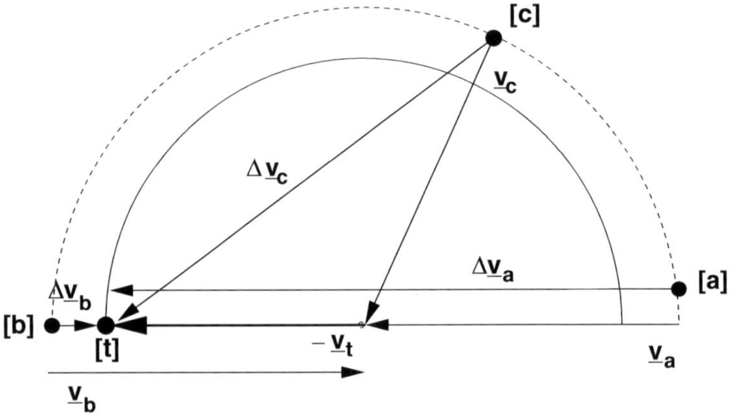

Fig. 4.2. Collision geometry within the horizontal plane ($h \approx 0°$), as a function of the impact azimuth angle $A = \angle(\underline{v}_t, -\Delta\underline{v}_c)$, in the case where the target velocity magnitude $v_t = |\underline{v}_t|$ is less than the debris velocity magnitude $v_d = |\underline{v}_a| = |\underline{v}_b| = |\underline{v}_c|$.

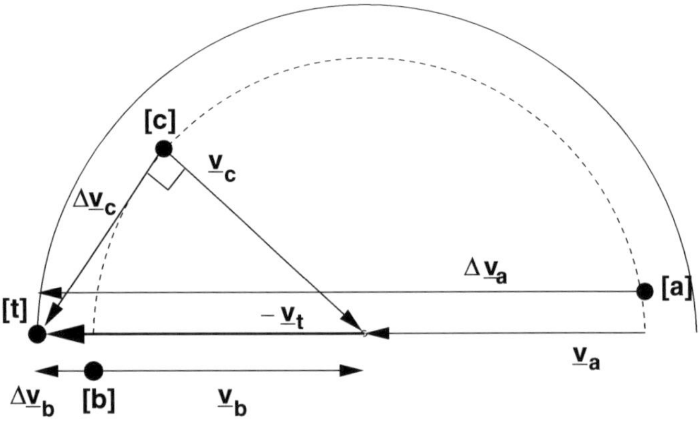

Fig. 4.3. Collision geometry within the horizontal plane ($h \approx 0°$), as a function of the impact azimuth angle $A = \angle(\underline{v}_t, -\Delta\underline{v}_c)$, in the case where the target velocity magnitude $v_t = |\underline{v}_t|$ exceeds the debris velocity magnitude $v_d = |\underline{v}_a| = |\underline{v}_b| = |\underline{v}_c|$.

Sometimes it is useful to know from which azimuth direction impacts can be expected at a given orbit position of the target. Let i_t be the orbit inclination and u_t the argument of true latitude of the target position, then the corresponding declination δ is defined through

$$\sin\delta = \sin u_t \, \sin i_t \tag{4.21}$$

At a collision position the declination of the target and impactor object must be identical ($\delta = \delta_t = \delta_d$). For an impactor orbit inclination i_d this leads to two possible orbit positions u_d, for the ascending (↑) and descending pass (↓).

$$u_{d,\uparrow} = \arcsin\left(\sin u_t \frac{\sin i_t}{\sin i_d}\right) \tag{4.22}$$

$$u_{d,\downarrow} = \pi - u_{d,\uparrow} \tag{4.23}$$

Let $\tilde{A}$ be the flight path azimuth of an orbit, measured in the local horizontal plane from north, to the velocity vector, around $-\underline{U}$. At the collision declination δ the following relations apply for the target (and similarly for the impactor).

$$\sin\tilde{A}_t = \frac{\cos i_t}{\cos\delta} \tag{4.24}$$

$$\cos\tilde{A}_t = \frac{\sin i_t \, \cos u_t}{\cos\delta} \tag{4.25}$$

The difference $\Delta\tilde{A} = \tilde{A}_d - \tilde{A}_t$ between the flight azimuths of the impactor $\tilde{A}_d$ and target $\tilde{A}_t$ can be determined from

$$\sin\left(\Delta\tilde{A}_{\uparrow\downarrow}\right) = \sin i_t \cos i_d \cos u_t - \cos i_t \sin i_d \cos u_{d,\uparrow\downarrow} \tag{4.26}$$

$$\cos\left(\Delta\tilde{A}_{\uparrow\downarrow}\right) = \cos i_t \cos i_d - \sin i_t \sin i_d \cos u_t \cos u_{d,\uparrow\downarrow} \tag{4.27}$$

These equations can be used to determine the two possible approach azimuths $A = \angle(\underline{v}_t, -\Delta\underline{v})$ of the debris object relative to the target direction of motion $\underline{v}_t$, for the ascending pass (↑), and for the descending pass (↓) of the impactor orbit. Let the ratio of the impactor orbit velocity v_d and the target orbit velocity v_t be given as $\eta = v_d/v_t$ (covering all cases in Fig. 4.2 and Fig. 4.3), then the azimuths $A_{\uparrow\downarrow}$ for the two possible impactor approach directions are defined as

$$\sin A_{\uparrow\downarrow} = \frac{-\eta \, \sin\left(\Delta\tilde{A}_{\uparrow\downarrow}\right)}{\sqrt{1 + \eta^2 - 2\eta\cos\left(\Delta\tilde{A}_{\uparrow\downarrow}\right)}} \tag{4.28}$$

$$\cos A_{\uparrow\downarrow} = \frac{1 - \eta \, \cos\left(\Delta\tilde{A}_{\uparrow\downarrow}\right)}{\sqrt{1 + \eta^2 - 2\eta\cos\left(\Delta\tilde{A}_{\uparrow\downarrow}\right)}} \tag{4.29}$$

$$A_{\uparrow\downarrow} = 2 \, \arctan\left(\frac{1 - \cos A_{\uparrow\downarrow}}{\sin A_{\uparrow\downarrow}}\right) \tag{4.30}$$

In accordance with Eq. 4.7 the angle $A_{\uparrow\downarrow} \in [-\pi, +\pi]$ is counted positive around the inward radial vector $-\underline{U}$. The ratio of velocities is constrained to $0 < \eta < \sqrt{2}$.

4.3 COLLISION FLUX ASSESSMENT FOR TYPICAL TARGET ORBITS

The space debris environment is an extremely complex dynamic system, with strong spatial and temporal variability. Since the MASTER-2001 model is based on quasi-deterministic principles (though with considerable uncertainties on the input data), all observed features can be traced back to a unique source. This allows for a very detailed analysis of the impact flux which is generated at certain size regimes on a user-defined target orbit, and it permits the exploration of results which sometimes are surprising.

In this section some important target orbits in the LEO, GTO, and GEO regime will be investigated with respect to their debris flux from objects with $d > 1$ cm. A detailed analysis will only be performed for this size threshold, since it covers the high-risk impactors, and since it offers a good compromise between the size of the population sample (which increases with smaller d) and details in the graphical results (which increase with larger d). According to Table 3.2 the MASTER population of $d > 1$ cm consists of 573,945 objects, of which 189,461 reside in LEO. In the LEO regime, fragments (78.8%) are the dominant contribution, followed by SRM slag (11.6%) and NaK droplets (7.9%). In the MEO and GEO region, with 384,484 objects of $d > 1$ cm, fragments dominate (60.7%), followed by SRM slag (38.7%). Other sources (e.g. mission-related objects) provide only minor contributions.

For the LEO regime two sample orbits shall be used to sense the MASTER-2001 space debris environment: the orbit of the International Space Station (ISS), and the orbit of ESA's ERS-2 satellite (see Table 4.1). For both target objects the same type of flux diagrams will be compared side-by-side to highlight the different collision flux signatures. All targets will be idealized as spheres, with uniform cross-section. The analysis epoch for all orbits is chosen identical to the reference epoch of the MASTER model on May 1, 2001.

Table 4.1. Target orbits for a collision flux analysis in LEO (for ERS and ISS), in GTO, and in GEO. All data sets refer to May 1, 2001. Due to nodal averaging of flux results, information on Ω is not considered.

		ERS	ISS	GTO	GEO
a	[km]	7159.5	6738.1	24551.1	42200.0
e	[–]	0.0011	0.0006	0.7174	0.0006
i	[°]	98.6	51.6	7.0	0.1
ω	[°]	90.5	11.4	178.0	352.3

The ERS-2 satellite (abbreviated as "ERS" hereafter) is operating on a near-circular, Sun-synchronous orbit at a mean altitude of 780 km. Fig. 4.7 shows that the peak flux along the orbit can be expected at high latitudes, and particularly during the crossing of small circles of latitude at $u = 90° \pm 20°$ and $u = 270° \pm 20°$, where the highest concentration of objects from the inclination band $65° \leq i \leq 75°$ is encountered (compare Fig. 4.15). Fig. 4.6 resolves the flux contributions in orbit position u and collision azimuth A. Due to the dominance of fragments from high inclination orbits at $i \approx 65°$ to $75°, 82°, 90°$ and $100°$, the by far dominant flux contributions on the ascending ERS pass ($-90° \leq u \leq +90°$) are received from the left side of the orbit. The few impacts from the right-hand side are caused by near head-on collisions from orbits of $i_c < i < 110°$, where $i_c = 180° - i_{ERS} = 81.4°$ is the complementary inclination band, causing collisions under $A \approx 0°$, if the orbit nodes have a 180° separation with respect to ERS. The closed band structures in Fig. 4.6 each represent a certain debris inclination band, which leads to impacts under two possibles azimuth angles for an ascending and descending pass of the debris orbit. For the inclination bands shown in Fig. 4.15 such azimuth pairs can be determined for a given orbit position by means of Eq. 4.21 – 4.30. As ERS enters its descending pass ($90° \leq u \leq +270°$), the flux directionality is reversed, and most impactors are approaching from the right. The closed band structures are restored as soon as the relevant inclination bands are reached again. Over a full orbit, the rather asymmetric azimuth directionality with u maps onto a fairly symmetric azimuth flux pattern in Fig. 4.11. The majority of the debris approach ERS from $-30° \leq A \leq +30°$ at most probable velocities of $\Delta v \approx 15\,\mathrm{km/s}$ (see Fig. 4.4), corresponding to almost twice the orbital speed. This is in agreement with Eq. 4.19, which also predicts a cosine-law for the collision velocities as a function of the impact azimuth, as illustrated in Fig. 4.10. Since the dominating fragment population is mostly in orbits of small eccentricities (see Fig. 4.18), there is only a minor dispersion of the Δv's from the ideal cosine-law. The contributions to the collision velocity spectrum from different inclination bands are indicated in Fig. 4.14. The dominant inclinations $i \approx 65°$ to $75°, 82°, 90°$ and $100°$ are clearly discernible by their flux levels. Debris at $i \approx 98°$ spans the widest range of possible velocities from $\Delta v \approx 0$ (for a "rendezvous" type collision) to $\Delta v \approx 2 \times v_t$ (for a near head-on collision). For all other inclinations the Δv ranges (particularly the lower limits), and hence the collision azimuth ranges are increasingly constrained with decreasing inclinations of the debris orbits. This is particularly noted for SRM slag objects at $i \approx 28°$ which impact ERS at $\Delta v \approx 12 \pm 2\,\mathrm{km/s}$ from $-60° \leq A \leq -30°$ on the ascending pass, and from $30° \leq A \leq 60°$ on the descending pass (see Fig. 4.6).

The second LEO example to be analyzed is the International Space Station (ISS) in a near-circular orbit of mean altitude 360 km, with a moderate inclination of 51.6° (see Table 4.1). For this target orbit the debris environment is dominated by solid rocket motor (SRM) slag particles, which were mainly released into orbits of large eccentricities (see Fig. 4.19), at inclinations near 28°, due to KSC based insertions into GEO transfer trajectories. The inclination bands which were dominant for ERS are still noticeable in the ISS flux-versus-inclination diagram in Fig. 4.17, but they only play a minor role. Since the SRM slag orbits have a smaller inclination than ISS, most of the impact flux in Fig. 4.8 is approaching ISS from the left side on the ascending arc, and from the right side on the descending arc. As compared to the ERS case, the role of the contributing debris inclination bands has been completely inverted, with the dominant SRM slag population impacting from $-90° \leq A \leq -60°$ on the ascending pass, and from $60° \leq A \leq 90°$ on the descending pass. The peak flux contributions along the orbit are accumulated within $\pm 40°$ of the equator crossings. The sine-shaped flux-free corridor in Fig. 4.8

is due to the changing flight azimuth of ISS with respect to the high-inclination fragment orbits (see Eq. 4.24 and Eq. 4.25). The extension of the azimuth flux gap across the corridor is determined by the lack of debris in the vicinity of the complementary ISS inclination of $i_c = 180° - i_{ISS} = 128.4°$. The nearest populated debris band is at 100°, leaving a "dead water" zone of $\Delta A \approx \pm 28.4°$, centered on $A(u) \approx 28° \times \cos u$. Fig. 4.13 shows how the local azimuth profiles add up over the entire orbit. Distinct, spiky flux peaks appear at $A = \pm 60°$, with a flux-free corridor remaining within $-15° \leq A \leq +15°$. There are also tail-on collisions for ISS, but at an insignificant level. In principle, ISS could also intercept slower, co-planar debris, which would impact from $A \approx 0°$. Such objects, however, would need to have low pericenters, which are filtered out by atmospheric capture. Fig. 4.12 shows how the collision azimuth A correlates with the approach velocity Δv. As in the ERS scenario, the azimuth distribution follows a basic cosine-law according to Eq. 4.19. Due to prevailing large eccentricities of the SRM orbits the resulting pericenter velocities are on the order of $v_d \approx 9.9\,\mathrm{km/s}$, as compared to the lower ISS velocity of $v_t \approx 7.7\,\mathrm{km/s}$. This allows for collision geometries as illustrated in Fig. 4.2, and for a wider spreading of approach velocities Δv with varying eccentricities (contraction or expansion of the dashed outer circle in Fig. 4.2), particularly around $A \approx \pm 90°$. The flux peaks in the azimuth distribution at $A \approx \pm 60°$ directly map onto most probable ISS impact velocities of $9\,\mathrm{km/s} \leq \Delta v \leq 11\,\mathrm{km/s}$ in Fig. 4.5. In contrast with ERS, the ISS collision velocities cover a much wider range, with fairly uniform flux contributions from $4\,\mathrm{km/s} \leq \Delta v \leq 11\,\mathrm{km/s}$. This is related to the equal-sized, flux producing Δv bands versus inclination in Fig. 4.16. The lowest possible impact velocities in this chart are resulting from fragments which are co-orbiting with ISS at $i \approx 51.6°$. The V-shape border-line of $\Delta v_{min}(i)$ for ISS lies in the center of the populated inclination bands. When moving away from the ISS target orbit inclination, $\Delta v_{min}(i)$ increases with a slope of ~0.12 km/s per degree to either side of the inclination band (the same slope applies for ERS, though starting at the right-most edge of the populated inclination bands). Fig. 4.16 also gives an indication of the debris source which causes the highest impact velocities on ISS. These are SRM slag objects from LEO insertions at $i \approx 100°$, leading to $\Delta v \approx 14.5\,\mathrm{km/s}$ (see Fig.4.5) from approach directions of $A \approx \pm 15°$ (see Fig. 4.13).

The determination of collision fluxes on highly eccentric orbits is a particularly challenging benchmark for a space debris environment model. As an example, the GEO transfer orbit (GTO) of an Ariane 5 shall be investigated. It has a pericenter altitude of $H_{pe} \approx 560\,\mathrm{km}$, an apocenter of $H_{ap} \approx 35{,}786\,\mathrm{km}$ at the GEO altitude, an eccentricity of $e = 0.7174$, an inclination of $i = 7.0°$, and a pericenter position at $\omega = 178°$, near the descending node. Fig. 4.21 shows that most of the impacts of debris with $d > 1\,\mathrm{cm}$ result from fragments, with peak fluxes at $u \approx \omega \pm 30°$, when passing through the densely populated altitude bands at 800 km to 900 km, and at 1,400 km, under elevations of $h \approx \pm 12°$ on space-bound and Earth-bound passes through the LEO regime (see Fig. 4.29, and Eq. A.20). Fig. 4.21 shows that only in the close vicinity of the pericenter the flux from SRM slag is equal-ranking to the fragment contribution. Fig. 4.20 correlates orbit position and impact azimuth. Peak fluxes occur near pericenter passes at $u \approx \omega \pm 30°$ under azimuth angles of $A \approx \pm 45°$. A single, secondary flux peak is visible near the apocenter pass through the GEO regime, almost from the anti-flight direction, at $A \approx -160°$ (see also Fig. 4.23). The offset $\Delta A \approx +20°$ relative to the tail-on direction of $A = -180°$ is resulting from Eq. 4.30 for $\eta = v_d/v_t \approx 2$ and $\Delta\tilde{A} = i_d - i_t \approx 10°$. Fig. 4.28 identifies these impacts as fragments and SRM slag on near-GEO orbits of $0 \leq i_d \leq 15°$ (see Fig. 4.24), catching up with the slower GTO target at relative velocities of $v_{GEO} - v_{GTO,ap} \approx 1.5\,\mathrm{km/s}$. Most of the flux, however, is encountered at velocities of $8.5\,\mathrm{km/s} \leq \Delta v \leq 14\,\mathrm{km/s}$ during LEO passes. Due to the low relative velocities near GEO the dispersion in impact elevations h in Fig. 4.22 is much more pronounced than during LEO passes where impact velocities are high. Collision flux contributions from $A \approx \pm 70°$ under $h \approx \pm 80°$ are caused by low-inclination, co-orbiting SRM slag objects on highly eccentric orbits (see Fig. 4.26 and 4.27). The prevailing flux contributions, however, result from orbits of low to moderate eccentricities $0.001 \leq e \leq 0.1$ (both in LEO and GEO), and from the densely populated LEO inclination bands between 65° and 100° (see Fig. 4.25), with relative velocities of $\Delta v \approx 11 \pm 3\,\mathrm{km/s}$ (see Fig. 4.24 and 4.28).

Finally, the debris impact flux on a typical GEO mission shall be analyzed, with GEO orbital parameters as indicated in Table 4.1. Fig. 4.32 shows that most of the impacts occur under azimuth angles of $A \approx \pm 80°$. In Fig. 4.33 these are identified as collisions with nearly co-orbiting GEO fragments at velocities of $0.0 \leq \Delta v \leq 1.0\,\mathrm{km/s}$, with a peak at $0.8\,\mathrm{km/s}$ due to impacts from debris orbit clusters near 15° (the upper limit of the 53-year cyclic inclination variation in GEO). Minute "peaks" near $\Delta v \approx 1.5\,\mathrm{km/s}$ and $3.0\,\mathrm{km/s}$ can be correlated with head-on collisions with SRM slag on GTO orbits, and with oblique impacts resulting from fragments on highly eccentric Molnija-type orbits of inclinations $i \approx 65°$. Most flux contributing orbits, however, are of eccentricities $0.001 \leq e \leq 0.05$ as shown in Fig. 4.31. Due to the dominant near-circular impactor orbits with small relative velocities, the impact elevation for the prevailing side-on impacts is spread across a relatively broad band of $-15° \leq h \leq +15°$ (see Fig. 4.30).

Flux predictions not only change with the chosen target orbit, but also with the analysis epoch (see Section 3.9, and Chapters 5 and 6), and with the lowest size threshold of the debris population model considered. The latter effect is quite evident when analyzing the ISS orbit for impact flux from debris with diameters $d > 10\,\mathrm{cm}$. In this case, fragmentation debris overrule SRM slag. As a result, in Fig. 4.17 the fragment inclination bands between 65° and 100° become dominant, and the flux distributions with orbit position, impact azimuth, and impact velocity change considerably.

For orbits of a few representative space missions (see Table 4.2), the variation of the collision flux with the debris population size threshold shall be looked at. For debris of $d > 0.1\,\mathrm{mm}$ to $d > 10\,\mathrm{cm}$ Table 4.3 lists the mean times between collisions of any population member with a spherical target of $1\,\mathrm{m}^2$ cross-section. For the same target Table 4.4 provides corresponding results in terms of relative flux levels as compared to the high-risk ERS orbit. The figures indicate that the risk of receiving a destructive impact from an object of $d > 1\,\mathrm{cm}$ is acceptable for most missions, even when assuming a target cross-section of a few tens of square meters. The mean time between catastrophic collisions of any two catalog objects is presently on the order of 10 years. Some more debris flux results are summarized in (Klinkrad et al., 2003).

The flux scenarios which were deployed and analyzed here cover only a small part of the MASTER-2001 capabilities. The reader is invited to explore more features in the flux charts provided, and to investigate further target orbit scenarios with a personal copy of the MASTER model.

Table 4.2. Representative target orbits used in a debris flux analysis with MASTER-2001 (GStar = Globalstar constellation).

	H_{pe} [km]	H_{ap} [km]	i [°]
ISS	356.0	364.1	51.6
ERS	773.5	789.2	98.6
GStar	1,400.0	1,400.0	52.0
GPS	20,000.0	20,000.0	55.0
GTO	560.0	35,786.0	7.0
GEO	35,786.0	35,786.0	0.1

Table 4.3. Mean time $\Delta\bar{t}_c$ between collisions of a 1 m^2 spherical target with the MASTER-2001 population of diameters larger that a given size threshold, on orbits as specified in Table 4.2.

	$d > 0.1$ mm	$d > 1$ mm	$d > 1$ cm	$d > 10$ cm
ISS	221 d	217 y	149,253 y	1.872×10^6 y
ERS	12.3 d	11.56 y	6,993 y	120,773 y
GStar	13.0 d	16.21 y	18,115 y	248,756 y
GPS	4.52 y	10,090 y	11.2×10^6 y	990×10^6 y
GTO	1.05 y	532 y	465,000 y	8.85×10^6 y
GEO	4.67 y	2,267 y	2.44×10^6 y	24.6×10^6 y

Table 4.4. MASTER-2001 collision flux on a 1 m^2 spherical target, from space debris larger that a given size threshold, on orbits as specified in Table 4.2.

	quantity [units]	$d > 0.1$ mm	$d > 1$ mm	$d > 1$ cm	$d > 10$ cm
ERS	F_{ERS} [m^{-2} y^{-1}]	29.6	8.65×10^{-2}	1.43×10^{-4}	8.28×10^{-6}
ISS	F_{ISS}/F_{ERS} [–]	1/17.9	1/18.8	1/21.3	1/15.5
GStar	F_{GStar}/F_{ERS} [–]	1/1.05	1/1.40	1/2.59	1/2.06
GPS	F_{GPS}/F_{ERS} [–]	1/133.9	1/872.9	1/1,601	1/8,198
GTO	F_{GTO}/F_{ERS} [–]	1/29.9	1/46.0	1/66.5	1/73.3
GEO	F_{GEO}/F_{ERS} [–]	1/138.3	1/196.1	1/348.8	1/203.9

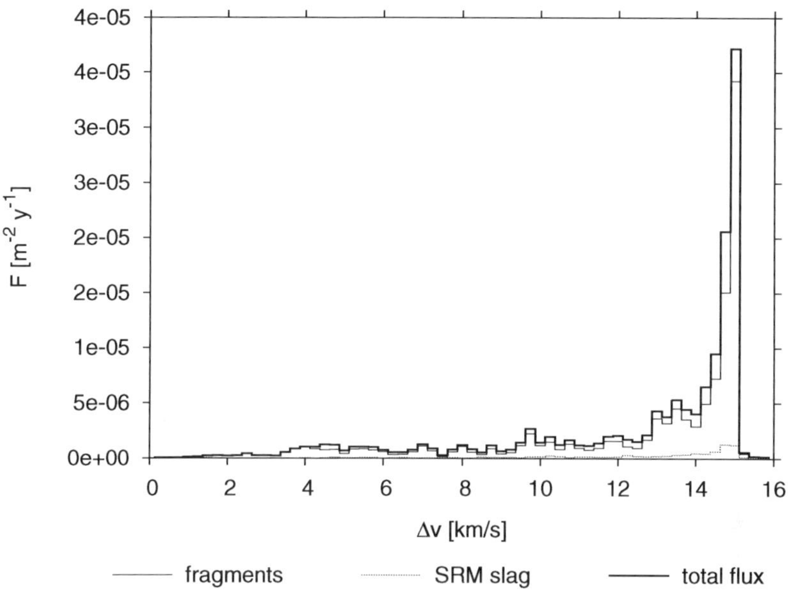

Fig. 4.4. Source-wise and total debris flux for $d > 1$ cm on an ERS orbit, as a function of the impact velocity Δv (class width: $\Delta(\Delta v) = 0.25\,\mathrm{km/s}$).

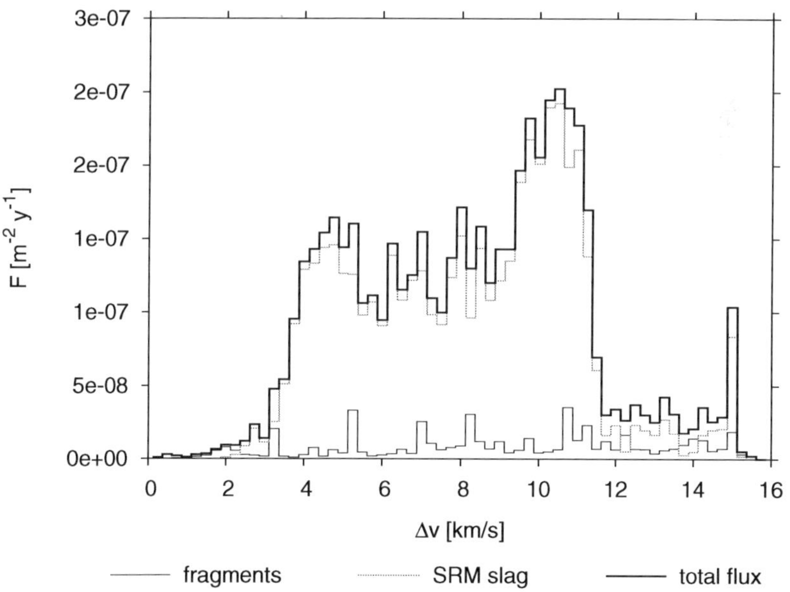

Fig. 4.5. Source-wise and total debris flux for $d > 1$ cm on an ISS orbit, as a function of the impact velocity Δv (class width: $\Delta(\Delta v) = 0.25\,\mathrm{km/s}$).

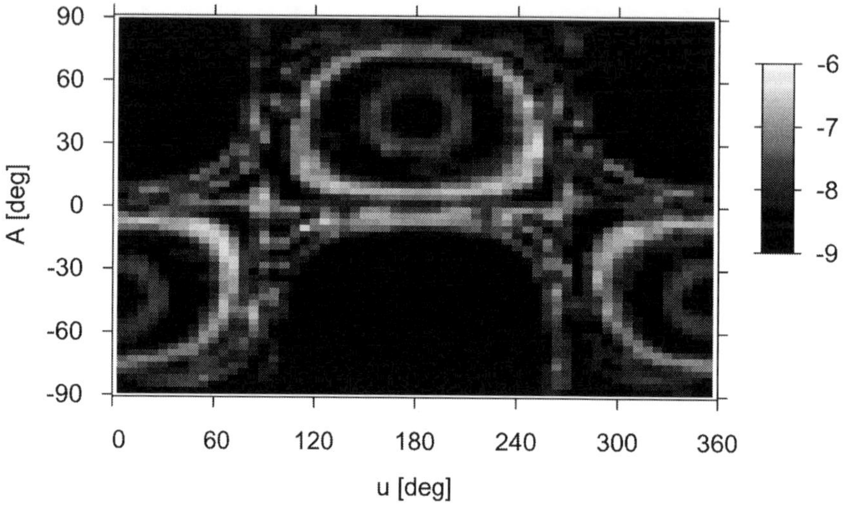

Fig. 4.6. Debris flux distribution for $d > 1$ cm on an ERS orbit, as a function of impact azimuth A and orbit position $u = \omega + f$. Contour levels are given as $\log_{10}$ of the impact flux $F\,[\mathrm{m}^{-2}\,\mathrm{y}^{-1}]$ (class widths: $\Delta u = 6°$, $\Delta A = 3°$; see color plate on page 378).

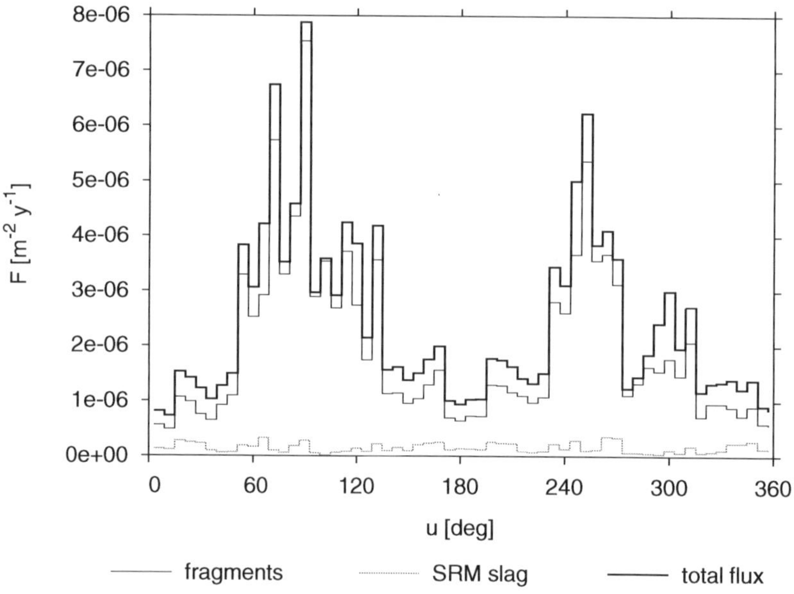

Fig. 4.7. Source-wise and total debris flux for $d > 1$ cm on an ERS orbit, as a function of the orbit position $u = \omega + f$ (class width: $\Delta u = 6°$).

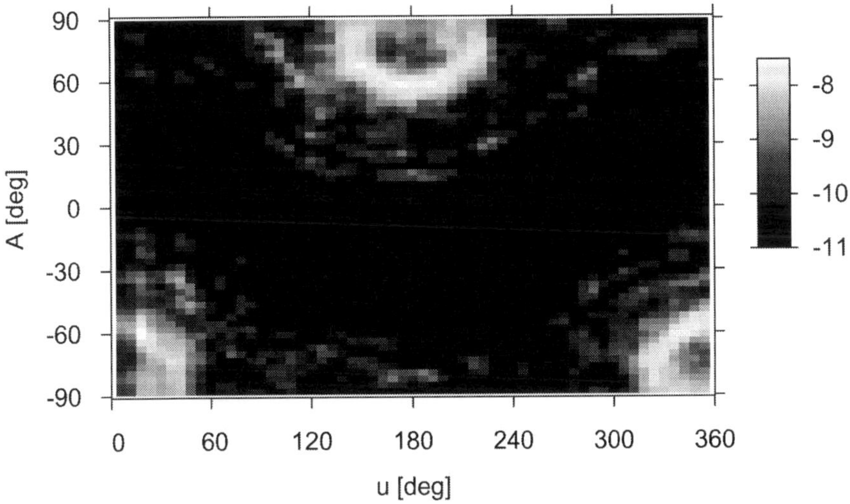

Fig. 4.8. Debris flux distribution for $d > 1$ cm on an ISS orbit, as a function of impact azimuth A and orbit position $u = \omega + f$. Contour levels are given as $\log_{10}$ of the impact flux F [$\mathrm{m}^{-2}\,\mathrm{y}^{-1}$] (class widths: $\Delta u = 6°$, $\Delta A = 3°$; see color plate on page 378).

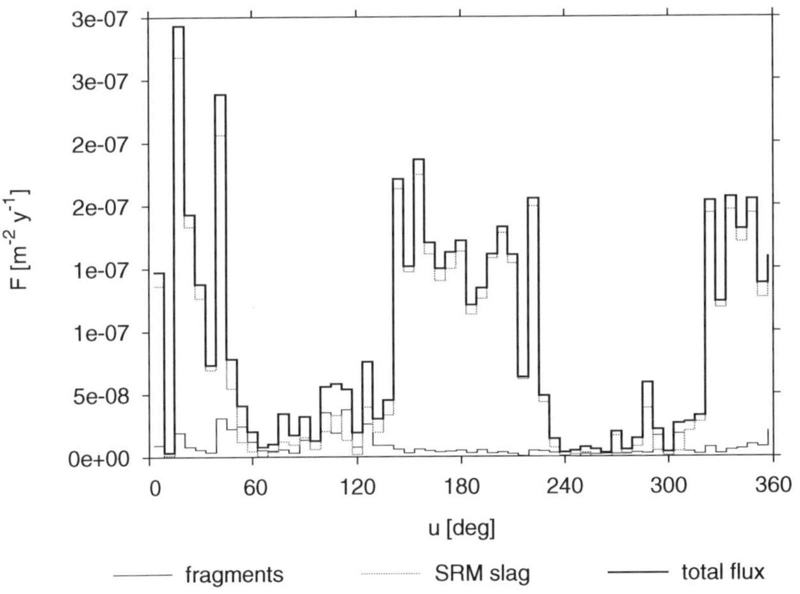

Fig. 4.9. Source-wise and total debris flux for $d > 1$ cm on an ISS orbit, as a function of the orbit position $u = \omega + f$ (class width: $\Delta u = 6°$).

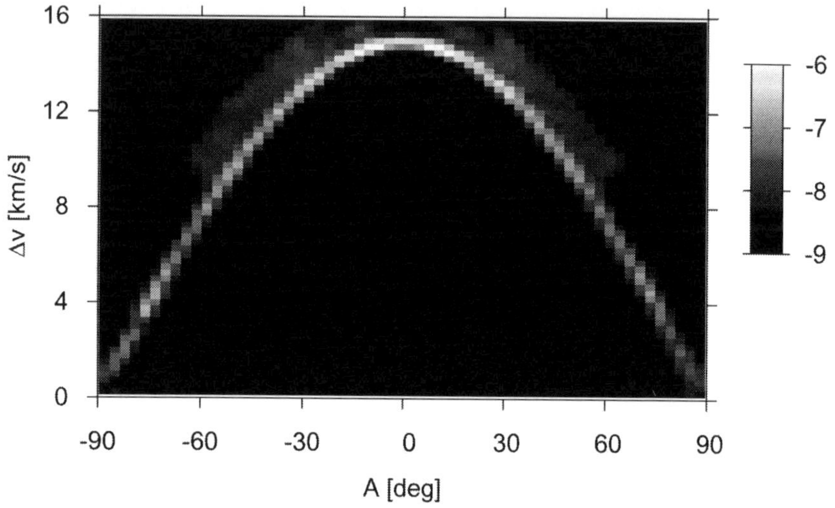

Fig. 4.10. Debris flux distribution for $d > 1$ cm on an ERS orbit, as a function of impact azimuth A and impact velocity Δv. Contour levels are given as $\log_{10}$ of the impact flux $F\,[\mathrm{m}^{-2}\,\mathrm{y}^{-1}]$ (class widths: $\Delta A = 3°$, $\Delta(\Delta v) = 0.25$ km/s; see color plate on page 379).

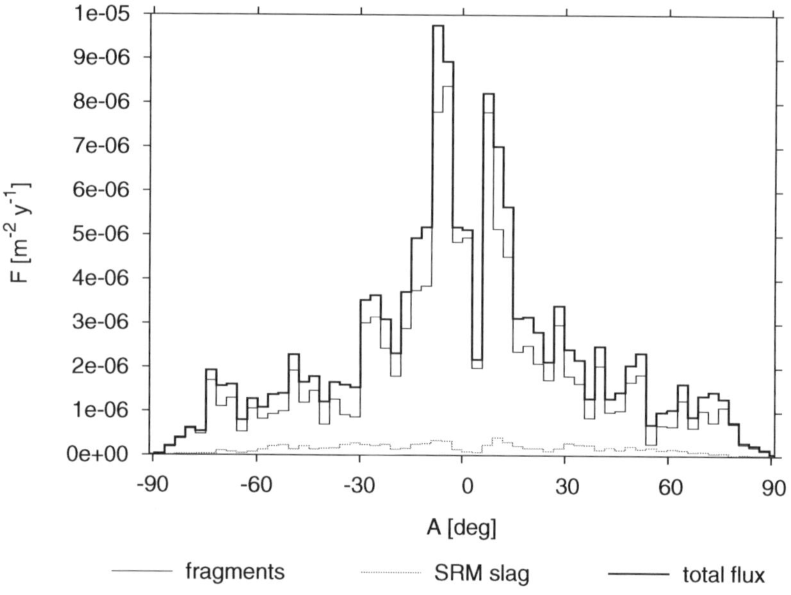

Fig. 4.11. Source-wise and total debris flux for $d > 1$ cm on an ERS orbit, as a function of the impact azimuth A (class width: $\Delta A = 3°$).

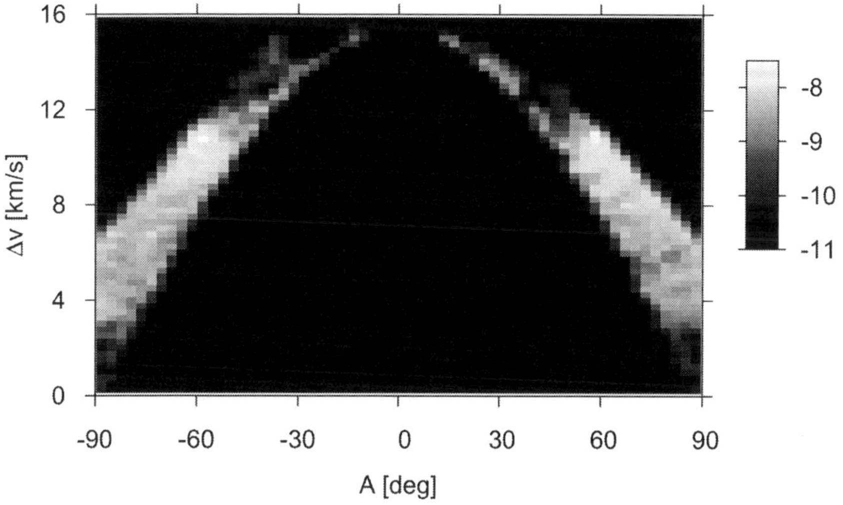

Fig. 4.12. Debris flux distribution for $d > 1$ cm on an ISS orbit, as a function of impact azimuth A and impact velocity Δv. Contour levels are given as $\log_{10}$ of the impact flux $F\,[\mathrm{m}^{-2}\,\mathrm{y}^{-1}]$ (class widths: $\Delta A = 3°$, $\Delta(\Delta v) = 0.25$ km/s; see color plate on page 379).

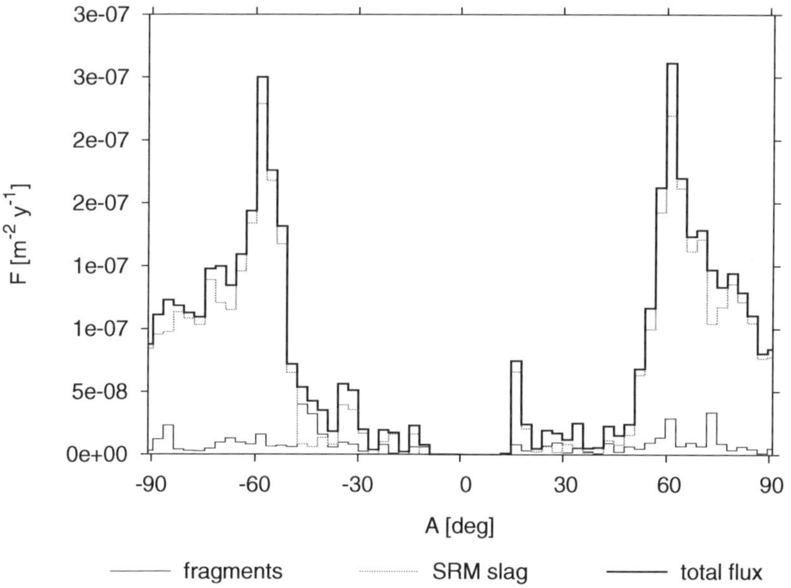

Fig. 4.13. Source-wise and total debris flux for $d > 1$ cm on an ISS orbit, as a function of the impact azimuth A (class width: $\Delta A = 3°$).

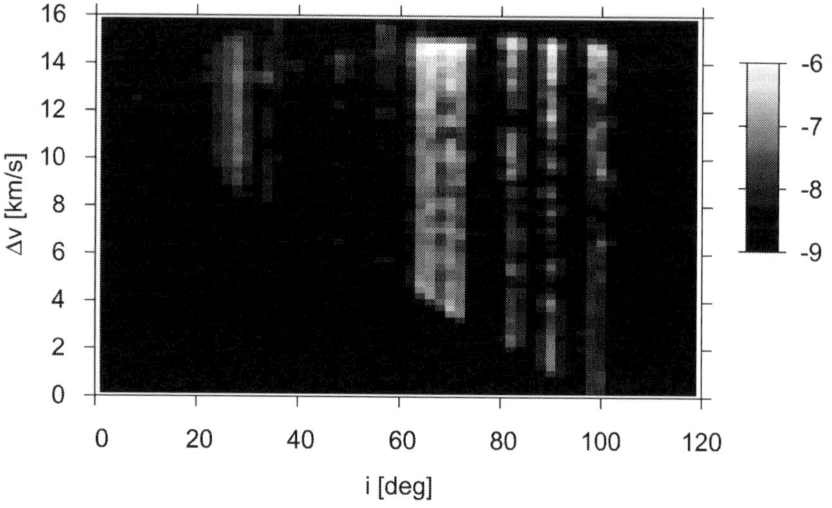

Fig. 4.14. Debris flux distribution for $d > 1\,\text{cm}$ on an ERS orbit, as a function of the impactor orbit inclination i and the impact velocity Δv. Contour levels are given as $\log_{10}$ of the impact flux $F\,[\text{m}^{-2}\,\text{y}^{-1}]$ (class widths: $\Delta i = 2°$, $\Delta(\Delta v) = 0.25\,\text{km/s}$; see color plate on page 380).

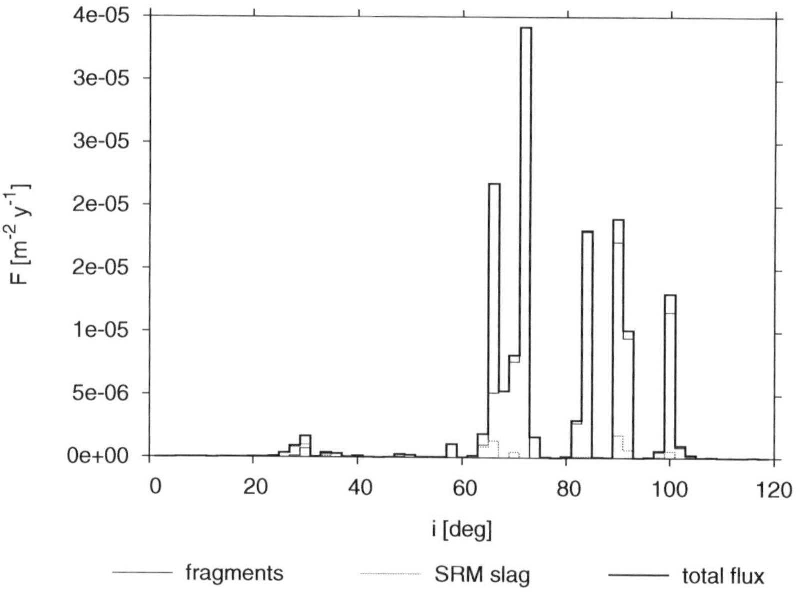

Fig. 4.15. Source-wise and total debris flux for $d > 1\,\text{cm}$ on an ERS orbit, as a function of the impactor orbit inclination i (class width: $\Delta i = 2°$).

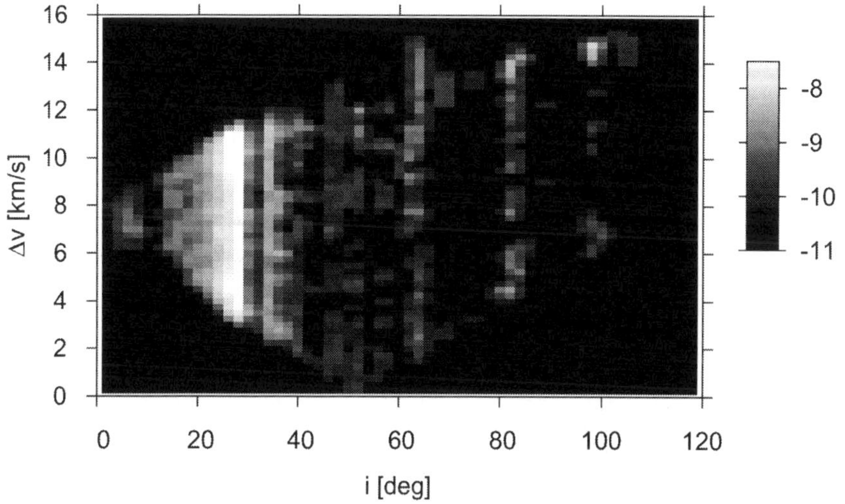

Fig. 4.16. Debris flux distribution for $d > 1\,\mathrm{cm}$ on an ISS orbit, as a function of the impactor orbit inclination i and the impact velocity Δv. Contour levels are given as $\log_{10}$ of the impact flux $F\,[\mathrm{m}^{-2}\,\mathrm{y}^{-1}]$ (class widths: $\Delta i = 2°$, $\Delta(\Delta v) = 0.25\,\mathrm{km/s}$; see color plate on page 380).

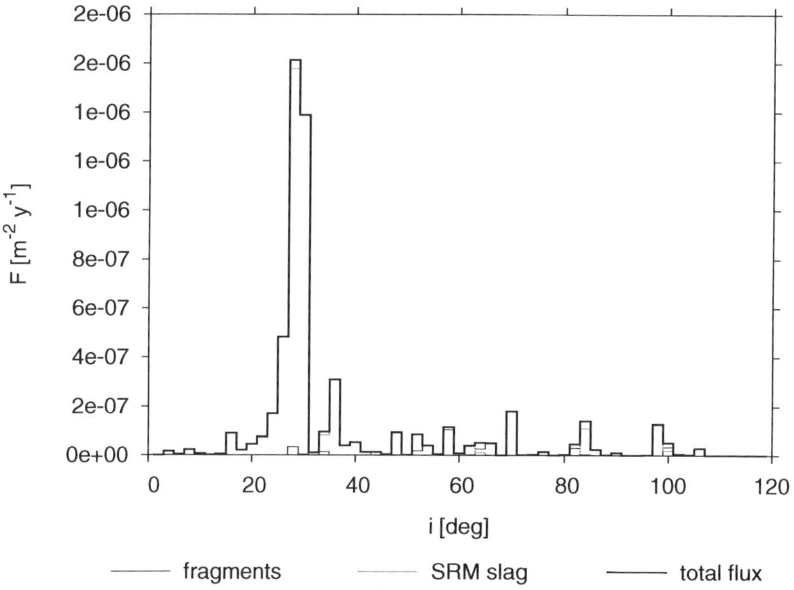

Fig. 4.17. Source-wise and total debris flux for $d > 1$ cm on an ISS orbit, as a function of the impactor orbit inclination i (class width: $\Delta i = 2°$).

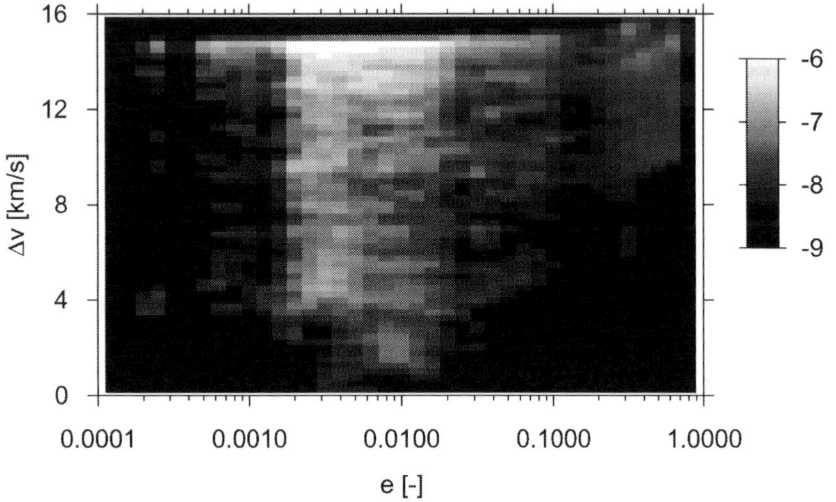

Fig. 4.18. Debris flux distribution for $d > 1$ cm on an ERS orbit, as a function of the impactor orbit eccentricity e and the impact velocity Δv. Contour levels are given as $\log_{10}$ of the impact flux $F\,[\mathrm{m}^{-2}\,\mathrm{y}^{-1}]$ (class widths: $\Delta(\Delta v) = 0.25$ km/s, 40 equal-size classes in $log_{10}\, e$; see color plate on page 381).

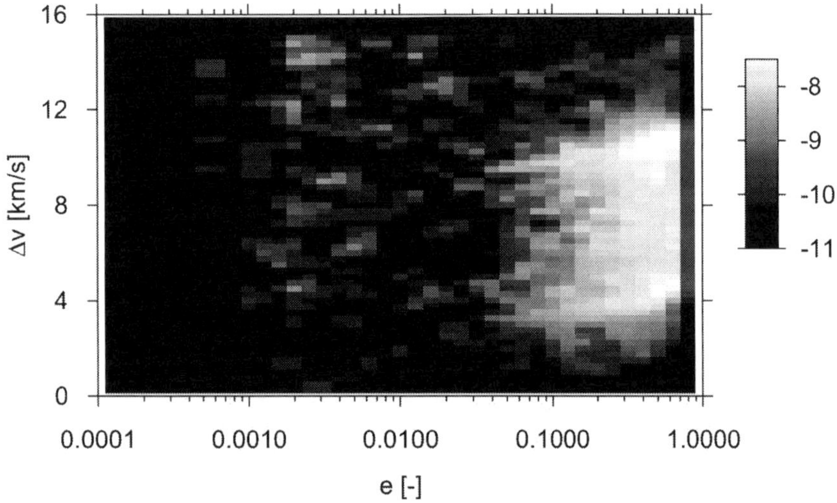

Fig. 4.19. Debris flux distribution for $d > 1$ cm on an ISS orbit, as a function of the impactor orbit eccentricity e and the impact velocity Δv. Contour levels are given as $\log_{10}$ of the impact flux $F\,[\mathrm{m}^{-2}\,\mathrm{y}^{-1}]$ (class widths: $\Delta(\Delta v) = 0.25$ km/s, 40 equal-size classes in $log_{10}\, e$; see color plate on page 381).

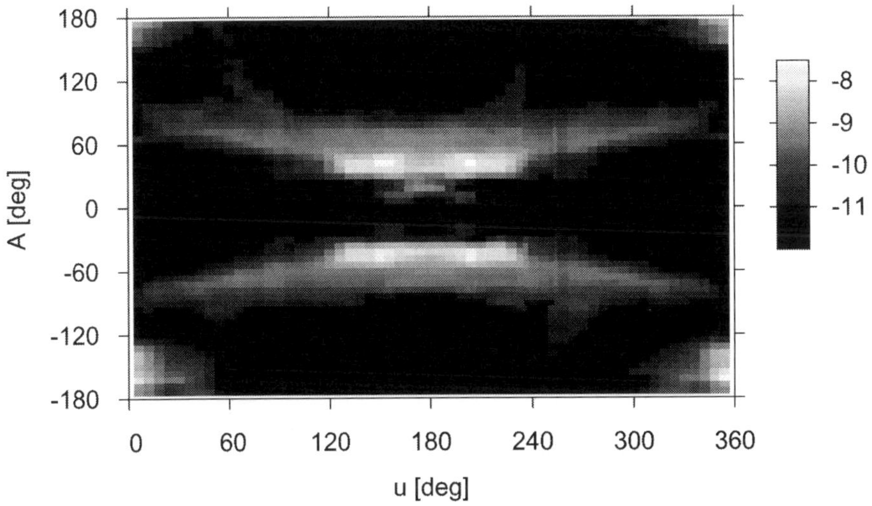

Fig. 4.20. Debris flux distribution for $d > 1$ cm on a GTO orbit, as a function of impact azimuth A and orbit position $u = \omega + f$. Contour levels are given as $\log_{10}$ of the impact flux $F\,[\mathrm{m}^{-2}\,\mathrm{y}^{-1}]$ (class widths: $\Delta u = 6°$, $\Delta A = 6°$; see color plate on page 382).

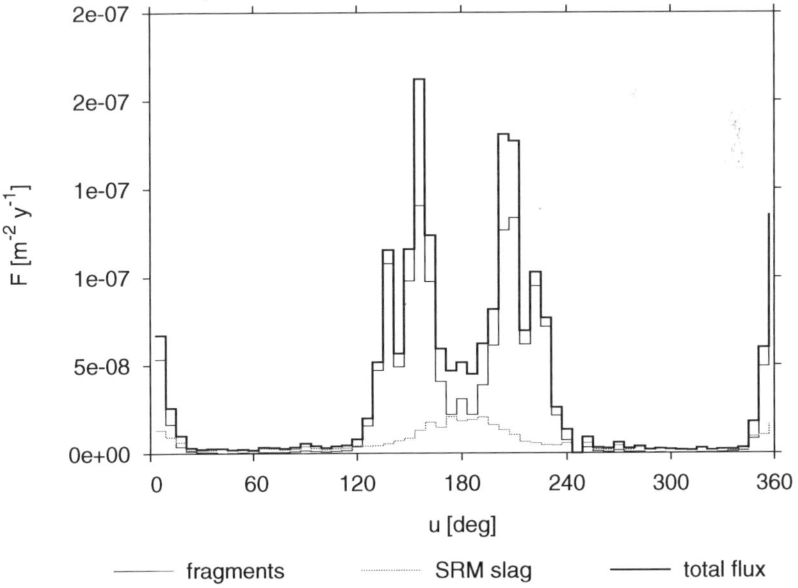

Fig. 4.21. Source-wise and total debris flux for $d > 1$ cm on a GTO orbit, as a function of the orbit position $u = \omega + f$ (class width: $\Delta u = 6°$).

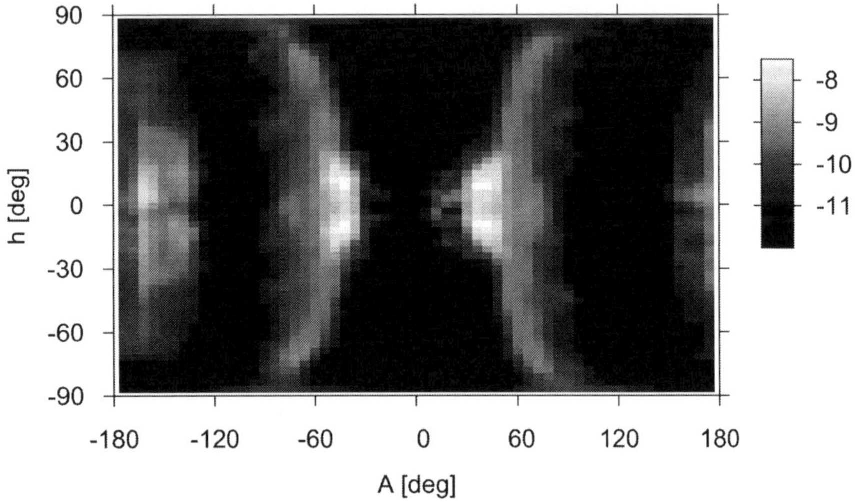

Fig. 4.22. Debris flux distribution for $d > 1$ cm on a GTO orbit, as a function of impact azimuth A and impact elevation h. Contour levels are given as $\log_{10}$ of the impact flux F [m^{-2} y^{-1}] (class widths: $\Delta A = 6°$, $\Delta h = 3°$; see color plate on page 382).

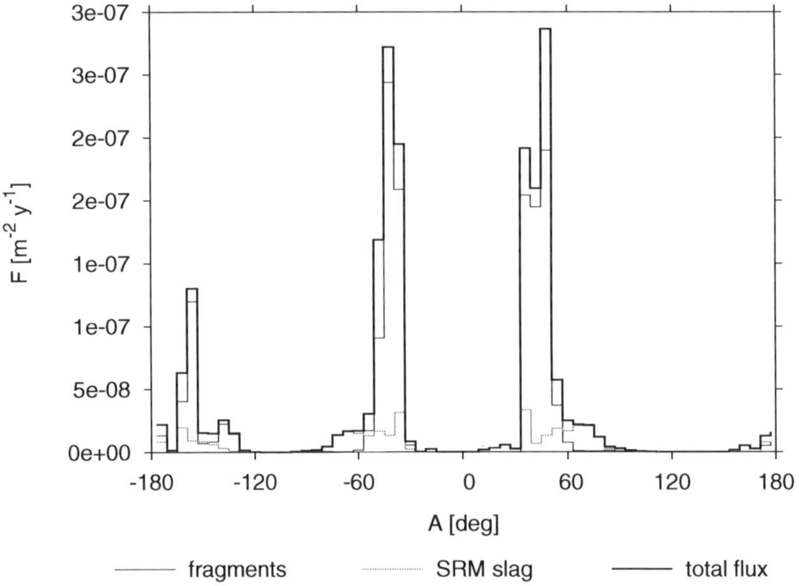

Fig. 4.23. Source-wise and total debris flux for $d > 1$ cm on a GTO orbit, as a function of the impact azimuth angle A (class width: $\Delta A = 6°$).

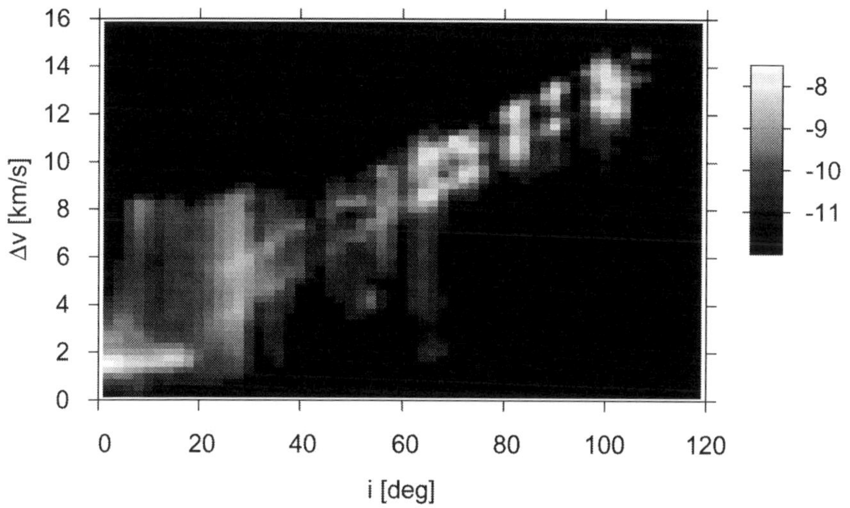

Fig. 4.24. Debris flux distribution for $d > 1$ cm on a GTO orbit, as a function of impact velocity Δv and impactor orbit inclination i. Contour levels are given as $\log_{10}$ of the impact flux $F\,[\mathrm{m}^{-2}\,\mathrm{y}^{-1}]$ (class widths: $\Delta i = 2°$, $\Delta(\Delta v) = 0.25$ km/s; see color plate on page 383).

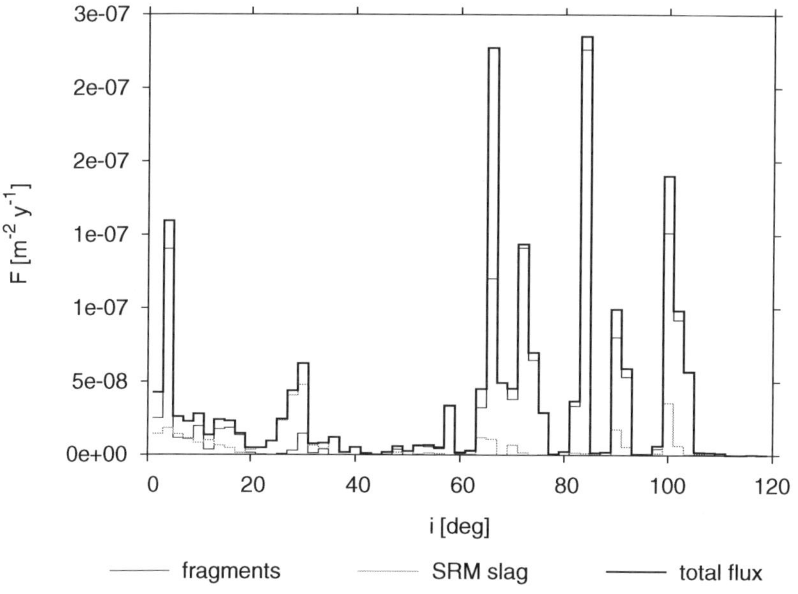

Fig. 4.25. Source-wise and total debris flux for $d > 1$ cm on a GTO orbit, as a function of the impactor orbit inclination i (class width: $\Delta i = 2°$).

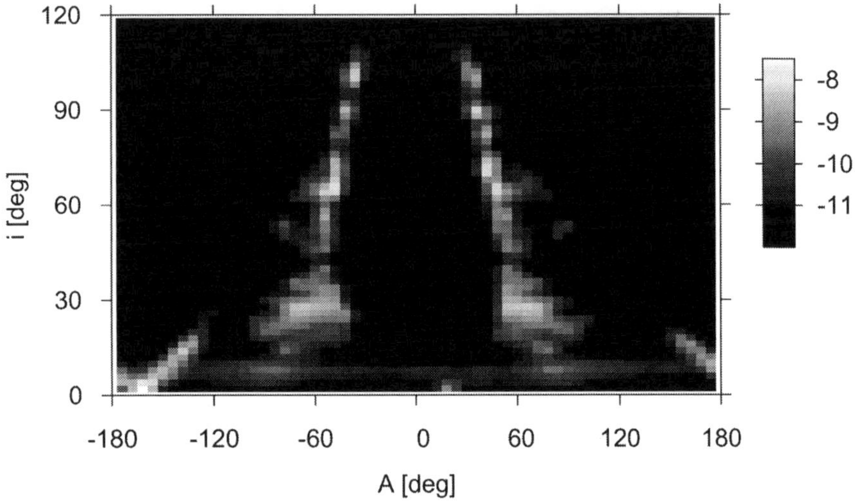

Fig. 4.26. Debris flux distribution for $d > 1\,\mathrm{cm}$ on a GTO orbit, as a function of the impact azimuth A and the impactor orbit inclination i. Contour levels are given as $\log_{10}$ of the impact flux $F\,[\mathrm{m}^{-2}\,\mathrm{y}^{-1}]$ (class widths: $\Delta A = 6°$, $\Delta i = 2°$; see color plate on page 383).

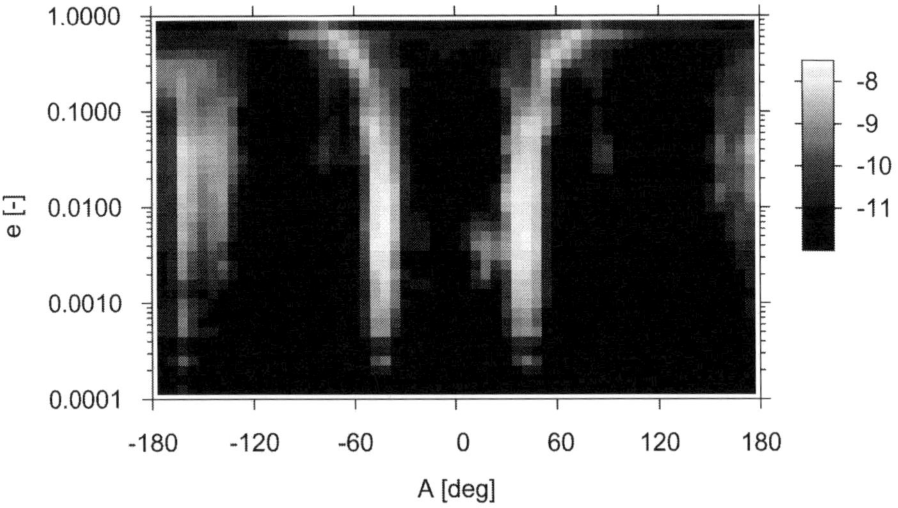

Fig. 4.27. Debris flux distribution for $d > 1\,\mathrm{cm}$ on a GTO orbit, as a function of the impact azimuth A and the impactor orbit eccentricity e. Contour levels are given as $\log_{10}$ of the impact flux $F\,[\mathrm{m}^{-2}\,\mathrm{y}^{-1}]$ (class widths: $\Delta A = 6°$, 40 equal-size classes in $log_{10}\,e$; see color plate on page 384).

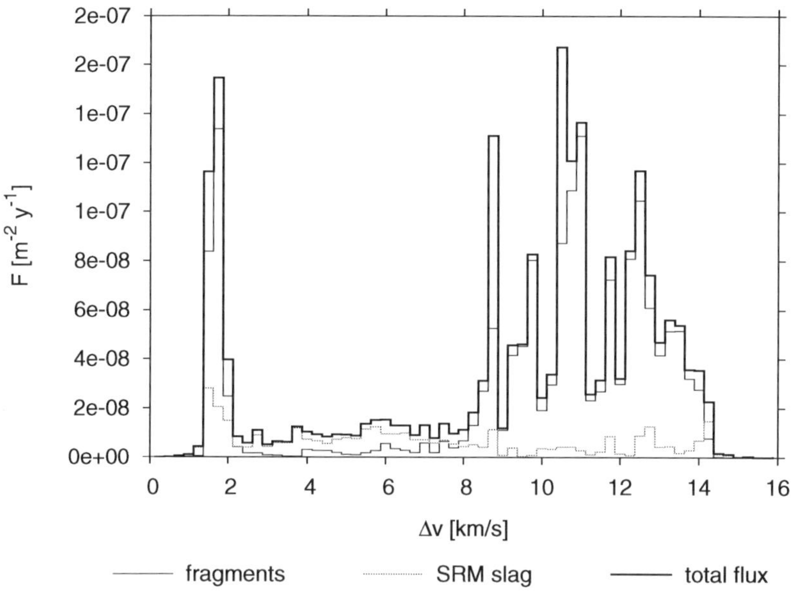

Fig. 4.28. Source-wise and total debris flux for $d > 1\,\text{cm}$ on a GTO orbit, as a function of the impact velocity Δv (class width: $\Delta(\Delta v) = 0.25\,\text{km/s}$).

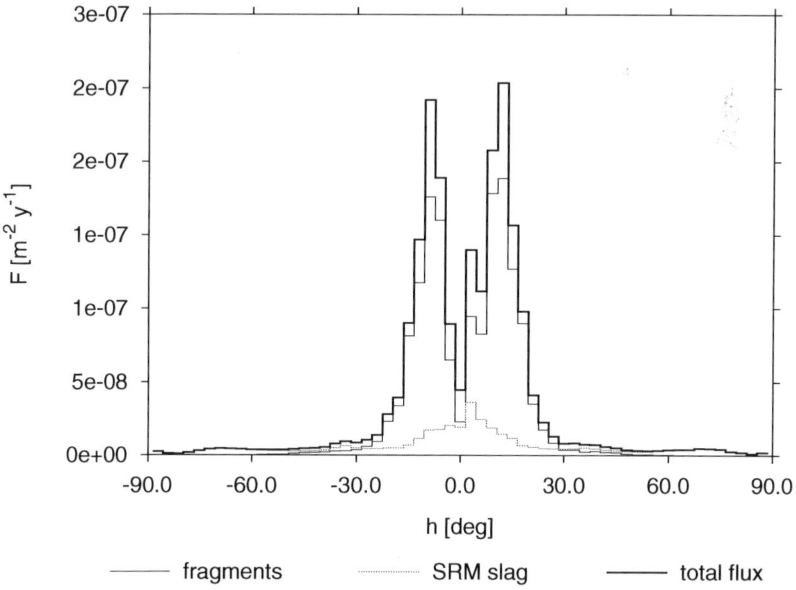

Fig. 4.29. Source-wise and total debris flux for $d > 1\,\text{cm}$ on a GTO orbit, as a function of the impact elevation h (class width: $\Delta h = 3°$).

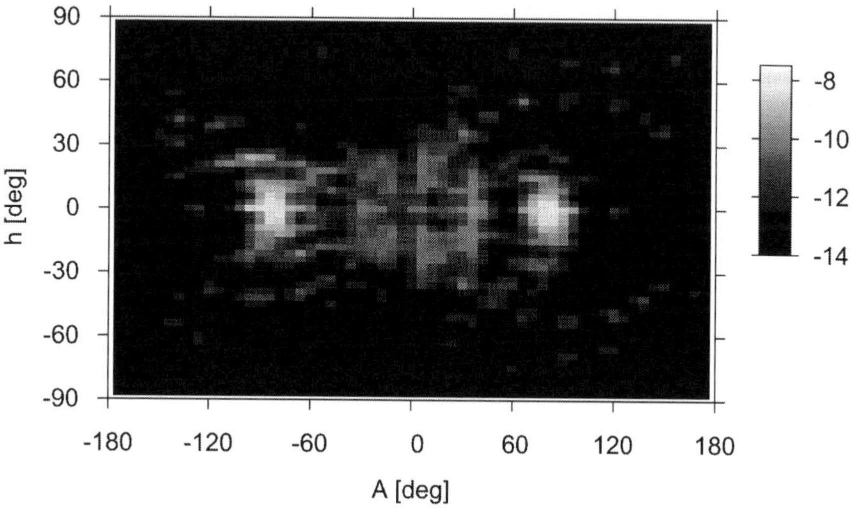

Fig. 4.30. Debris flux distribution for $d > 1\,\mathrm{cm}$ on a GEO orbit, as a function of the impact elevation h and the impact azimuth A. Contour levels are given as $\log_{10}$ of the impact flux $F\,[\mathrm{m}^{-2}\,\mathrm{y}^{-1}]$ (class widths: $\Delta h = 3°$, $\Delta A = 6°$; see color plate on page 384).

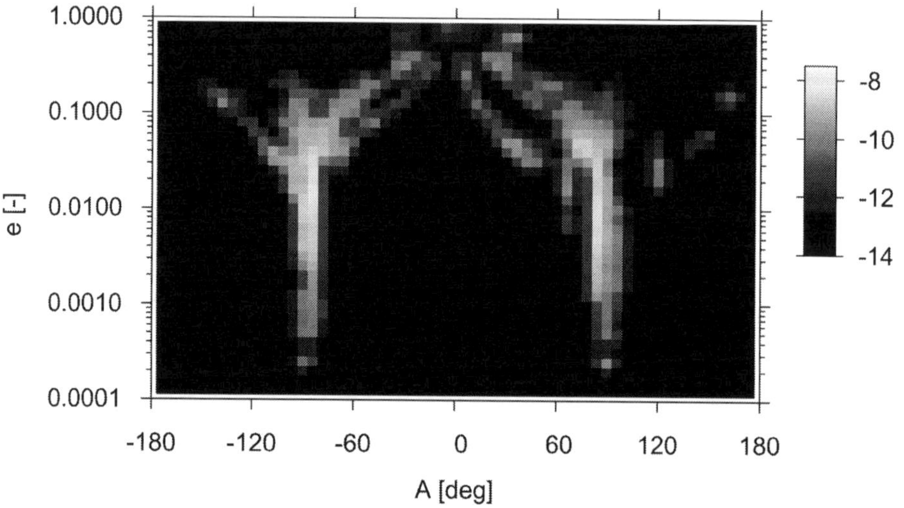

Fig. 4.31. Debris flux distribution for $d > 1\,\mathrm{cm}$ on a GEO orbit, as a function of the impactor orbit eccentricity e and the impact azimuth A. Contour levels are given as $\log_{10}$ of the impact flux $F\,[\mathrm{m}^{-2}\,\mathrm{y}^{-1}]$ (class widths: $\Delta A = 6°$, 40 equal-size classes in $log_{10}\, e$; see color plate on page 385).

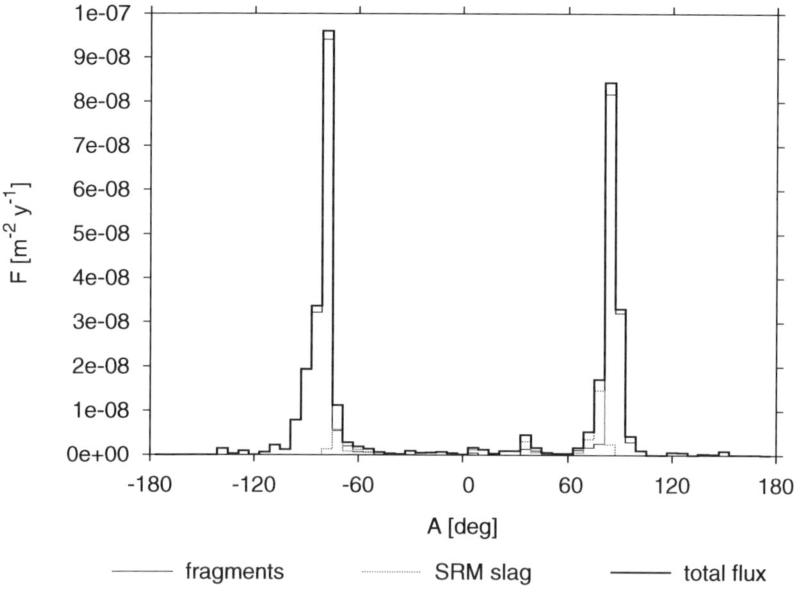

Fig. 4.32. Source-wise and total debris flux for $d > 1\,\mathrm{cm}$ on a GEO orbit, as a function of the impact azimuth A (class width: $\Delta A = 6°$).

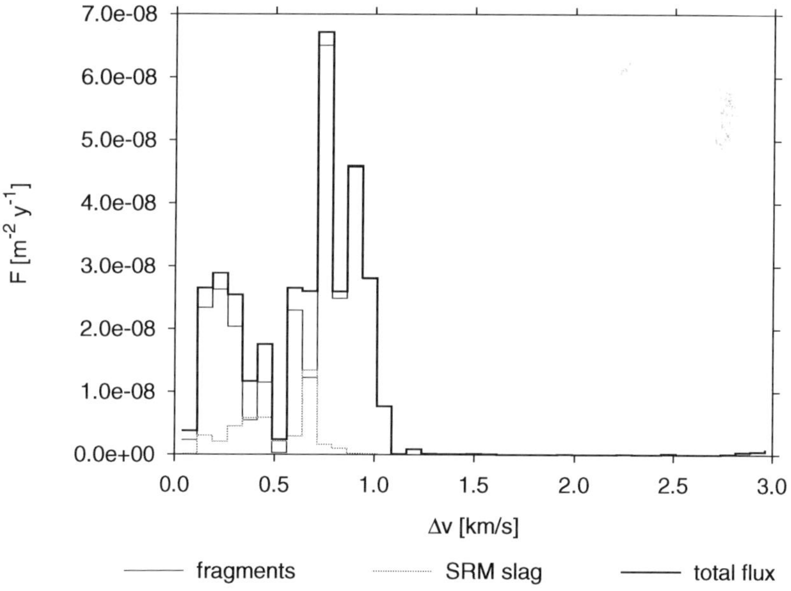

Fig. 4.33. Source-wise and total debris flux for $d > 1\,\mathrm{cm}$ on a GEO orbit, as a function of the impact velocity Δv (class width: $\Delta(\Delta v) = 75\,\mathrm{m/s}$).

4.4 REFERENCES

Bendisch, J., Bunte, K., Sdunnus, H., Wegener, P., Walker, R., and Wiedemann, C. (2002). Upgrade of ESA's MASTER Model. Technical Report (final report) ESA contract 14710/00/D/HK, ILR/TUBS, Braunschweig.

Klinkrad, H. (1993). Collision Risk Analysis for Low Earth Orbits. *Advances in Space Research*, vol.13:551–557.

Klinkrad, H. et al. (2003). *ESA Space Debris Mitigation Handbook*. ESA/ESOC, second edition. issue 1.0.

5

Modeling of the Future Space Debris Environment

H. Klinkrad, C. Martin and R. Walker

In Chapters 3 and 4 the historic evolution of the space debris environment up to the present was developed. All deployment and release events were largely considered as deterministically known, with traceable release epochs, release types, and release orbits. When trying to forecast the future evolution of the space debris environment, this deterministic knowledge base needs to be replaced by a statistical model of deployment and release events, based on a thorough analysis of past activities and recent trends. The current chapter will explain corresponding "traffic models" and their effect on the long-term evolution of the space debris environment in the case of unchanged operational practices (a so-called "business-as-usual" scenario). The expression "long-term" in this context refers to forecasting timespans of up to 100 years. Traffic models, prediction methods, and resulting trends, which are discussed hereafter refer to the DELTA 2.0 software (Debris Environment Long-Term Analysis (Walker et al., 2000)).

5.1 ORBIT PROPAGATION METHODS FOR LONG-TERM PREDICTIONS

Section 3.1 outlined the importance of efficient orbit prediction methods for the propagation of huge space debris populations over time spans of several decades. For repetitive, long-term forecasts up to 100 years, as required by models of the future debris environment, even more efficient orbit theories are required to obtain acceptable CPU times in corresponding software implementations.

The DELTA 2.0 program uses as a starting point an initial population provided by MASTER-2001 for its reference epoch of May 1, 2001 (Bendisch et al., 2002). Since DELTA is mainly concerned with long-term stability and proliferation aspects of the future space debris environment, only particles with a damage potential are considered. The related size threshold is adopted as $d > 1$ mm. Hence, only a small fraction of the initial MASTER population of $d > 1\,\mu\text{m}$ meets this criterion, and is maintained for the long-term analysis.

The MASTER-2001 orbit propagator described in Section 3.1 uses singly averaged orbital elements (see curve [2] in Fig. A.5), allowing for propagation time steps of typically 2 days in context with a 4th-order multi-step predictor-corrector method. For DELTA 2.0 larger time steps are required to save CPU time. This is accomplished by propagating singly averaged Kepler elements by means of a simple "point-slope formula", where $x_i(t) = x_i(t_o) + \dot{x}_i(t_o) \cdot (t - t_o)$, and $x_i = a, e, i, \Omega, \omega$ for $i = 1, \dots, 5$. Though less accurate than higher-order methods, this algorithm is very stable and robust in numerical applications. In comparative tests against verified orbit propagators prediction time steps of up to 8 days could be used, while maintaining numerical stability and acceptable accuracy.

The perturbation model used in the DELTA Orbit Propagator (DELTOP) is compatible with the FOCUS-1 propagator described in Section 3.1, considering secular and long-periodic zonal harmonic perturbations due to J_2 and J_3, GEO resonant perturbations due to $J_{2,2}$, luni-solar perturbations according to Cook's theory (Cook, 1962), airdrag perturbations based on King-Hele's algorithms (King-Hele, 1987), and solar radiation pressure perturbations as derived by Aksnes for a cylindrical Earth shadow (Aksnes, 1976). Predictions of mean solar activity, as input to the atmosphere model, are produced as concatenated replicates of the past two solar cycles.

In a comparison with the FOCUS-1 orbit generator, the DELTOP propagator is 2.5 to 5 times faster, as a result of the simplified integration algorithm with larger time steps (Walker et al., 2000).

5.2 CONCEPTS OF A LONG-TERM DEBRIS ENVIRONMENT PROJECTION

The DELTA model is designed according to the same principles as the MASTER model. Consequently, it uses a volume discretization approach which characterizes a space debris population at a certain epoch by means of cell-passage events, as described in Section 3.2, for control volume partitions in LEO, MEO, and GEO as listed in Table 5.1.

Table 5.1. DELTA-2001 discretization of the control volumes LEO, MEO and GEO. All ranges in right ascension are from $\alpha = -180°$ to $+180°$. Ranges in declination are from $\delta = -90°$ to $+90°$ in LEO and MEO, and from $\delta = -20°$ to $+20°$ in GEO. Altitudes refer to a spherical Earth of radius $a_e = 6378.137$ km.

	H_{min} [km]	H_{max} [km]	ΔH [km]	$\Delta\alpha$ [°]	$\Delta\delta$ [°]	Cells
LEO	125	2,000	25	360	5	2,700
MEO	2,000	34,775	475	360	5	2,484
GEO	34,775	36,775	40	10	4	18,000

The DELTA model combines an automated pre-processing of the MASTER-2001 space debris population with the long-term simulation of evolution dynamics, and with the subsequent post-run analysis of results. The occurrence of launch, explosion, and solid rocket motor (SRM) firing events is queried from event data files at regular points in time during the simulation. Separate pre-processing programs, a launch event model, an explosion event model and an SRM firing event model randomly predict future object releases based upon average event rates. The future event data are stored in chronologically ordered explosion and SRM firing event files. Due to the statistical nature of such events, a full DELTA simulation of the long-term environment evolution must be repeated several times, with each run using different random conditions for the generation of event sequences. Collisions are generated autonomously by the debris environment evolution model, based upon spatial densities, statistically sampled collision probabilities (see Eq. 3.1–3.3), and the mass-related kinetic energies of the events. DELTA typically requires 10 complete Monte Carlo runs to arrive at confidence intervals in the forecast results of about $\pm 25\%$ (not considering errors in the model theory itself). The concept of the main program's environment evolution engine is based upon a synchronous processing of sources and sinks at each propagation time step, starting with the initial MASTER-2001 population of $d > 1$ mm on May 1, 2001, which comprises $\sim$120,000 representative objects. The only population sources which are covered by this size regime are launch and mission related objects (MROs), fragments, solid rocket motor slag, and sodium–potassium (NaK) droplets. The latter one is a historic source for which no further release events need to be simulated. For fragments and SRM slag objects DELTA uses the same release models as applied by MASTER (see Sections 3.4 and 3.5). For each object in a DELTA simulation the complete information is retained on its source, object type, orbit, mass, size, representation factor, start and end of operational life (for orbit-maintained systems), and lethality ratio (specific kinetic energy of a collision in kJ/kg which, for LEO objects of $m > 1$ kg, would result in a catastrophic fragmentation). These detailed object labels, for instance, allow feedback collisions (the generation of collisions by collision fragments) to be traced.

5.3 MODELING FUTURE LAUNCH TRAFFIC AND RELEASE EVENTS

The number of collisions in the future is strongly dependent upon the number of launches, which could significantly increase the total mass and cross-section in orbit. This would give an increased likelihood of catastrophic collisions involving large intact objects, larger explosion fragments, or larger collision fragments ("feedback" effect). Furthermore, an increase in the on-orbit mass would exacerbate the feedback effect, and act as a reservoir for fueling a collision cascading process. Therefore, the future traffic model used by DELTA is critical to the predictive accuracy of long-term future environments. Indeed, the estimation of future activity can be a source of great uncertainty, along with the initial conditions (the current environment), if they are not validated against measurements.

The launch rate fluctuates from year to year and is extremely difficult to predict, because launches are dictated by political, economical, and technological considerations. Between the mid 1960s and early 1980s the launch rate was at a peak of ~120 per year, but then declined down to ~70 in the 1990s (see Fig. 2.2), where it reached another stable level.

There are two different types of future launch traffic model: the steady-state model and the mission model. The steady-state model is based on the assumption that the historical launch activity of the past few years is typical of future activity, with a constant, average overall launch rate. This overall launch rate can then be scaled up and down year-by-year to investigate the effects of different growth scenarios. Most long-term evolution models use some sort of steady-state model, because it offers stability, flexibility, and established information on the orbit and mass distributions of launch-related objects. However, this type of model does not consider changes in this "steady state", such as certain historic activities ceasing and new activities commencing in the future. The mission model is purely based on estimates of the future uses of space such as new technology over the next 25 years, new space architectures, new launchers, and speculations of civil, military, and commercial objectives and benefits. There are two main mission models that predict US civil and military activity during the period 1990 to 2010. The Civil Needs Data Base (CNDB), a NASA-sanctioned analysis of future architectures necessary to meet the US governmental and scientific needs, and the US Department of Defense (DoD) mission model. European mission scenarios up to the year 2020 have been estimated on the basis of the capability and availability of future European launchers. Unfortunately, mission models rely on historically poor predictions by analysts of the future uses of space (which tend to be over-optimistic). Both civil and military requirements change from year to year, in an almost unpredictable manner, according to the political and economic situation at the time.

Neither a pure steady state nor a pure mission model is ideally suited for a good prediction of future launch traffic. Attempts to reproduce the past decade of space history show that a better solution is to combine with the mission model reliable and stable steady-state data, which has mature and certain information on the future uses of space. This includes the deployment, operation, and support of multiple-satellite constellations, and the International Space Station infrastructure. Most long-term simulation models, such as NASA's EVOLVE (Opiela and Krisko, 2001), the Italian SDM (Rossi et al., 1995b), and the UK IDES model (Walker et al., 1999) are currently using this type of combined traffic model. The DELTA 2.0 model (Walker et al., 2000) adopts the same approach, with a steady-state part reflecting a continuation of recent historic launch activity, and a mission-model part including a few realistic satellite constellations.

The derivation of DELTA's future launch traffic model involved the extraction and analysis of data on more than 2,100 launch- and mission-related objects for the period May 1993 to May 2001 from ESA's DISCOS database (Hernández et al., 2001). The data include information on launch epoch, COSPAR identifier, launch country, object type, initial orbit (semimajor axis, eccentricity, inclination), mass, and cross-sectional area. Each object is assigned a classification code to combine all characterization criteria, and all objects with a common classification code are grouped together in order to derive summary orbit, mass, and cross-section data. For each payload class, a mean launch rate is established. Objects are also grouped launch-wise in order to establish links and relationships between different classification codes. In this way, typical historical launch patterns can be replicated by the model of future launch traffic, and can thus be carried forward into the long-term future.

Historic launch activities show characteristic patterns, for instance when plotting distributions of semimajor axis versus eccentricity, or eccentricity versus inclination. With the high-resolution classification scheme used, a total of 484 different object classes were identified from 2,100 objects in the historical launch data of the 8 years prior to May 1, 2001. The projected orbit distributions of the future launch traffic model strongly resemble the distributions of the historical data, both in terms of the locations and scattering of different clusters. All key orbit regimes are well represented, such as near-circular LEO, MEO, and GEO, and also high eccentricity GTO and Molniya orbits. Overall, the model predicts an average annual input of $\sim$315 launch- and mission-related objects (representing 340 tons of mass) to the existing DELTA population. This corresponds to a mean overall annual launch rate of 70.5, covering the entire regime of Earth orbits.

The DELTA model of future explosions is derived from a list of 172 historical on-orbit explosion events (see also Table 2.2), which was also used to derive the MASTER-2001 reference population serving as input to the DELTA program. Combined with a harmonized fragmentation model as outlined in Section 3.4, this ensures that there is a smooth transition at the reference epoch between the historical population and its projected evolution. Over the 8-year reference period of historical data, 47 on-orbit fragmentation events were analyzed and processed, corresponding to a mean annual explosion rate of 5.25 (see Fig. 2.16). In total, over 30 different object classes were identified for the future explosion traffic model. Their orbit distribution reveals distinct patterns with respect to semimajor axis, eccentricity, and inclination. Most explosions occur in near-circular, low-Earth orbits, at high orbital inclinations, with clusters at $i \approx 65^\circ$, 82°, and 98°. However, explosions of objects decaying from highly eccentric orbits (particularly the small Proton SOZ ullage motors) also make a noticeable contribution. The MASTER-2001 reference population includes 8 fictitious GEO break-up events for the time-span 1993 to 2001 (mean annual explosion rate of 1.0) to match GEO observation data. These unconfirmed events are not considered in the DELTA traffic model, and since the last confirmed GEO fragmentation event lies outside the 8 year reference timespan, no future GEO explosions are simulated by DELTA in the long-term population projections.

Similarly to the model of future explosions, the DELTA model of future SRM firings has been derived from a historical SRM firing list used by MASTER-2001 to generate the reference population. Combined with a harmonized slag generation model (see Section 3.5), this ensures a smooth transition between the historic population evolution up to 2001, and its future projection. Over the 8-year period of historical data prior to the MASTER-2001 reference epoch, 176 SRM firing events were analyzed and processed in order to produce the model of future SRM firings, with an average annual firing rate of 19 (see Fig. 2.19). In total, 90 different source object classes were identified for the firing model, with characteristic orbit distributions with respect to semimajor axis, eccentricity, and inclination. Among these a number of distinct firing scenarios can be identified: direct insertions into near-circular, high-inclination LEO (particularly at 82° and 98° inclination); transfers from near-circular LEO to highly eccentric GTO (e.g. Ariane launches from Kourou with $i = 7°$, and US launches from Kennedy Space Center with $i = 28.5°$); and transfers from GTO to GEO with a delivery orbit of near-zero eccentricity and inclination, and $a \approx 42{,}000$ km.

Since the future traffic models have been derived from historical activities over the past 8 years, they represent a "business-as-usual" scenario of the future traffic. The DELTA program has the capability to up- or down-scale this reference scenario for pre-defined timespans in the prediction.

5.4 DEPLOYMENTS OF CONSTELLATIONS AND NANO-SATELLITES

Continuous coverage from space of certain areas on the Earth is of great importance for meteorology, remote sensing, communications, navigation, and defense applications. The unique advantages of the GEO region in this context were identified by Hermann Noordung (1929) and Arthur C. Clarke (1945) long before the first satellite was launched. The ideal geostationary orbit (GEO) is circular ($e \approx 0$), within the equatorial plane ($i \approx 0$), and at an altitude of $H \approx 35{,}786$ km. It has an orbit period of $T = 23^h56'04.09''$, exactly corresponding to one sidereal day. Hence, it moves synchronously with the underlying Earth. Several satellites in GEO, when uniformly spaced in geographic longitude, can jointly obtain near-global coverage of the Earth's surface. This capability is, for instance, used by an international network of meteorological satellites (e.g. GOES, Meteosat, Insat, and GMS). Such combinations of GEO spacecraft can be regarded as first satellite constellations.

GEO "constellations" have the drawback of incomplete Earth coverage, and of deteriorated observation geometry with increasing latitude of the target area. These shortcomings can be overcome by uniformly distributing several satellites in co-planar, near-circular orbits of moderate inclinations, and by deploying several such orbit planes at equidistant node separations. Such configurations are often denoted as "Walker constellations" (Walker, 1971). In many cases the orbit periods of such constellations are synchronized with the Earth rotation to have repeating ground track patterns after an integer number of days and orbit revolutions. The best-known examples of such constellations are the US GPS/Navstar and the Russian Glonass constellations (see Table 5.2). Both operate in or near the semi-synchronous orbit region, with orbit periods of 719.94 min (GPS) and 675.73 min (Glonass), and with repeating ground tracks after 2 orbits and 1 sidereal day (GPS), and after 17 orbits and 8 sidereal days (Glonass). Later, as of 2008, they

Table 5.2. Operational, partially operational, and planned satellite constellation designs (a more comprehensive list is provided by (Springmann and de Weck, 2004), based on data from the US Federal Communications Commission).

Constellation	H_{pe} [km]	H_{ap} [km]	i [°]	Satellites + spares	No. of planes	Satellites per plane
Galileo	23,616	23,616	56.0	27 + 3s	3	9 (+1s)
GPS/Navstar	20,182	20,182	55.0	24 + 3s	6	4 (+1/2s)
Glonass	19,132	19,132	64.8	24 + 3s	3	8 (+1s)
ICO	10,390	10,390	45.0	12	2	6
Ellipso Concordia	8,050	8,050	0.0	6 + 1s	1	6 (+1s)
Ellipso Borealis	7,605	633	116.6	8 + 2s	2	4 (+1s)
Skybridge	1,469	1,469	53.0	80	20	4
GlobalStar	1,414	1,414	52.0	48 + 4s	8	6 (+1/2s)
Teledesic	1,375	1,375	84.7	288 + 36s	12	24 (+3s)
Orbcomm	825	825	45.0	28 + 8s	4	7 (+2s)
Iridium	780	780	86.4	66 + 6s	6	11 (+1s)

will be joined by the European Galileo system, which will be operating 3,434 km above the GPS constellation, on an orbit with a period of 861.48 min, and a ground track repetition within 5 revolutions and 3 sidereal days.

Between February 1978 and July 2002 there were 44 GPS payloads launched and deployed, of which 28 were still functional in 2002 (the total number of associated mission-related objects and rocket bodies was on the order of 150). For the Glonass constellation 39 payloads were launched between October 1982 and July 2002, of which 10 were still functional in 2002 (the total number of associated mission-related objects and rocket bodies was on the order of 300). With these object releases, the GPS and Glonass systems are dominating the environment near semi-synchronous altitudes.

Table 5.3. Satellite data and deployment status of LEO constellation designs. The mass m, cross-section A, and operational lifetime t_{oper} refer to a single satellite (latest generation). The time of deployment start $t_{o,depl}$ and service start $t_{o,serv}$ refer to the entire constellation. The status relates to 2004.

Constellation	m [kg]	A [m^2]	t_{oper} [y]	$t_{o,depl}$	$t_{o,serv}$	Status
ICO	2,600	10	10	2001	2002	only one launch in 2001
Ellipso	1,000	8	5	2003	2004	– pending –
Skybridge	1,250	12	8	2003	2004	– pending –
GlobalStar	450	10	7.5	1998	1999	on-orbit and in operation
Teledesic	1,400	12	10	2003	2004	– halted –
Orbcomm	42	3	4	1997	1998	on-orbit and in operation
Iridium	689	7	7–9	1997	1998	on-orbit and in operation

Table 5.4. Launched and planned nano- and pico-satellite missions (mass: $m \leq 30\,\text{kg}$).

Name	No.	m [kg]	Dimensions [cm] × [cm] × [cm]	Orbit: $H_{pe} \times H_{ap}$ at i [km] × [km] at [°]
Launched (1994–2004):				
Astrid	1	27	29 × 45 × 45	968 × 1,023 at 83°
Astrid-2	1	30	95 × 40 × 40	1,000 × 1,000 at 83°
ASUSAT-1	1	5.8	32 × 24 × 24	750 × 800 at 100°
DARPA	2	0.5	10 × 7.5 × 2.5	700 × 700 at 90°
LatinSat-A,B	2	11.35		650 × 650 at 65°
MASAT	1	0.5	20 × 7.5 × 2.5	700 × 700 at 90°
MEPSI-2	2	0.25	10 × 7.5 × 2.5	511 × 539 at 97.8°
Munin	1	6	21 × 21 × 21	698 × 1,800 at 96°
PCSAT	1	17.5		794 × 794 at 67°
Reflector	1	6	150 × 50 × 50	996 × 1,050 at 99.7°
Saudi-Sat	2	12		650 × 650 at 65°
SNAP-1	1	6.5	20 × 17.5 × 17.5	684 × 707 at 98°
STENSAT	1	0.25	9 × 9 × 2.5	700 × 700 at 90°
SQUIRT-1	1	16	45 × 40 × 38	794 × 794 at 67°
SQUIRT-2	1	23.1	21 × 23.5 × 23.5	700 × 700 at 90°
Thelma/Louise	2	0.5	20 × 7.5 × 2.5	700 × 700 at 90°
TUBSAT-N	1	8	32 × 32 × 10.4	400 × 776 at 78.9°
TUBSAT-N1	1	3	32 × 32 × 3.4	400 × 776 at 78.9°
UNAMSAT	1	17	25 × 25 × 25	988 × 1,010 at 82.9°
UniSat	1	10		650 × 650 at 65°
UniSat-2	1	11.8	15 × 25 × 25	650 × 650 at 65°
Westpac	1	28.8	24.5 (spherical)	835 × 835 at 98.7°
XSS-10	1	28		518 × 805 at 39.7°
Planned (after 2004):				
3CornerSat	3	15	45 (spherical)	380 × 380 at 51°
BLUEsat	1	< 15	22 (cube)	LEO
CUBEsat	≤ 24	1	10 (cube)	LEO
ION-F	3	15	30 × 50 × 50	380 × 380 at 51°

Table 5.3 shows that since the late 1990s LEO constellations have become increasingly important, mainly for communication purposes. Of the multitude of registered projects (Springmann and de Weck, 2004) only a few have reached an operational status: Globalstar, Orbcomm, and Iridium. All these constellations are deployed in near-circular orbits within a narrow altitude band, at moderate to high inclinations. Hence, they produce a noticeable step increase of the catalog population at their release altitudes (see Fig. 2.14). The orbit altitudes of Orbcomm and Iridium near 800 km, and of Globalstar near 1,400 km are located in regions of peak concentrations of space debris of destructive sizes (see Fig. 3.21 and 3.22). As a consequence, the three currently operational constellations with future re-

placements shall be considered in a modified baseline scenario for a "business-as-usual" forecast of the space debris environment (see Section 5.6). Table 5.3 lists the relevant spacecraft quantities which are used as input to the DELTA simulation.

Apart from constellations of full-size spacecraft, low-cost micro-, nano-, and pico-satellites have become of interest since the mid 1990s. So far, such missions have mostly been launched with single or dual payloads, into orbits below 1,000 km altitude (see Table 5.4). However, conceptual studies exist to use large swarms of nano- or pico-satellites and to coordinate their payloads to perform communication, reconnaissance, astronomy or high-resolution Earth observation with large synthetic apertures. Such swarms, deployed in narrow bands of altitude and inclination, can lead to strong increases in local spatial densities, and thus may lead to collisions with large-size, intact objects. With masses mainly between 5 and 10 kg, nano- and pico-satellites would reach sufficient impact energies to lead to catastrophic break-ups of the targets. The sensitivity of long-term space debris environment evolution to swarms of small satellites will be addressed in Section 5.6.

5.5 DEFINITION OF A BUSINESS-AS-USUAL FORECAST SCENARIO

With the model of launch traffic and release events outlined in Section 5.3 a long-term projection of the space debris environment with the DELTA 2.0 program can be performed. The prediction timespan shall be 100 years, starting on May 1, 2001, with an initial MASTER-2001 space debris population composed of 120,000 representative objects larger than 1 mm. The assumptions for the traffic model shall follow a "business-as-usual" scenario with the following characteristics:

- constant annual launch rate of 70.5, with a mean input of 315 objects and 340 tons of mass per year, into orbits described by the deployment model
- constant annual rate of on-orbit explosions of 5.25, distributed across orbit families and object masses correlated with past events
- constant annual rate of 19 SRM firings, correlated with orbit families and motor sizes of past events
- existing LEO constellations (Iridium, Globalstar, Orbcomm) and MEO constellations (GPS/Navstar, Glonass) adopted from the input population; no constellation upgrades or replacements considered
- existing nano- and pico-satellites adopted from the input population; no deployments of satellite swarms considered
- lethality threshold for impacts set to $E_i/m_t \geq 40\,\mathrm{kJ/kg}$ (kinetic energy of impactor related to the mass of the target object)

The whole population is propagated with 8-day time steps, considering all relevant perturbations, and taking into account orbit maintenance of operational LEO and GEO spacecraft (by suppressing perturbations during pre-defined operational lifetimes). Launch, release, and collision events are triggered statistically, by a procedure outlined in Eq. 5.2 to 5.4.

Let c from Eq. 3.1 be the mean number of impacts of debris particles larger than a certain size threshold, on a target of a given collision cross-section A_c, in a time interval Δt. Furthermore, let n be the total number of debris objects of the identified size regime which reside in the same control volume (e.g. altitude shell), and which could collide with the target. Then the probability P_j that the target is hit by exactly j out of $n \geq j$ possible objects is given by a binomial distribution.

$$P_j = \binom{n}{j} \left(\frac{c}{n}\right)^j \left(1 - \frac{c}{n}\right)^{n-1} \tag{5.1}$$

For large n and small ratios c/n this binomial distribution can be well approximated by a Poisson distribution (see Eq. 3.2), which is no more explicitly dependent on n.

$$P_j = \frac{c^j}{j!} \exp(-c) \tag{5.2}$$

The probabilities P_j for $j = 0, \ldots, k$ are then determined until a small threshold value of ϵ is reached such that $P_{k+1} \leq \epsilon$. The resulting probabilities can be normalized, such that their sum equals 1.

$$\hat{P}_j = \frac{P_j}{\sum\limits_{i=0}^{k} P_i} \quad \rightarrow \quad \sum_{j=0}^{k} \hat{P}_j = 1 \tag{5.3}$$

A reliable random number generator is then used to draw a number $\hat{r} \in [0,1]$. The corresponding number of successful events in the analyzed time interval is determined by the largest j for which the sum of the normalized probabilities $\hat{P}_j$ is still less than $\hat{r}$, such that

$$\sum_{i=0}^{j} \hat{P}_i \leq \hat{r} < \sum_{i=0}^{j+1} \hat{P}_i \quad \rightarrow \quad j = \text{fct}(\hat{r}) \tag{5.4}$$

This procedure is applied to all launch and on-orbit release event types which are covered by the DELTA traffic model. The parameter c for the different traffic model sources is defined as follows:

- launch traffic model: $c =$ (mean annual launch rate of an object and orbit sub-class [y^{-1}]) $\times$ (propagation time step [y])
- explosion traffic model: $c =$ (mean annual explosion rate of an object and orbit sub-class [y^{-1}]) $\times$ (propagation time step [y])
- SRM slag traffic model: $c =$ (mean annual firing rate of a solid motor and orbit sub-class [y^{-1}]) $\times$ (propagation time step [y])

DELTA's classification scheme comprises 75 classes of orbit type, 20 classes of object type, and 9 classes of object mass. If an event is identified to occur, based on the statistical Monte Carlo procedure, then its event time is selected randomly inside the current propagation timespan Δt. Likewise the positions of the orbit node, of the pericenter, and of the orbit position are randomly selected from $\Omega, \omega, f \in [0, 2\pi]$. The semimajor axis a and inclination i are slightly randomized within $\Delta a = \pm 20\,\mathrm{km}$ and $\Delta i = \pm 2°$ to reduce selection effects from the discrete classifications in the traffic model database.

All population source terms that are considered by the model of projected traffic can be regarded as "open loop" inputs, with no feedback from spatial and temporal changes in population densities. The latter effects, however, are of great importance when trying to forecast future collision rates. In contrast with the previously mentioned sources, collisions are "closed loop" inputs to the overall population, driven by the population state itself.

Collision fluxes are determined in a quasi-deterministic manner described in Eq. 4.1 to 4.9. The total number of expected collisions c between projectiles from the i-th size class of cross-section A_i, with a target of size A_t, during a timespan Δt is determined from

$$c_i = F_i \left(\sqrt{A_i} + \sqrt{A_t}\right)^2 \Delta t \tag{5.5}$$

For each size class $i = 1, ..., I$ Eq. 5.2 to 5.4 are applied to determine the expected number of collision events in the forecast time interval Δt. If one or more collisions are predicted, then the lethality of the event is assessed, based on the ratio ψ_i of the projectile's kinetic energy to the mass of the target object.

$$\psi_i = \frac{1}{2} \frac{m_i}{m_t} (\Delta v_i)^2 \tag{5.6}$$

with m_i the mass of the projectile in the i-th size class, m_t the mass of the target and Δv_i the projectile impact velocity.

An event is considered to be lethal, with a total disintegration of the target, if $\psi_i \geq \psi_*$, where the default threshold for a catastrophic fragmentation is set to $\psi_* = 40\,\mathrm{kJ/kg}$. This value has been established from ground-based hypervelocity impact tests on full-size targets.

If a fragmentation event is predicted (either a collision, or an explosion), then the mass of the involved object(s) is transferred into fragments according to the NASA break-up model (see Section 3.4). The source object is subsequently deleted from the population, and the fragments are inserted and binned into the different orbit, object and mass classes.

Due to the assumed traffic model, with constant annual launch rate $\dot{n}_l$, the number N of large-size launch- and mission-related objects (LMROs), which have masses capable of producing catastrophic collision events, increases almost linearly with time according to Eq. 5.7 (see also Fig. 5.1), with the slope of the LEO population growth undergoing oscillations due to atmospheric drag variations with the 11-year solar cycle.

$$N(t) \approx N(t_o) + \dot{n}_l\,(t - t_o) \tag{5.7}$$

If one assumes that at some epoch t a population of $N(t)$ space objects (large debris and intact objects), of an average collision cross-section $\overline{A}_c$, randomly move through a control volume V, at mean relative velocities of $\Delta\overline{v}_c$, then the mean collision rate $\dot{n}_c$ per volume can be predicted according to the laws of kinetic gas theory.

$$\frac{\dot{n}_c(t)}{V} \approx \frac{1}{\sqrt{2}} \left(\frac{N(t)}{V} \right)^2 \overline{A}_c \, \Delta\overline{v}_c \tag{5.8}$$

This theoretical assessment, suggesting a progressive increase of catastrophic collisions with a linear population growth, is confirmed by Fig. 5.2. It shows the evolution of the cumulative number of catastrophic collisions in LEO over 100 years, with a discrimination between background events (collisions involving only launch- and mission-related objects, and explosion fragments), and uncontrolable feedback events (collisions involving previously generated collision fragments)[5.1]. Fig. 5.2 indicates that background collisions, which (and only which) are predicted by Eq. 5.8, will be outnumbered by feedback collisions after ~2085. This collision type will dominate the long-term space debris environment, if space flight activities are continued on a business-as-usual basis. All 52 collisions which are forecast by the DELTA model for the 100-year prediction timespan are produced in the LEO region.

All release events which are caused by the traffic model are triggered by statistical Monte Carlo techniques, with different random number seeds for each run. In order to arrive at a stable mean forecast, 10 DELTA runs must be performed for the same prediction timespan to arrive at 1σ dispersions which are within $\pm 25\%$ of the mean value. The corresponding confidence intervals are indicated in Fig. 5.1, 5.3 and 5.5. These uncertainties are only reflecting statistical noise in the event triggering process. They do not reflect errors in the traffic model, or in the dynamic population model.

Fig. 5.3 shows the future evolution of the critical population of $d > 10\,\text{cm}$ in LEO, MEO, and GEO. Any member of this population is likely to cause a catastrophic fragmentation when impacting on a large-size intact object. The related object numbers in LEO are predicted to increase by a factor of ~6 over the next 100 years. The corresponding evolution of the spatial density altitude profiles in LEO is given in Fig. 5.4. Most of the collisions occur near the densely populated altitude bands of 800 ± 200 km and $1,400 \pm 100$ km, and in inclination bands between 65° and 100°. Out of the total predicted number of ~50,000 objects of $d > 10\,\text{cm}$ after 100 years, ~30,000 are expected to be collision fragments. The long-term MEO environment is dominated by explosion fragments (with many of them generated on highly elliptic orbits), and the growth in the GEO region is attributed to launch traffic only.

The evolution of the space debris population of $d > 1\,\text{cm}$ is summarized in Fig. 5.5 and 5.6. All of these population members have the potential to disable a functional payload. Fig. 5.5 shows the predicted number of objects larger than 1 cm for the LEO, MEO, and GEO region, as a function of time. Due to the absence of GEO fragmentations in the traffic model, the GEO population in this size regime remains almost unchanged. Dominated by future explosion events, the

[5.1]collisions with constellation members are insignificant in the business-as-usual scenario

MEO population increases from ~200,000 in 2001 to ~875,000 in 2100, at a slightly progressive rate. Since all collisions are forecast in the LEO region, the LEO population has a strongly progressive growth from ~190,000 in 2001 to ~1,600,000 in 2100. Fig. 5.6 indicates that up to the year 2040 a near-linear population increase due to explosions is prevailing. Thereafter the accelerated population growth due to collisions dominates the environment. All other sources (e.g. SRM slag and NaK droplets) are of minor importance.

A business-as-usual scenario for future space activities is likely to deteriorate the space object environment to a level which could dramatically increase the risk on presently preferred operational orbits, in particular within the LEO region. Probably in less than 50 years collision fragments will start dominating the space debris environment, and probably in less than 100 years feedback collisions with previously generated collision fragments will prevail as the cause of catastrophic break-ups. This is the time when some orbit regions may become unstable. A stability criterion for the space debris environment can be formulated as an equilibrium condition, where for a given control volume (e.g. a spherical shell) the object sources (launches, explosions, collisions, release events, and objects entering from adjacent control volumes) are inferior or equal to the object sinks (objects entering into adjacent control volumes, e.g. due to airdrag or active removal). If this condition is violated, the stability in this region is at risk. Through reduced traffic rates and inhibited release events, however, there is still a possibility of control. An orbital region becomes truly unstable, with no possibility of reversing the trend, if the collisions within the already existing population, exceed the sink terms. Several authors have defined stability metrics for the space debris environment in a mathematical form (Kessler, 1991; Rossi et al., 1995a). Most of these formulations, however, use simplifying assumptions for the debris environment, such as purely altitude-dependent population distributions, with dynamics according to the kinetic gas theory, and with simple, mathematically tractable expressions for sources and sinks. Based on these assumptions, Kessler postulates that at LEO altitudes near 900 km and 1,400 km regions of instability may have already developed today (Kessler, 1991).

Due to its semi-deterministic concept of modeling the space debris environment the DELTA program uses its own stability metric for each spherical sector of the LEO, MEO, and GEO control volumes. DELTA maintains a time history of spatial densities $D_{i,j,k}$ for critical particle sizes of $d > 10\,\text{cm}$, and for each volume cell indexed by its geocentric radius, declination, and right ascension (see Eq. 3.23). The recorded time histories can be fitted by a least-squares technique to a second-order polynomial, which can reflect trends ($\propto t$), and progressions ($\propto t^2$; according to Eq. 5.8).

$$D_{i,j,k}(t) = b_{0,(i,j,k)} + b_{1,(i,j,k)}\,(t - t_0) + b_{2,(i,j,k)}\,(t - t_0)^2 \tag{5.9}$$

where $b_{0,(i,j,k)}$, $b_{1,(i,j,k)}$, $b_{2,(i,j,k)}$ are fitted coefficients. Due to its low order, the fit suppresses statistical noise and higher-frequency harmonics (e.g. due to the 11-year solar cycle).

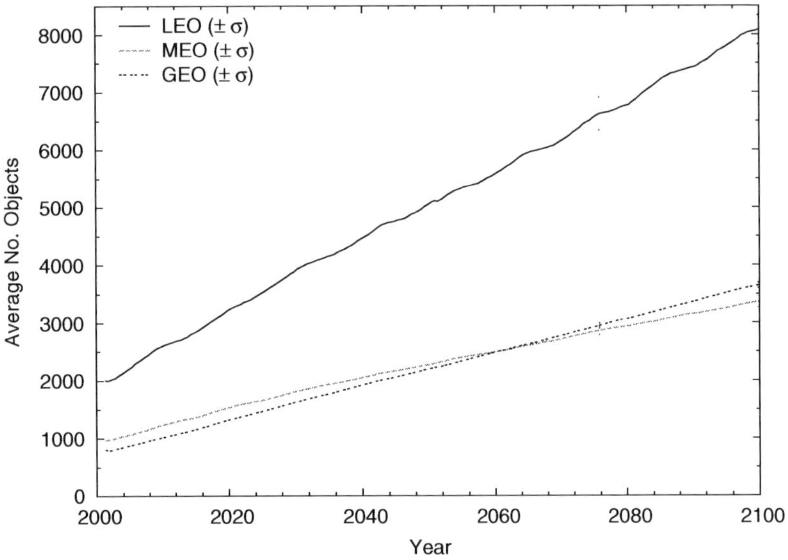

Fig. 5.1. Evolution of the space object population of $d > 1$ m in the LEO, MEO, and GEO regimes, for a business-as-usual scenario, over a 100-year prediction timespan. The 1σ confidence intervals are based on 10 Monte Carlo runs.

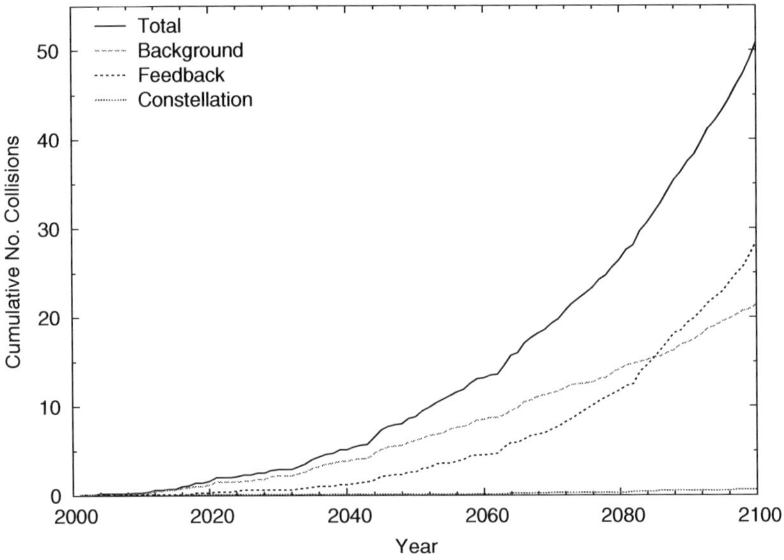

Fig. 5.2. Cumulative number of catastrophic collisions in LEO for a business-as-usual scenario, over a 100-year prediction timespan, discriminated by collision type.

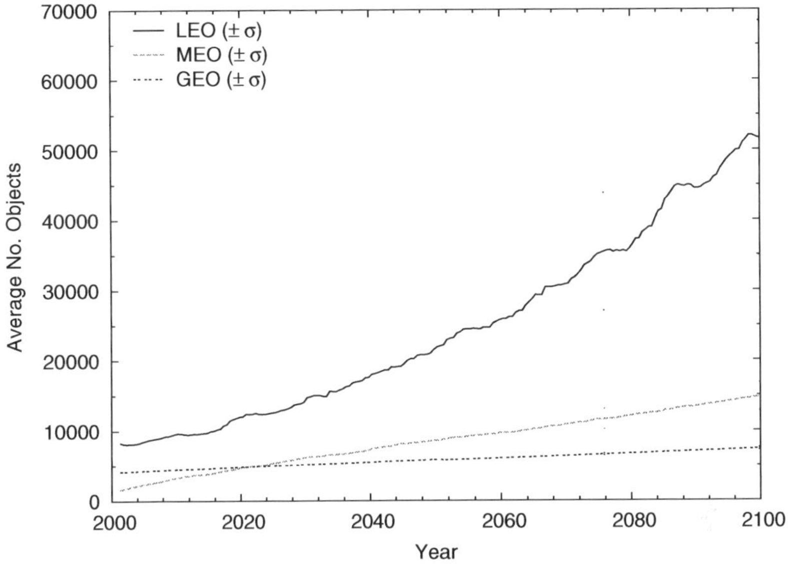

Fig. 5.3. Evolution of the space object population of $d > 10\,\text{cm}$ in the LEO, MEO, and GEO regimes, for a business-as-usual scenario, over a 100-year prediction timespan. The 1σ confidence intervals are based on 10 Monte Carlo runs.

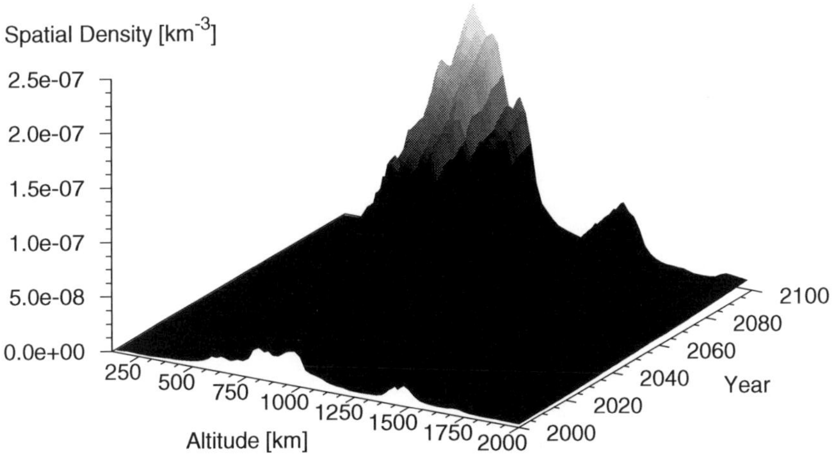

Fig. 5.4. Evolution of spatial density altitude profiles in LEO for objects with $d > 10\,\text{cm}$, for a business-as-usual scenario, during a 100-year prediction timespan.

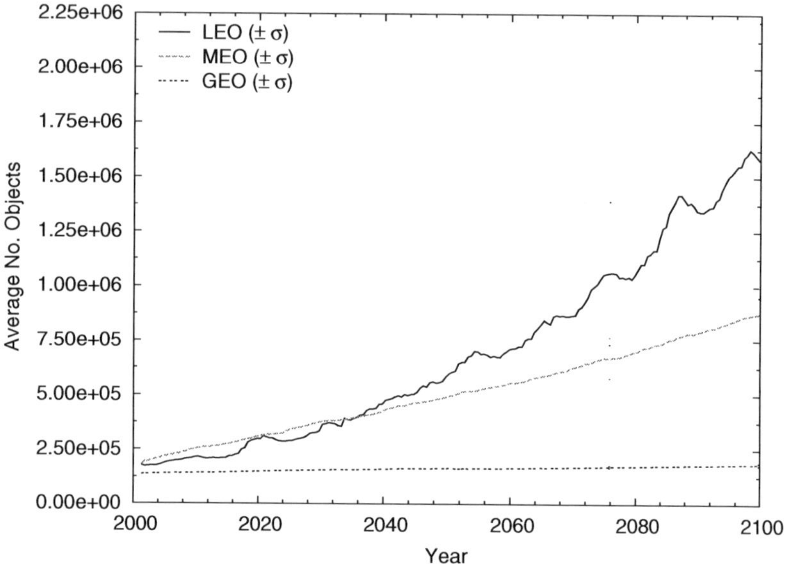

Fig. 5.5. Evolution of the space object population of $d > 1$ cm in the LEO, MEO, and GEO regimes, for a business-as-usual scenario, over a 100-year prediction timespan. The 1σ confidence intervals are based on 10 Monte Carlo runs.

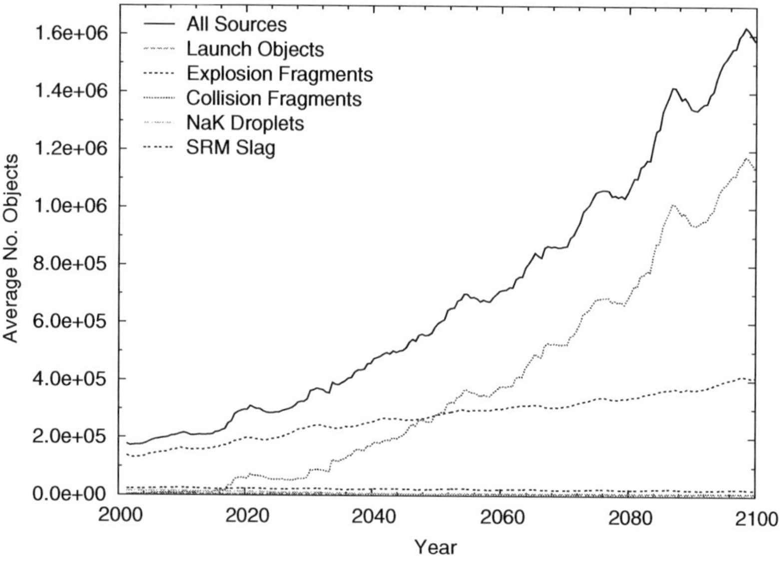

Fig. 5.6. Evolution of the LEO space object population of $d > 1$ cm, for a business-as-usual scenario, over a 100-year prediction timespan, discriminated by contributing sources.

The debris environment stability in a given volume cell can be monitored via the time derivative of its spatial density.

$$\frac{\mathrm{d}}{\mathrm{d}t}\Big(D_{i,j,k}(t)\Big) = \dot{D}_{i,j,k}(t) = b_{1,(i,j,k)} + 2\,b_{2,(i,j,k)}\,(t - t_0) \tag{5.10}$$

At a given time t, a certain region can be regarded as stable, if there is no net increase in spatial density, such that $\dot{D}_{i,j,k}(t) \leq 0$. Else, if $\dot{D}_{i,j,k}(t) > 0$ for objects of $d > 10$ cm, then a local instability is evolving. For a business-as-usual scenario this is first occurring in some isolated volume cells at LEO altitudes of 800 ± 200 km, and at declinations between 65° and 90°. Towards the end of the 100-year prediction timespan several regions of local instabilities begin to merge, and complete altitude shells and declination bands reach debris concentrations which pose an unacceptably high risk to space operations.

The ultimate stage of uncontrolled debris growth would be a self-supporting process of collisional cascading, when collisions in between collision fragments start to play an increasingly important role. In this phase the debris population would slowly grind itself to ever-decreasing sizes. Such a development, however, is a very long-term perspective, which will only take place several hundred years from now, if no remedial action in the form of debris mitigation measures is taken in the near future.

5.6 VARIATIONS OF A BUSINESS-AS-USUAL FORECAST SCENARIO

Since the business-as-usual scenario will serve as benchmark for the assessment of the effectiveness of debris mitigation measures, it is important to assess the sensitivity of this reference forecast with respect to varying traffic model assumptions. Three different aspects shall be investigated: (1) sensitivity with respect to changes in the annual deployment, explosion, and SRM firing rates, (2) sensitivity with respect to constellation deployments and nano-satellite swarm releases, and (3) sensitivity with respect to adopted fatality thresholds of catastrophic collision events.

For traffic rate variations two scenarios are analyzed. In the first one, "doubling traffic", current launch, explosion, and SRM firing rates are doubled relative to the business-as-usual case. On aggregate, the corresponding deployed objects are increased by 50%. In a second scenario of "halving traffic", the scaling of deployment rates is inverted, leading to a total number of released objects which is 25% below the business-as-usual case. The resulting number of catastrophic collisions when doubling traffic increases from 50 to 90, with a growth of the population of $d > 10$ cm of ~80% over 100 years. When halving the traffic rates, the number of catastrophic collisions reduces to 39, causing a 10 cm population after 100 years which is ~20% smaller than in the business-as-usual case. Hence, though increasing the traffic rates will aggravate the problem, a reduction of the traffic will not effectively improve the long-term space debris environment.

The deployment of satellite constellations and nano-satellite swarms in narrow altitude and inclination bands can strongly increase local spatial densities and resulting collision risk levels. This effect shall be investigated for three scenarios: (1) maintenance of the existing constellations, (2) deployment of a new 1,000 satellite constellation, and (3) release of a swarm of 1,000 nano-satellites.

In the first case the currently existing Orbcomm, Iridium, and Globalstar constellations (see Table 5.2 and 5.3) shall be maintained through three complete replacement generations, with no disposal at their end of operation. This leads to a maximum of 108 Orbcomm satellites by 2006, 216 Iridium satellites by 2008, and 156 Globalstar satellites by 2014 (including spares). With no orbit maintenance after mission completion, the Orbcomm constellation will completely decay due to airdrag before 2060. Due to their higher operational altitude (Globalstar), or due to their smaller area-to-mass ratio (Iridium), the other two constellations remain in orbit throughout the simulation time span, with Iridium encountering two spacecraft losses due to collisions. Analysis shows, however, that the maintenance of the existing constellations does not cause a noticeable long-term risk increase.

In a second scenario a fictitious Walker-type constellation of 1,000 satellites, each with 700 kg mass and $10\,\mathrm{m}^2$ cross-section, is deployed in orbits of 800 km altitude and 82° inclination. Three generations are simulated, each with a 7-year operational lifespan, starting in 2010, and terminating their mission in 2031 with no active disposal of the total fleet of 3,000 payloads. A DELTA simulation over 100 years predicts that more than 900 of these payloads will be involved in a collision event, at an increasing rate as time progresses. Another 700 catastrophic collision are forecast with the background population. Out of the aggregate number of ~1,600 events more than 90% can be identified as feedback collisions, caused by fragments from previous collisions. By the year 2100 a total of 750,000 objects of $d > 10\,\mathrm{cm}$ are predicted, of which ~93.5% are collision fragments. This is an outlook of a possible run-away situation for a pessimistic deployment and mitigation scenario, involving the release of large amounts of additional mass ($\Delta m = 700,000\,\mathrm{kg} \approx 0.14 \times m_{2001}$) and cross-section ($\Delta A = 10,000\,\mathrm{m}^2 \approx 0.25 \times A_{2001}$) into today's most densely packed altitudes and inclinations.

The fact that on-orbit mass plays a dominant role in the long-term debris environment projections is further illustrated by another deployment scenario, where 1,000 nano-satellites of spherical shape with diameters of 10 cm and masses of 1 kg are released simultaneously in 2010 into a Sun-synchronous orbit of altitude 800 km and inclination 98°. As for the constellation scenario, a densely populated altitude and inclination band is chosen. In this case, however, the total emitted mass and collision cross-section are smaller by a factor $1/700$ and $1/1,273$, respectively. As a result, the number of catastrophic collisions over a 90-year prediction timespan through 2100 increases only by ~10% (+5 events) with respect to the business-as-usual scenario. Due to the relatively low kinetic energies associated with collisions caused by the nano-satellites, the effect of the few additional break-ups has little influence on the overall population growth in the critical size regime of $d > 10\,\mathrm{cm}$, with a relative increase of less than 10% (which is less than the standard deviation of the randomized traffic model itself).

In a final scenario the sensitivity of the DELTA forecasts shall be examined with respect to the assumed fatality threshold ψ_* (see Eq. 5.6) of the break-up model. Apart from the experimentally verified reference value of $\psi_* = 40\,\mathrm{kJ/kg}$, alternative thresholds of 30 kJ/kg and 60 kJ/kg shall be analyzed. As compared with the business-as-usual scenario, the number of cumulative collisions after 100 years increases by ~35% for the lower break-up threshold of 30 kJ/kg, and it decreases by ~25% for the higher threshold of 60 kJ/kg. These offsets are slightly beyond the $\pm 1\sigma$ confidence interval of the business-as-usual (BAU) reference scenario with $\psi_* = 40\,\mathrm{kJ/kg}$.

5.7 ALTERNATIVE DEBRIS ENVIRONMENT PROJECTION MODELS

So far, only ESA's DELTA 2.0 projection tool for the long-term space debris environment has been described. However, several other programs with similar functionalities have been developed by different research teams since the mid 1980s. The most prominent of these are (in chronological order of appearance) NASA's EVOLVE (Opiela and Krisko, 2001), RSA's SDPA (Nazarenko and Menshikov, 2001), ISTI's SDM (Rossi et al., 1995b; Rossi et al., 1995a), QinetiQ's IDES (Walker et al., 1999), and NASA's LEGEND software (Liou et al., 2004).

EVOLVE has been developed at NASA for almost 20 years, going back to 1986. It has been among the first and leading tools for long-term environment forecasts and stability analysis. EVOLVE describes a time-varying, one-dimensional, solely altitude-dependent space debris environment evolution in the LEO regime. Special versions of the software address environment stability near the GEO ring, and close to constellation deployment altitudes. NASA plans to replace EVOLVE by the new LEGEND software (LEO-to-GEO Environment Debris Model), which provides a full 3D, time-dependent LEO-to-GEO environment model, with efficient LEO and GEO orbit propagators, and with state-of-the-art traffic and break-up models.

The SDPA model (Space Debris Prediction and Analysis) was originally developed at the Center for Programme Studies (CPS) of the Russian Space Agency (RSA). SDPA is a semi-analytic, stochastic model for the near- and long-term prediction of the space debris environment in LEO and GEO, with altitude- and latitude-dependent, time-varying spatial density distributions. The model takes into account launch and release traffic, but no feedback collisions.

The IDES software (Integrated Debris Evolution Suite) has been developed at DERA, and later at QinetiQ, under UK funding. IDES uses its own, fast orbit propagator, and applies quasi-deterministic principles based on the MASTER concept to establish spatial density and transient velocity distributions, and to determine collision fluxes with target orbits. For the long-term projection of the space debris environment, ESA's DELTA program employs the concepts which were developed within IDES. Since both models were developed by the same core team, the functionalities of DELTA and IDES are very similar, and they can exchange initial population files and traffic models.

The SDM tool (Semi-Deterministic Model) has been developed by an Italian team at CNUCE, and later ISTI (Institute for Information Science and Technology), under ASI and ESA funding. SDM uses fast orbit propagation techniques (either object-wise or statistically). It determines spatial density distributions based on an input population which is specified by object-wise entries of mass, diameter, semimajor axis, eccentricity and inclination (Ω, ω, M are randomly generated). While collision probabilities consider spatial density peaks at high latitudes, SDM is mainly a 1D, altitude- and time-dependent model, which can also accept an initial MASTER population as input.

The models EVOLVE, SDM, IDES, and LEGEND all cover the combined modes of MASTER-2001 (for building up a present population from historic launch and release traffic), and of DELTA 2.0 (for propagating a present population under different assumptions of traffic and mitigation scenarios). Most of the tools mentioned (including MASTER and DELTA), in their latest releases, apply similar principles to define their traffic models, and they use similar techniques and mathematical algorithms to generate release events (explosions, collisions, and SRM slag ejections). An important improvement was the introduction of the EVOLVE 4.0 break-up model (Johnson et al., 2001), as described in Section 3.4. This state-of-the-art model, which was rigorously verified against on-orbit fragmentation events and ground test data, tends to produce lower imparted velocities and more fragments of critical masses than predecessor models. These changes lead to more catastrophic collisions in long-term forecasts, with earlier developments of critical object concentrations in some LEO regions. Hence, after introduction of the EVOLVE 4.0 break-up model, the need for space debris mitigation measures as a means to control the debris environment became even more evident.

Long-term forecasts by the previously mentioned tools are important inputs to decision processes towards international agreements on space debris mitigation. There are several international forums where such issues are discussed. Two leading, technically oriented bodies are the United Nations' Technical Sub-Committee of UNCOPUOS (Committee on the Peaceful Uses of Outer Space), who published a Technical Report on Space Debris (anon., 1999), and the Inter-Agency Space Debris Coordination Committee (IADC). At both forums international information exchange on long-term debris environment simulations regularly takes place to support the development of globally agreed space debris mitigation guidelines.

5.8 REFERENCES

Aksnes, K. (1976). Short-Period and Long-Period Perturbations of a Spherical Satellite Due to Direct Solar Radiation Pressure. *Celestial Mechanics*, vol.13:89–104.

anon. (1999). *Technical Report on Space Debris*. United Nations Publications, New York, USA.

Bendisch, J., Bunte, K., Sdunnus, H., Wegener, P., Walker, R., and Wiedemann, C.

(2002). Upgrade of ESA's MASTER Model. Technical Report (final report) ESA contract 14710/00/D/HK, ILR/TUBS, Braunschweig.

Cook, G. (1962). Luni-Solar Perturbations of the Orbit of an Earth Satellite. *Geophysical Journal of the Royal Astronomical Society*, vol.6, no.3:271–291.

Hernández, C., Pina, F., Sánchez, N., Sdunnus, H., and Klinkrad, H. (2001). The DISCOS Database and Web Interface. In *Proceedings of the Third European Conference on Space Debris*, ESA SP-473, pages 803–807.

Johnson, N., Krisko, P., Liou, J., and Anz-Meador, P. (2001). NASA's New Break-Up Model of EVOLVE 4.0. *Advances in Space Research*, vol.28, no.9:1377–1384.

Kessler, D. (1991). Collisional Cascading: The Limits of Population Growth in Low Earth Orbit. *Advances in Space Research*, vol.11, no.12:63–66.

King-Hele, D. (1987). *Satellite Orbits in an Atmosphere: Theory and Applications*. Blackie, Glasgow, UK.

Liou, J.-C., Hall, D., Krisko, P., and Opiela, J. (2004). LEGEND – A Three-Dimensional LEO-to-GEO Debris Evolutionary Model. *Advances in Space Research*, vol.34, no.5:981–986.

Nazarenko, A. and Menshikov, I. (2001). Engineering Model of the Space Debris Environment. In *Proceedings of the Third European Conference on Space Debris*, ESA SP-473, pages 293–298.

Opiela, J. and Krisko, P. (2001). Evaluation of Orbital Debris Mitigation Practices Using EVOLVE 4.1. *51st IAF Congress*, IAA-01-IAA.6.6.04.

Rossi, A., Cordelli, A., Pardini, C., Anselmo, L., Farinella, P., and Jehn, R. (1995a). Long-Term Evolution of Earth Orbiting Objects. *47th IAF Congress*, IAA-95-IAA.6.4.06.

Rossi, A., Cordelli, A., Pardini, C., Anselmo, L., and Parinella, P. (1995b). Modeling of the Space Debris Evolution: Two New Computer Models. *Advances in the Astronautical Sciences (Space Flight Mechanics)*, pages 1217–1231.

Springmann, P. and de Weck, O. (2004). Parametric Scaling Model for Non-Geosynchronous Communication Satellites. *Journal of Spacecraft and Rockets*, vol.41, no.3:472–477.

Walker, J. (1971). Some Circular Orbit Patterns Providing Continuous Whole Earth Coverage. *Journal of the British Interplanetary Society*, vol.24:269–384.

Walker, R., Stokes, P., Wilkinson, J., and Swinerd, G. (1999). Enhancement and Validation of the IDES Orbital Debris Environment Model. *Space Debris Journal*, vol.1, issue 1:1–20.

Walker, R., Swinerd, G., Wilkinson, J., and Martin, C. (2000). Long Term Space Debris Environment Prediction. Technical Report (final report) ESA contract 12808/98/D/IM(SC), DERA, Farnborough, UK.

6

Effects of Debris Mitigation Measures on Environment Projections

H. Klinkrad, C. Martin, R. Walker and R. Jehn

Chapter 5 outlined the consequences of continuing space activities in a business-as-usual fashion, with the conclusion that a timely change of operational practices is required in order to maintain a stable space debris environment, which will permit safe space operations in the long-term future. The main driver for future debris proliferation was found to be the on-orbit mass reservoir, predominantly of LEO objects beyond 100 kg, and potential large-size colliders (mission-related, or due to on-orbit explosions), capable of producing catastrophic break-ups. These break-ups may lead to enhanced feedback collisions and to the onset of a self-sustained runaway situation due to collisional cascading. In the current chapter debris mitigation measures will be identified that can effectively tackle the main causes of an uncontrolled population growth.

6.1 SPACE DEBRIS MITIGATION OPTIONS

All major tools for long-term projections of the space debris environment lead to similar recommendations for the most effective debris mitigation measures in view of technical, operational, and economic feasibility versus improvements in environmental stability. The following main categories of measures can be distinguished:

- reduction of mission-related objects
- prevention of on-orbit explosions (see Section 6.2)
- prevention of non-explosive release events
- collision avoidance between trackable objects (see Chapter 8)
- post-mission disposal of space systems (see Section 6.3)
- removal of passive on-orbit objects

Mission-related objects (MROs) contribute ~10.5% of the trackable catalog population, with ~60% of these related to launch systems, and ~40% related to payloads. MROs, also referred to as operational debris, are defined as objects released during nominal operations by both spacecraft and rocket bodies. This includes debris from launcher staging and payload separation (such as adapters, shrouds, and clamp bands), and objects released during spacecraft deployment and commissioning (such as parts of explosive bolts, solar array latches, and lens covers). Most of these objects are released with low relative velocities, and so they remain in close proximity to the operational orbit of the source object.

The release of MROs can be limited by system design. The best method of reducing the population of MROs is not to produce the objects in the first instance. This is reflected in most debris mitigation standards through recommendations to minimize, or to avoid the use of debris-generating systems (e.g. yo-yo de-spinners, nozzle closures of propulsion systems, protective lens covers, etc.). System design is also encouraged to ensure that released parts (e.g. antenna deployment mechanisms, protective covers, explosive bolts, ullage motors, heat shields, etc.) are retained with the primary object. This can be achieved through the use of lanyards, sliding or hinged covers, and special catchment devices. Moreover, materials and basic system technologies (e.g. tanks, surface materials, structures, etc.) should be selected such that they are resistant to environmental degradation (e.g. aging by radiation, atomic oxygen and micro-particle impact erosion, and thermal cycling). Finally, solid rocket motor designs and the composition of the solid propellant should be chosen such that SRM slag ejection is minimized.

For debris objects that are released in a planned manner (such as MROs), and which pass through LEO, NASA guidelines (anon., 1995) request that:

- the total area–time product should be no larger than 0.1 m^2 y, where the area–time product is defined as the sum over all cross-sections of operational debris of $d > 1$ mm, with each multiplied by its total orbital lifetime spent below 2,000 km altitude
- the total object–time product should be no larger than 100 y, where the object–time product is defined as the sum over all operational debris of $d > 1$ mm, with each multiplied by its total orbital lifetime spent below 2,000 km altitude

These values result in a probability of less than 10^{-6} that debris from a single release event will impact and potentially damage an average operational spacecraft. For operational debris of $d > 5$ cm, planned to be released within 300 km of GEO, NASA requests that their apogee heights be 300 km or more below GEO within 25 years.

The CNES standard (anon., 1999) and the draft European Space Debris Mitigation Standard (EDMS) (anon., 2001) request that a maximum of one debris object may be created from a single payload launch, or a maximum of two objects from a multiple payload launch. Separated upper stages and adapters are to be considered as debris objects within these constraints. No lifetime limit is imposed on MROs for such releases. US Government Mitigation Practices (anon., 1997) require an evaluation and justification for each planned LEO release event of a debris object of $d > 5$ mm, if it will remain on orbit for more than 25 years. Currently, no quantitative guidelines on mission-related objects and release events are specified by Russia (anon., 2000) and Japan (anon., 1996).

There is an important class of release event which represents ~27% of all known on-orbit fragmentations: deliberate explosions and collisions, which caused ~2,200 cataloged fragments by January 2002[6.1]. Most of these events are associated with classified payloads, sometimes in context with ASAT (anti-satellite) tests. The NASA Standard (anon., 1995) leaves room for such tests, as long as they comply with the area-lifetime and object-lifetime limits defined for MROs. One example is an ASAT test performed on September 12, 1985, when an aircraft-launched ASAT missile destroyed the Solwind P78-1 satellite on an ascending pass along the coast of California. Of 285 fragments cataloged, only 2 remained on orbit by January 2002.

Non-explosive release events include slag particle emissions from solid rocket motor (SRM) burns, the release of NaK droplets from RORSAT reactor core ejections, and the detachment of surface degradation products. Of these sources, only SRM slag may have effects on the long-term debris environment stability, because NaK droplets are a decaying, non-reproducing population, and surface degradation products are too small to affect environment stability. With maximum slag diameters of a few centimeters, their specific energy input during collisions will generally be below 40 kJ/kg, which is the adopted threshold for catastrophic breakups of a target. However, typical SRM slag particles can cause mission-critical damage on operational space systems. Fig. 6.2 illustrates the combined effects of passivation measures and SRM slag suppression, applied as of 2005, on the evolution of the debris population of $d > 1$ cm over 100 years.

There have been numerous proposals for the use of ground-based (Bekey, 1997) or space-borne (Bondarenko and Lyagushin, 1997) laser systems to remove small-size debris from orbit. These systems are intended to fire a laser beam at a debris object, ablate parts of the surface, and cause a plasma blow-off. This ejection of debris material acts like a propulsive jet, perturbing the orbit and reducing its lifetime. The energy that the laser can couple to a debris particle limits the size of objects which can be efficiently de-orbited. There are many other problems associated with the use of such laser systems, including requirements on pointing accuracy, sustained power levels, and sufficient contact times.

A clean way of removing large masses from LEO would be their return by means of systems designed to survive re-entry. So far, only the US Space Shuttle has this capability. However, due to safety considerations and cost aspects, just a small fraction of its 113 flights through 2003 have contributed to this mitigation option, with a negligible effect on the overall on-orbit mass.

[6.1]33% of the historic on-orbit fragmentations are of unknown origin; some of these are likely to be deliberate events as well

6.2 EXPLOSION PREVENTION BY END-OF-LIFE PASSIVATION

Explosions of spacecraft and upper stages in orbit have been the major source of debris in the past, with 175 such events up to January 2002, at a mean annual rate of ~4.5 (see Section 2.2). With ~30%, the majority of these are believed to be due to critical propulsion failures, and another 4% are expected to be due to battery explosions on upper stages and satellites. These failures, which caused more than 3,500 cataloged fragments, might have been avoided, if on-board passivation techniques had been employed. Such procedures are a standard on many of today's launchers, and so far there are no recorded explosions of successfully passivated orbital stages. End-of-life (EOL) passivation was first considered as a design requirement at the beginning of the 1980s. All upper stages and spacecraft which were launched before then, and which are still in orbit, continue to pose an explosion hazard (e.g. a Titan III-C transtage launched in 1967 exploded after 27 years in orbit). Of ~5,000 orbital rocket bodies launched since 1957 (including solid boost motors), 2,780 were still on orbit by mid 2004. Of these, ~50% were launched before 1990, and ~25% were launched before 1982. Hence, there is a significant number of latent explosion sources still on orbit.

Space debris mitigation standards recommend that all on-board reservoirs of stored energy (e.g. propellants, pressurants, batteries, momentum control gyros) should be permanently depleted when they are no longer required for any nominal or post-mission operations. The draft EDMS and CNES guidelines (anon., 2001; anon., 1999) define a 1-year period after EOL within which to complete passivation measures. The draft EDMS also requests a system availability of better than 0.9 to successfully perform the passivation procedure. The NASA Standard and the US Government Practices (anon., 1995; anon., 1997) propose that passivation be performed as soon as it does not pose an unacceptable risk to the payload. The following passivation aspects should be considered:

- idle burn or venting of residual propellants, with fuel valves left open
- venting of all pressure systems, and/or activation of pressure relief mechanisms to avoid explosions due to external heating
- discharge of batteries, shut down of charging lines, and maintenance of a permanent discharge state
- deactivation of range safety systems
- dissipation of energy contained in momentum control gyros

All of these passivation measures should be analyzed to verify that they do not introduce extra failure mechanisms. Ariane launchers, for instance, maintain a differential pressure between the oxidizer and fuel tank to avoid a reversal and rupture of their common bulkhead (this was identified as a possible cause for the Ariane 1 H-10 upper stage explosion in November 1986, see Table 2.2).

Fuel depletion or "idle" burns of orbital stages may be performed such that the resulting thrust leads to a braking maneuver, leaving the stage in a reduced-lifetime orbit. The residual lifetime should be less than 25 years to be compliant with international recommendations for space debris mitigation.

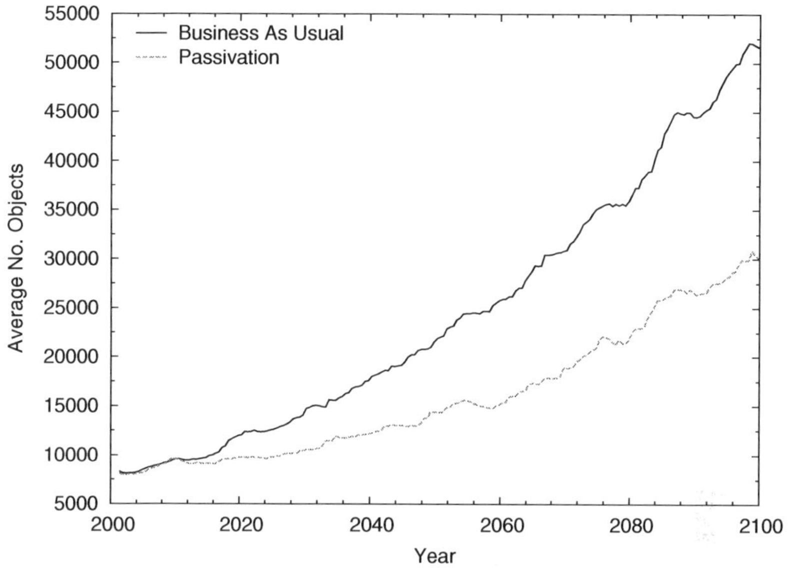

Fig. 6.1. Evolution of the LEO space object population of $d > 10$ cm over a 100-year prediction timespan, for a business-as-usual scenario (BAU), and for a passivation scenario (applied to all spacecraft and upper stages after 2005).

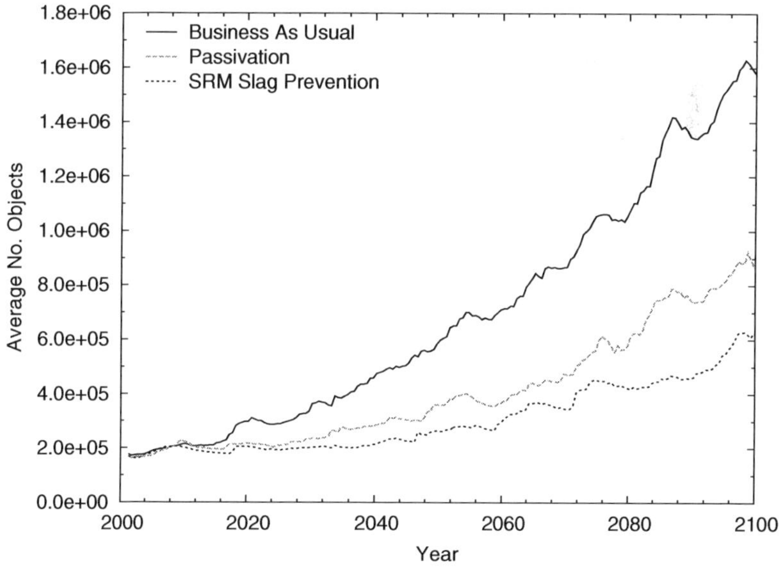

Fig. 6.2. Evolution of the LEO space object population of $d > 1$ cm over a 100-year prediction timespan, for a business-as-usual scenario (BAU), for a passivation scenario (applied to all spacecraft and upper stages after 2005), and for a SRM slag prevention scenario (no SRM slag after 2005).

A recent world-wide survey by the Inter-Agency Space Debris Coordination Committee (IADC) of commercial launcher manufacturers has indicated that depletion of residual propellants at EOL is already implemented by many launch systems, including H-II, Delta, Titan, Ariane, Long March 4, Soyuz, and Proton. This is reflected in several publications reporting on extensive debris mitigation research activities and implemented operational procedures, for instance for the Ariane series of launchers (Bonnal et al., 1997). The Ariane measures include controlled fuel and pressurant venting, enhanced propellant tank protection, and battery passivation. Debris mitigation techniques are also investigated for use on Zenit launch systems in the framework of the Sea Launch program (Kompanietz et al., 2000). They include several design solutions to prevent explosions of the second stage through propellant venting, and a proposed pneumatic system as a replacement for solid propellant retro-rockets used for de-orbiting. China has reported on the development of a residual propellant venting system for the Long March 4 upper stage (Zhang, 1997), confirming that this system has been integrated and employed in the nominal launcher design since 1999.

The passivation of batteries at EOL is not implemented yet with high priority. This is mainly due to spacecraft operators considering the risks associated with battery explosions to be low, as a result of inherent design features and associated control systems. Moreover, in the case of launch vehicles, the batteries are completely discharged at EOL, and have no means of recharging. However, the design and cost impact of including battery passivation is not considered significant in comparison with fuel venting or pressure relief.

Fig. 6.1 shows that global implementation of passivation measures by 2005 will lead to a reduction of critical-size objects with $d > 10$ cm by almost 50% within a 100-year prediction timespan. Fig. 6.2 supports this trend for smaller objects of $d > 1$ cm. These positive effects of explosion prevention by passivation go along with a reduction of catastrophic collisions from 52 for a BAU scenario to 29. While the absolute growth of potentially hazardous objects is reduced, a trend towards increasing growth rates is maintained, indicating that the suppression of on-orbit explosions by EOL passivation is a necessary but not sufficient condition to provide a stable space debris environment for future generations. Passivation measures can delay collisional cascading to a more distant future, but they cannot guarantee long-term stability of the environment.

6.3 POST-MISSION DISPOSAL FROM LOW-EARTH ORBITS

The most effective means of stabilizing the space debris environment is the removal of mass, particularly with $m > 100$ kg, from altitude shells and inclination bands with high spatial densities. Such regions presently exist in the LEO regime and near the GEO ring.

Orbits at LEO altitudes are mainly perturbed by the non-spherical geopotential, by luni-solar attraction, by solar radiation pressure, and by aerodynamic forces. Of these, only airdrag, acting opposite to the direction of motion, is a non-conservative, energy dissipating perturbation. Since the orbital energy $E = -\mu/(2\,a)$ is only a function of the semimajor axis, airdrag leads to a secular decrease in a. At the same time the orbital eccentricity decreases.

$$\frac{da}{dt} = -B\,\rho\,v\,a\,\frac{1 + e^2 + 2e\cos f}{1 - e^2} \tag{6.1}$$

$$\frac{de}{dt} = -B\,\rho\,v\,(e + \cos f) \tag{6.2}$$

In these equations, $B = c_D\,A/m$ is the ballistic parameter of the spacecraft, v is the orbital velocity, and ρ is the local air density. In the present context, a rotationally symmetric, mean exponential atmosphere shall be assumed, which only varies with geodetic altitude H (see also Annex B).

$$\rho = \rho_{pe}\,\exp\left(-\frac{H - H_{pe}}{H_{\rho,pe}}\right) \tag{6.3}$$

Here, the index "pe" denotes pericenter conditions, and $H_{\rho,pe}$ is the density scale height, with a mean value of 9 km at 120 km altitude, increasing with higher altitudes (e.g. 100 km at 800 km altitude). Hence, at $H = 120\,\text{km}$ a decay by $\Delta H = -9\,\text{km}$ leads to a density increase by $\exp(1) \approx 2.71828$. As a consequence of these steep density gradients, drag perturbations on eccentric orbits can be idealized as braking impulses $-\Delta v_{pe}$ during pericenter passes. Due to the laws of orbit mechanics, as manifested by Eq. A.21, such impulses first lower the apogee, while not affecting the perigee. Only after the eccentricity is reduced to zero, does the drag act on the entire orbit and cause it to spiral down into the atmosphere. The remaining orbital lifetime t_L can be estimated according to (King-Hele, 1987).

$$t_L = \frac{2}{3\,\rho_{pe}\,B}\sqrt{\frac{a}{\mu}}\,X(e,z) \tag{6.4}$$

$$X(e,z) = \frac{3e\,\exp(z)}{4\,I_0(z) + 8e\,I_1(z)}\left\{1 + \frac{7e}{6} + \frac{5e^2}{16} + \frac{1}{2z}\left(1 + \frac{11e}{12} + \frac{3}{4z} + \frac{3}{4z^2}\right) + O\left(e^3, \frac{1}{z^4}\right)\right\}$$

In these equations $I_k(z)$ are integer Bessel functions of order $k = 0$ and 1, and argument $z = ae/H_{\rho,pe}$ (Abramowitz and Stegun, 1984). In the case of near-circular orbits, $\exp(z) \rightarrow 1$, $I_0 \rightarrow 1$, and $I_1 \rightarrow 0$. Fig. 6.3 shows the resulting normalized lifetimes $\hat{t}_L = t_L \cdot A/m$ as a function of eccentricity and perigee altitude of the initial orbit. As an example, a spacecraft with $A/m = 0.004\,\text{m}^2/\text{kg}$ (or $m/A = 250\,\text{kg/m}^2$), on a near-circular orbit of 575 km altitude would have a normalized lifetime of $\hat{t}_L = 0.1\,\text{y}\,\text{m}^2/\text{kg}$ and a true lifetime of $t_L = \hat{t}_L \cdot m/A = 25$ y. Alternatively, Fig. 6.4 shows the initial apogee, and the perigee required to reach a remaining orbital lifetime of $t_L \approx 25$ y, as a function of the mass-to-area ratio m/A of a satellite. As for the previous scenario, with $H_{ap} = H_{pe} = 575\,\text{km}$

and $m/A = 250\,\mathrm{kg/m^2}$, the 25 y lifetime limit is matched exactly, and no de-orbit maneuver is necessary. For the same m/A ratio, an initially circular orbit at $\overline{H} = 1{,}000\,\mathrm{km}$ would require lowering the perigee to $H_{pe} \approx 400\,\mathrm{km}$ for a remaining lifetime of $t_L \leq 25\,\mathrm{y}$.

Table 6.1. Required Δv_{ap} and corresponding propellant mass fraction $\Delta m_p/m_o$, for a direct LEO de-orbit from an initially circular orbit of altitude H to an eccentric orbit with perigee altitude $H_{pe} \approx 80\,\mathrm{km}$, as a function of the initial altitude and propulsion system (I_{sp} = specific impulse; w_e = exhaust velocity).

H	Δv_{ap}	$\Delta m_p/m_o$ [%] for chemical propulsion systems			
[km]	[m/s]	N_2H_4	solid motors	N_2H_4/N_2O_4	LOX/LH
200	35.9	1.65	1.30	1.17	0.83
300	65.0	2.97	2.34	2.11	1.49
400	93.3	4.23	3.34	3.02	2.14
500	120.8	5.44	4.30	3.90	2.76
600	147.7	6.61	5.23	4.74	3.36
700	173.9	7.74	6.13	5.56	3.95
800	199.4	8.83	7.00	6.35	4.51
900	224.3	9.87	7.84	7.11	5.06
1000	248.6	10.88	8.65	7.85	5.60
1100	272.3	11.85	9.44	8.56	6.11
1200	295.4	12.79	10.20	9.26	6.61
1300	317.9	13.70	10.93	9.93	7.10
1400	339.9	14.57	11.64	10.58	7.57
1500	361.5	15.42	12.33	11.21	8.03
1600	382.5	16.24	13.00	11.82	8.48
1700	403.0	17.03	13.65	12.41	8.91
1800	423.0	17.80	14.27	12.99	9.34
1900	442.6	18.54	14.88	13.55	9.75
2000	461.8	19.26	15.47	14.09	10.15
I_{sp} [s]		220.0	280.0	310.0	440.0
w_e [m/s]		2,158	2,747	3,041	4,316

The purpose of an orbit lifetime reduction maneuver is to lower the altitude and increase the airdrag to the maximum possible extent for the given propulsion system, propellant efficiency, and fuel capacity.

The performance of any propulsion system, for a given maneuver size Δv, can be expressed in terms of the required propellant mass fraction Δm_p relative to the spacecraft initial mass m_o. This performance index is solely dependent on the energy content of the propellant used, and on the characteristic exhaust velocity w_e of its combustion products.

$$\frac{\Delta m_p}{m_o} = 1 - \exp\left(-\frac{\Delta v}{w_e}\right) \tag{6.5}$$

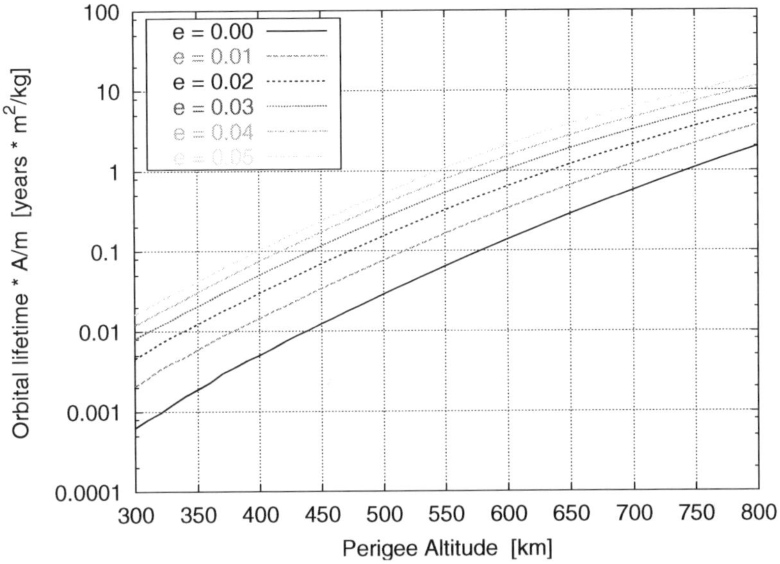

Fig. 6.3. Normalized orbital lifetime $\hat{t}_L = t_L \cdot (m/A)$ for a static, mean atmosphere, as a function of the initial perigee altitude and eccentricity.

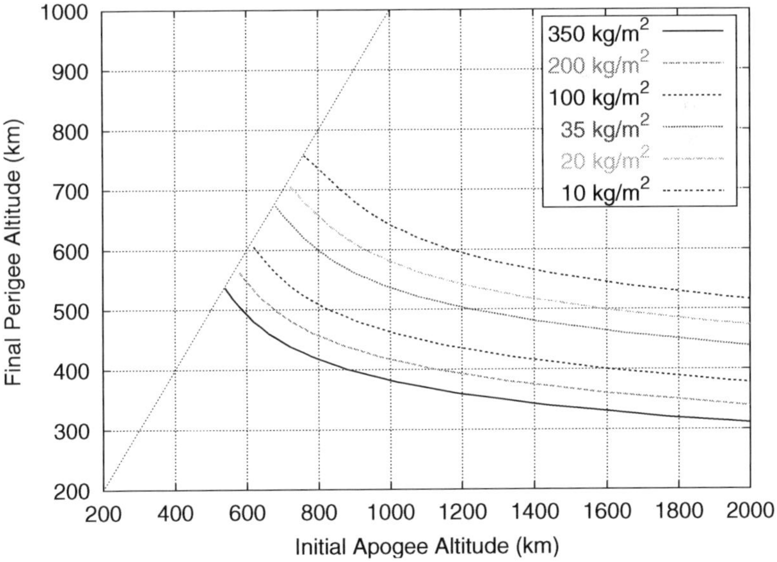

Fig. 6.4. Required perigee altitude for a remaining lifetime of $t_L \approx 25$ years, as a function of the initial apogee altitude and the area-to-mass ratio. A static, mean atmosphere model is assumed.

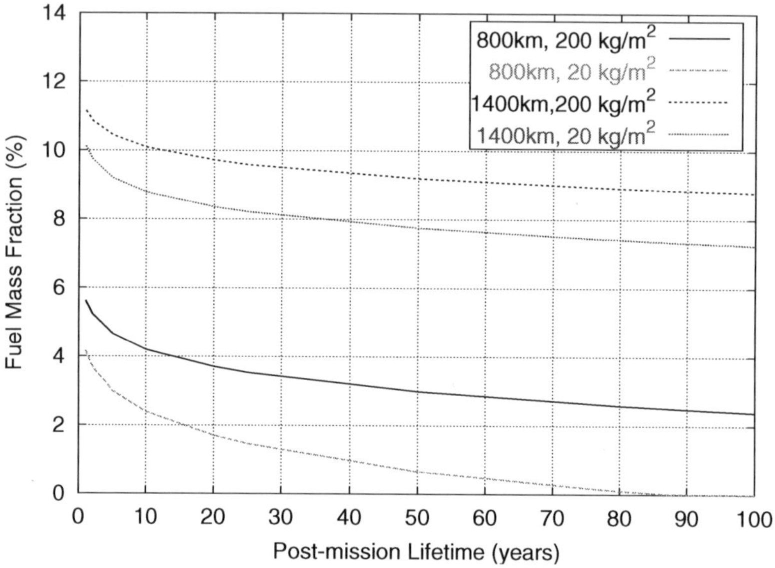

Fig. 6.5. De-orbit efficiency of a chemical motor ($I_{sp} = 280\,\text{s}$) as a function of initial circular orbit altitude, area-to-mass ratio, and remaining orbital lifetime.

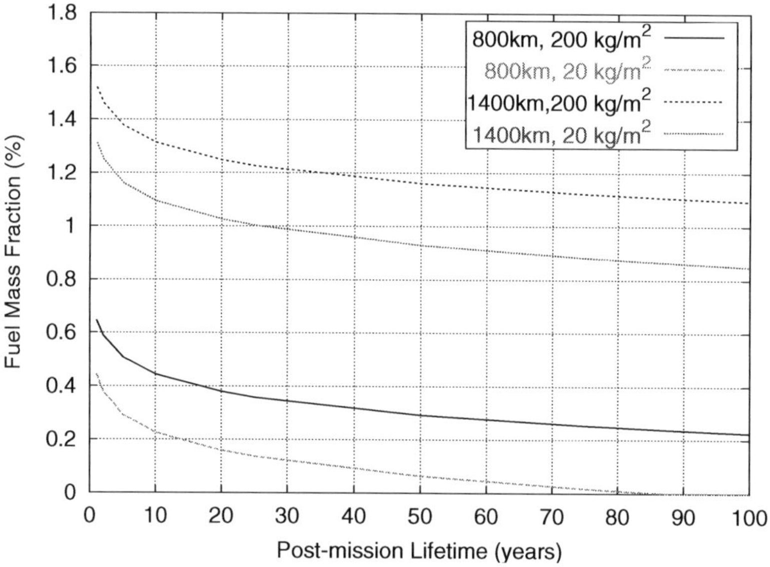

Fig. 6.6. De-orbit efficiency of ion propulsion ($I_{sp} = 3{,}400\,\text{s}$) as a function of initial circular orbit altitude, area-to-mass ratio, and remaining orbital lifetime.

Instead of w_e, the specific impulse I_{sp} is often used as an alternative performance index. It is defined as $I_{sp} = w_e/g$, in units of seconds (where $g = 9.81\,\mathrm{m/s^2}$ is the normalizing gravitational acceleration). Table 6.1 shows that for commonly used chemical motors I_{sp} ranges from 220 s for mono-propellant hydrazine systems, to 440 s for high-energy hydrogen/oxygen bi-propellant systems. Electrical propulsion systems with high-performance ion thrusters can reach specific impulses on the order of 3,000 s.

Chemical motors have the advantage of relatively high thrust levels of typically 1 N to 10 N for maneuver engines, and ≤ 0.05 N for attitude control thrusters. Their optimal maneuver strategy for orbit lifetime reductions is an apogee burn Δv_{ap}, opposite to the direction of motion $\underline{V}$.

$$\Delta v_{ap} = \sqrt{\frac{\mu}{r_{ap}}}\left(1 - \sqrt{\frac{2}{1+\beta}}\right) \tag{6.6}$$

$$\Delta \underline{v}_{ap} = -\Delta v_{ap}\,\underline{V} \tag{6.7}$$

with $\beta = r_{ap}/r_{pe}$. This corresponds to an inverse Hohmann transfer according to Eq. A.22. For a direct de-orbit, with immediate re-entry within half a revolution, the perigee should be lowered to $H_{pe} \leq 80$ km ($r_{pe} \leq 6{,}458$ km)[6.2]. Table 6.1 indicates the required Δv_{ap} as a function of the initial altitude of a circular orbit. The achievable thrust level F_T for chemical attitude and orbit control thrusters often requires extended burn durations Δt_T to generate the desired velocity increment according to

$$\Delta t_T = \frac{m_o - \Delta m_p/2}{F_T}\,\Delta v \tag{6.8}$$

For extended burn durations with constant thruster orientation the optimum impulsive, tangential maneuver is not achieved, and the maneuver efficiency η_T drops with the length of the burn arc Δf_T.

$$\eta_T = \frac{\sin(\Delta f_T/2)}{\Delta f_T/2} \tag{6.9}$$

where Δf_T is an arc of true anomaly, given in radians, and centered on the apogee location. For instance, a burn arc length of $\Delta f_T = 40°, 80°$ and $120°$ would cause a maneuver efficiency reduction and a wasted fuel fraction of 2.0%, 7.9% and 17.3%, respectively. Hence, if necessary, a Δv should be split into a sufficient number of individual maneuvers with acceptable burn arc lengths.

A good example of a multiple-maneuver strategy is the end-of-life disposal of the French SPOT-1 satellite (Alby, 2003), in compliance with the 25-year lifetime limitation rule requested by the CNES space debris mitigation standard (anon., 1999). After 17 years of successful operations SPOT-1 was de-orbited in November 2003 from its initial near-circular orbit of altitude 822 km and inclination 98°.

[6.2]note that $\sim 99.99\%$ of the Earth's atmospheric density is concentrated below $H \approx 82$ km

In a first step, as a safety measure, the orbit of the 2-ton, 3-axis-stabilized spacecraft was lowered to 808 km by 2 burns. Between November 19 and 27, 2003, the pericenter of this orbit was reduced by 8 successive apogee maneuvers to $H_{pe} = 619\,\text{km}$ (with an almost unchanged final apocenter at $H_{ap} = 804\,\text{km}$). Each burn of the mono-propellant hydrazine thrusters had a duration of $\Delta t_T = 1,000\,\text{s}$ (equivalent to an arc length of $\Delta f_T \approx 60°$), consuming 4.4 to 5.4 kg of fuel per burn, and lowering the perigee by about 24 km in each step. In a final depletion burn of 2,400 s, on November 28, 2003, the remaining 8.4 kg of propellant was used to further reduce the pericenter to $H_{pe} = 580\,\text{km}$, before SPOT-1 was passivated and shut down. The achieved disposal orbit has an estimated lifetime of 16.5 years, which is well within the recommended value $t_L \leq 25$ years.

For the SPOT-1 de-orbit a propellant mass fraction of ~2.5% of the initial spacecraft mass was used to obtain a remaining orbital lifetime of ~16.5 years. These numbers (for $I_{sp} \approx 220\,\text{s}$, and $m/A \approx 100\,\text{kg/m}^2$) match well with the estimates shown in Fig. 6.5 for an initial orbit at $\overline{H} = 800\,\text{km}$ (for $I_{sp} \approx 280\,\text{s}$). Fig. 6.5 also illustrates one reason for selecting a 25-year lifetime limit: there is a progressive increase in the required fuel mass fraction when going to shorter orbital lifetimes.

Disposal strategies with perigee altitude reduction may be assisted by drag augmentation, once the target orbit is reached. Inflatable structures like ballutes are, however, constrained to use at lower altitudes. Also, their necessarily large surface area can pose an additional collision risk to operational systems. Before deploying drag augmentation devices it should be demonstrated that their orbital area–time product is lower than for a naturally decaying object. Correspondingly, the accelerated decay of the spacecraft or upper stage must outweigh the introduced short-term increase in collision risk.

An alternative to chemical thrusters are electric propulsion systems. They use inert propellants which are ionized and accelerated by spacecraft-provided electrical energy, instead of chemical energy released in a combustion process. Depending on the physical principles of pre-processing and accelerating the propellant, it is possible to distinguish between resisto-jets, arc-jets, pulsed electro-thermal thrusters, magneto-plasma-dynamic thrusters, and ion thrusters. The most efficient ones, with the widest applications in spaceflight, are ion thrusters with specific impulses of typically 3,000 s ($w_e = 29,400\,\text{m/s}$), and efficiencies of up to 90% (as compared with <45% for chemical thrusters). Preferred propellants are noble gases (e.g. Xe), or metals with a low evaporation temperature and ionization energy (e.g. Hg or Cs). Common thrust levels range from a few milli-newtons to several newtons, with typical values in the range of $10\,\text{mN} \leq F_T \leq 200\,\text{mN}$. Due to these small accelerations, ion thrusters tend to be operated in a continuous mode. Hence, when used for de-orbiting, their thrust is constantly aligned with the orbit velocity vector to have a maximum effect on the orbit energy, and thus on the orbit altitude reduction. In contrast with impulsive maneuvers from chemical thrusters which use Hohmann transfers, continuous de-orbit burns with ion thrusters lead to an inward motion on a spiral trajectory. The Δv required to go

Table 6.2. History of space tether deployments (1967–1996).

Mission	Date	Orbit	Length	Comments
Gemini 11	1967	LEO	30 m	spin stabilised (0.15 rpm)
Gemini 12	1967	LEO	30 m	gravity gradient; stable swing
H-9M-69	1980	sub-orbital	500 m	partial deployment
S-520-2	1981	sub-orbital	500 m	partial deployment
Charge-1	1983	sub-orbital	500 m	full deployment
Charge-2	1984	sub-orbital	500 m	full deployment
Echo-7	1988	sub-orbital	—	magnetic field alignment
Oedipus-A	1989	sub-orbital	958 m	spin stabilised (0.7 rpm)
Charge-2B	1992	sub-orbital	500 m	full deployment
TSS-1	1992	LEO	20 km	conductive, deployed/retrieved
SEDS-1	1993	LEO	20 km	downward deployment/cut
PMG	1993	LEO	500 m	conductive, upward deployment
SEDS-2	1994	LEO	20 km	downward deployment
Oedipus-C	1995	sub-orbital	1 km	spin stabilized (0.7 rpm)
TSS-1R	1996	LEO	19.6 km	conductive, severed after 4 days
TiPS	1996	LEO	4 km	at 1022 km/63° since 1996

from an initial circular orbit of radius r_1 to a circular destination orbit of radius r_2 is simply the difference between the final and initial orbit velocity.

$$\Delta v = \sqrt{\frac{\mu}{r_2}} - \sqrt{\frac{\mu}{r_1}} \tag{6.10}$$

The corresponding propellant mass fraction can be determined from Eq. 6.5, using $w_e = 29,400\,\mathrm{m/s}$ for ion thrusters. This ejection velocity is an order of magnitude larger than that for chemical systems, and hence the required propellant mass fraction $\Delta m_p/m_o$ is reduced by the same amount. This is also illustrated in Fig. 6.6. The required time span Δt_T of continuous thruster operation is indicated by Eq. 6.8 for the previously defined thrust levels of $10\,\mathrm{mN} \leq F_T \leq 200\,\mathrm{mN}$.

While electrical propulsion systems are very efficient, their thrust levels are insufficient for controlled, direct de-orbits. Moreover, they require the availability of on-board resources throughout the maneuver duration of several days or weeks. Resisto-jets and arc-jets, however, have the advantage of accepting almost any propellant, including waste products.

Another elegant way of removing objects from orbit is by means of electrodynamic tethers, which consist of a long conductive wire with a small end-mass, possibly including an electron emitter or collector (see Table 6.2 for past tether deployments). When moving such a conductive wire through the Earth's magnetic field, a voltage is induced which is proportional to the magnetic induction B_m (see Eq. 6.11–6.13), the velocity v_t, and the tether length l_t. This voltage causes electrons to move through the insulated wire, leaving it through the end-mass, and closing the circuit through the surrounding space plasma. Such a current-carrying wire causes a Lorentz force which is proportional to the vector product $\underline{v}_t \times \underline{B}_m$, with non-vanishing contributions only from the horizontal component $B_{m,\theta}$ of the magnetic induction (see Eq. 6.12). The direction of this force is opposite to the spacecraft motion, thereby acting like an energy-dissipating drag force, and reducing the orbit altitude. Electrodynamic tethers provide optimal performance at near-equatorial inclinations where the Earth's magnetic field is perpendicular to the spacecraft motion. For more inclined orbits, such systems will take longer to de-orbit. This is because the contributing meridian component $B_{m,\theta}$ of the magnetic induction B_m at a geocentric distance r is a function of the magnetic co-latitude θ_m (angular distance from the magnetic north pole), with a maximum for $\theta_m = 90°$ at the magnetic equator, and close to the geographic equator.

$$B_m(r,\theta_m) = \frac{\mu_m M_m}{4\pi r^3}\sqrt{1 + 3\cos^2\theta_m} = \sqrt{B_{m,\theta}^2 + B_{m,r}^2} \tag{6.11}$$

$$B_{m,\theta}(r,\theta_m) = \frac{\mu_m M_m}{4\pi r^3}\sin\theta_m \quad \text{with} \quad \underline{B}_{m,\theta} = B_{m,\theta}\,\underline{N} \tag{6.12}$$

$$B_{m,r}(r,\theta_m) = -\frac{\mu_m M_m}{4\pi r^3}\,2\cos\theta_m \quad \text{with} \quad \underline{B}_{m,r} = B_{m,r}\,\underline{U} \tag{6.13}$$

Here, $\mu_m = 1.256 \times 10^{-6}\,\mathrm{V\,s\,A^{-1}\,m^{-1}}$ is the magnetic field constant, $M_m \approx 8 \times 10^{22}\,\mathrm{A\,m^2}$ is the Earth's magnetic dipole moment, $B_{m,\theta}$ and $B_{m,r}$ are the northward meridian and radial components of $\underline{B}_m$, and B_m is its magnitude.

Besides a latitude dependence, Eq. 6.11 to 6.13 also show a change in B_m due to geocentric distance. This decrease of B_m and of the resulting Lorentz force with $1/r^3$ is much more moderate than a corresponding decrease of airdrag forces according to an exponential density law (see Eq. 6.3). As a consequence, conductive tethers can be used for de-orbit operations from much higher altitudes. The required time Δt_t for a tether-induced descent from an initial circular orbit of radius r_1 to a circular destination orbit of radius r_2 can be approximated by an expression derived in (Hoyt and Forward, 1999).

$$\Delta t_t = \frac{m_o\,R_t}{12\,l_t^2\,a_e^6\,\cos^2\zeta_t\,\langle\cos^2 i_m\rangle\,B_{m,\theta=90°}^2}\left(r_1^6 - r_2^6\right) \tag{6.14}$$

In this equation l_t is the tether length, m_o is the overall system mass (spacecraft plus tether), R_t is the tether resistance, ζ_t is the angle between the tether axis and the local vertical, $B_{m,\theta=90°} = B_m(\theta_m = 90°) \approx 31\,\mu\mathrm{T}$ is the magnitude of the magnetic induction at the geomagnetic equator, and $\langle\cos^2 i_m\rangle$ is the averaged geomagnetic inclination of the spacecraft orbit according to

$$\begin{aligned}\left\langle\cos^2 i_m\right\rangle = \frac{1}{16}\,(&6 + 2\cos 2i + 3\cos(2i - 2\vartheta_m) \\ &+ 2\cos(2\vartheta_m) + 3\cos(2i + 2\vartheta_m))\end{aligned} \tag{6.15}$$

with the orbital inclination i, and $\vartheta_m \approx 11.5°$ the tilt of the geomagnetic dipole axis with respect to the Earth's rotation axis. Eq. 6.14 does not take into account

Table 6.3. De-orbiting capabilities of conductive tethers. Assumptions: 10 m^2 airdrag cross-section; conductive aluminum tether with 2.5% of the spacecraft mass.

Constellation	$\overline{H}_o$ [km]	i [°]	Natural re-entry	Tether-induced re-entry
Orbcomm	825	45	150 years	11 days
Iridium	780	86	100 years	7.5 months
GlobalStar	1,414	52	9,000 years	37 days
Skybridge	1,469	53	11,000 years	46 days
Teledesic	1,375	85	7,000 years	17 months

airdrag effects, and it only applies under the previously mentioned simplifying model assumptions. The vertical offset ζ_t of the tether axis is normally established by a balance between electrodynamic, gravity gradient, and centrifugal forces. In order to avoid dynamic instabilities in its attitude, while maintaining high electrodynamic drag, the tether properties should be chosen such that ζ_t is maintained in the vicinity of 35° (assuming a straight tether).

Table 6.3 highlights the effectiveness of conductive tethers for the end-of-life de-orbiting of constellation satellites, assuming aluminum tethers with 2.5% of the spacecraft mass (Hoyt and Forward, 1999). As compared to natural re-entry, tether-induced decays are several orders of magnitude faster, and all of those listed comfortably comply with a 25-year lifetime rule.

The thinness of tethers (with typical diameters close to 1 mm) makes them extremely susceptible to being severed by debris or micrometeoroid impactors, which need only be a fraction of the tether diameter to cause destruction. The resulting tether fragments may lead to an increased overall threat to operational spacecraft as compared with the intact tether. To reduce the risk of severing, single strand tethers can be replaced by more robust ribbon-like or interconnected multiple-thread tethers.

An alternative form of space tether exploits the principle of momentum exchange to de-orbit spacecraft or upper stages. This procedure implies an orbit rendezvous, with the active spacecraft subsequently maneuvering to a higher altitude, while roping down the passive object by means of an attached, non-conductive tether. Assuming that the tethered system behaves like a gravity-gradient stabilized dumbbell, the orbital motion of both objects is now coupled to the common center of mass, with the lower, passive object moving too slow, and the upper, active satellite moving too fast for its geocentric distance. When the tether is cut, the lower object attains an orbit with a lower perigee and the active satellite enters into an orbit with a higher apogee to conserve the angular momentum and the orbital energy. With a tether of sufficient length this principle can be used to de-orbit decommissioned satellites, spent orbital stages, or mission-related objects, while at the same time the active "remover satellite" is propelled to a higher orbit, from which it can continue its mission. Such sequential de-orbits can be performed most efficiently within densely populated inclination bands.

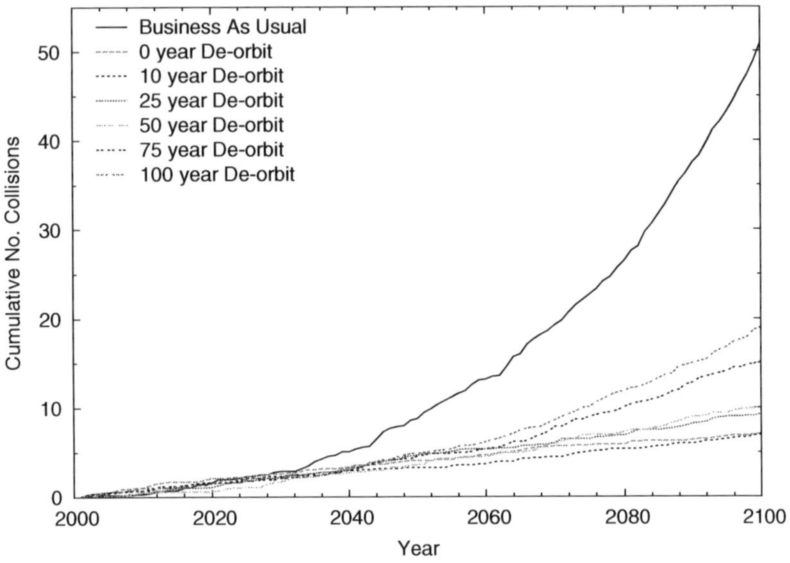

Fig. 6.7. Evolution of catastrophic collision rates in LEO over a 100-year prediction timespan, for a business-as-usual scenario (BAU), and for a combined passivation and delayed de-orbit scenario with different remaining lifetimes.

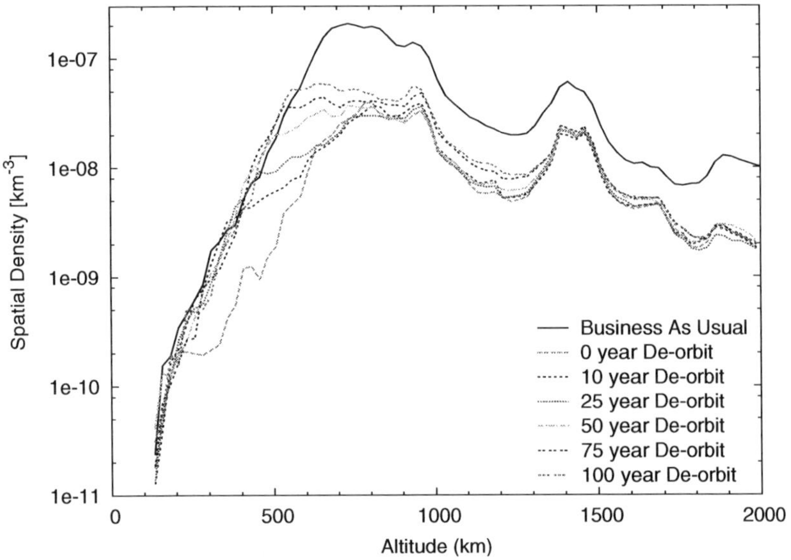

Fig. 6.8. Spatial density altitude profiles of objects with $d > 10$ cm after a 100-year prediction timespan, for a business-as-usual scenario (BAU), and for a combined passivation and delayed de-orbit scenario with different remaining lifetimes.

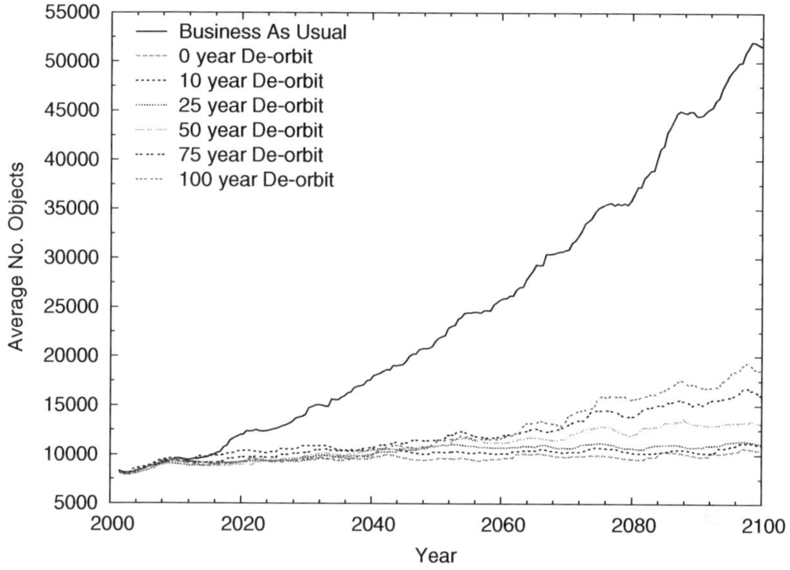

Fig. 6.9. Evolution of object counts of $d > 10\,\text{cm}$ in LEO over a 100-year prediction timespan, for a business-as-usual scenario (BAU), and for a combined passivation and delayed de-orbit scenario with different remaining lifetimes.

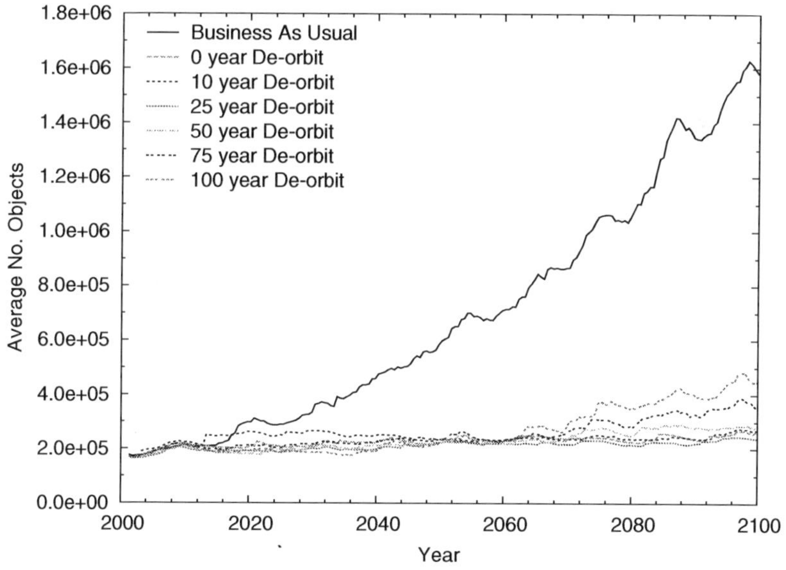

Fig. 6.10. Evolution of object counts of $d > 1\,\text{cm}$ in LEO over a 100-year prediction timespan, for a business-as-usual scenario (BAU), and for a combined passivation and delayed de-orbit scenario with different remaining lifetimes.

The natural decay from a low-Earth orbit under the influence of airdrag according to Eq. 6.1 can be accelerated by the deployment of drag augmentation devices at the end of the mission. A ballute is one example of such a system. It consists of a large inflatable balloon, mostly deployed by an orbital stage. The research of Russian experts suggests that a ballute of 10 m diameter, deployed by Soyuz stages at mission completion, can be capable of reducing their remaining orbital lifetimes by up to a factor of 5, with only a small mass penalty on the launcher system. However, as with tether systems, drag augmentation devices must prove their value by creating a smaller area–time product than for the naturally decaying system, resulting in a decay rate which is fast enough to outweigh the increased collision risk during the remaining orbital lifetime (Klinkrad et al., 2003).

In the case of highly eccentric orbits (HEO) and GEO transfer orbits (GTO), luni-solar perturbations play an important role. Periodic changes of the perigee altitude are particularly driven by solar gravitational attraction, as a function of the orbit parameters e, i, and Ω, with larger amplitudes for larger values of e and i. The minima of the periodic changes in H_{pe} are obtained for $\Omega = 0°$ and $180°$, while the maxima are reached for $\Omega = 90°$ and $270°$. The corresponding variations are $\Delta H_{pe} \approx \pm 60$ km for $i = 7.0°$ (Kourou launch), and $\Delta H_{pe} \approx \pm 100$ km for $i = 28.5°$ (KSC launch). The density scale height at 300 km is such that the air density increases by a factor of 2 for a descent by $\Delta H_{pe} = -40$ km, and it reduces to about 1/2 for an ascent by $\Delta H_{pe} = +40$ km. The orbital lifetime is thus strongly affected by the air density at perigee pass, and by the resulting airdrag. A proper selection of the launch epoch can reduce the lifetime of a HEO or GTO trajectory significantly. In the case of an injection stage, a directed depletion burn after payload deployment can be used to further lower the perigee altitude and reduce the orbit lifetime.

Independent of the method employed to achieve a required post-mission lifetime limit (propulsive de-orbit by chemical or ion thrusters, tether-induced de-orbit, or natural decay), the benefit on the long-term evolution of the LEO regime can be analyzed by a comparison against a business-as-usual (BAU) scenario as described in Section 5.5.

For post-mission lifetime limits of 100, 75, 50, 25, 10, and 0 years (immediate de-orbit) Fig. 6.7 shows that the predicted cumulative number of catastrophic collisions over a timespan of 100 years is significantly reduced, from 52 for business-as-usual to 18 for a 100-year rule, and to 7 for a lifetime limit of $t_L \leq 10$ years. An even stronger reduction is noted for the number of environment-critical objects with $d > 10$ cm according to Fig. 6.9, and for the corresponding number of mission-critical objects with $d > 1$ cm according to Fig. 6.10. In both of these diagrams a post-mission lifetime limitation of $t_L \leq 25$ years leads to a stable space debris environment in LEO, with no net growth of the population, despite continuing space operations at present deployment rates. This result confirms the recommendation for a 25-year lifetime rule, which is supported by most existing national and international space debris mitigation standards.

Fig. 6.8 shows a snapshot of the LEO altitude profile of spatial densities for $d > 10$ cm after a forecast of 100 years. All lifetime limitations successfully reduce the concentrations above 700 km by at least a factor of 5. Below altitudes of 700 km the different concentration profiles diverge according to the adopted lifetime rule, and according to the corresponding pericenter altitudes of disposal orbits.

At high LEO altitudes the de-orbiting fuel requirements for the most common chemical propulsion systems may become prohibitive (see Fig. 6.5). Hence, several debris mitigation standards allow re-orbiting to super-LEO altitudes at the end of a mission. The lower altitude limit of the target disposal area is prescribed as $H \geq 1,700$ km by JAXA (formerly NASDA), as $H \geq 2,000$ km by CNES, EDMS, and US Government, and as $H \geq 2,500$ km by NASA. Long-term simulations show that when applying such measures on all LEO missions with end-of-life altitudes of $H_{pe} \geq 1,400$ km, then spatial densities in the disposal region will grow by a factor of $\sim$2 over a 100-year timespan, with a small likelihood of collision. While this appears to be a tolerable risk level, all possible alternatives for de-orbit should be exploited to avoid the creation of such a LEO graveyard region with almost infinite orbital lifetimes, and with possible long-term effects on the safety of HEO and GTO missions.

6.4 POST-MISSION DISPOSAL FROM GEO-SYNCHRONOUS ORBITS

A particularly valuable region for space operations is the geostationary orbit (GEO), which is ideally circular ($e \approx 0$), within the equatorial plane ($i \approx 0$), at an altitude of $H \approx 35,786$ km, and of an orbit period $T \approx 23^h56'04.09''$, corresponding to one sidereal day.

Orbits in GEO and its vicinity are mainly perturbed by luni-solar gravitational attraction and by the geopotenial (solar radiation pressure plays a secondary role). Luni-solar forces tend to increase the orbital inclination, which would grow up to 46° and return to 0° within 170 years if the Earth was spherical. However, due to Earth oblateness and its corresponding zonal harmonic J_2, the mass concentration near the equator limits the luni-solar effect on the inclination amplitude to $i_{max} \approx 15°$, and at the same time reduces the period of the cyclic motion of the inclination vector to $T_i \approx 53$ years. Fig. 6.11 shows positions $\underline{i}(i, \Omega)$ of the inclination vectors $\underline{i} = (\sin i \cos \Omega, \sin i \sin \Omega) \approx (i \cos \Omega, i \sin \Omega)$ of cataloged near-GEO objects in 2002. Since launches into GEO only began in the mid 1960s, with the first failures occurring shortly thereafter, the initial generation of abandoned spacecraft will not complete their 53 years inclination cycle before the late 2010s. Given the historic evolution of the GEO population, as illustrated in Fig. 2.4 and Fig. 2.5, most GEO objects were abandoned in the past 27 years, and they are hence still in a phase of inclination build-up. "Banana"-shaped clusters in the inclination vector distribution $\underline{i}(i, \Omega)$, which were detected in optical observations of near-GEO objects down to 15 cm sizes (see Fig. 3.35), are a strong indication of undetected GEO break-up events.

With varying constellations of the perturbing bodies, the initial inclination drift rate cyclicly changes from 0.747°/y to 0.943°/y within 9.3 years. Depending on the permitted inclination excursion Δi, the inclination maintenance maneu-

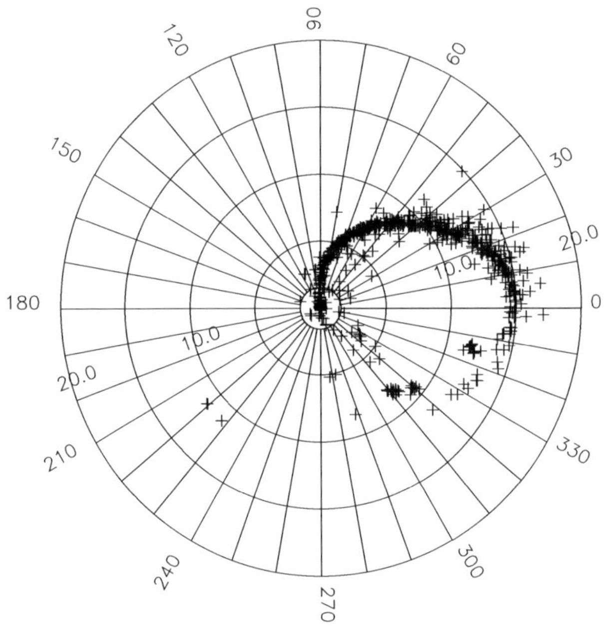

Fig. 6.11. Status in 2002 of the orbit inclination i (radial coordinate) and right ascension of the ascending node Ω (angular coordinate) of objects which were abandoned near the GEO ring. The orbit momentum vectors precess about a pole at $i \approx 7.5°$ and $\Omega = 0°$ with a period of $T_i \approx 53$ years.

vers (north–south station-keeping) of operational GEO satellites require an out-of-plane correction by

$$\frac{\Delta v_{\mathrm{N/S}}}{[\mathrm{m/s}]} = 6,148 \ \sin \Delta i \approx 107.3 \ \frac{\Delta i}{[°]} \tag{6.16}$$

which must be performed in intervals of $\Delta t/[\mathrm{d}] \approx 861.4\,\Delta i/[°]$ (Agrawal, 1986).

Along-track perturbations of a GEO orbit are dominated by resonant tesseral harmonics of the geopotential, especially by the terms $J_{2,2}$ and $J_{3,1}$ (where $J_{n,m} = \sqrt{C_{n,m}^2 + S_{n,m}^2}$). Fig. 6.14 shows the resulting energy levels in an equatorial cross-section of the GEO orbit, as a function of the geographic longitude λ. A satellite positioned at a longitude with no further orbit maintenance would move towards the closest energy minimum at an increasing drift rate according to Fig. 6.13. Since no energy is dissipated, the spacecraft would regain the same (potential) energy level on the opposite side of the potential well, and then reverse the oscillation process about the stable longitude at $\lambda_{L1} \approx 73.9°$E or at $\lambda_{L2} \approx 103.7°$W. In order to avoid a longitude drift, which takes a spacecraft outside its conventional longitude slot of $\lambda_o \pm 0.1°$, east–west station-keeping maneuvers must be applied with magnitudes depending on the nominal GEO position.

$$\frac{\Delta v_{\mathrm{E/W}}}{[\mathrm{m\,s^{-1}\,y^{-1}}]} = 1.74\ \sin\left(2\lambda_o - 2\lambda_L\right) \tag{6.17}$$

where $\lambda_L = \lambda_{L1}$ or λ_{L2}. The minimum time interval between east–west maneuvers is $\Delta t \approx 31$ days (Agrawal, 1986).

The GEO orbit region has a steadily increasing population of catalog objects, which grows at a mean rate of ~30 new entries per year, and which had a total count of ~850 objects by 2003 (see Fig. 2.4 and 2.5), with ~330 thereof controlled spacecraft. Inside, and close to the GEO ring approximately 29% of the on-orbit mass and 41% of the on-orbit cross-section are concentrated. So far, there have been two recorded fragmentations in GEO, both of which were explosions (the Ekran-2 satellite in 1978, and a Titan III-C Transtage in 1992). Optical observations, however, indicate that there is a large un-cataloged GEO population, probably caused by undetected fragmentation events (see Section 3.10).

To protect the precious GEO environment, consensus was reached between commercial space operators, national space agencies, and international entities to maneuver GEO spacecraft at the end of their operational lifetime to a disposal orbit which is near-circular, and which has a pericenter clearance of the GEO ring according to the so-called "IADC formula".

$$\frac{\Delta H_{geo}}{[\mathrm{km}]} \geq 235 + 1,000\ c_R\ \frac{A/m}{[\mathrm{m^2/kg}]} \quad . \tag{6.18}$$

In this equation $c_R \in [0,2]$ is the reflectivity coefficient (0.0 for total absorption, and 2.0 for total reflection), and A/m is the spacecraft area-to-mass ratio at the end of life. The first term on the right-hand side is composed of a minimum GEO clearance distance of 200 km (including a corridor for re-positioning maneuvers of operational spacecraft), and a 35 km amplitude to account for periodic luni-solar perturbations on the orbit eccentricity. The second term considers periodic effects due to solar radiation pressure. Long-term stability studies suggest that the initial eccentricity of the disposal orbit should be sufficiently small (e.g. $e < 0.005$) to avoid an amplitude build-up which could lead to a violation of the IADC threshold.

The effectiveness of GEO re-orbiting can be demonstrated in a long-term, 100-year simulation. Fig. 6.12 compares the GEO environment projections for a business-as-usual scenario (BAU, see Section 5.5), for applied passivation measures at end-of-life, and for re-orbiting according to the IADC guideline (plus passivation) after 2010. Clearly, only the end-of-life re-orbiting strategy guarantees a stable GEO population of critical diameters $d > 10$ cm, with tolerable collision flux levels beyond this century. At the same time the spatial densities in the disposal region grow. The related collision risk, however, is considerably lower than that for the BAU scenario in GEO, because dispersions in the disposal orbits lead to much lower concentration levels. No collisions were detected during 10 Monte Carlo runs with DELTA 2.0 over a 100-year prediction timespan.

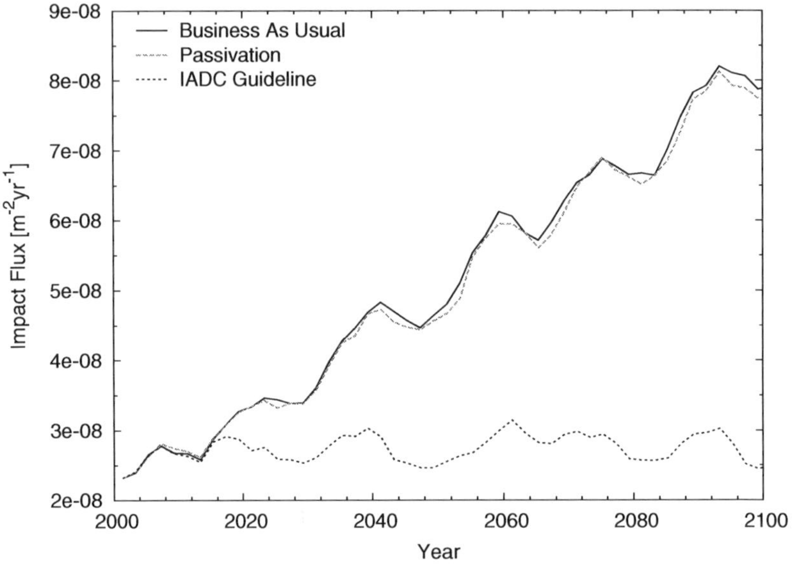

Fig. 6.12. Evolution of object flux of $d > 10$ cm in GEO over a 100-year prediction time-span, for a business-as-usual scenario (BAU), for passivation at end-of-life, and for combined passivation and re-orbiting according to the IADC guideline.

A typical GEO end-of-life re-orbit requires a Δv of about 11 m/s. This corresponds to a propellant mass fraction of about 2.3% of the entire north–south and east–west station keeping propellant budget for a 10-year mission. The decision by a GEO spacecraft operator as to when a re-orbit maneuver should be initiated should be based on the estimated fuel remaining, and on the spacecraft health status. The assessment of the remaining propellant is normally done by "book-keeping" of performed and subsequently calibrated orbit maintenance maneuvers. Due to variations in the thruster performances, the translation of the well-known cumulated Δv into a derived fuel mass is a difficult task, with considerable uncertainties associated. For the end-of-life disposal of GEO spacecraft conservative assumptions for the fuel estimate should be used (e.g. the lower 1σ value of the confidence interval), and the transfer maneuver should be split into at least 3 burns. The first Δv should take the spacecraft to an intermediate orbit, with a perigee near GEO and an apogee above GEO (e.g. at $\Delta H_{geo}/2$), the second burn at this apocenter should attain a new apogee above ΔH_{geo}, and the third maneuver should circularize the disposal orbit in a depletion burn. Such a strategy reduces the risk that the spacecraft gets stranded on an elliptic orbit, with a perigee close to GEO, if the propellant is exhausted during the first part of a Hohmann transfer, according to Eq. A.21 and A.22.

For spacecraft that were controlled by ESA's Space Operations Center (ESOC), Table 6.4 lists their end-of-life GEO disposal details. Except for Olympus-1 and

GEOS-2 (which marginally violates the IADC rule), all re-orbits were compliant with the later established IADC guideline. Olympus-1 had to be de-orbited below GEO, as a result of spacecraft contingencies. For the timespan 1997 to 2003 Tables 6.5 and 6.6 show the world-wide statistics of GEO end-of-life disposal, discriminated according to operator and disposal practice (Jehn et al., 2004; Samson, 1999).

Of 103 spacecraft retirements, 34 resulted in a disposal orbit which was compliant with the IADC rule of Eq. 6.18, 35 resulted in a drift orbit with insufficient GEO clearance, and 34 spacecraft were left in libration orbits about the stable longitudes $\lambda_{L1} \approx 73.9°$E (22 disposals), or $\lambda_{L2} \approx 103.7°$W (10 disposals), or about both λ_{L1} and λ_{L2} (2 disposals). The latter case can be reproduced by placing a spacecraft near the unstable equilibrium longitude of 162.5°E (see Fig. 6.14), from where it will drift towards the closest stable equilibrium longitude. Since the initial (potential) energy at the starting point was close to the global maximum, the spacecraft momentum is large enough to cross the second unstable equilibrium longitude at 12.0°W, and drop into the second potential well, before the satellite reaches its initial potential energy level again and reverses its motion to complete the cycle. The longitude oscillation can be thought of as traveling in a roller-coaster on a track which is described by the energy function in Fig. 6.14. The corresponding velocities as a function of the starting condition are provided in Fig. 6.13. The drift rates in libration orbits can reach up to $\dot{\lambda} = \pm 0.45°/\mathrm{d}$ during passes of the stable equilibrium longitudes. The libration periods can be determined from

$$T = \frac{4}{k} \int_0^{\pi/2} \frac{1}{\sqrt{1 - \sin^2(\Delta\lambda_L)\sin^2\psi}}\, d\psi = \frac{4}{k} K(\Delta\lambda_L) \tag{6.19}$$

$$k = 6 \frac{a_e}{a_{geo}} \sqrt{\frac{\mu}{a_{geo}^3}} \sqrt{J_{2,2}} \approx 8.91 \times 10^{-8} \left[\mathrm{s}^{-1}\right] \tag{6.20}$$

where $a_{geo} = 42{,}164$ km, $\Delta\lambda_L$ is the libration amplitude, $J_{2,2} = 1.812 \times 10^{-6}$, and $K(\Delta\lambda_L)$ is the elliptic integral of the first kind, which takes values between $K(0.0) = \pi/2$ and $K(\pi/2) = \infty$. The corresponding libration periods range from 816 days for small amplitudes to infinity for amplitudes of $\pm\pi/2$. Due to the simplifying assumption of a $J_{2,2}$ perturbing potential, Eq. 6.19 cannot be applied to spacecraft which start their libration near the eastern unstable longitude of 162.5°E, and which subsequently cross the potential wells of both stable longitudes, mainly under the influence of the harmonics $J_{3,1}$, $J_{3,2}$ and $J_{3,3}$ (see Table C.1 and Fig. C.1). The resulting shape of the potential in Fig. 6.14 can be very well correlated with the geoid in Fig. C.2.

For most spacecraft which were abandoned in libration orbits it may be assumed that they were operated until total depletion of their propellant resources, with no intention of post-mission disposal. Such undesirable practices can often be correlated with certain spacecraft operators, leading to higher concentrations of defunct GEO spacecraft, particularly at eastern longitudes.

Table 6.4. GEO end-of-life disposal history of ESA controlled spacecraft (1984 to 2002).

ESA spacecraft	Launch date	Disposal date	Disposal ΔH_{geo}
GEOS-2	1978-Jul-14	1984-Jan-25	+260 km
OTS-2	1978-May-11	1991-Jan-03	+318 km
MET-2 (F2)	1981-Jun-19	1991-Dec-03	+334 km
OLYMPUS-1	1989-Jul-12	1993-Aug-30	−213 km
ECS-2	1984-Aug-04	1993-Nov-10	+335 km
MET-4 (MOP-1)	1989-Mar-06	1995-Nov-08	+833 km
MET-3 (P2)	1988-Jun-15	1995-Nov-23	+941 km
MARECS-A	1981-Dec-20	1996-Aug-22	+1,536 km
ECS-1	1983-Jun-16	1996-Dec-16	+377 km
ECS-4	1987-Sep-16	2002-Dec-01	+414 km

Table 6.5. GEO end-of-life re-orbiting practices during the years 1997 to 2003. The retirement summaries are sorted by disposal type and end-of-life year.

	1997	1998	1999	2000	2001	2002	2003	Total
left around L_1	1	7	5	3	5	1	–	22
left around L_2	2	3	1	1	1	1	1	10
left around L_1/L_2	–	–	–	2	–	–	–	2
moved to drift orbit (too low perigee)	6	6	4	2	6	5	6	35
moved to drift orbit (IADC compliant)	6	6	5	3	2	4	8	34
Total	15	22	15	11	14	11	15	103

Table 6.6. GEO end-of-life re-orbiting practices during the years 1997 to 2003. The retirement summaries are sorted by disposal type and spacecraft operator.

	China	Intelsat	Japan	Russia	USA	Others	Total
left in libration orbit L_1, L_2, L_1/L_2	3	–	–	26	4	1	34
moved to drift orbit (too low perigee)	1	3	2	2	12	15	35
moved to drift orbit (IADC compliant)	–	8	6	3	5	12	34
Total	4	11	8	31	21	28	103

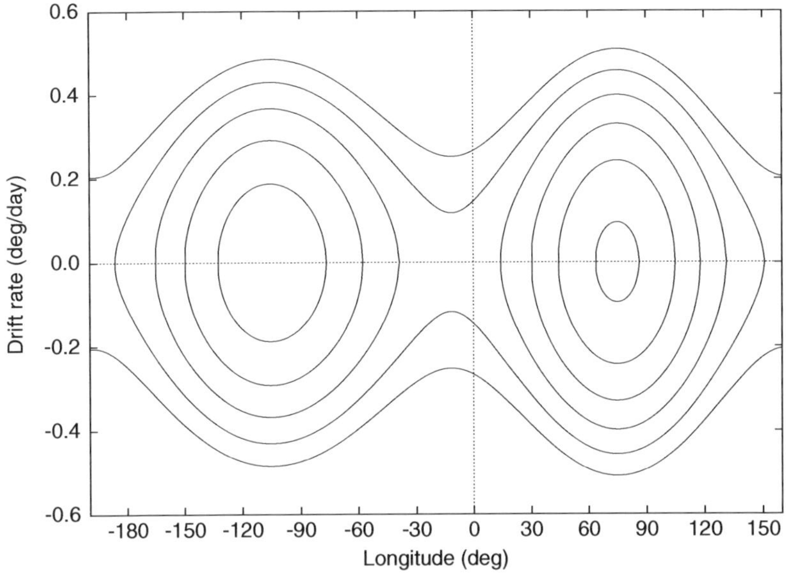

Fig. 6.13. Phase diagram of the longitude drift rate $\dot{\lambda}$ of a near-geosynchronous orbit, as a function of its instantaneous longitude λ, and of its instantaneous orbital energy. Contours relate to energy levels in Fig. 6.14.

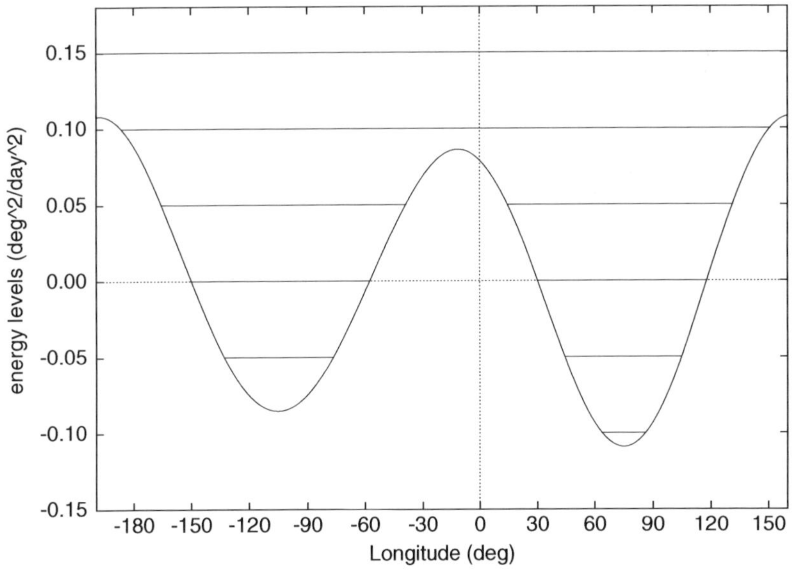

Fig. 6.14. Physical explanation of drift orbits, and of libration motion around $\lambda_{L1} \approx 73.9°$E, around $\lambda_{L2} \approx 103.7°$W, or around λ_{L1} and λ_{L2}. Energy levels relate to contours in Fig. 6.13.

6.5 DE-ORBIT OF GTO UPPER STAGES

Upper stages of rockets which inject GEO spacecraft into a GEO transfer orbit can pose a small, but long-term risk to both the LEO and GEO environments. Natural orbit perturbations due to the Sun and Moon, in concert with airdrag during the perigee pass, can reduce the orbit lifetime, but often not sufficiently to comply with a post-mission 25-year lifetime limit. This is a problem, for instance, for Ariane 5 GTOs, which have perigees typically at $H_{pe} \approx 600$ km. One way to improve the situation is to use residual propellant from the statistical reserve of the Ariane 5 EPS upper stage to perform an idle burn until depletion. Assuming that the propellant reserve is $\Delta m_p = 100$ kg, then for an empty EPS mass of $m = 3,000$ kg a maneuver size of $\Delta v \approx 100$ m/s can be achieved (hence, for EPS the relation $\Delta m_p/[\text{kg}] \approx \Delta v/[\text{m/s}]$ holds).

The best way of applying a Δv to reach the maximum reduction in GTO perigee is by an apogee burn (true anomaly $f = 180°$), opposite to the direction of motion (thrust direction $\phi = 180°$). Unfortunately, the time to reach apogee is on the order of 5.3 hours, which exceeds the availability of the EPS orbit and attitude control system. Hence, the maneuver needs to be performed during the ascending arc. One can show from the Gauss perturbation equations (Eq. A.33 and A.34) that for an impulsive maneuver of size $\Delta v \ll v$, applied at an orbit position f, the change in perigee altitude can be determined from

$$\frac{\Delta r_{pe}}{r_{pe}} = \frac{2}{e}\left(1 - \frac{r_{pe}}{r}\right) \cos\phi \, \frac{\Delta v}{v} - \sin f \, \frac{r}{r_{pe}} \, \sin\phi \, \frac{\Delta v}{v} \tag{6.21}$$

where $r_{pe} = r_{pe,o}$, $r = r(f)$, and $v = v(f)$ are defined in Eq. A.7, A.6, and A.9. The maneuver thrust direction ϕ is defined as the in-plane rotation angle around the inverse orbit normal vector $-\underline{W}$, measured from the tangential in-flight direction $\underline{T}$. The corresponding angle $\tilde{\phi}$ relative to the transversal direction $\underline{V}$ is given by $\tilde{\phi} = \phi + \gamma$, with the flight path angle γ from Eq. A.20.

With methods of standard calculus Eq. 6.21 can be used to solve for the optimum maneuver orientation ϕ_{opt}, which causes a maximum perigee reduction for a given orbit position $0 < f \leq \pi$ on the outbound orbit arc.

$$\phi_{opt} = \pi - \arctan\left[\frac{e}{2}\frac{r}{r_{pe}}\left(1 - \frac{r_{pe}}{r}\right)^{-1} \sin f\right] \tag{6.22}$$

For the parameters of an Ariane 5 GTO ($H_{pe} = 600$ km, $a = 24,571$ km, $e = 0.716$) the resulting function $\phi_{opt}(f)$ is shown in Fig. 6.15 (note that $\phi_{opt}(f)$ is only a weak function of H_{pe}). Applying the optimum maneuver directions in Eq. 6.21 for different initial perigee altitudes, and for maneuver sizes of $\Delta v = 25$ m/s to 100 m/s, results in perigee altitude reductions as listed in Table 6.7, as a function of the maneuver position on the orbit and time delay Δt_{pe} since the perigee pass. For maximum efficiency, a perigee-reducing Δv should be applied at the maximum possible delay Δt_{pe} that complies with system availability.

Table 6.7. Achievable reduction in GTO post-maneuver perigee altitudes H_{pe} for different initial perigee altitudes $H_{pe,o}$, and different Δv levels, as a function of the true anomaly f of the firing position, assuming an optimum thrust azimuth ϕ_{opt} according to Fig. 6.15.

	Post-Δv perigee reduction H_{pe} for an initial perigee at $H_{pe,o}$ = 200 km									
	f [deg]	15	30	45	60	75	90	120	150	180
	Δt_{pe} [min]	2.83	5.83	9.21	13.25	18.40	25.45	52.18	127.66	315.53
	ϕ_{opt} [deg]	98.53	106.29	112.77	117.77	121.37	123.75	126.31	132.52	180.00
$\Delta v = 25$ m/s	H_{pe} [km]	195.70	190.83	184.82	177.09	166.91	153.30	109.19	28.83	-38.03
$\Delta v = 50$ m/s	H_{pe} [km]	191.40	181.65	169.64	154.17	133.82	106.60	18.38	–142.34	–276.05
$\Delta v = 75$ m/s	H_{pe} [km]	187.10	172.48	154.46	131.26	100.73	59.90	–72.43	–313.51	–514.08
$\Delta v = 100$ m/s	H_{pe} [km]	182.80	163.31	139.28	108.34	67.64	13.20	–163.24	–484.68	–752.11
	Post-Δv perigee reduction H_{pe} for an initial perigee at $H_{pe,o}$ = 400 km									
	f [deg]	15	30	45	60	75	90	120	150	180
	Δt_{pe} [min]	2.97	6.11	9.65	13.87	19.25	26.59	54.22	130.90	317.48
	ϕ_{opt} [deg]	98.57	106.36	112.87	117.90	121.54	123.97	126.69	133.23	180.00
$\Delta v = 25$ m/s	H_{pe} [km]	395.49	390.39	384.09	376.00	365.36	351.17	305.48	223.61	156.89
$\Delta v = 50$ m/s	H_{pe} [km]	390.98	380.77	368.19	351.99	330.72	302.33	210.95	47.22	–86.21
$\Delta v = 75$ m/s	H_{pe} [km]	386.48	371.16	352.28	327.99	296.08	253.50	116.43	–129.17	–329.32
$\Delta v = 100$ m/s	H_{pe} [km]	381.97	361.55	336.37	303.99	261.44	204.67	21.90	–305.57	–572.43
	Post-Δv perigee reduction H_{pe} for an initial perigee at $H_{pe,o}$ = 600 km									
	f [deg]	15	30	45	60	75	90	120	150	180
	Δt_{pe} [min]	3.10	6.39	10.09	14.50	20.11	27.75	56.27	134.10	319.43
	ϕ_{opt} [deg]	98.60	106.43	112.97	118.04	121.71	124.18	127.06	133.92	180.00
$\Delta v = 25$ m/s	H_{pe} [km]	595.28	589.94	583.35	574.89	563.78	549.00	501.73	418.39	351.82
$\Delta v = 50$ m/s	H_{pe} [km]	590.56	579.88	566.71	549.78	527.57	498.00	403.46	236.79	103.64
$\Delta v = 75$ m/s	H_{pe} [km]	585.85	569.81	550.06	524.66	491.35	447.00	305.19	55.18	–144.54
$\Delta v = 100$ m/s	H_{pe} [km]	581.13	559.75	533.41	499.55	455.14	396.01	206.92	–126.43	–392.72

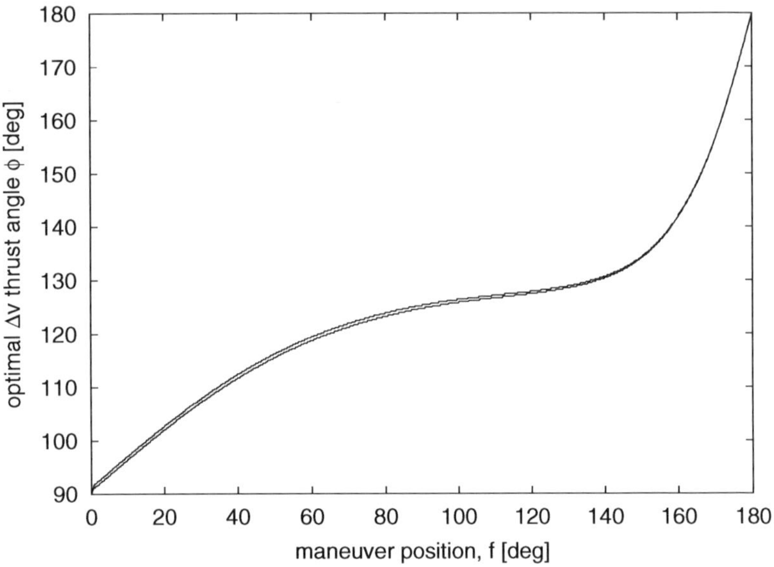

Fig. 6.15. Optimum orientation of the in-plane Δv maneuver thrust angle ϕ for a maximum reduction of the perigee altitude of a GEO transfer orbit. ϕ_{opt} is given as a function of the true anomaly f of the firing position, for an initial perigee of $H_{pe,o} = 200\,\text{km}$ (lower curve) and 600 km (upper curve).

The orbital lifetime t_L of GEO transfer orbits is a complex function of luni-solar, airdrag, and solar radiation pressure perturbations. For perigees of $H_{pe} = 200\,\text{km}$ it is on the order of a few years, for $H_{pe} = 400\,\text{km}$ it increases to several hundred years, and for $H_{pe} = 600\,\text{km}$ it reaches a few thousand years. The air density at perigee plays a decisive role in this context. Since it increases exponentially with decreasing altitude, a perigee lowering by $H_\rho = 33.5\,\text{km}$ relative to $H_{pe} = 200\,\text{km}$, and a reduction by $H_\rho = 68.1\,\text{km}$ from $H_{pe} = 600\,\text{km}$, causes a lifetime reduction by a factor of $1/\exp(1) \approx 1/3$. For the Ariane 5 EPS stage, in most instances a perigee reduction to $H_{pe} < 300\,\text{km}$ will lead to conformance with the 25-year lifetime limit stipulated by several mitigation guidelines.

6.6 DEFINITION OF PROTECTED REGIONS

In the previous Sections 6.3 and 6.4 a reduction of spatial density in the LEO and GEO regions was identified as key to the long-term stability control of the space debris environment. Another region of increasing importance are the near-circular, semi-synchronous orbits of approximately 12-h periods, with inclinations of 55° to 65° (hereafter denoted as "12-h circular MEO"), which are used, or will be used by the navigation satellite constellations GPS, GLONASS, and Galileo (see Table 5.2).

Due to their importance for spacecraft operations, the LEO environment, the 12-h circular MEO vicinity, and the GEO ring are identified as protected regions in most space debris mitigation standards. Details on the definition of these regions are provided by the following standards and guidelines (Klinkrad et al., 2003):

- NASA Safety Standard (anon., 1995)
- US Government Orbital Debris Mitigation Standard Practices (anon., 1997)
- NASDA Space Debris Mitigation Standard (anon., 1996)
- CNES Safety Requirements – Space Debris (anon., 1999)
- RASA Branch Standard – Space Technology Items (anon., 2000)
- European Space Debris Mitigation & Safety Standard (EDMS) (anon., 2001)

A self-standing synthesis, based on the listed debris mitigation standards, was produced by the Inter-Agency Space Debris Coordination Committee (IADC), which comprises the Italian Space Agency (ASI), the British National Space Centre (BNSC), the French Centre National d'Etudes Spatiales (CNES), the China National Space Administration (CNSA), the Deutsches Zentrum für Luft- und Raumfahrt (DLR), the European Space Agency (ESA), the Indian Space Research Organization (ISRO), the Japan Aerospace Exploration Agency (JAXA), the National Aeronautics and Space Administration (NASA) of the USA, the National Space Agency of the Ukraine (NSAU), and the Russian Federal Space Agency (RSA ROSCOSMOS). The IADC Debris Mitigation Guidelines (anon., 2002) are a consensus document, based on commonly accepted core recommendations of existing standards, including (1) reduction of mission-related objects, (2) suppression of on-orbit explosions, and (3) removal of spacecraft and upper stages from protected orbital regions after their completed mission. A major output of the IADC discussions is the rule on GEO end-of-life re-orbiting according to Eq. 6.18. This recommendation was also adopted by the International Telecommunications Union (ITU). The IADC Guidelines were drafted by IADC member organizations and presented to the Scientific and Technical Subcommittee (S&T SC) of UNCOPUOS (United Nations Committee on the Peaceful Uses of Outer Space) in 2003.

Within the USA, several entities operate satellites, for instance NASA, DoD (Department of Defense), and NOAA (National Oceanic and Atmospheric Administration). All of them have accepted a common set of US Government Orbital Debris Mitigation Standard Practices (while specific standards of these entities may be more comprehensive and restrictive).

Mitigation practices can be enforced through the issue of launch licenses. In the USA, this is done by the Federal Aviation Administration (FAA) of the Department of Transportation (DOT), by the National Oceanic and Atmospheric Administration (NOAA) of the Department of Commerce (DOC), and by the Federal Communications Commission (FCC). In Europe, national space legislation governs the regime of launch licenses for the UK, via the British National Space Centre (BNSC), for the Netherlands, and for Sweden.

All space debris mitigation guidelines and standards share a core set of mitigation principles, and all of them define certain orbital regions as protected areas, from which objects must be removed at their end of life, and within which

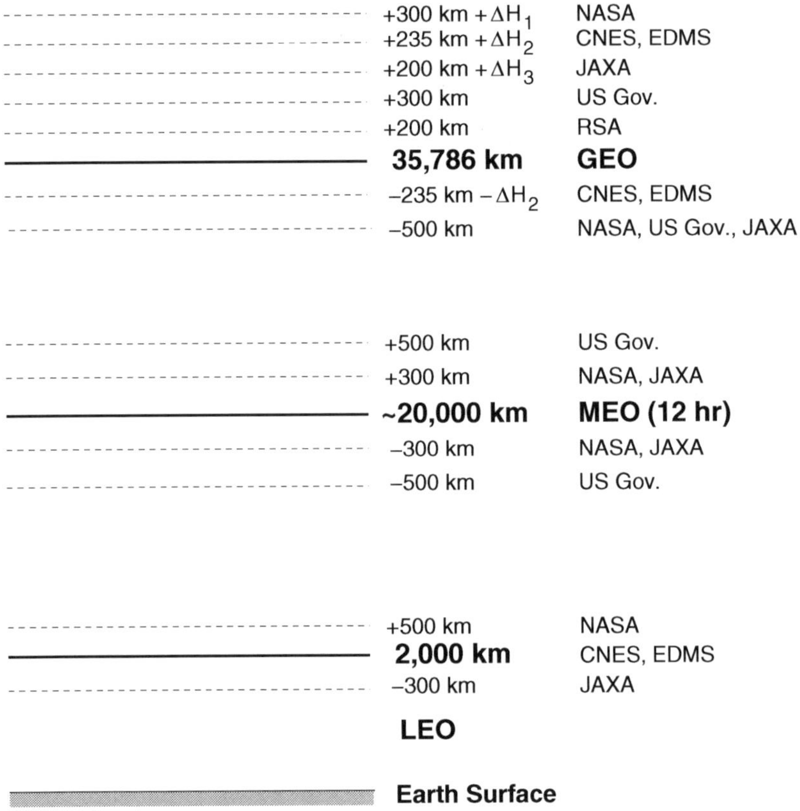

Fig. 6.16. Schematic view of protected orbital regions in LEO, around GEO, and around semi-synchronous, near-circular MEO altitudes according to different space debris mitigation standards (see Table 6.8 and 6.9 for the definition of thresholds and recommended practices).

their aggregate post-mission resident time (contiguous or periodic) must be limited. Fig. 6.16 illustrates the lower and upper altitude boundaries of the LEO, 12-h circular MEO, and GEO protected regions, as defined by the different standards. Table 6.8 (for LEO and 12-h circular MEO), and Table 6.9 (for GEO and GTO) provide more detail on the applicable numbers, and on different post-mission disposal procedures.

Most guidelines identify the LEO regime, which contains almost 70% of all cataloged objects, as a valuable operational resource which requires special protection. The applicable LEO altitude range is defined from the Earth's surface, described by the Earth ellipsoid (see Annex A.3), with an upper limit set by a spherical shell of altitude $H = 2{,}000\,\mathrm{km}$ with respect to the equatorial radius

Table 6.8. Summary of space debris mitigation practices recommended by different existing guidelines for the end-of-life LEO and MEO orbit disposal (see Table 6.9 for a definition of $\Delta H_{2,geo}$).

	LEO	12 h circular MEO
NASA	• retrieval within 10 years • re-entry within 25 years • re-orbit so that [$H_{pe} > 2,500$ km and $H_{ap} < 35,288$ km]	• re-orbit so that either [$H_{pe} > 2,500$ km and $H_{ap} < 19,900$ km] or [$H_{pe} > 20,500$ km and $H_{ap} < 35,288$ km]
US Gov.	• direct retrieval after EOL • re-entry within 25 years • re-orbit so that [$H_{pe} > 2,000$ km and $H_{ap} < 19,700$ km]	• re-orbit so that either [$H_{pe} > 2,000$ km and $H_{ap} < 19,700$ km] or [$H_{pe} > 20,700$ km and $H_{ap} < 35,300$ km]
CNES and EDMS	• direct re-entry • re-entry within 25 years • re-orbit so that [$H_{pe} > 2,000$ km and $H_{ap} < H_{geo} - \Delta H_{2,geo}$]	• direct re-entry • no requirement for a disposal maneuver
JAXA (NASDA)	• direct retrieval • re-entry within 25 years • re-orbit so that $H_{pe} > 1,700$ km ($> 2,500$ km preferred)	• re-entry within 25 years • re-orbit so that either [$H_{pe} > 1,700$ km and $H_{ap} < 19,900$ km] or [$H_{pe} > 20,500$ km and $H_{ap} < 35,288$ km]
RSA	only a general recommendation for EOL lifetime limitation	

$a_e = 6,378.135$ km (with $H = 2,500$ km for the NASA standard, and 1,700 km for the JAXA standard). To regulate post-mission disposal, alternative options are outlined for direct retrieval (CNES, EDMS, JAXA, US Government), delayed retrieval within 10 years (NASA), direct or delayed re-entry within 25 years, or re-orbit to super-LEO altitudes. In the latter case, the disposal perigees must be outside the LEO region, and the disposal apogees must remain below a protected 12-h circular MEO altitude shell (US Government), or below a clearance zone around the GEO ring (NASA, CNES, EDMS).

A protected 12-h MEO altitude shell is defined by NASA ($19,900\,\text{km} \leq H \leq 20,500\,\text{km}$), by the US Government ($19,700\,\text{km} \leq H \leq 20,700\,\text{km}$), and by JAXA ($19,900\,\text{km} \leq H \leq 20,500\,\text{km}$). For the post-mission disposal of spacecraft and upper stages which service these orbits, a lifetime reduction according to the 25-year rule, a de-orbit, or a re-orbit is requested. In the latter two cases the disposal orbits may not interfere with other protected regions (LEO or GEO), nor with the MEO region itself. The CNES, EDMS, and RSA standards do not identify the environment of MEO constellation orbits as a protected region. This is mainly due

Table 6.9. Summary of space debris mitigation practices recommended by different existing guidelines for the end-of-life GTO and GEO orbit disposal.

	GTO (with $H_{pe} < 2,000$ km)	GEO
NASA	• re-orbit so that [$H_{pe} > 2,500$ km and $H_{ap} < 35,288$ km]	• re-orbit above GEO so that $\Delta H_{1,geo}$ [km] > $300 + 1,000\,A/m$
US Gov.	• re-orbit so that either [$H_{pe} > 2,000$ km and $H_{ap} < 19,700$ km] or [$H_{pe} > 20,700$ km and $H_{ap} < 35,300$ km] or [$H_{pe} > 36,100$ km]	• re-orbit ~300 km above GEO so that $H_{pe} > 36,100$ km
CNES and EDMS	• direct de-orbit after EOL • re-entry within 25 years • re-orbit so that [$H_{pe} > 2,000$ km and $H_{ap} < H_{geo} - \Delta H_{2,geo}$]	• re-orbit above GEO so that $\Delta H_{2,geo}$ [km] > $235 + 1,000\,c_R\,A/m$
JAXA (NASDA)	• for orbital stages obtain $H_{ap} < H_{geo} - 500$ km within 25 years	• re-orbit above GEO so that $\Delta H_{3,geo}$ [km] > $200 + 0.022\,a\,c_R\,A/m$
RSA	• general recommendation for EOL lifetime limitation	• re-orbit above GEO so that $\Delta H_{4,geo} > 200$ km

to two reasons: (1) the periodic presence of retired GTO and HEO objects in a protected MEO region is difficult to avoid, and (2) re-orbited MEO constellation satellites are likely to revisit the protected MEO region. An analysis of circular disposal orbits indicates that under the combined influence of Earth oblateness (J_2) and luni-solar perturbations the eccentricities start to build up, in particular for certain combinations of ω and Ω, with periods governed by 2ω and $2\omega + \Omega$ (Chao, 2001; Merz, 2003). These cyclic changes have periods of a few centuries, leading to a continuous growth of the eccentricity to peak values of up to $e \approx 0.5$ within the next 180 years. The variational amplitudes of e, and hence the "instability" of the disposal orbits, largely depend on the orbit inclination of a constellation (see Table 5.2), with larger amplitudes of eccentricity perturbations for higher inclinations.

Highly eccentric orbits (HEOs) cover almost 10% of all cataloged objects. The most frequently used HEOs for space missions are the Geostationary Transfer Orbits (GTOs), which have a pericenter within the LEO regime, and an apocenter at $H_{ap} \approx 35,786$ km, close to the GEO ring. GTOs have eccentricities on the order of $e \approx 0.73$, and inclinations of typically $0° \leq i \leq 28.5°$, depending on the launch site latitude, and on the launcher injection strategy. For the post-mission disposal of GTO spacecraft and upper stages, most mitigation standards require a final orbit with a pericenter above the LEO region and an apocenter below the

GEO ring, with a sufficient clearance margin (NASA, CNES, EDMS, US Government). US Government guidelines also request that the destination orbits do not cross a safety region around the 12-h circular MEO altitude. As an alternative to the re-orbit option the CNES and EDMS standards allow a direct de-orbit, or a disposal in an orbit of less than 25 years' lifetime. This mitigation measure is also requested by the JAXA standard, with no re-orbit alternative.

The geostationary orbit (GEO) and its vicinity contains more than 9% of all cataloged objects. The tight margins on orbit altitude, eccentricity, and inclination which operational spacecraft in the GEO ring must maintain, make this a particularly vulnerable region in space, where explosions or collisions can dramatically increase spatial densities and corresponding risks of damage to valuable space assets. This is why all debris mitigation standards and guidelines request post-mission disposal of GEO spacecraft by re-orbiting and subsequent passivation in a near-circular disposal orbit, with a pericenter sufficiently above the GEO ring to stay clear of operational spacecraft which are station-maintained, or which are drifting to a newly assigned longitude. Some standards require an additional altitude separation to account for long-period variations of the disposal orbit, as a result of luni-solar and solar radiation pressure perturbations. Table 6.9 lists the different GEO end-of-life re-orbiting rules of NASA, US Government, CNES, EDMS, JAXA, and RSA ROSCOSMOS. The CNES and EDMS recommendation is identical to the GEO re-orbit formula of IADC, as described by Eq. 6.18. This rule was also adopted by the International Telecommunications Union (ITU). As one of its major activities, the ITU assigns allotments of longitude slots and frequencies for future GEO payloads during regular World Administrative Radio Conferences (WARCs). Accordingly, ITU is in a good position to impose debris mitigation measures on GEO spacecraft operators, some of whom are not yet compliant with the commonly agreed end-of-life disposal measures (see Tables 6.5 and 6.6).

6.7 REFERENCES

Abramowitz, M. and Stegun, A. (1984). *Pocketbook of Mathematical Functions.* Verlag Harri Deutsch, Frankfurt a.M.

Agrawal, B. N. (1986). *Design of Geosynchronous Spacecraft.* Prentice-Hall, Inc., Englewood Cliffs, NJ.

Alby, F. (2003). SPOT-1 End of Life Disposal Manoeuvres. *35th COSPAR Scientific Assembly*, PEDAS1/B1.6-0035-04.

anon. (1995). *NASA Safety Standard – Guidelines and Assessment Procedures for Limiting Orbital Debris.* NSS 1740.14.

anon. (1996). *Space Debris Mitigation Standard.* NASDA-STD-18.

anon. (1997). *US Government Orbital Debris Mitigation Standard Practices.*

anon. (1999). *CNES Exigence de Securité – Débris Spatiaux: Méthode et Procedure.* CNES, MPM-51-00-12, issue 1, rev. 0.

anon. (2000). *Russian Aviation & Space Agency (RASA) Branch Standard – General Requirements for Space Debris Mitigation.* RASA.

anon. (2001). *European Space Debris Safety and Mitigation Standard – Draft.* issue 1, rev. 3.

anon. (2002). *IADC Space Debris Mitigation Guidelines.* issue 1, rev. 1.

Bekey, I. (1997). Project Orion: Orbital Debris Removal Using Ground-Based Sensors and Lasers. In *Proceedings of the Second European Conference on Space Debris*, ESA SP-393, pages 699–701.

Bondarenko, S. and Lyagushin, S. (1997). Prospects of Using Military Lasers and Military Space Technology for Space Debris Removal. In *Proceedings of the Second European Conference on Space Debris*, ESA SP-393, pages 703–706.

Bonnal, C., Sanchez, M., and Naumann, W. (1997). Ariane Debris Mitigation Measures. In *Proceedings of the Second European Conference on Space Debris*, ESA SP-393, pages 681–688.

Chao, C. (2001). MEO Disposal Orbit Stability and Direct Re-Entry Strategy. *11th AAS/AIAA Space Flight Mechanics Conference*, AAS-01-244.

Hoyt, R. and Forward, R. (1999). Performance of the Termination Tether^{TM} for Autonomous De-Orbit of LEO Spacecraft. *50th IAF Congress*, IAF-99-IAA.6.6.09.

Jehn, R., Hernández, C., Klinkrad, H., Flury, W., and Agapov, V. (2004). Space Traffic Management in the Geostationary Ring. *55th IAF Congress*, IAA-04-IAA.5.12.4.05.

King-Hele, D. (1987). *Satellite Orbits in an Atmosphere: Theory and Applications.* Blackie, Glasgow, UK.

Klinkrad, H. et al. (2003). *ESA Space Debris Mitigation Handbook.* ESA/ESOC, second edition. issue 1.0.

Kompanietz, E. et al. (2000). Minimization of Space Debris Formation on Zenith-3SL ILV Launches Within the Sea Launch Program. *51st IAF Congress*, IAA-00-IAA.6.6.05.

Merz, K. (2003). On the Stability of Near Semi-Synchronous, Near Circular Orbits. Technical Report no. 459, issue 2, rev. 0, ESA/ESOC/MAO.

Samson, P. (1999). Classification of Geostationary Objects – Users Guide. Technical Report no. 420, ESA/ESOC/MAO.

Zhang, W. (1997). Elimination of the Potential Hazard of the Long March 4 Launch Vehicle On-Orbit Break-Up. In *Proceedings of the Second European Conference on Space Debris*, ESA SP-393, pages 689–691.

7

Hypervelocity Impact Damage Assessment and Protection Techniques

H. Klinkrad and H. Stokes

Spacecraft which operate in densely populated altitude regimes are experiencing a steady debris and meteoroid particle flux which strongly increases with decreasing particle sizes (see Table 3.2 and 3.3, and Fig. 2.38). The consequences of resulting impacts can range from small surface pits for µm-size impactors, via clear hole penetrations for mm-size objects, to partial or complete destruction via shockwaves for projectiles larger than a few centimeters. The most probable impact velocities are in the range from 0 to 15 km/s for space debris, and between 5 km/s and 30 km/s for meteoroids (denoted as hypervelocity impacts or HVI). At such speeds, the impact of an aluminum sphere of 1 cm diameter deploys the same energy as an exploding hand-grenade, with equally devastating consequences, unless special protection measures are applied.

7.1 HYPERVELOCITY ACCELERATORS AND HYDROCODE SIMULATIONS

To understand the physical processes during hypervelocity impacts it is essential to develop experimental accelerators which under well-controlled laboratory conditions propel particles to orbital velocities. For well defined projectile materials, shapes, and velocities, and for given target materials, such accelerators can be used to produce calibration shots which may help interpret impacts on retrieved space hardware. Likewise, they can be used to optimize shield designs and their capability to defeat impacting objects.

The most frequently used accelerators are based on the principle of a gun. So called two-stage light-gas guns (LGG), as illustrated in Fig. 7.1 and 7.2, are fired by igniting a low-explosive powder in a high-pressure chamber. The developing gas propels a piston down a pump tube, towards a high-pressure section which is closed by a diaphragm. As the piston moves down the barrel, a light gas (H_2

or He) contained in the pump tube is compressed until the cylindrical piston is trapped in the conical entrance of the high-pressure section. At this time the burst pressure of the diaphragm on the opposite side is exceeded. It ruptures, and the light gas escapes into the launch tube, accelerating a sabot which contains the projectile. As the sabot leaves the launch tube it disintegrates under aerodynamic forces, releasing the projectile towards the target. The sabot is subsequently caught by a screening orifice, while the motion of the unaffected projectile, and its impact on the target are monitored by diagnostic equipment, such as laser beams, X-ray tubes, and high-speed cameras.

The achievable projectile velocity of a light-gas gun can be estimated by assuming an isentropic expansion (frictionless, with no heat exchange) of the light gas from the high-pressure section into the launch tube. For the large gun of the Ernst-Mach-Institute (EMI), which is schematically shown in Fig. 7.1, let the pump-tube be filled with pre-heated hydrogen of pressure $p_o = 5\,\text{bar}$ and temperature $T_o = 500\,\text{K}$. During its compression by the piston, the content of the pump tube reaches a peak pressure of $p_{max} = 10,000\,\text{bar}$, which corresponds to the burst pressure of the diaphragm. The resulting temperature increase to $T_{max} \approx 4,400\,\text{K}$ can be determined from

$$T_{max} = T_o \left(\frac{p_{max}}{p_o} \right)^{\frac{\kappa-1}{\kappa}} \tag{7.1}$$

where $\kappa = (n+2)/n$ is the ratio of specific heats for a gas of n rotational and translational degrees of freedom. Hence for atomic helium $\kappa = 1.66$, and for molecular hydrogen $\kappa = 1.4$.

The speed of sound in the high-pressure section at the time of maximum compression is given by

$$a = \sqrt{\kappa \frac{R}{M} T_{max}} \tag{7.2}$$

where $R = 8,314.41\,\text{J}\,\text{kmol}^{-1}\,\text{K}^{-1}$ is the universal gas constant, and M is the molecular mass, with $M = 4\,\text{kg/kmol}$ for He, and $2\,\text{kg/kmol}$ for H_2. The resulting escape velocity after rupture of the diaphragm can be determined from the equation of de Saint-Venant and Wantzel.

$$w_e = a \sqrt{\frac{2}{\kappa - 1} \left\{ 1 - \left(\frac{p}{p_{max}} \right)^{\frac{\kappa-1}{\kappa}} \right\}} \tag{7.3}$$

The maximum possible velocity according to Eq. 7.3 can be obtained for gases with low molecular weight (since $w_e \propto 1/\sqrt{M}$), by an expansion into vacuum ($p \to 0$).

$$w_{e,max} = a \sqrt{\frac{2}{\kappa - 1}} = \sqrt{\frac{2\kappa}{\kappa - 1} \frac{R}{M} T_{max}} \tag{7.4}$$

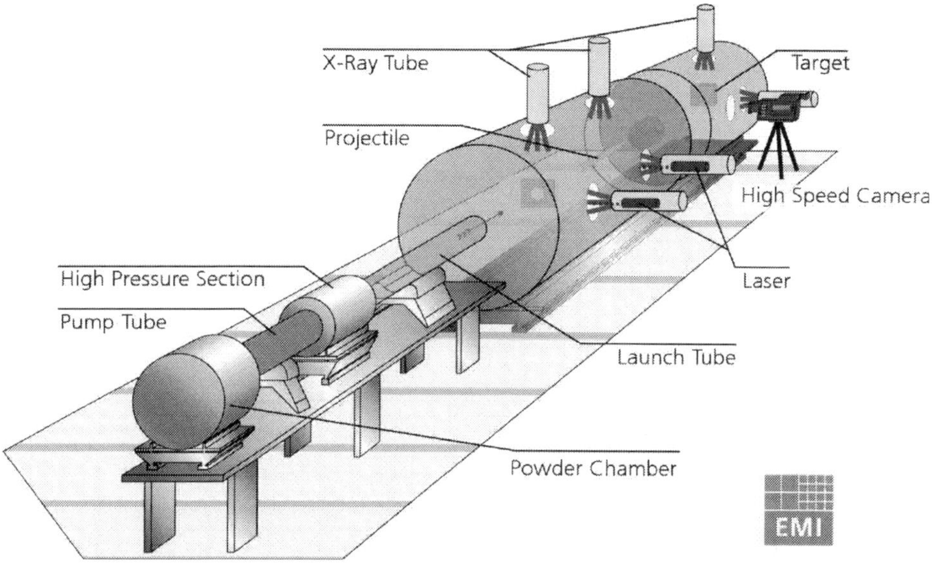

Fig. 7.1. System components of the large 2-stage light-gas gun and its associated diagnostic equipment at the Ernst-Mach-Institute (source: EMI).

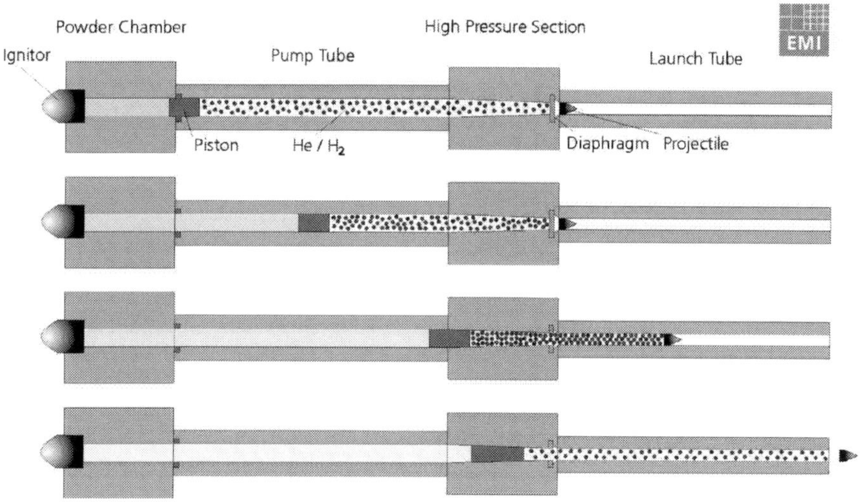

Fig. 7.2. Functional principle of a 2-stage light-gas gun (source: EMI).

Based on the simplified assessment by means of Eq. 7.3 and 7.4, neglecting friction losses and projectile mass dependencies, the large EMI 2-stage light-gas

Table 7.1. Performance ranges of different types of hypervelocity accelerators, based on data published in (anon., 1995).

Accelerator type	Mass and velocity range covered		
powder guns	1,200 g at 1.0 km/s	⟶	160 g at 3.8 km/s
single-stage light gas guns	1,200 g at 0.8 km/s	⟶	160 g at 2.0 km/s
two-stage light gas guns	80 g at 3.0 km/s	⟶	0.2 g at 9.5 km/s
modified light gas guns	5 g at 10.0 km/s	⟶	0.1 g at 15.0 km/s
electromagnetic rail guns	20 g at 1.5 km/s	⟶	1.0 g at 7.5 km/s
electrostatic guns	6 g at 2.0 km/s	⟶	0.05 g at 18.0 km/s
shaped charges	10 g at 10.0 km/s	⟶	1.0 g at 11.5 km/s

gun would theoretically be able to reach velocities of up to $w_{e,max} = 11.3\,\mathrm{km/s}$ with hydrogen[7.1], and of up to $w_{e,max} = 6.7\,\mathrm{km/s}$ for helium. These limits comply with experimentally verified performances for this type of accelerator, as listed in Table 7.1.

Table 7.1 summarizes performances of the most common types of hypervelocity accelerators, ranging from single-stage gas and powder guns, which propel 1 kg masses to 2 km/s, to electrostatic accelerators, which reach velocities of up to 18 km/s for masses on the order of 0.01 g. The latter type of equipment is often used to simulate impacts of meteoroids, which typically have masses of no more than a few micrograms, and most probable relative velocities between 10 and 20 km/s. At larger masses of up to 20 g, electromagnetic "rail guns" are more effective. They use the Lorentz force to accelerate a projectile which is driven by a mobile metal body or plasma layer, which closes an electric circuit between two conductive rails, and propels the mass under the influence of a generated magnetic field.

Hypervelocity accelerators, which translate chemical energy of an explosive into kinetic energy of a projectile, are most widely used to simulate damage scenarios and to optimize shield designs. Shaped charges belong to this class of devices. They use a high-explosive which is cast in a conical shape around a likewise conical aluminum liner, both with a half-cone angle of about 11°. When the explosive is ignited the aluminum liner is compressed, melted and accelerated as a jet of liquid metal to speeds of up to 11.5 km/s (for masses of about 1 g). Special techniques are required to remove a trailing slug, and to transform the elongated jet into cylindrical shapes of technically useful length-to-diameter aspect ratios. However, the shape of the projectile, and the reproducibility of test conditions cannot be well controlled. This is why light-gas guns are generally preferred for HVI tests. They cover a wide range of masses and velocities (see Table 7.1), ranging from more than 1 kg at 1 km/s for single-stage guns to 0.2 g at 9.5 km/s for two-stage guns. Most of these guns use light-weight plastic sabots to carry and accelerate the projectiles, which are then aerodynamically separated after leaving

[7.1]actual limit of the large EMI light-gas gun: ~8 km/s for masses of 1 g the launch tube. In this way the shape of projectiles can be arbitrarily chosen, and their orientation from release to impact can be well controlled. Consequently, HVI test conditions for light-gas guns are reproducible within acceptable tolerances.

The main objective of hypervelocity accelerators is to perform impact tests for projectiles of a certain shape, material, mass, and velocity, on a target of a monolithic or composite material and structure. The response of the target is often monitored by high-speed cameras and X-ray equipment, in order to analyze the evolution of the generated ejecta cloud and the dynamics of the projectile penetration into the target. At the current state of accelerator technology it is not yet possible to perform routine HVI experiments at speeds which correspond to the most probable on-orbit collision velocities, which for 1-cm objects in LEO can range from 10.5 km/s (ISS at $i = 51.5°$, see Fig. 4.5) to 14.5 km/s (Sun-synchronous orbits at $i = 98.5°$, see Fig. 4.4), and which are hence beyond the velocity limit of 2-stage light-gas guns (see Table 7.1). The extrapolation of HVI results into so far unreachable domains for experiments can be performed by means of hydrocode computer simulations.

Hydrocode programs solve a set of conservation equations for mass, momentum, and energy, together with adjoining equations of state (EOS), which include shock phenomena under consideration of material response with change of state. The resulting partial differential equations are integrated in time, across a spatial grid which can be defined as body-fixed (Lagrange grid) or inertially fixed (Euler grid). The Lagrange method has advantages in the traceability of body-related grid-point histories, and in the definition of material boundaries, at the expense of large grid distortions during an impact. On the other hand, the Euler method allows one to compute diffusion and mixing of materials, which is not possible with Lagrange codes. As an alternative approach, meshless techniques were developed, which typically use Lagrangian nodes that are not physically connected by a grid, but instead use mathematical interpolation functions to track their relative motion. In such a way the grid distortion is avoided, and the combined benefits of the Lagrange and Euler methods can be exploited. The most prominent meshless method is known as smoothed particle hydrodynamics (SPH).

Hydrocodes must be validated and calibrated with experimental data at projectile velocity and mass ranges which can be covered by both approaches. An overview of hydrocode techniques and implementations is provided in (anon., 2004) and (anon., 2001).

7.2 EFFECTS OF HYPERVELOCITY IMPACTS

When a spherical, metallic projectile hits a ductile target of infinite thickness, then at low velocities plastic deformation prevails. With increasing speeds the projectile starts to break up, leaving a residual crater on the target. Beyond 5 km/s (depending on the materials), the hypervelocity impact results in a complete projectile break-up and melting, and in a mass ejection from a crater that typically has a depth D_c, which is 2 to 5 times the diameter d_p of the projectile (see Fig. 7.3, right-

Table 7.2. Properties of common space system materials, relevant for impact damage assessments (material density ρ, speed of sound a_o in the material, change of speed of sound $\Delta a/\Delta v_p$ with impact particle velocity).

	Al alloy	Ti alloy	Steel alloy	Quartzglass	Plexiglass
$\rho\ [\mathrm{g/cm^3}]$	2.785	4.528	7.896	2.204	1.186
$a_o\ [\mathrm{km/s}]$	5.328	4.877	4.569	0.794	2.598
$\Delta a/\Delta v_p\ [-]$	1.490	1.049	1.490	1.695	1.516

most image). As the impact velocity v_p increases, elastic and plastic stress–strain behavior prevails, density and compressibility effects set in, and finally thermal properties become important. For hypervelocity impacts, the projectile velocity exceeds the speed of sound of the target and projectile material, which further increases with an increase in the projectile impact velocity (see Table 7.2 for material-dependent coefficients).

$$a\left(v_p\right) = a_o + \frac{\Delta a}{\Delta v_p}\, v_p \qquad (7.5)$$

After impact of the projectile the compression wave steepens until it starts propagating as a concentric shock wave, with jumps in density, pressure, grid-point velocity, and internal energy across the shock front. In the phase of impulsive pressure loading the volume decreases in a linear relationship $|\Delta p/\Delta V| = \text{const.}$, along the so-called Rayleigh line. The response of the material during unloading is governed by a non-linear dependency $|\Delta p/\Delta V| \neq \text{const.}$, which follows the so-called Hugoniot function (close to the release isentrope). The area which is included by the Rayleigh line and the Hugoniot function represents the waste heat which is released into the material. In a target of semi-infinite thickness, a given grid location experiences a pressure peak at the transit of the shock wave, followed by a steady-state phase of elevated pressure, and a recovery phase of declining pressure. Compressive and tensile waves in the longitudinal direction are accompanied by torsional and shear waves in the transverse direction. If the target has a finite thickness, the free boundaries reflect the shock, turning an incoming compressive wave into a reflected tensile wave. The superimposition of these waves can locally lead to a yield stress violation, and cause spallation of the material along a well defined fracture line (see Fig. 7.3, second image from the right). For fixed impact conditions a decrease in the target thickness will first lead to a spall detachment, then cause a perforation, and finally result in a clear hole puncture. Fig. 7.3 shows a combination of possible failure modes which include plastic flow when exceeding elastic limits, and fracture when exceeding cohesive material strength.

Fig. 7.4 indicates that for the same impact velocity and mass, the projectile shape and orientation is an important parameter for the characterization of the impact damage. Elongated, cylindrical projectiles tend to have a higher penetration capability than spheres or plates. This is why low-cost shaped charges, which

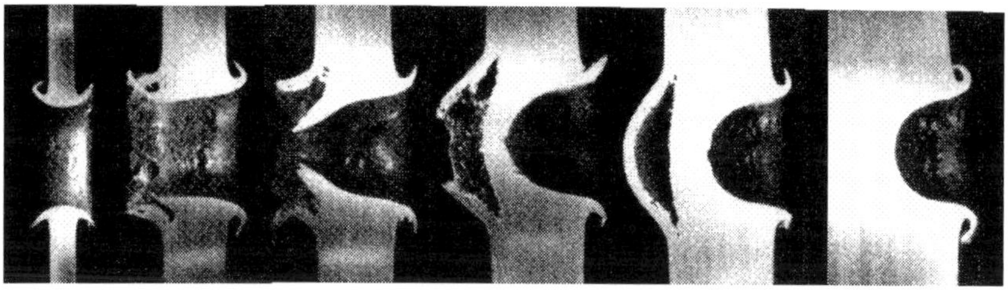

Fig. 7.3. Single-wall impact damage characteristics on Al 1100 targets, as a function of the relative target thickness $t_t/d_c = 1.0, 2.4, 2.8, 3.4, 3.9, 10.0$ (left to right), for an impact by a glass projectile of diameter $d_p = 3.2\,\text{mm}$ at a velocity of $v_p = 6\,\text{km/s}$ (source: (Grün et al., 2001); note: display of rightmost target is truncated in thickness).

Fig. 7.4. Impact crater geometry of an aluminum plate, sphere, and cylinder on a semi-infinite aluminum target (source: (anon., 1995)).

produce high-speed, elongated jets of liquid metal, are used in military applications. Interestingly, the resulting impact damage for an aluminum projectile is almost unaffected by its solid or liquid state.

For high velocities and slender projectiles the peak pressures and temperatures shortly after impact can reach levels of 100 GPa and 10,000 K, which are comparable to conditions at the center of the Earth (~365 GPa and ~6000 K). The time scale during which the structural impact damage evolves is on the order of a few microseconds.

7.3 SINGLE-WALL DAMAGE EQUATIONS

The consequences of hypervelocity impacts are commonly described by means of damage equations. These are mainly used to determine whether particles exceed the ballistic limit of a target, and therefore provide a means to assess the probability of penetration. Care must be taken when using damage equations, to understand their underlying assumptions and physical limitations. For instance, all equations listed hereafter have been derived for spherical impactors, ignoring particle shape effects. For a given mass, such a spherical impactor is likely to cause less damage than an elongated object (see Fig. 7.4, right image). It should also be noted that all damage equations have only been validated in a confined range of velocities and impactor masses, and for certain shield and impactor materials. Ac-

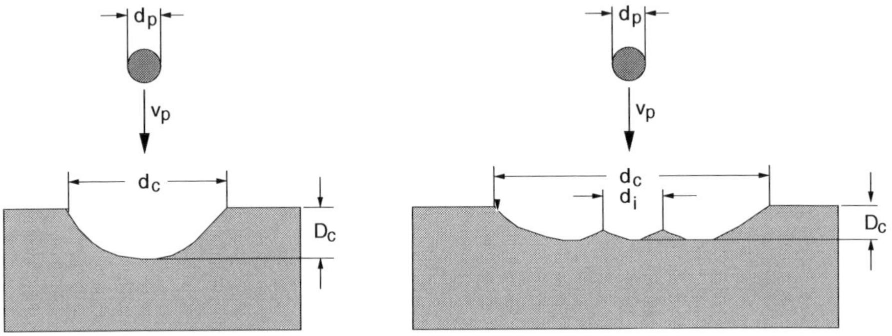

Fig. 7.5. Characteristic quantities of single-wall impacts on ductile materials (left) and brittle materials (right) of a thick plate target.

cordingly, the empirical relationships should only be used for preliminary assessments. The validation of a shield design should in any case be performed by HVI tests, and/or by numerical hydrocode simulations. Recurring parameters which are used in the following damage equations are listed in Table 7.3. The damage equations quoted hereafter are based on information compiled in (Klinkrad et al., 2003).

A non-perforating spherical projectile impacting under a normal angle on a ductile target produces a near hemispherical crater of diameter d_c and depth D_c (see Fig. 7.5, left). The same projectile impacting a brittle target will normally produce an interior crater with diameter d_i, inside a larger outer crater of diameter d_c and depth D_c (see Fig. 7.5, right). Most spacecraft and orbital stages comprise a large percentage of aluminum alloy, together with titanium, stainless steel, and other monolithic metal structures. All these fall into the category of ductile materials. Typical brittle surfaces include solar arrays, mirrors on telescopes, and Shuttle window panes.

The diameter of an impact crater on a semi-infinite single wall of monolithic material is characterized by Eq. 7.6, with coefficients listed in Table 7.4.

$$d_c = K_1\, d_p^{\lambda}\, \rho_p^{\beta}\, \rho_t^{\kappa}\, v_p^{\gamma}\, \left(\cos\alpha_p\right)^{\xi} \tag{7.6}$$

For a spherical projectile impacting on a planar aluminum target, pure crater damage will occur, if the target thickness t_t is much larger than the crater depth D_c (typically $t_t > 3\,D_c$). With decreasing target thickness deformation and separation of the rear wall sets in at $t_t \approx 2.2\,D_c$, followed by spall separation at $t_t \approx 1.8\,D_c$, and wall puncture at $t_t < 1.8\,D_c$.

For a fully perforated target the impact damage is described by the following hole diameter equation, with coefficients listed in Table 7.5.

$$d_c = d_p \left[K_1 \left(\frac{t_t}{d_p}\right)^{\lambda} \rho_p^{\beta}\, \rho_t^{\gamma}\, v_p^{\gamma}\, \left(\cos\alpha_p\right)^{\xi} + K_o \right] \tag{7.7}$$

Table 7.3. Definition of parameters used in HVI damage equations.

Symbol	Units	Description
t_t, t_w, t_s	cm	thickness of target, back wall, shield (total)
$t_{t,lim}$	cm	ballistic limit (min. required single-wall thickness)
$t_{w,lim}$	cm	ballistic limit (min. required back wall thickness)
C_w	–	ballistic limit parameter for back wall
K_1, K_2	–	calibration factors
K_o	–	calibration constant
$\lambda, \beta, \gamma, \xi, \kappa,$		
$\delta, \mu, \upsilon, \upsilon_1, \upsilon_2$	–	calibration exponents
d_c	cm	crater or hole diameter
D_c	cm	crater depth
d_p	cm	particle (impactor) diameter
$d_{p,lim}$	cm	max. defeatable particle diameter
S	cm	spacing between shield and back wall
ρ_t, ρ_p, ρ_s, ρ_w	$\mathrm{g\ cm^{-3}}$	density of target, particle, shield, back wall
v_p, v_n	$\mathrm{km\ s^{-1}}$	particle impact velocity and its normal component
α_p	deg	particle impact angle (with respect to surface normal)
τ	Pa	yield stress of back wall
τ_1^*	Pa	ref. yield stress, 276×10^6 Pa (= 40,000 lb in^{-2})
τ_2^*	Pa	ref. yield stress, 483×10^6 Pa (= 70,000 lb in^{-2})

Table 7.4. Calibration parameters for the impact crater equation of ductile and brittle materials according to different sources.

Equation	Material	K_1	λ	β	γ	ξ	κ
ESA	ductile	0.8–1.32	1.056	0.519	2/3	2/3	0
Gault	brittle	1.08	1.071	0.524	0.714	0.714	–0.5
McHugh et al.	brittle	1.28	1.2	0	2/3	2/3	0.5
Cour-Palais	brittle	1.06	1.06	0.5	2/3	2/3	0

Table 7.5. Calibrated parameters for the clear hole impact equation.

Equation	K_1	λ	β	γ	ξ	υ	K_o
Maiden	0.88	2/3	0	1	1	0	0.9
Nysmith	0.88	0.45	0.5	0.5	0.5	0	0
Sawle	0.209	2/3	0.2	0.2	0.2	–0.2	1
Fechtig	5.24e–5	0	1/3	2/3	2/3	0	0

Table 7.6. Calibrated parameters for a single-wall ballistic limit equation, according to different sources.

Equation	Target	K_1	λ	β	γ	ξ	κ
ESA	thick plate	0.36–0.99	1.056	0.519	2/3	2/3	0
ESA	thin plate	0.26–0.64	1.056	0.519	0.875	0.875	0
Pailer & Gruen	any	0.77	1.212	0.737	0.875	0.875	–0.5
Frost	any	0.43	1.056	0.519	0.875	0.875	0
Naumann et al.	any	0.65	1.056	0.5	0.875	0.875	–0.5
Naumann	any	0.326	1.056	0.499	2/3	2/3	0
McHugh et al.	thick glass	1.18–4.48	1.2	0	2/3	2/3	0.5
Cour-Palais	thick glass	0.98–3.17	1.06	0.5	2/3	2/3	0

In between the thick target case of Eq. 7.6, and the thin target case of Eq. 7.7, there are intermediate damage scenarios, where the capability of a single-wall shield to withstand a perforation is described by the ballistic limit. This corresponds to the minimum target thickness $t_{t,lim}$, which can defeat an incoming object, for given impactor and shield properties, and for a given impact velocity and impact incident angle.

$$t_t \geq t_{t,lim} = K_1\, d_p^{\lambda}\, \rho_p^{\beta}\, \rho_t^{\kappa}\, v_p^{\gamma}\, \left(\cos\alpha_p\right)^{\xi} \tag{7.8}$$

Coefficients for this ballistic limit equation are provided in Table 7.6. Eq. 7.8 can be re-arranged to indicate the maximum defeatable, critical projectile diameter $d_{p,lim}$.

$$d_p \leq d_{p,lim} = \left[\frac{t_t}{K_1\, \rho_p^{\beta}\, \rho_t^{\kappa}\, v_p^{\gamma}\, \left(\cos\alpha_p\right)^{\xi}}\right]^{\frac{1}{\lambda}} \tag{7.9}$$

The shieldable impactor diameter defined in Eq. 7.9 declines with increasing impact velocity according to $d_{p,lim} \propto v^{-2/3}$. This dependency is plotted in Fig. 7.7 for a single-wall shield ($t_s = 0.0$).

7.4 MULTIPLE-WALL DAMAGE EQUATIONS

Single-wall shields do not offer efficient protection against hypervelocity impacts. For a given structural mass per shield area multi-wall protection systems turn out to be much more effective. A thin outer shield exploits the kinetic energy of the impactor to break it up, and a back wall defeats the dispersed fragments.

The concept of multi-wall shields can be attributed to the astronomer Fred Whipple. He devised a mass-efficient protection against HVI projectiles which consists of a thin bumper shield at a sufficient stand-off distance from a thicker back wall. The effect of such a "Whipple shield" is illustrated in Fig. 7.6 (left image), for an impact of a 5-mm Al particle at 8 km/s on a 0.5 mm shield, and the

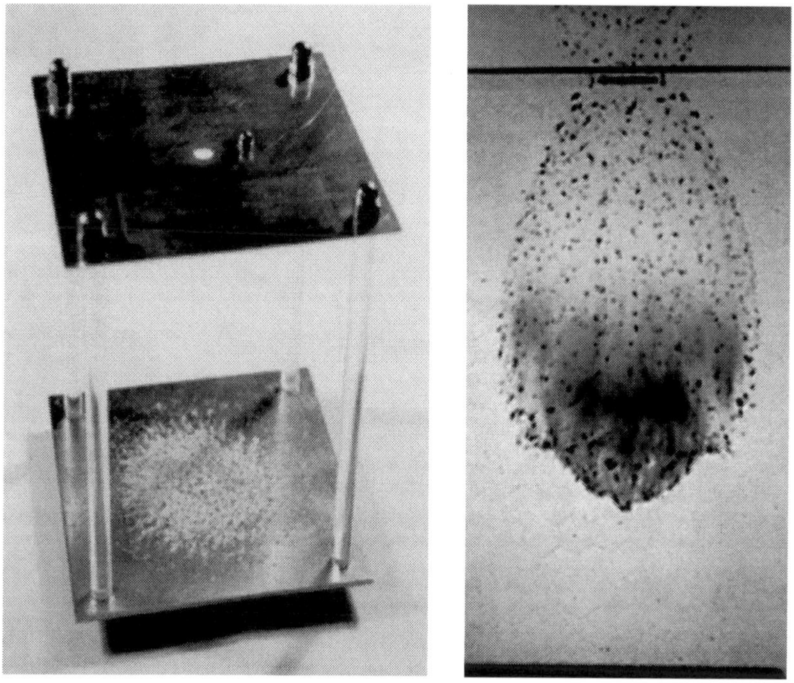

Fig. 7.6. Experimental verification of the effectiveness of Whipple shields as debris impact protection. Left: hole produced by an Al-sphere of $d_p = 5\,\mathrm{mm}$ at $v_p = 8.0\,\mathrm{km/s}$ on a shield of $t_s = 0.5\,\mathrm{mm}$, with fragment dispersion on the back wall of $t_w = 2\,\mathrm{mm}$, at a distance of $S = 100\,\mathrm{mm}$ (source: EMI). Right: shadowgraph of dispersed impactor and shield fragments after shield puncture (separate experiment; source: EMI).

spreading of fragments on a back wall at 10 cm distance. As a projectile hits the bumper, compression and tensile waves are generated which cause a complete disintegration of the impactor (provided that $t_s/d_p \geq 0.1$ and $v_p \geq 7\,\mathrm{km/s}$). Apart from a few surface ejecta particles, a cloud of mainly liquid projectile and shield material forms, which disperses laterally from the impact hole, and progresses longitudinally at almost unaltered speed, before hitting the back wall. Due to the lateral dispersion, and due to the time-delayed arrival of different cloud particles, the surface loading on the back wall is considerably reduced as compared to the point load caused by the intact projectile at the time of shield impact. Hence, the back wall can defeat the cloud of projectile fragments. The right image in Fig. 7.6 shows a shadowgraph of a dispersed impactor cloud, with the projectile mass mainly concentrated in the dark area, in the lower part of the cloud, and shield material concentrated in the spotted regions on the cloud periphery, and in the ejecta cone. The core of the cloud is almost void of any material.

For a classical Whipple shield made of Al 7075-T6 alloy, the minimum thickness of the back wall can be estimated from

$$t_w \geq t_{w,lim} = \frac{C_w \, \rho_p \, d_p^3 \, v_p}{S^2} \tag{7.10}$$

where $C_w = 21.7 \pm 7.3$ for $\sigma_{0.2}$ yield conditions, and $C_w = 4.3 \pm 0.7$ for fracture conditions. For a given back wall thickness t_w (typically $t_w \geq 0.2\,d_p$), this equation can also be used to determine the necessary spacing S between bumper and back wall (typically $S \geq 30\,d_p$).

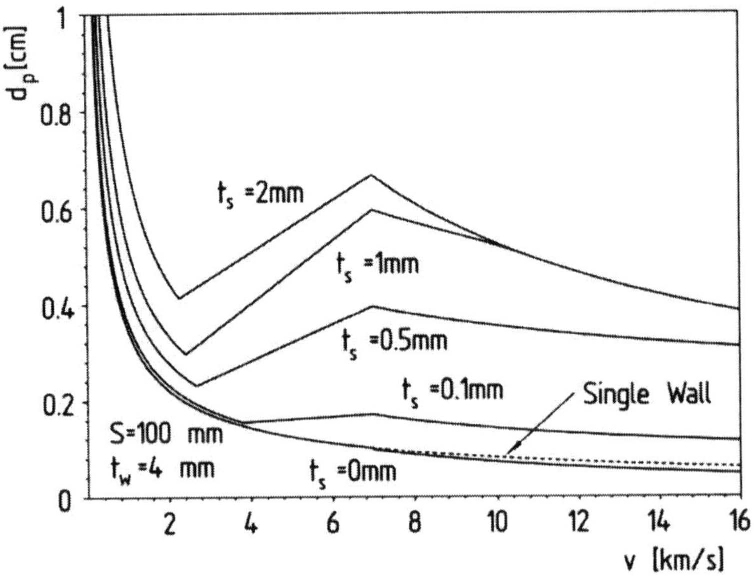

Fig. 7.7. Ballistic limit curves for an aluminum Whipple shield with a wall thickness of $t_w = 4$ mm and a separation distance of $S = 100$ mm, as a function of projectile diameter d_p, shield thickness t_s, and impact velocity v_n (source: RWTH Aachen).

Fig. 7.7 shows the ballistic limit curve, in terms of maximum defeatable projectile diameter d_p, for a Whipple shield with back wall thickness $t_w = 4$ mm, stand-off distance $S = 100$ mm, and varying bumper shield thickness t_s. With the relative bumper thickness increasing to $t_s/d_p > 0.05$, the shield starts to become effective. At velocities $v_n < 3$ km/s, in the ballistic region, the material strength of the projectile exceeds the dynamic pressure at impact, and so the impactor is only poorly fragmented by the bumper. At $3\,\text{km/s} \leq v_n \leq 7\,\text{km/s}$, in the shatter or transition region, the projectile is broken up into a finite number of pieces, which may be solid or liquid. At $v_n > 7$ km/s, in the true hypervelocity range, shock-induced pressure and tensile strain by far exceed the material strength of the projectile, thus causing it to completely break up into a cloud of numerous liquid particles, and vaporized material.

The ballistic limit of a multiple-wall shield can be expressed in terms of the

Table 7.7. Calibration parameters for the multiple-wall ballistic limit equation, according to different sources (with $\hat{\tau}_1 = \sqrt{\tau_1^*/\tau}$; $\hat{\tau}_2 = \sqrt{\tau_2^*/\tau}$; $\gamma = 2/3$, $\xi = 5/3$, and $\mu = 1$ for $v_n < 3$ km/s; $\gamma = 1$, $\xi = 1$, and $\mu = 0$ for $v > 6(7)$ km/s).

Equation	v_n [km/s]	K_1	K_2	λ	β	κ	δ	υ_1, υ_2
ESA (tripple	$v_n < 3$	$0.312\,\hat{\tau}_1$	$1.667\,K_1$	1.056	0.5	0	0	0, 0
wall eq.)	$v_n > 7$	$0.107\,\hat{\tau}_2$	0	1.5	0.5	0	–0.5	0.167, 0
NASA (mod.	$v_n < 3$	$0.6\,\hat{\tau}_1$	$1.667\,K_1$	1.056	0.5	0	0	0, 0
Cour-Palais)	$v_n > 7$	$0.129\,\hat{\tau}_2$	0	1.5	0.5	0	–0.5	0.167, 0
NASA	$v_n < 3$	$0.3\,\hat{\tau}_1$	$1.233\,K_1$	1.056	0.5	0	0	0, 1
(shock eq.)	$v_n > 6$	$22.545\,\hat{\tau}_1$	0	3	1	–1	–2	0, 0
NASA	$v_n < 3$	$0.4\,\hat{\tau}_1$	$0.925\,K_1$	1.056	0.5	0	0	0, 1
(bumper eq.)	$v_n > 6$	$18.224\,\hat{\tau}_1$	0	3	1	–1	–2	0, 0

minimum required back wall thickness $t_{w,lim}$, or in terms of the maximum defeatable projectile diameter $d_{p,lim}$.

$$t_w \geq t_{w,lim} = K_1\, d_p^{\lambda}\, \rho_p^{\beta}\, \rho_w^{\kappa}\, \rho_s^{\upsilon_1}\, v_p^{\gamma}\, \left(\cos\alpha_p\right)^{\xi}\, S^{\delta} - K_2\, t_s^{\mu}\, \rho_s^{\upsilon_2} \tag{7.11}$$

$$d_p \leq d_{p,lim} = \left[\frac{t_w + K_2\, t_s^{\mu}\, \rho_s^{\upsilon_2}}{K_1\, \rho_p^{\beta}\, \rho_w^{\kappa}\, \rho_s^{\upsilon_1}\, v_p^{\gamma}\, \left(\cos\alpha_p\right)^{\xi}\, S^{\delta}}\right]^{\frac{1}{\lambda}} \tag{7.12}$$

These ballistic limits are defined for the two velocity ranges $0 < v_n < v_1 = 3$ km/s and $6(7)$ km/s $= v_2 < v_n < \infty$, with corresponding model coefficients as listed in Table 7.7. In the intermediate velocity regime, at $v_1 \leq v_n \leq v_2$, results are linearly interpolated between the left and right boundaries according to

$$t_{w,lim}\left(v_n\right) = t_{w,lim}\left(v_1\right)\, f_1\left(v_n\right) + t_{w,lim}\left(v_2\right)\, f_2\left(v_n\right) \tag{7.13}$$

$$d_{p,lim}\left(v_n\right) = d_{p,lim}\left(v_1\right)\, f_1\left(v_n\right) + d_{p,lim}\left(v_2\right)\, f_2\left(v_n\right) \tag{7.14}$$

with the weighting functions $f_1\left(v_n\right)$ and $f_2\left(v_n\right)$.

$$f_1\left(v_n\right) = \frac{v_2 - v_n}{v_2 - v_1} \qquad f_2\left(v_n\right) = \frac{v_n - v_1}{v_2 - v_1} \tag{7.15}$$

These multiple-wall ballistic limit equations may also be used to estimate the behavior of honeycomb structures, which normally consist of aluminum cells enclosed in face-sheets made of aluminum or CFRP (carbon fiber reinforced plastic). Such structures provide an excellent material stiffness for a small mass-per-area ratio. However, since the honeycomb cells are aligned parallel to the surface normal, an impact of a hypervelocity projectile results in a channeled fragment cloud, with a corresponding concentration of the pressure load on the back wall. To reduce this problem multiple layers of honeycombs can be used, with horizontal displacements by half the cell width between different layers.

For Al/Al single and double honeycomb panels (Turner et al., 2002) recommend the ESA triple-wall equation for near-vertical impacts, with coefficients listed in Table 7.7. In this case the dependency on the incident angle is adjusted via $\xi = 8/3$ (instead of 5/3), to represent channeling effects. Beyond impact velocities of 7 km/s the same authors suggest introducing an equivalent back wall thickness which mimics the combined behavior of the heterogenic structure.

7.5 HVI SHIELD DESIGNS AND IMPLEMENTATIONS

The effectiveness of Whipple shields can be improved in different ways. One option is to increase the number of bumper shields, which in succession cause further shock waves to disrupt and disperse the projectile, and to raise its thermal state, causing it to melt or vaporize within a short penetration distance. Another option, often used in combination with multiple shields, is to deploy different bumper materials. For the outer bumper, which is exposed to the harsh space environment, sufficiently inert materials are preferred, such as aluminum alloys, corrugated aluminum, metal matrix composites, CFRP composites, or Kevlar. For the inner bumper layers Nextel fabrics can be applied.

A mass and performance optimized shielding concept, which was devised for the manned Columbus module of the International Space Station (ISS), uses a "stuffed Whipple shield" with an Al 6061-T6 outer bumper of 2.5 mm thickness, followed at 71.1 mm distance by a 4 mm sheet of four Nextel 312-AF62 fabric layers on top of a 6 mm Kevlar 129-812 layer in an epoxy resin 914. This internal bumper is separated by 42 mm from an Al 2219-T851 back wall of 4.8 mm thickness (anon., 2004). The performance of the Columbus shield was analyzed with hydrocodes for spherical Al 2007 projectiles of $d_p = 12$ to 17 mm at $v_p = 3$ to 11 km/s. For the lower size and velocity regime these tests were verified against experiments with light-gas guns. The shield is designed to withstand impacts at orbital speeds by objects of up to 1 cm size.

The protection of manned space systems is a high priority for several reasons. The penetration of HVI projectiles may cause immediate injury or death of crew members due to the kinetic energy of fragments, heat, light-flash, and over-pressurization. Time-delayed consequences may include hypoxia-induced crew unconsciousness and suffocation, especially if the impact hole is large and the depressurization is rapid. Since manned modules are pressurized, special precautions must be taken in the structural design to avoid propagating crack formations after local failures (holes) in the pressure shell. Moreover, the air escaping from the leak can cause perturbing torques which may exceed the control capabilities of the spacecraft and lead to a loss of attitude, entailing power degradation and (for large objects like ISS) possible structural damage due to centrifugal forces.

For unmanned spacecraft the acceptable risk level due to hypervelocity impacts is higher than for manned systems. However, with minor adaptations in the spacecraft designs their survivability can be significantly enhanced. One such example is the Canadian Radarsat (Terrillon et al., 1991; anon., 1995). Radarsat was launched in November 1995 into a Sun-synchronous orbit of altitude 788 km $\times$ 794 km at 98.59° inclination. In this orbit regime most of the critical impacts occur at velocities of 13 km/s $\leq v_p \leq$ 15 km/s under azimuth angles of $-30° \leq A \leq +30°$ (see Fig. 4.4 and 4.11). Since Radarsat is 3-axis-stabilized, with constant yaw, roll, and pitch relative to an orbit-related coordinate system, the deployment of debris impact protection could be concentrated on surfaces close to the ram direction. An analysis by NASA indicated that hypervelocity impacts by 1 mm aluminum projectiles could seriously damage unprotected 24-gauge wire bundles, and impactors of 0.4 mm size could perforate unprotected hydrazine lines. Further tests showed that the deployment of low-weight Nextel shields effectively protected the wiring and piping. As a consequence, a layer of Nextel was added to the multi-layer insulation (MLI) blankets which covered the external electronic boxes and harnesses. Moreover, critical boxes received thicker walls, or were moved away from the debris approach direction, behind less critical hardware. With a total mass penalty of 0.6% (17 kg) for debris protection measures the Radarsat designers achieved an improvement in system survivability from 50% to 87% over a planned 5-year mission lifetime.

The survivability of a space object is a function of its size and geometry, its structure, its orbit and attitude, and its exposure time to the space debris and meteoroid environment. A measure of survivability is the probability of no failure (PNF), which is often synonymously denoted as probability of no penetration (PNP). Let an individual system component have a ballistic limit $d_{p,lim}$, in terms of defeatable projectile diameter. If the most probable impact velocities are high (e.g. $\overline{v}_p \approx 10.5$ km/s for ISS, or $\overline{v}_p \approx 14.5$ km/s for Sun-synchronous orbits), then one may assume a constant mean value of $d_{p,lim}$ across the relevant velocity range with the highest flux contribution (compare Fig. 7.7). Based on the concepts of Poisson statistics (see Eq. 3.2) the probability of no penetration can be determined from the impact flux $F(d \geq d_{p,lim})$ of debris and meteoroid objects which exceed the ballistic limit, from the collision cross-section A of the target, and from the time Δt of exposure to the space debris environment.

$$P_{\text{PNP}} = \exp\Big(-F(d \geq d_{p,lim})\,A\,\Delta t\Big) \qquad (7.16)$$

The probability $P_{n\geq 1}$ of one or more failures due to shield penetration is given by the complement of Eq. 7.16.

$$P_{n\geq 1} = 1 - P_{\text{PNP}} \qquad (7.17)$$

For complex spacecraft a combined failure analysis of all critical sub-system components should be performed, also taking into account geometric shielding effects by up-stream equipment boxes.

The overall system reliability and survivability should be verified against well defined failure criteria, which should be based on extensive HVI tests and hydrocode simulations. For the International Space Station design requirements call for a probability of less than 0.5% per year that a debris or meteoroid impact causes a critical failure (anon., 1995). This requirement is in line with an annual impact probability of 0.07% for an object of $d_p \geq d_{p,lim}^{(ISS)} \approx 1\,\text{cm}$, and for a projected ISS cross-section of $\overline{A} = 100\,\text{m}^2$, according to the MASTER-2001 model (see also Table 4.4).

7.6 REFERENCES

anon. (1995). *Orbital Debris – A Technical Assessment*. National Academy Press, Washington, DC.

anon. (2001). *Position Paper on Orbital Debris*. International Academy of Astronautics, Paris, France.

anon. (2004). Protection Manual. Technical Report IADC-WD-00-03, rel. 3.3, Inter-Agency Space Debris Coordination Committee (IADC).

Grün, E., Gustafson, B., Dermott, S., and Fechtig, H. (2001). *Interplanetary Dust*. Springer Verlag, Berlin/Heidelberg/New York.

Klinkrad, H. et al. (2003). *ESA Space Debris Mitigation Handbook*. ESA/ESOC, second edition. issue 1.0.

Terrillon, F., Warren, H., and Yelle, M. (1991). Orbital Debris Shielding Design of the RADARSAT Satellite. *42nd IAF Congress*, IAF-91-283.

Turner, R., Taylor, E., McDonnell, J., Stokes, H., Marriot, P., Wilkinson, J., Catling, D., Vignjevic, R., Berthoud, L., and Lambert, M. (2002). Cost-Effective Debris Shields for Unmanned Spacecraft. In *International Journal of Impact Engineering*, volume vol.28, no.2, pages 365–378.

8

Operational Collision Avoidance with Regard to Catalog Objects

H. Klinkrad, J. Alarcón and N. Sánchez

In previous chapters all collision flux estimates and corresponding collision probabilities were purely based on stochastic methods which did not consider the estimated orbital positions of objects. While orbits of the whole debris population were propagated across the historic evolution of the environment, the coarse first-order prediction methods were only used to derive time histories of spatial object densities and transient velocities in a gridded 3D control volume. Such statistical assessments are justified for small debris objects which were generated by stochastic release models into orbits with poorly known initial conditions. For larger-size objects, however, USSTRATCOM is maintaining a catalog of tracked objects (see Chapter 2), with routinely updated orbit information in a so-called Two-Line Element (TLE) format. When used with care, such data can be employed to improve the safety of operational spacecraft by means of a conjunction prediction and collision warning service for missions in densely populated orbital regions.

8.1 ORBIT PREDICTION AND ASSOCIATED UNCERTAINTIES

The subsequent conjunction event analysis shall be illustrated at the example of ESA's ERS-2 and Envisat satellites, which are operated at mean altitudes of $\overline{H} \approx$ 780 km, on near-circular, Sun-synchronous orbits. Different levels of accuracy are available for the orbit determination products of these satellites, depending on the user requirements, and on the accepted turn-around time. For quick-look, near real-time data, the orbit determination position accuracy over fitted observation arcs at mean atmospheric conditions is on the order of $\Delta r_{U,1\sigma} \approx 0.5$ m in radial direction, $\Delta r_{W,1\sigma} \approx 1.0$ m in out-of-plane direction, and $\Delta r_{V,1\sigma} \approx 3.0$ m in along-track direction. The corresponding velocity errors are $\Delta v_{U,1\sigma} \approx 3.0$ mm/s, $\Delta v_{W,1\sigma} \approx 1.0$ mm/s, and $\Delta v_{V,1\sigma} \approx 1.0$ mm/s. Though these accuracies are almost two orders of magnitude worse than the best possible fits, they are still two orders of magnitude better than the presumable accuracy of the TLE data which are provided by the USSTRATCOM catalog (see Table 8.1). Hence, when orbit determination and orbit prediction accuracy is dealt with hereafter, the focus will be on the TLE data sets.

TLE data are provided as mean, doubly averaged Kepler states (with mean motion instead of semimajor axis), according to Brouwer's theory, with adaptations by Lane, Crawford, and Hujsak as described in (Hoots and Roehrich, 1980). The data are formatted in two lines (hence "Two-Line Elements"), with additional information on the reference epoch, on a drag related parameter, and on the international designation codes of the tracked objects. TLE orbits are described in a true equator and mean equinox reference frame (TEME), where the orientation of the equator plane accounts for precession and nutation of the Z-axis under the influence of Sun and Moon (see Fig.A.2), and the X-axis points to the meridian of the mean vernal equinox in a system which accounts for precession only. To recover the best possible accuracy of fitted, osculating orbits, TLE data must be processed by the SGP-4 or SDP-4 software (Simplified General Perturbations theory, and Simplified Deep-Space Perturbations theory, rev.4, (Hoots and Roehrich, 1980)).

The SGP-4 theory is applied for all orbits with periods of $T \leq 225\,\text{min}$. It performs a propagation in time of doubly averaged elements according to their secular rates of change due to the zonal harmonics J_2 and J_4 of the Earth potential, and due to drag perturbations in an atmosphere with a power-law altitude profile of air density. The propagated, doubly averaged elements at epoch are subsequently converted into singly averaged elements, by overlaying long-periodic perturbations due to J_3, before a final conversion step to osculating elements by superimposition of first-order, short-period perturbation amplitudes due to J_2. The underlying principle is illustrated in Fig. A.5 (see also Eq. A.46 to A.50).

The SDP-4 theory applies for all orbits with a period of $T > 225\,\text{min}$. It considers secular and long-periodic perturbations due to J_2, J_3, and J_4, resonant geopotential perturbations for 12-hour and 24-hour orbits, luni-solar perturbations, and a non-conservative perturbation term which can accommodate airdrag and solar radiation pressure effects. The reconstruction of osculating elements is (as for SGP-4) limited to first-order accuracy, taking into account J_2 short-periodic amplitudes. Due to a concentration on low-Earth orbits, the SDP-4 theory will only play a minor role in the following investigations.

TLE data per se are not meant for precise orbit analyses. Furthermore, a metric of the quality of the orbit fit by TLE sets is not available to the general user. To some extent, however, the latter problem can be overcome. If one assumes that a TLE data set is the best possible fit to an observed, osculating orbit over a timespan of typically 36 hours (Foster, 2001b), then one may invert the process, and conclude that a first-order replicate of the original orbit (accurate to J_2) can be reconstructed by performing a least-squares fit to an analytically generated SGP-4 arc with a numerical integrator, using an accurate and realistic model of the effective perturbations. This concept is applied to 14 catalog objects with perigee

Table 8.1. Assessed accuracy of TLE data for Catalog orbits with eccentricities of $e < 0.1$ and perigees of $H_{pe} < 800$ km. Position and velocity uncertainties are provided as 1σ values in the radial (U), transversal (V), and out-of-plane direction (W) of an orbit-related coordinate system (see Fig. A.2).

	Position error [m]			Velocity error [mm/s]		
	$\Delta r_{U,1\sigma}$	$\Delta r_{V,1\sigma}$	$\Delta r_{W,1\sigma}$	$\Delta v_{U,1\sigma}$	$\Delta v_{V,1\sigma}$	$\Delta v_{W,1\sigma}$
$i < 30°$	102	419	122	404	112	118
$30° \leq i \leq 60°$	129	434	163	428	142	186
$i > 60°$	104	556	139	559	110	148

altitudes of $338\,\text{km} \leq H_{pe} \leq 35{,}781\,\text{km}$, orbit eccentricities of $0.0006 \leq e \leq 0.728$, and inclinations of $4.32° \leq i \leq 98.27°$. For each of the 14 sample objects a 24-hour SGP-4 orbit is numerically fitted by adjusting an initial osculating state at epoch, and a drag calibration parameter. As a by-product of the batch least-squares orbit determination, an error covariance matrix is produced and stored in a look-up table for 12 different orbit categories, partitioned into 3 perigee altitude classes ($H_{pe} < 800$ km, $800\,\text{km} \leq H_{pe} \leq 25{,}000$ km, and $H_{pe} > 25{,}000$ km), 3 inclination classes ($i < 30°$, $30° \leq i \leq 60°$, and $i > 60°$), and 2 eccentricity classes ($e \leq 0.1$ and $e > 0.1$). Table 8.1 summarizes the 1σ uncertainties in position and velocity for near-circular low-Earth catalog orbits, which are the main source of conjunction objects for ERS-2 and Envisat. The process of deriving covariance information shall now be briefly outlined.

Let H be the Jacobian of the partial derivatives of the modeled observations $\underline{h}(\underline{x}_\circ)$ (here: TLE derived osculating states) with respect to the state vector $\underline{x}_\circ = (\underline{r}_\circ, \underline{v}_\circ)$ at initial epoch $t_\circ$.

$$\mathsf{H} = \left. \frac{\partial \underline{h}(\underline{x}_\circ)}{\partial \underline{x}_\circ} \right|_{\underline{x}_\circ = \hat{\underline{x}}_\circ} \tag{8.1}$$

where $\hat{\underline{x}}_\circ$ is the initial state estimate. Furthermore, let W be a weighting matrix, inversely proportional to the squares of the standard deviations σ_n of the uncorrelated measurement errors.

$$\mathsf{W} = \operatorname{diag}\left(\sigma_1^{-2}, \ldots, \sigma_n^{-2}, \ldots, \sigma_N^{-2}\right) \tag{8.2}$$

The error covariance matrix $\mathsf{C}_\circ$ of the fitted initial state $\underline{x}_\circ$ at epoch $t_\circ$ is then given by the inverse of the normal matrix of the batch least-squares process (Montenbruck and Gill, 2000).

$$\mathsf{C}_\circ = \left(\mathsf{H}^T \mathsf{W} \mathsf{H}\right)^{-1} \tag{8.3}$$

This covariance matrix can be mapped to a different epoch t by means of the state transition matrix $\Phi\,(t_\circ, t)$.

$$\mathsf{C}(t) = \Phi\,(t_\circ, t)\;\mathsf{C}_\circ(t_\circ)\;\Phi^T\,(t_\circ, t) \tag{8.4}$$

Φ describes the sensitivity of the predicted state $\underline{x}(t)$ in response to a variation of the initial state $\underline{x}_\circ(t_\circ)$.

$$\Phi\,(t_\circ,t) = \frac{\partial \underline{x}(t)}{\partial \underline{x}_\circ(t_\circ)} \tag{8.5}$$

The variational equations on the right-hand side can be expressed analytically for unperturbed Kepler orbits and for J_2 perturbed orbits. For more refined perturbation models, they must be integrated numerically, together with the equations of motion (see Eq. A.30 and A.31). Partial derivatives of the perturbing forces, which are required for the integration in time of Eq. 8.5, are provided in (Montenbruck and Gill, 2000) and (Alarcón, 2002).

The integration of the variational equations can cause a large computational burden, if too detailed perturbation models are used, particularly for the geopotential. Trial computations demonstrated that a sufficient accuracy is obtained for a perturbation model consisting of an Earth gravity field with spherical harmonics up to degree and order 5, airdrag based on an MSISe-90 atmosphere, direct solar radiation pressure, and luni-solar attraction.

All conjunction event computations hereafter shall be referred to a mean equatorial coordinate system of 2000.0 (denoted as J2000), with coordinates X, Y, and Z (see Fig. A.2). In this system, the orientation of an orbit-related coordinate system is described by the unit vectors in radial ($\underline{U}$), transversal ($\underline{V}$), and out-of-plane direction ($\underline{W}$). For a given orbit state $\underline{x} = (\underline{r}, \underline{v})$ at time t, they can be derived from

$$\underline{U} = \frac{\underline{r}}{|\underline{r}|} \qquad \underline{W} = \frac{\underline{r} \times \underline{v}}{|\underline{r} \times \underline{v}|} \qquad \underline{V} = \underline{W} \times \underline{U} \tag{8.6}$$

These direction vectors define a transformation matrix $\mathsf{R}_{U,V,W}$ which can be used to map the covariance matrix $\mathsf{C} = \mathsf{C}_{X,Y,Z}$ into the orbit-related frame.

$$\mathsf{C}_{U,V,W} = \mathsf{R}_{U,V,W}\ \mathsf{C}\ \mathsf{R}^T_{U,V,W} \tag{8.7}$$

$$\mathsf{R}_{U,V,W} = \begin{pmatrix} U_X & U_Y & U_Z \\ V_X & V_Y & V_Z \\ W_X & W_Y & W_Z \end{pmatrix} \tag{8.8}$$

The diagonal elements of the covariance matrix $\mathsf{C}_{U,V,W}$ now describe the 1σ variations of the radial, transversal, and out-of-plane components of position and velocity, which are more directly linked to the underlying orbit mechanics, and which are thus easier to interpret than in the X, Y, Z frame. In Table 8.1, corresponding standard deviations of the covariance matrix $\mathsf{C}_{U,V,W}(t_\circ)$ are listed for different fits to TLE-generated orbits. In these fits, given a satellite specific area-to-mass ratio A/m, the drag coefficient c_D of the ballistic parameter $B = c_D\,A/m$ was adjusted as part of an extended initial state vector, $\hat{\underline{x}}_\circ = (\underline{r}_\circ, \underline{v}_\circ, c_D)$. For the following analysis it will be assumed that this drag calibration term is uncorrelated with the orbit state (which may not be true in high-drag environments, as noted in (Matney et al., 2002)). The 1σ uncertainty in the c_D coefficient shall be adopted as 20% of its nominal value, reflecting combined errors in the ambient air density, and in the interaction of the satellite with the free-molecular flow.

8.2 DETERMINATION OF NEAR-MISS CONJUNCTION EVENTS

As part of a collision risk assessment one first needs to determine events of near misses between pairs of space objects from the trackable catalog population. The focus hereafter will be on the collision risk of ESA's ERS-2 and Envisat satellites with objects of the USSTRATCOM catalog. The orbits of the ESA satellites will be assumed to be available from an operational orbit determination at epoch t_o, with resulting orbit files containing fitted and predicted states for a timespan of $t \in [t_o - 2\,\mathrm{d}, t_o + 8\,\mathrm{d}]$, to cover (with safety margin) a typical prediction timespan of +7 days. The operational satellites will hereafter be denoted as targets (index "t"). They have operational orbit determination accuracies as mentioned in Section 8.1. The potential conjunction objects are extracted from the USSTRATCOM catalog, with recent orbit information provided in TLE format. These TLE data are propagated across the forecast time interval of +7 days, providing first-order osculating states by means of the SGP-4 orbit theory (the SDP-4 theory is used, if the orbital period exceeds 225 minutes). The conjunction counterparts will be denoted as risk objects (index "r"). Their orbits have deduced standard deviations relative to reconstructed "true" orbits as listed in Table 8.1.

Different methods have been devised to determine close conjunction events between pairs of objects during a pre-defined time interval. (Hoots et al., 1984) developed a theory which is based on geometric considerations (the same concept was applied for instance by (Bérend, 1997) and (Klinkrad, 1997)). The conjunction detection in this case is performed by the successive application of an altitude filter, a plane geometry filter, and a phase filter. This approach is particularly useful, if the two orbits are available through an analytic orbit theory.

The altitude filter rejects all risk objects whose orbits do not intersect the spherical altitude shell between the perigee and apogee of the target orbit, taking into account the orbital decay and short-periodic altitude variations of both objects within the forecast interval. Objects which pass this filter are subjected to an orbit geometry check, where the intersection line of the planes of the risk and target orbit is determined. For the given intersection geometry the closest approach distance at the ascending and descending locations of the nodal line are computed. Conjunction event predictions which are within a user-defined, spherical control volume of radius R_c (see Eq. 8.23 and Eq. 8.9ff), centered on the target position, are maintained as potential candidates for a close fly-by. For the orbits of both, the target and risk object, the times at passes of the common lines of nodes of the orbit plane intersections are stored, with a symmetric time window to account for a safety distance of $|\Delta \underline{r}| \leq R_c$. In a final step, an orbit phase filter is applied, which checks for overlaps of the two separate time window patterns within the forecast timespan. If overlaps are detected, then an iterative Newton scheme is applied to find the zero-transition of the range-rate between the two objects ($\Delta \underline{v}\, \Delta \underline{r} / \Delta r = \dot{\rho} = 0.0$), which marks the time of closest approach, and the local minimum of the distance between the orbits. Special techniques must be applied in the case of near-coplanar orbits.

The outlined detection concept for proximity events entails complex filter algorithms, with CPU-time demanding iterative root finders, and with a non-negligible risk of missing close fly-bys due to perturbations affecting the conjunction geometries. With ever-increasing computer capacities alternative algorithms for conjunction event detections became a viable alternative. The following sieve algorithm has been implemented in ESA's CRASS software (Alarcón, 2002).

As first step of the sieve, again an altitude filter is applied to the two orbit histories of the target and risk object. In a subsequent step, the time history of ranges $\rho(t)$ between the target orbit $\underline{r}_t(t)$ and the risk orbit $\underline{r}_r(t)$ is analyzed at equidistant time steps Δt across the prediction interval (for instance $t \in [t_\circ, t_\circ + 7\,\mathrm{d}]$). At each time step the coordinates of the range vector are checked against an adjusted safety distance

$$R_{c,1} = R_c + v_e\,\Delta t \tag{8.9}$$

where $v_e = \sqrt{2\mu/r}$ is the escape velocity. A close conjunction within a time interval Δt is not possible, if $\rho_X > R_{c,1}$, or $\rho_Y > R_{c,1}$, or $\rho_Z > R_{c,1}$, or $\rho > R_{c,1}$ (checked in this sequence). For those orbits which passed this step, the reference control volume of radius R_c around the target object cannot be violated, if a curvature-adjusted safety distance $R_{c,2} = R_c + g\,(\Delta t)^2$ is not penetrated (where g is the gravity acceleration of the Earth). This corresponds to the following condition.

$$R_{c,2} = R_c + g\,(\Delta t)^2 < \sqrt{(\Delta r)^2 - (\Delta\underline{r}\cdot\Delta\underline{v}/\Delta v)^2} \tag{8.10}$$

The threshold $R_{c,1}$ can be further refined, considering the effective relative velocity, rather than the escape velocity, to obtain

$$R_{c,3} = R_{c,2} + \frac{1}{2}\,|\,\Delta\underline{v}\cdot\Delta\underline{r}/\Delta r\,|\,\Delta t \tag{8.11}$$

For those orbits passing the previous range sieves a root finder is started, which determines the time t_{tca} of closest approach $\Delta r_{tca} = \rho_{tca}$ from a zero-transition of the range-rate time history.

$$\Delta\underline{v}_{tca}\,\Delta\underline{r}_{tca}/\Delta r_{tca} = \dot{\rho}_{tca} = 0.0 \qquad \rightarrow \qquad t_{tca} \tag{8.12}$$

Table 8.2 shows the effectiveness of different steps in the conjunction event sieve. For each of the identified conjunctions a final test is performed to verify that $\Delta r_{tca} \le R_c$. At the established time t_{tca} and for the associated target state $\underline{x}_t = (\underline{r}_t, \underline{v}_t)$ a target-centered, orbit-related coordinate system $\underline{U}_t$, $\underline{V}_t$, $\underline{W}_t$ can be deployed according to Eq. 8.6. This leads to a transformation matrix $(\mathsf{R}_t)_{U,V,W}$ as described in Eq. 8.8, which allows a mapping $(X,Y,Z) \mapsto (U,V,W)$ of the conjunction distance $\Delta\underline{r}_{tca}$, and approach velocity $\Delta\underline{v}_{tca}$ in the case of a direct hit.

$$(\Delta\underline{r}_{tca})_{U,V,W} = (\Delta\underline{r}_t)_{U,V,W} = (\mathsf{R}_t)_{U,V,W}\,(\Delta\underline{r}_{tca})_{X,Y,Z} \tag{8.13}$$

$$(\Delta\underline{v}_{tca})_{U,V,W} = (\Delta\underline{v}_t)_{U,V,W} = (\mathsf{R}_t)_{U,V,W}\,(\Delta\underline{v}_{tca})_{X,Y,Z} \tag{8.14}$$

where $\Delta\underline{r}_{tca}$ and $\Delta\underline{v}_{tca}$ are defined as

$$(\Delta\underline{r}_{tca})_{X,Y,Z} = (\,\underline{r}_r(t_{tca}) - \underline{r}_t(t_{tca})\,)_{X,Y,Z} \tag{8.15}$$

$$(\Delta\underline{v}_{tca})_{X,Y,Z} = (\,\underline{v}_r(t_{tca}) - \underline{v}_t(t_{tca})\,)_{X,Y,Z} \tag{8.16}$$

Table 8.2. Performance of successive sieve steps for the detection of close conjunction events in the CRASS software (see Eq. 8.9 to 8.11 for a definition of the threshold radii $R_{c,1}$, $R_{c,2}$, and $R_{c,3}$).

Conjunction sieve	Rejected/analyzed (%)	Passed/total (%)
altitude range filter	59.4	40.6
$\rho_X \leq R_{c,1}$	74.3	10.4
$\rho_Y \leq R_{c,1}$	72.9	2.8
$\rho_Z \leq R_{c,1}$	50.8	1.4
$\rho \leq R_{c,1}$	34.5	0.92
$R_{c,2} \leq \sqrt{(\Delta r)^2 - (\Delta \underline{r} \cdot \Delta \underline{v}/\Delta v)^2}$	4.3	0.88
$\rho \leq R_{c,3}$	41.9	0.51
$\rho_{tca}\,(\dot{\rho} = 0.0) \leq R_c$	99.0	0.005

Using Eq. 4.6, 4.7 and 4.8, the vector $(\Delta \underline{r}_{tca})_{U,V,W}$ can be expressed in terms of distance Δr_{tca}, azimuth $A_{r,tca}$, and elevation $h_{r,tca}$ of the conjunction, as illustrated in Fig. A.3. Correspondingly, the approach velocity $(\Delta \underline{v}_{tca})_{U,V,W}$ of the risk object can be expressed in terms of its magnitude Δv_{tca}, and the azimuth $A^{\star}_{v,tca} = A_{v,tca} + \pi$, and elevation $h^{\star}_{v,tca} = -h_{v,tca}$ of the direction *from* which the risk object would approach in the case of a direct hit on the target.

So far, a spherical threshold surface has been deployed around the target. Due to the secular effects of drag perturbations on the orbit period, with a resulting dominance of along-track position uncertainties, it is advantageous to define an ellipsoidal threshold surface, with its major axis in the along-track direction. The CRASS program uses an ellipsoid aligned with the $\underline{U}_t, \underline{V}_t, \underline{W}_t$ axes, with dimensions $R_{c,U} = 10\,\text{km}$, $R_{c,V} = 25\,\text{km}$, and $R_{c,W} = 10\,\text{km}$, for a spherical threshold radius of $R_c \geq \max(R_{c,U}, R_{c,V}, R_{c,W})$ used in the sieve algorithms. A conjunction event falls into the ellipsoidal control volume, if $k_c^2 \leq 1$ in the following equation. Else, the event is rejected from further analysis.

$$k_c^2 = \left(\frac{\Delta r_U}{R_{c,U}}\right)^2 + \left(\frac{\Delta r_V}{R_{c,V}}\right)^2 + \left(\frac{\Delta r_W}{R_{c,W}}\right)^2 \qquad (8.17)$$

Owing to more frequent calls to the sieve algorithms, the computer run-time for the conjunction event sieve increases for steps that are too small. The run-time also increases for time steps that are too large, since fewer orbits can be rejected by the CPU-time-efficient initial sieve steps. In between these extremes one can find an optimal analysis time step, which is on the order of $\Delta t \approx 180\,\text{s}$ (about 3% of the orbital period of ERS-2 and Envisat).

Due to their simplicity and robustness the outlined sieve algorithms are found to be superior in their CPU time efficiency, and in their conjunction event detection capability, as compared with traditional methods which are based on filters considering altitude, orbit plane intersection, orbit phasing, and minimum approach distance.

8.3 COLLISION RISK ESTIMATE FOR NEAR-MISS CONJUNCTIONS

Each event which passes the criteria of the conjunction event sieve can be assessed for its collision probability relative to the target object. Several authors have developed methods to achieve this objective (Foster, 1992; Khutorovsky et al., 1993; Bérend, 1997; Alfriend et al., 1999; Klinkrad, 1997; Patera, 2001). All these approaches have the following assumptions in common:

- The position uncertainty can be described by a 3D Gaussian distribution, with a probability density function as given by Eq. 8.22.
- The target and risk object move along straight lines at constant velocities.
- The uncertainties in the velocities can be neglected.
- The target and risk object position uncertainties are not correlated.
- The position uncertainties during the encounter are constant, with corresponding covariances as at the time of closest approach.

The collision probability assessment scheme described in the following is based on the formulations by Alfriend and Akella (Alfriend et al., 1999). It uses as inputs the relative position $\Delta\underline{r}_{tca}$ (Eq. 8.15) and relative velocity $\Delta\underline{v}_{tca}$ (Eq. 8.16) of the conjunction risk object relative to the target, at the time t_{tca} of the closest approach. In the vicinity of the event the relative risk object position $\Delta\underline{r}$ can be expressed as a function of the time offset $\Delta t_{tca} = t - t_{tca}$.

$$\Delta\underline{r}(t) = \Delta\underline{r}_{tca} + \Delta\underline{v}_{tca}\,(t - t_{tca}) \tag{8.18}$$

At the conjunction time t_{tca}, in accordance with the assumptions made, the propagated (7×7) error covariances of the extended target and risk object state vectors $\underline{\hat{x}}_{t,tca} = (\underline{r}_{t,tca}, \underline{v}_{t,tca}, (c_D)_{t,tca})$ and $\underline{\hat{x}}_{r,tca} = (\underline{r}_{r,tca}, \underline{v}_{r,tca}, (c_D)_{r,tca})$ can be written as

$$\hat{C}_t(t_{tca}) = \Phi_t\,(t_o, t_{tca})\;\hat{C}_{t,o}(t_o)\;\Phi_t^T\,(t_o, t_{tca}) \tag{8.19}$$

$$\hat{C}_r(t_{tca}) = \Phi_r\,(t_o, t_{tca})\;\hat{C}_{r,o}(t_o)\;\Phi_r^T\,(t_o, t_{tca}) \tag{8.20}$$

with state transition matrices as defined in Eq. 8.5. Since the errors in the orbit states of both objects are considered to be uncorrelated, both contributions can be combined into a common covariance matrix.

$$\hat{C} = \hat{C}(t_{tca}) = \hat{C}_t(t_{tca}) + \hat{C}_r(t_{tca}) \qquad \left(\hat{C}_{7\times 7} \mapsto C_{3\times 3}\right) \tag{8.21}$$

Of this (7×7) matrix, only the position uncertainties of the upper left (3×3) sub-matrix shall be taken into account, which correspond to the position error covariance matrix $C = C(t_{tca})$, with a 1σ error ellipsoid as illustrated in Fig. 8.1. Since the position error is assumed to have a 3D normal distribution, the probability density function $p(\Delta\underline{r})$ in the vicinity of the point of closest approach can be described as

$$p\,(\Delta\underline{r}) = \frac{1}{\sqrt{(2\pi)^3\det(C)}}\,\exp\left[-\frac{1}{2}\,\Delta\underline{r}^T\,C^{-1}\,\Delta\underline{r}\right] \tag{8.22}$$

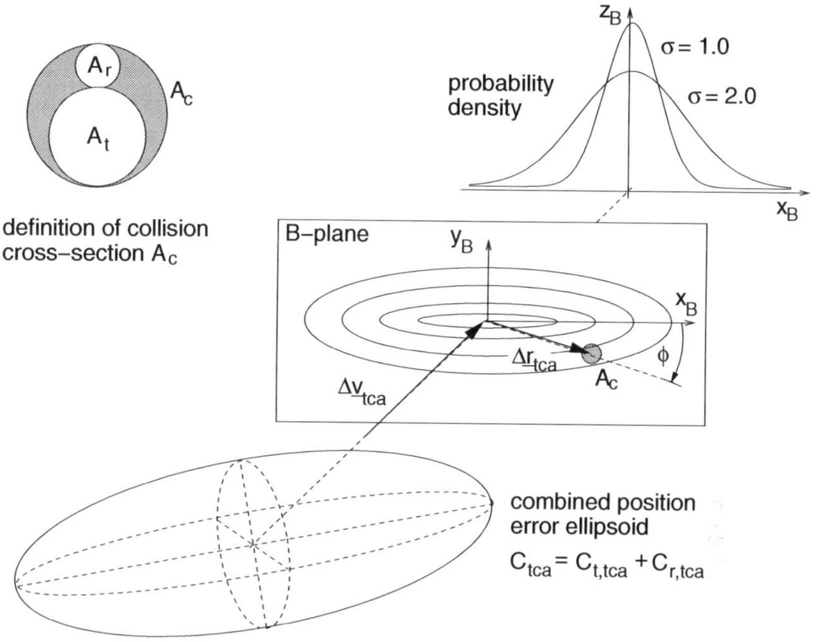

Fig. 8.1. 2D mapping onto the B-plane of the 3D combined position uncertainty of the target and risk object at the time of closest approach (tca). The target position is centered at $x_B, y_B = 0$, and the risk object transits the B-plane at a fly-by distance of Δr_{tca}, (where $\Delta \underline{r}_{\mathrm{tca}} \parallel$ B-plane) with a velocity of Δv_{tca} (where $\Delta \underline{v}_{\mathrm{tca}} \perp$ B-plane). The collision probability is determined by the size of the combined collision cross-section A_c, and by the conjunction location within the 2D probability density distribution (see also Fig. 8.2).

Let R_t and R_r be the radii of spheres which enclose the target and risk object, respectively. A collision will occur, if these two spheres intersect, or in other words, if a closest approach occurs within the volume described by a sphere of collision radius R_c, cross-section A_c, and volume V_c.

$$R_c = R_t + R_r \qquad A_c = \pi R_c^2 \qquad V_c = \frac{4}{3} \pi R_c^3 \tag{8.23}$$

The probability of collision P_c can be determined from a volume integral of the 3D probability density function in Eq. 8.22, over the spherical region V_c, centered on the risk object position.

$$P_c = \frac{1}{\sqrt{(2\pi)^3 \det(\mathrm{C})}} \int_{(V_c)} \exp\left[-\frac{1}{2} \Delta \underline{r}^T \, \mathrm{C}^{-1} \, \Delta \underline{r} \right] \mathrm{d}V \tag{8.24}$$

One can demonstrate that this volume integral can be reduced to a surface integral, by mapping the position error ellipsoid onto elliptical contours of constant probability on the B-plane (Foster, 1992). The B-plane[8.1] is perpendicular to the relative velocity vector $\Delta \underline{v}_{tca}$ at the time of closest approach (see Fig. 8.1). Due to Eq. 8.12 the range-rate at this point has a zero transit, which means that the conjunction range vector $\Delta \underline{r}_{tca}$ lies within the B-plane. Let the unit direction vectors $\underline{X}_B$ and $\underline{Y}_B$ be defined as

$$\underline{X}_B = \frac{\Delta \underline{r}_{tca}}{|\Delta \underline{r}_{tca}|} \qquad \underline{Y}_B = \frac{(\Delta \underline{r}_{tca}) \times (\Delta \underline{v}_{tca})}{|(\Delta \underline{r}_{tca}) \times (\Delta \underline{v}_{tca})|} \tag{8.25}$$

These two unit vectors define a transformation matrix R_{X_B,Y_B} which maps the 3D covariance matrix $\mathsf{C} = \mathsf{C}_{X,Y,Z}$ onto a 2D covariance matrix C_B in the B-plane.

$$\mathsf{C}_B = \mathsf{C}_{X_B,Y_B} = \mathsf{R}_{X_B,Y_B}\, \mathsf{C}\, \mathsf{R}^T_{X_B,Y_B} \tag{8.26}$$

$$\mathsf{R}_{X_B,Y_B} = \begin{pmatrix} X_{B,X} & X_{B,Y} & X_{B,Z} \\ Y_{B,X} & Y_{B,Y} & Y_{B,Z} \end{pmatrix} \tag{8.27}$$

The main axes orientation of the elliptic contours of equal probability can be determined from the eigenvalues $\lambda_{i,B} (i = 1, 2)$ and their corresponding eigenvectors $\underline{e}_{i,B}$ which solve the equation

$$(\mathsf{C}_B - \lambda_{i,B} \mathsf{I})\, \underline{e}_{i,B} = \underline{0} \tag{8.28}$$

where I is the (2×2) identity matrix. The semimajor axis of the 1σ position error ellipse in the B-plane is defined as $a_{1\sigma,B} = \sqrt{\max(\lambda_{1,B}, \lambda_{2,B})}$. The semiminor axis is defined as $b_{1\sigma,B} = \sqrt{\min(\lambda_{1,B}, \lambda_{2,B})}$. The direction of the semimajor axis of the ellipse shall be denoted as $\underline{e}_{a,B}$. The corresponding unit vector is $\underline{x}_B = \underline{e}_{a,B}/|\underline{e}_{a,B}|$, and its orientation with respect to the conjunction direction $\underline{X}_B$ is given by the rotation angle ϕ_B (see Fig. 8.1).

$$\phi_B = \arccos(\underline{x}_B\, \underline{X}_B) \tag{8.29}$$

Let $\Delta \underline{r}_B$ be a conjunction position within the B-plane, then the 3D collision probability integral in Eq. 8.24 can be reduced to a 2D integral over the projected, circular collision cross-section of radius R_c, centered at the predicted fly-by location $\Delta \underline{r}_{tca}$.

$$P_c = \frac{1}{2\pi \sqrt{\det(\mathsf{C}_B)}} \int_{-R_c}^{+R_c} \int_{-\sqrt{R_c^2 - x_B^2}}^{+\sqrt{R_c^2 - x_B^2}} \exp[-A_B]\, dy_B\, dx_B \tag{8.30}$$

$$A_B = \frac{1}{2} \Delta \underline{r}_B^T\, \mathsf{C}_B^{-1}\, \Delta \underline{r}_B \tag{8.31}$$

Fig. 8.1 and 8.2 illustrate that with increasing position errors the probability density function is spread over a larger area. Since its total integral is always 1, there must be a certain scaling factor k_σ to the 1σ error distribution which yields a

[8.1] the term B-plane is adopted from the nomenclature of planetary fly-bys

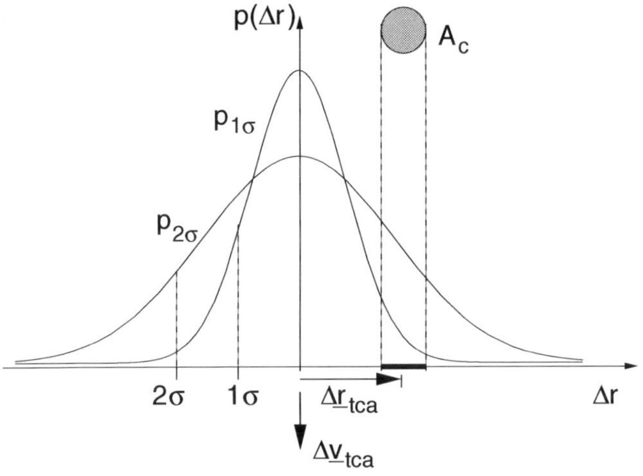

Fig. 8.2. Determination of the collision risk by integration of the position probability density over the combined collision cross-section A_c at a location $\Delta r_{\rm tca}$ of the risk object, relative to the target (cut along $\Delta \underline{r}_{\rm tca}$ in Fig. 8.1, vertical to the B-plane).

maximum probability density $p(\Delta \underline{r}_{tca})$, and hence a maximum collision probability P_c for a given position offset $\Delta \underline{r}_{tca}$ in the B-plane. If the 1σ dispersion is scaled with k_σ, then the covariance is scaled according to

$$C_B^\star = k_\sigma^2 \, C_B \tag{8.32}$$

Analogous to Eq. 8.22 the scaled probability density distribution function in the B-plane can be defined.

$$p_B\left(\Delta \underline{r}_B\right) = \frac{1}{2\pi \, k_\sigma^2 \, \sqrt{\det(C_B)}} \exp\left[-\frac{1}{k_\sigma^2} A_B\right] \tag{8.33}$$

The resulting derivative $\partial p_B / \partial k_\sigma$ reads as follows (Bérend, 1997):

$$\frac{\partial p_B}{\partial k_\sigma} = \frac{1}{2\pi \, k_\sigma^2 \, \sqrt{\det(C_B)}} \left[-\frac{2}{k_\sigma^3} + \frac{2A_B}{k_\sigma^5}\right] \exp\left[-\frac{1}{k_\sigma^2} A_B\right] \tag{8.34}$$

This function has a zero transition, with a corresponding maximum in p_B for

$$k_{\sigma,max}^2 = \frac{1}{2} \Delta \underline{r}_B^T \, C_B^{-1} \, \Delta \underline{r}_B \tag{8.35}$$

Hence, Eq. 8.33 has a maximum for the scaling parameter k_σ which makes the argument of the exponent equal to -1. For a given distance $\Delta \underline{r}_{tca}$ of the closest approach, and for a fixed geometry ($a_{1\sigma,B}/b_{1\sigma,B}$ = const) and orientation (ϕ_B = const) of the error ellipse in the B-plane, the maximum possible collision probability $P_{c,max}$ is obtained by applying $k^2_{\sigma,max}$ in Eq. 8.32, and by using this scaled covariance matrix $C^\star_B$ to evaluate Eq. 8.30. If one assumes that the probability is constant[8.2] within the integration radius R_c, with its magnitude adopted at the integration center $\Delta \underline{r}_{tca}$, then the maximum collision probability according to Eq. 8.30 can be expressed in a compact way (Alfriend et al., 1999).

$$P_{c,max} = \frac{R_c^2}{\exp(1)\,\sqrt{\det(C_B)}\;\Delta \underline{r}_{tca}^T\, C_B^{-1}\, \Delta \underline{r}_{tca}} \tag{8.36}$$

The outlined algorithms for a statistical collision risk assessment can be validated by means of Monte Carlo techniques. A detailed summary of corresponding evaluations for the methods of (Alfriend et al., 1999) and (Bérend, 1997), (Khutorovsky et al., 1993), and (Klinkrad, 1997) can be found in (Alarcón, 2002). The solution concepts of (Khutorovsky et al., 1993) and (Klinkrad, 1997) showed limitations for close fly-bys with small position errors and large collision cross-sections. Consequently, the most generally applicable algorithms by (Alfriend et al., 1999), which are similar to those of (Bérend, 1997), will be adopted as baseline in the subsequent analysis.

Throughout this chapter a circular collision cross-section is assumed which is defined by two equally circular cross-sections of a target object and risk object as illustrated in Fig. 8.1. The corresponding collision risk results may be quite conservative, since "collision" circles are defined by the maximum span of both objects, which may lead to a large fraction of empty space. In (Patera, 2001) a special technique is described which employs line integrals to consider more general target geometries. This refined approach can often reduce the collision probabilities and the number of necessary avoidance maneuvers.

8.4 STATISTICAL FORECAST OF AVOIDANCE MANEUVER FREQUENCY

Mission planners and spacecraft operators should have an idea of the expected collision risk with cataloged and uncataloged objects before launching their satellite into a certain orbital regime. In order to protect the space environment (with first priority), and in order to protect the projected space mission (with second priority), catastrophic collisions should be avoided whenever possible. In Section 5.5 such catastrophic collision events were defined as having specific energy inputs into the target object exceeding 40 kJ/kg. This is the case for nearly every collision with an object of $d > 10$ cm, and hence for almost all USSTRATCOM catalog objects.

For the following analysis the collision probability algorithms derived in Section 8.3 will be combined with statistical flux results for objects of $d > 10$ cm obtained from ESA's MASTER-2001 model (see Section 4.3). The solution concept was developed in (Foster, 2001b), and has been implemented in ESA's Debris Risk Assessment and Mitigation Analysis software DRAMA (Sánchez-Ortiz et al.,

[8.2]this assumption is justified, if $R_c \ll b_{1\sigma,B}$

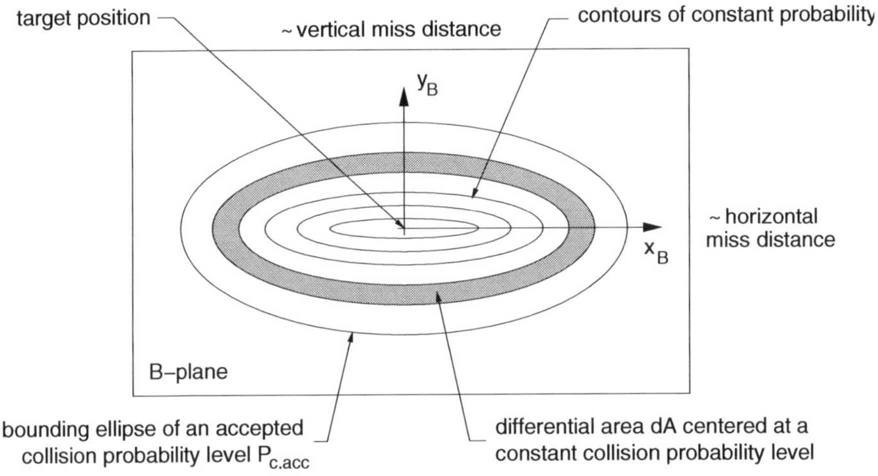

Fig. 8.3. Illustration of the collision risk estimation concept, based on the integration of statistical objects fluxes over the collision probability density function in the B-plane, centered at the target object.

2004). In a first step, the flux contributing debris objects are grouped into different orbit categories, with J_H perigee altitude classes, J_e eccentricity classes, and J_i orbit inclination classes. The orbit categories are chosen such that they match with those for which typical error covariance matrices of catalog objects were derived (see Section 8.1). For each of these orbit classes a mean approach direction and collision velocity is determined in the target-centered, orbit-related coordinate system, for J_A different azimuth classes, and J_h different elevation classes (with one class centered on $A_v = 0°$ and $h_v = 0°$). Impacts in the horizontal plane, between near-circular orbits, are prevailing by far for large-size objects in the LEO environment. As a consequence, one may assume equal orbit velocity magnitudes for both the target and risk object. With the resulting information on approach velocity and direction, corresponding B-planes for $J = J_H \times J_e \times J_i \times J_A \times J_h$ mean collision events of all orbit classes and approach directions can be defined, and the combined target and risk object position error covariance matrices can be mapped onto elliptical contours of equal collision probability in the class-specific B-plane (see Fig. 8.3).

Let $P_{c,acc}$ be an accepted collision probability level defined by a spacecraft operator, let the collision cross-section be defined as $A_c = (\sqrt{A_t} + \sqrt{A_r})^2$ (see Fig. 8.1), and let the collision radius R_c be small as compared with the 1σ dimensions of the position error ellipsoid in the B-plane, then for each of the $j = 1, \ldots, J$ event classes one may define contours of equal collision probability $P_c = p_c\, A_c$, which follow the contours of equal collision probability density p_c. If the spacecraft is maneuvered whenever the collision risk from a catalog object exceeds the accepted level $P_{c,acc}$, then the associated rate of collision avoidance maneuvers $\dot{N}_{c,man}$ can be determined from the rate of catalog object passes within an elliptic area described by $P_c > P_{c,acc}$ (Foster, 2001b).

$$\dot{N}_{c,man} = \sum_{j=1}^{J} \int_{0}^{A_j(P_{c,acc})} F_{j,cat} \, \mathrm{d}A \tag{8.37}$$

where $F_{j,cat}$ is the catalog object flux of the j-th contributing class, which is a subset of the MASTER-2001 object flux $F_{j,all}$ for $d > 10$ cm according to

$$F_{j,cat} \approx \frac{1}{\bar{c}_{TLE}(d \geq 10\,\mathrm{cm})} F_{j,all}(d \geq 10\,\mathrm{cm}) \tag{8.38}$$

with the size-dependent ratio $\bar{c}_{TLE}$ of expected, modeled objects in relation to trackable catalog objects observable by the US Space Surveillance Network (see Eq. 3.30 and Table 3.2). For LEO altitudes, with detection thresholds of $d_{SSN} \geq 10$ cm, one finds that $\bar{c}_{TLE} \approx 2.0$.

Similar to Eq. 8.37 the total collision rate $\dot{N}_{c,tot}$ due to catalog objects is obtained by an area integral out to infinity, weighted by a local collision probability $P_{c,j}(\Delta\underline{r}_{tca}, R_c)$ according to Eq. 8.30, with $\Delta\underline{r}_{tca} = (\Delta\underline{r}_{x_B}, \Delta\underline{r}_{y_B})$.

$$\dot{N}_{c,tot} = \sum_{j=1}^{J} \int_{0}^{\infty} P_{c,j} \, F_{j,cat} \, \mathrm{d}A \tag{8.39}$$

Eq. 8.39 can be split into two separate contributions, representing the collision rate $\dot{N}_{c,red}$ that should be reduced by means of evasive maneuvers whenever $P_c > P_{c,acc}$, and a residual collision rate $\dot{N}_{c,res}$ which is not attempted to be reduced due to $P_c \leq P_{c,acc}$.

$$\dot{N}_{c,red} = \sum_{j=1}^{J} \int_{0}^{A_j(P_{c,acc})} P_{c,j} \, F_{j,cat} \, \mathrm{d}A \tag{8.40}$$

$$\dot{N}_{c,res} = \sum_{j=1}^{J} \int_{A_j(P_{c,acc})}^{\infty} P_{c,j} \, F_{j,cat} \, \mathrm{d}A \tag{8.41}$$

$$\dot{N}_{c,tot} = \dot{N}_{c,red} + \dot{N}_{c,res} \tag{8.42}$$

Based on the previous definitions a false alarm probability P_{fa} can be determined, which indicates the likelihood of performing an evasive maneuver for a near-miss event that would not have caused a collision.

$$P_{\mathrm{fa}} = 1 - \frac{\dot{N}_{c,red}}{\dot{N}_{c,man}} \tag{8.43}$$

Table 8.3. Definition of sample target objects, associated orbits, and target cross-section radii R_t, as used for a statistical collision risk analysis.

Object	a [km]	e [–]	i [°]	R_t [m]
Envisat	7162.0	0.0012	98.52	13.0
ERS-2	7153.0	0.0011	98.52	5.9
ISS	6754.0	0.0007	51.63	60.0

Table 8.4. Flux F and annual collision rate $\dot{N}_c$ of cataloged objects (index "cat"), and all modeled objects of $d > 10$ cm (index "tot"), for sample spacecraft and orbits as defined in Table 8.3.

Object	F_{tot} [m^{-2} y^{-1}]	$\dot{N}_{c,tot}$ [y^{-1}]	F_{cat} [m^{-2} y^{-1}]	$\dot{N}_{c,cat}$ [y^{-1}]
Envisat	28.14×10^{-6}	0.01494	13.71×10^{-6}	0.00728
ERS-2	36.12×10^{-6}	0.00395	18.84×10^{-6}	0.00206
ISS	2.12×10^{-6}	0.02394	0.84×10^{-6}	0.00954

From the previous event rates one can define the relative magnitudes in terms of fractional risk reduction $\eta_{c,red}$ and fractional residual risk $\eta_{c,res}$ with respect to the trackable catalog population.

$$\eta_{c,red} = \frac{\dot{N}_{c,red}}{\dot{N}_{c,tot}} \qquad \eta_{c,res} = \frac{\dot{N}_{c,res}}{\dot{N}_{c,tot}} \tag{8.44}$$

Similarly, one can define $\hat{\eta}_{c,red}$ and $\hat{\eta}_{c,res}$ with respect to the entire modeled population of diameters $d > 10$ cm.

$$\hat{\eta}_{c,red} = \frac{1}{\bar{c}_{TLE}}\eta_{c,red} = \frac{\dot{N}_{c,red}}{\dot{N}_{c,all}} \qquad \hat{\eta}_{c,res} = \frac{1}{\bar{c}_{TLE}}\eta_{c,res} = \frac{\dot{N}_{c,res}}{\dot{N}_{c,all}} \tag{8.45}$$

where $\bar{c}_{TLE} \approx 2.0$ in the LEO regime.

Table 8.3 shows sample orbits and defines target cross-section radii R_t for the Envisat and ERS-2 satellites, and for the International Space Station (ISS). Table 8.4 shows related fluxes F_{cat} and F_{tot}, and corresponding target cross-section penetration rates $\dot{N}_{c,cat}$ and $\dot{N}_{c,tot}$ for cataloged objects (index "cat"), and for modeled objects (index "tot") of diameters $d > 10$ cm according to the MASTER-2001 model (Bendisch et al., 2002). The radius $R_t \approx R_c$ (for $R_t \gg R_r$), defining the collision cross-section A_c, marks an area within which a passing object *may* cause a collision, though the actual projected cross-section of the target covers only a fraction of this circular safety zone (for instance ~30% in the case of ERS-2). The underlying conservative assumptions lead to an extra safety margin in collision risk assessments when applying the methods outlined in this text.

Based on the collision flux of catalog orbits, Eq. 8.37 to 8.44 can be used to derive collision avoidance maneuver rates $\dot{N}_{c,man}$, corresponding absolute and fractional collision rate reductions $\dot{N}_{c,red}$ and $\eta_{c,red}$, absolute and fractional residual collision rates $\dot{N}_{c,res}$ and $\eta_{c,res}$, and false alarm probabilities P_{fa}, as a function of the target orbit, the reaction timespan Δt_{tca} before conjunction, the accepted collision probability level $P_{c,acc}$, and the accuracy of the risk object orbit determination. The effect of different levels of risk object orbit determination accuracies can be analyzed by applying a scale factor $k_{\sigma,r}$ to the estimated TLE orbit uncertainties listed in Table 8.1. The orbit of the target object in this context is assumed to be perfectly known.

Table 8.5 lists predicted maneuver statistics and related risk reductions for Envisat. If a spacecraft operator is prepared to accept any collision risk $P_{c,acc} \leq 1.0$, then no maneuvers need to be performed ($\dot{N}_{c,man} = 0.0$), and the residual collision rate is equal to the collision rate of catalog objects $\dot{N}_{c,res} = \dot{N}_{c,cat} = 0.00727\,\mathrm{y}^{-1}$ (see also Table 8.4). In this case, the risk is invariant with pre-event warning times and orbit prediction uncertainties. Since no maneuvers are performed, the probability of false alarms is zero. The false alarm rate P_{fa} grows with increasing uncertainty in the orbit determination (expressed in terms of $k_{\sigma,r}$), with decreasing levels of accepted collision probability $P_{c,acc}$, and (in general) with increasing time-to-go Δt_{tca}. Hence, the false alarm rate shows an opposite trend to the avoidance maneuver frequency $\dot{N}_{c,man}$ (see Eq. 8.43). For the time-to-go or reaction time Δt_{tca} three values are listed: 8, 24, and 48 hours. For a certain accepted collision probability the maximum number of maneuvers may, in contrast to normal expectations, be required closer to the event (e.g. at $\Delta t_{tca} = 12$ h or 8 h). This is due to the dependency of collision probability on the error dispersion (see Fig. 8.2 and the discussion on page 225 to 226), and due to the finite size of the collision cross-section A_c in comparison with the 1σ position probability density distribution in the B-plane.

Under all circumstances the predicted mean number of maneuvers in Table 8.5 will lead to a collision risk reduction. The effectiveness of this risk reduction measure increases with decreasing time-to-go ($\eta_{c,red} = 61.87\%$, 82.24%, and 93.15% for $\Delta t_{tca} = 48$ h, 24 h, and 8 h, where $P_{c,acc} = 10^{-4}$ and $k_{\sigma,r} = 1.0$), with increasing orbit determination accuracy of the risk object ($\eta_{c,red} = 93.15\%$, 97.82%, and 98.83% for $k_{\sigma,r} = 1.0$, 0.1, and 0.01, where $P_{c,acc} = 10^{-4}$ and $\Delta t_{tca} = 8$ h), and with a decreasing level of accepted collision probability ($\eta_{c,red} = 57.14\%$ and 93.15%, for $P_{c,acc} = 10^{-3}$ and 10^{-4}, where $k_{\sigma,r} = 1.0$ and $\Delta t_{tca} = 8$ h). Hence, in order to achieve a maximum risk reduction for a small number of avoidance maneuvers, the orbit of the risk object should be improved to the same level as the well-known target orbit (typically $k_{\sigma,r} \approx k_{\sigma,t} \approx 0.01$), based on acquired tracking data, and the maneuver decision should be taken at the latest possible time which is compatible with operational requirements (typically $\Delta t_{tca} \approx 8$ h in LEO). An independent orbit determination, based on traditional range and angular measurements, can result in $k_{\sigma,r} \approx 0.01$, which reduces the TLE-based number of avoidance maneuvers by more than one order of magnitude from $\dot{N}_{c,man} \geq 9$ per year to $\dot{N}_{c,man} \leq 1$ in four years, for $P_{c,acc} = 10^{-4}$, with a resulting risk reduction of 98.83%. Accepting a large collision probability of $P_{c,acc} = 10^{-3}$ for this scenario only slightly affects the risk, but reduces the maneuver rate by a factor 6.7 (only applies for small $k_{\sigma,r}$).

Table 8.5. Collision risk event statistics for the Envisat orbit and cross-section (see Table 8.3), as a function of the accepted collision probability $P_{c,acc}$, the reaction timespan Δt_{tca} prior to the conjunction event, and the position uncertainty scaling factor $k_{\sigma,r}$ (relative to estimated TLE orbit uncertainties).

$k_{\sigma,r}$ [–]	$P_{c,acc}$ [–]	Δt_{tca} [h]	$\dot{N}_{c,man}$ [y^{-1}]	$\dot{N}_{c,red}$ [y^{-1}]	$\eta_{c,red}$ [–]	$\dot{N}_{c,res}$ [y^{-1}]	$\eta_{c,res}$ [–]	P_{fa} [–]
1.00	1.0e–4	8.0	8.9081900	0.0067785	0.9315035	0.0004984	0.0684965	0.9992390
		24.0	14.4605000	0.0059983	0.8242922	0.0012786	0.1757078	0.9995850
		48.0	8.8902600	0.0045023	0.6187145	0.0027746	0.3812855	0.9994940
	1.0e–3	8.0	1.8321300	0.0041579	0.5713799	0.0031190	0.4286201	0.9977310
		24.0	1.1292200	0.0030279	0.4160983	0.0042490	0.5839017	0.9973190
		48.0	1.0700500	0.0028496	0.3915994	0.0044273	0.6084006	0.9973370
	1.0	—	0.0000000	0.0000000	0.0000000	0.0072769	1.0000000	0.0000000
0.10	1.0e–4	8.0	0.3148470	0.0071185	0.9782366	0.0001584	0.0217634	0.9773900
		24.0	0.8143840	0.0071160	0.9778927	0.0001609	0.0221073	0.9912620
		48.0	1.4548400	0.0070864	0.9738159	0.0001905	0.0261841	0.9951290
	1.0e–3	8.0	0.1880860	0.0070789	0.9727961	0.0001980	0.0272039	0.9623630
		24.0	0.4516710	0.0069906	0.9606578	0.0002863	0.0393422	0.9845230
		48.0	0.7587870	0.0068167	0.9367550	0.0004602	0.0632450	0.9910160
	1.0	—	0.0000000	0.0000000	0.0000000	0.0072769	1.0000000	0.0000000
0.01	1.0e–4	8.0	0.2415500	0.0071921	0.9883525	0.0000848	0.0116475	0.9702250
		24.0	0.2793620	0.0071758	0.9861060	0.0001011	0.0138940	0.9743140
		48.0	0.1630680	0.0071295	0.9797448	0.0001474	0.0202552	0.9562790
	1.0e–3	8.0	0.0361450	0.0071440	0.9817339	0.0001329	0.0182661	0.8023520
		24.0	0.0484033	0.0071268	0.9793695	0.0001501	0.0206305	0.8527620
		48.0	0.0833164	0.0071101	0.9770794	0.0001668	0.0229206	0.9146610
	1.0	—	0.0000000	0.0000000	0.0000000	0.0072769	1.0000000	0.0000000

Table 8.6. Collision risk event statistics for the ISS orbit and cross-section (see Table 8.3), as a function of the accepted collision probability $P_{c,acc}$, the reaction timespan Δt_{tca} prior to the conjunction event, and the position uncertainty scaling factor $k_{\sigma,r}$ (relative to estimated TLE orbit uncertainties).

$k_{\sigma,r}$ [–]	$P_{c,acc}$ [–]	Δt_{tca} [h]	$\dot{N}_{c,man}$ [y^{-1}]	$\dot{N}_{c,red}$ [y^{-1}]	$\eta_{c,red}$ [–]	$\dot{N}_{c,res}$ [y^{-1}]	$\eta_{c,res}$ [–]	P_{fa} [–]
1.00	1.0e–4	8.0	1.7937600	0.0090785	0.9515948	0.0004618	0.0484052	0.9949390
		24.0	5.5439000	0.0089786	0.9411250	0.0005617	0.0588750	0.9983800
		48.0	11.3628000	0.0087578	0.9179797	0.0007825	0.0820203	0.9992290
	1.0e–3	8.0	1.0283100	0.0087790	0.9201989	0.0007613	0.0798011	0.9914630
		24.0	2.4766600	0.0077804	0.8155315	0.0017599	0.1844685	0.9968580
		48.0	3.2036900	0.0055705	0.5838855	0.0039699	0.4161145	0.9982610
	1.0	—	0.0000000	0.0000000	0.0000000	0.0095403	1.0000000	0.0000000
0.10	1.0e–4	8.0	0.3824720	0.0092052	0.9648768	0.0003351	0.0351232	0.9759320
		24.0	0.5676570	0.0091748	0.9616873	0.0003655	0.0383127	0.9838370
		48.0	0.5479990	0.0091209	0.9560317	0.0004195	0.0439683	0.9833560
	1.0e–3	8.0	0.0608863	0.0091199	0.9559340	0.0004204	0.0440660	0.8502140
		24.0	0.1829610	0.0090999	0.9538372	0.0004404	0.0461628	0.9502630
		48.0	0.3760670	0.0090539	0.9490094	0.0004865	0.0509906	0.9759250
	1.0	—	0.0000000	0.0000000	0.0000000	0.0095403	1.0000000	0.0000000
0.01	1.0e–4	8.0	0.7733580	0.0095233	0.9982169	0.0000170	0.0017831	0.9876860
		24.0	1.0159700	0.0093750	0.9826704	0.0001653	0.0173296	0.9907720
		48.0	0.9394820	0.0092964	0.9744291	0.0002440	0.0255709	0.9901050
	1.0e–3	8.0	0.0540630	0.0092740	0.9720820	0.0002663	0.0279180	0.8284600
		24.0	0.0423510	0.0092368	0.9681846	0.0003035	0.0318154	0.7818990
		48.0	0.0475306	0.0091405	0.9580912	0.0003998	0.0419088	0.8076920
	1.0	—	0.0000000	0.0000000	0.0000000	0.0095403	1.0000000	0.0000000

Table 8.6 lists predicted avoidance maneuver statistics for the ISS (the same comments and interpretations apply as for the Envisat case). In ISS operations $P_{c,acc} = 10^{-4}$ defines a "red alert" risk level. With 8 hours reaction time this accepted collision probability leads to an ISS maneuver rate of $\dot{N}_{c,man} \approx 1$, which corresponds well with an operational maneuver rate of $\dot{N}_{c,man} \approx 1.25$ per year, averaged over 6 years.

8.5 COLLISION AVOIDANCE FOR OPERATIONAL SATELLITES

The first confirmed, unintentional collision between two catalog orbits occurred on July 24, 1996, between the French Cerise satellite (95-033B) and a fragment (86-019RF) of an Ariane-1 H-10 upper stage which exploded on November 13, 1986, nine months after it delivered the SPOT-1 satellite into an orbit of $\overline{H} = 826$ km at $i = 98.73°$ (see second ranking fragmentation event in Table 2.2). The post-event reconstruction of the collision time matched well with the time of attitude loss that was logged in the recovered on-board data history of Cerise with a timing accuracy of $\Delta t \in [0, +5\,\mathrm{s}]$. Fig. 8.4 shows radar images of Cerise and its severed gravity gradient boom, as obtained by the TIRA radar with ISAR processing techniques (Inverted Synthetic Aperture Radar). The boom with its end-mass is shown in the left image. It was severed at ~4.20 m (~2/3 of its total length). The cut end of the remaining boom, and the satellite body are clearly discernible in the right image.

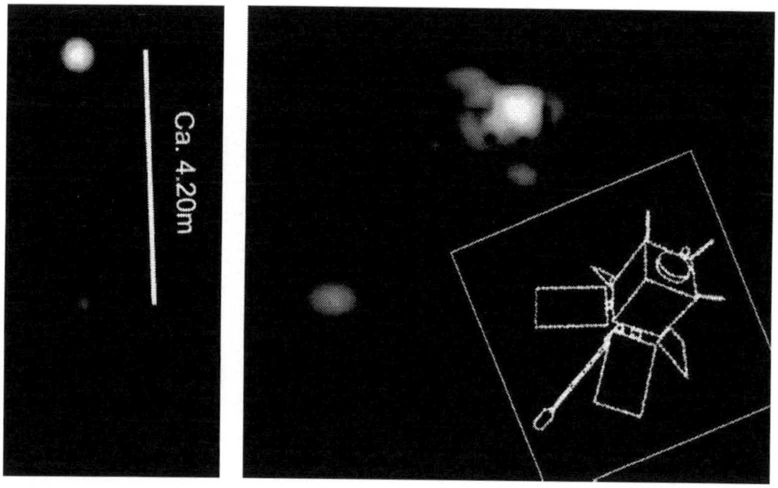

Fig. 8.4. Radar image of the damaged Cerise satellite (right), and of its severed gravity gradient boom (left), following the collision with a fragment of an Ariane-1 H-10 upper stage explosion (image source: FGAN).

The collision with Cerise occurred on July 24, 1996, at 09:48:02 UTC, at $H =$ 685.8 km, $\lambda = 59.75°$E, and $\phi = 38.22°$S above the Indian Ocean. The debris object hit Cerise with a relative velocity of 14.77 km/s, under an azimuth angle of 10.5° (almost head-on, from a 1 o'clock direction), at an elevation angle of $-0.1°$ (almost within the horizontal plane). At the time of collision Cerise was on an ascending pass, and the debris object was on a descending pass, with both orbits of very similar osculating elements $a \approx 7047.1$ km, $e \approx 0.002$, $i \approx 98.3°$, and $\omega \approx 93.0°$, only differing in their ascending node locations by $\Delta\Omega \approx 193.0°$. Due to the collision a momentum change could be detected in the post-event orbit determinations for both objects. Surprisingly, most of the Δv was noted in the radial and out-of-plane directions. The fragment even encountered a transversal velocity increase (and hence an increase in orbital energy), which is still unexplained (Alby et al., 1997). The occurrence of this unintentional collision between two catalog objects is in line with expected mean times between collision events according to Table 4.3 and Fig. 5.2.

The French Space Agency CNES is routinely monitoring conjunction events of spacecraft operated by them (which was not the case for Cerise), with objects from the USSTRATCOM catalog. In July 1997 they performed an evasive, 2-burn maneuver to raise the orbit of SPOT-2 (90-005A) by 400 m, in order to avoid a conjunction within $\rho_{tca} \approx 300$ m, on July 24, 1997, at 02:27:11 UTC, by an explosion fragment of a Thorad Agena D orbital stage (70-025JF). In the case of predicted high-risk events, CNES acquires and processes tracking data from the *Monge* radar (see Fig. 2.31) to improve the orbit of the risk object. If the risk remains above an accepted level of $P_c \leq 10^{-3}$, then evasive maneuvers are considered. The conjunction detection and risk analysis techniques which are used at CNES are described by (Bérend, 1997).

In the United States, NASA supports collision warnings for the manned missions of the Space Shuttle (STS), and of the International Space Station (ISS), based on Orbital Conjunction Messages (OCM) issued by USSTRATCOM. Different processes are applied for STS and ISS. USSTRATCOM uses general perturbation methods (analytical SGP-4 predictions) to screen the predicted orbit of ISS 72 hours ahead for catalog object conjunctions closer than 60 km. Special perturbation techniques (numerical orbit predictions), based on osculating state vectors from USSTRATCOM are used once the conjunction is within a box of ± 10 km $\times$ ± 40 km $\times$ ± 40 km ($\pm\Delta r_U \times \pm\Delta r_V \times \pm\Delta r_W$). If the event falls within ± 2 km $\times$ ± 25 km $\times$ ± 25 km, then covariance information on both orbits is included to assess the probability of collision (Foster, 2001a). The event forecasts are routinely updated every 8 hours, or upon special request. Avoidance maneuvers by ISS are considered once the fly-by event lies within a box of ± 0.75 km $\times$ ± 25 km $\times$ ± 25 km. In order to maneuver ISS, the time of closest approach must be within the following 30 hours, at least 3 consistent event predictions based on different orbit determinations must be available, and the uncertainties of the fitted orbit states must meet internal quality standards. If a maneuver decision is taken, then USSTRATCOM is informed of the post-maneuver trajectory, to verify that the new orbit does not pose a collision risk for the next 36 hours ahead.

Based on the described process, 6 maneuvers were initiated for ISS between June 1999 and May 2002 (of which 5 were actually performed). This corresponds to a mean collision avoidance maneuver rate of $\hat{\dot{N}}_{c,ISS} \approx 1.25\,\mathrm{y}^{-1}$. This operational maneuver rate matches NASA's predictions of $\dot{N}_{c,ISS} \approx 1.2\,\mathrm{y}^{-1}$ for $P_{c,ISS} \leq 10^{-4}$ (Foster, 2001a), and it is close to the forecasts in Table 8.6. If the threshold $P_{c,ISS} \leq 10^{-4}$ is applied as decision criterion, then the resulting avoidance maneuvers can lead to a collision risk reduction of up to 99% as compared with a no-maneuver scenario (see Table 8.6). The ISS procedure, based on a mixed conjunction geometry and collision probability assessment, reduces the number of maneuvers relative to a purely geometric, box-based approach by one order of magnitude. The resulting mean time between evasive maneuvers is compatible with operational requirements of ISS payloads.

The collision avoidance concept for the Space Shuttle (STS) differs from that for the ISS. Prior to the *Challenger* accident during its ascent phase of flight STS-51L, on January 28, 1986, no preventative measures were taken for collision avoidance, since the determination uncertainty of catalog orbits was deemed inadequate for this purpose. When Shuttle launches were resumed with *Discovery* (STS-26), on September 29, 1988, a new collision risk assessment concept was introduced. An extensive Monte Carlo analysis of catalog object conjunctions with STS led to the definition of an alert box with extensions $\pm 5\,\mathrm{km} \times \pm 25\,\mathrm{km} \times \pm 5\,\mathrm{km}$ ($\pm\Delta r_U \times \pm\Delta r_V \times \pm\Delta r_W$), in the STS orbital coordinate system. The dimensions of this box are compatible with an accepted collision probability per event of $P_c \leq 10^{-5}$ for a catalog population of 1988, and for typical orbit determination accuracies. Since STS-26 USSPACECOM and later USSTRATCOM were tasked to screen the STS orbits for 36 hours ahead, and raise an alert, if a conjunction event falls inside the alert box. For events within a reduced box of $\pm 2\,\mathrm{km} \times \pm 5\,\mathrm{km} \times \pm 2\,\mathrm{km}$ evasive maneuvers are considered, provided that payload or mission objectives are not compromised (NASA Flight Rule A 4.1.3-6). After the *Challenger* accident, 61 STS missions, covering STS-26 to STS-85 (1988 to 1997), with 568 days on orbit, led to 6 avoidance maneuvers. This corresponds to one maneuver on every tenth mission, and to a maneuver rate of $\hat{\dot{N}}_{c,STS} \approx 3.7\,\mathrm{y}^{-1}$ per year on orbit.

Since the mid 1990s ESA has been monitoring conjunction events of orbits of their remote sensing spacecraft ERS-1 (91-050A), ERS-2 (95-021A), and Envisat (02-009A) with respect to the USSTRATCOM catalog population. During this time frame mathematical methods and operational procedures were continuously refined. The most recent development status of ESA's collision risk analysis software CRASS is documented in (Alarcón, 2002). In total, 5 evasive maneuvers were performed by ESA spacecraft: 2 for ERS-1 (in 1996 and 1997), 1 for ERS-2 (in 2004), and 2 for Envisat (both in 2004). The ERS-1 maneuvers were performed for a conjunction with Cosmos-614 (73-098A) at a predicted distance of $\rho_{tca} \approx 200$ m at epoch 1997/06/25 13:24:47.76 UTC, and for a near fly-by of the Hilat satellite (83-063A) at $\rho_{tca} \approx 390$ m at epoch 1998/03/21 01:42:53 UTC. Following a power failure, the ERS-1 spacecraft ceased to be operational in March 2000, after exceeding the projected mission lifetime threefold. At the same time the conjunction prediction service was discontinued, since reliable orbit data for this target were no longer available.

Table 8.7. High-risk conjunctions of ERS-2 (95-021A) and Envisat (02-009A) between March 19 and December 27, 2004.

	Target object		Risk object		Prediction time		Time of close approach		Range & risk at t_{tca}	
No.	Obj. ID	$\Delta t_{t,tca}$ [d]	Obj. ID	$\Delta t_{r,tca}$ [d]	Date (0^h UTC)	Δt_{tca} [d]	t_{tca} Date & Time (UTC)		Δr_{tca} [km]	P_c [–]
1	95021A	1.536	85079B	2.619	2004/03/26	2.492	2004/03/28	11:48:28	0.166	2.630e–4
2	02009A	0.518	91082AE	0.940	2004/06/04	1.448	2004/06/05	10:44:39	0.754	1.781e–4
3	02009A	0.823	71067K	1.801	2004/06/08	1.741	2004/06/09	17:46:29	0.791	3.432e–4
4	02009A	0.869	81041A	1.832	2004/08/31	2.802	2004/09/02	19:14:11	1.297	2.186e–4
5	95021A	–0.048	86030A	1.342	2004/09/29	0.997	2004/09/29	23:56:02	0.067	1.546e–4
6	02009A	0.339	92093EE	1.038	2004/10/21	1.216	2004/10/22	05:10:36	0.081	5.488e–4
7	02009A	–0.656	76126AE	0.147	2004/12/15	0.274	2004/12/15	06:34:46	0.195	1.618e–3

	Conjunction position at t_{tca}			Approach velocity at t_{tca}				1σ Error ellipse in B-plane			
No.	Δr_{tca} [km]	$A_{r,tca}$ [°]	$h_{r,tca}$ [°]	Δv_{tca} [km/s]	$A_{v,tca}$ [°]	$h_{v,tca}$ [°]	A_c [m²]	ϕ_B [°]	$a_{1\sigma,B}$ [km]	$b_{1\sigma,B}$ [km]	Man.
1	0.166	102.62	53.71	14.867	–4.10	–0.02	265.038	59.96	0.655	0.116	**yes**
2	0.754	–159.17	4.24	5.613	–68.34	0.63	615.752	–3.52	3.598	0.143	no
3	0.791	76.26	–8.42	14.573	–11.86	0.19	581.069	–6.58	1.299	0.141	no
4	1.297	–75.12	–3.58	14.428	14.85	0.05	615.752	2.99	2.438	0.144	**yes**
5	0.067	–79.96	–54.11	14.523	13.35	0.09	149.571	52.74	1.011	0.143	no
6	0.081	53.26	49.17	14.778	7.38	–0.12	615.752	59.01	1.253	0.127	**yes**
7	0.195	–71.56	51.09	14.800	8.60	1.45	615.752	–55.96	0.165	0.117	no

	Risk object launch details			Osculating orbit at t_{tca}			Risk object characteristics & description		
No.	Launch date	Obj.	Owner	a [km]	e [–]	i [°]	m [kg]	A [m²]	Object description
1	1985/09/04	RB	SU	7156.03	0.00175	74.06	1421.21	12.920	Cosmos-3M stage 2 (S3M)
2	1991/11/28	PD	USA	7226.35	0.01292	98.96	2.00	0.010	DMSP F11 debris
3	1971/08/07	RB	USA	7148.43	0.00722	87.63	2.00	0.010	AKM (OV1 20, OV1 21 AKM)
4	1981/05/07	PL	SU	7154.90	0.00122	74.07	743.31	1.770	Cosmos 1269
5	1986/04/17	PL	SU	7152.86	0.00233	74.03	743.31	1.770	Cosmos 1741
6	1992/12/25	RD	RU	7158.89	0.00492	71.33	8.00	0.040	Zenit-2 fragmentation debris
7	1976/12/27	PD	SU	7255.83	0.05009	65.77	6.00	0.030	Cosmos 886 fragmentation debris

For all ESA missions analyzed here, an accepted collision probability of $P_c \leq P_{c,acc} = 10^{-4}$ is assumed as baseline[8.3]. In a selected time frame between March 19 and December 27, 2004, this threshold was violated twice by ERS-2 (as compared to a prediction of 3.5 in 9 months), and five times by Envisat (as compared to a prediction of 6 in 9 months according to Table 8.5). These events led to one avoidance maneuver for ERS-2, and two for Envisat (see Table 8.7). The ERS-2 maneuver was initiated after 4 consecutive predictions of a close conjunction with a Cosmos-3M second stage (85-079B) on days $t_{tca} - 7$ d through $t_{tca} - 3$ d. On day $t_{tca} - 3$ d the closest approach was determined for 2004/03/06 11:48:28 UTC at a distance of $\rho_{tca} \approx 170$ m, with a collision probability of $P_c = 2.63 \times 10^{-4} > P_{c,acc}$, using TLE data and applying general perturbation methods for the Cosmos-3M orbit. Based on osculating datasets and special perturbation techniques USSTRATCOM assessed the fly-by to be at a distance of 881 m, for the same epoch. In both results the conjunction location was above ERS-2. Hence, the ERS-2 altitude at the time of closest approach was lowered by $\Delta H_{tca} = -500$ m with two along-track burns, each of $\Delta v_V = -6.2$ cm/s, applied 2.5 and 1.5 orbits before the event, with restituting burns 0.5 and 1.5 orbits after the conjunction.

Fig. 8.5 illustrates the first of two Envisat avoidance maneuvers in 2004 (corresponding to event no. 4 in Table 8.7). On September 2, 2004, at 19:14:11 UTC, the Russian satellite Cosmos 1269 was predicted to pass Envisat at a distance of $\rho_{tca} \approx 1.30$ km, with a probability of collision $P_c = 2.186 \times 10^{-4}$, for a time-to-go of $\Delta t_{tca} = 1.83$ days. This high risk for a sizeable clearance is caused by a relatively oblique approach under $A_{v,tca} = 14.85°$, which causes a spread of the combined position error ellipsoid along the horizontal axis within the B-plane. Due to the oblique approach geometry a fuel-efficient avoidance strategy could be chosen, applying a small along-track maneuver of $\Delta V_V = +2$ cm/s, on September 1, at 23:52 UTC, to cause a delayed arrival of Envisat at the conjunction location. The resulting increase in the conjunction distance was +1.0 km in cross-track and +75 m in radial direction for a propellant consumption of 2×80 grams (avoidance maneuver and subsequent orbit restitution maneuver).

Fig. 8.6 illustrates the second Envisat avoidance maneuver in 2004 (corresponding to event no. 6 in Table 8.7). During 4 consecutive predictions on days $t_{tca} - 4$ d through $t_{tca} - 1$ d a close conjunction with an explosion fragment (92-093EE) of a Russian Zenith-2 orbital stage was predicted for 2004/10/22 at 05:10:35 UTC, at distances of $81\,\text{m} \leq \rho_{tca} \leq 316\,\text{m}$, with collision probabilities of $1.4 \times 10^{-4} \leq P_c \leq 5.5 \times 10^{-4}$, all of them exceeding the accepted level of $P_{c,acc} \leq 10^{-4}$. On day $t_{tca} - 1$ d NASA verified that also on the basis of more accurate osculating state vectors provided by the SSN the conjunction distance was predicted to be in the range $150\,\text{m} \leq \rho_{tca} \leq 300\,\text{m}$. On the same day the TIRA radar tracked the potential collider object. It was determined to be of an equivalent spherical size of ~15 cm, for a mean signal return of −15 dBsm, which showed strong amplitudes of ±40 dBsm, indicating a tumbling motion of a period of 12 s, with a superimposed spin of a period of 0.5 s.

[8.3]in late 2004 this threshold was lowered to $P_c \leq P_{c,acc} = 1/3,000$

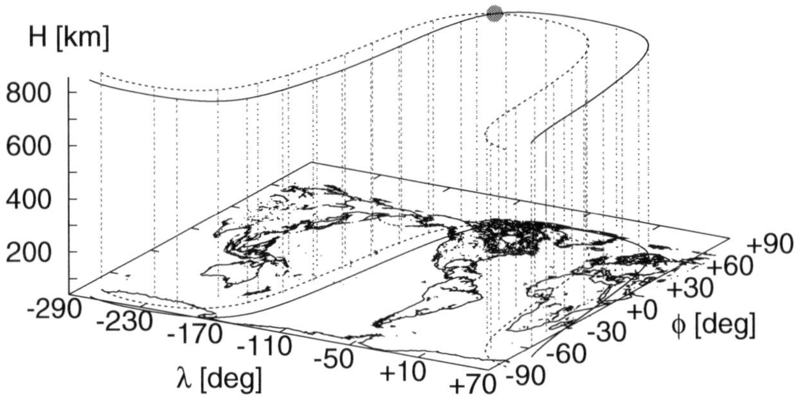

Fig. 8.5. Conjunction event scenario of Envisat (02-009A) with the Cosmos 1269 satellite (81-041A) at epoch 2004/09/02 19:14:11 UTC (see case no.4 in Table 8.7).

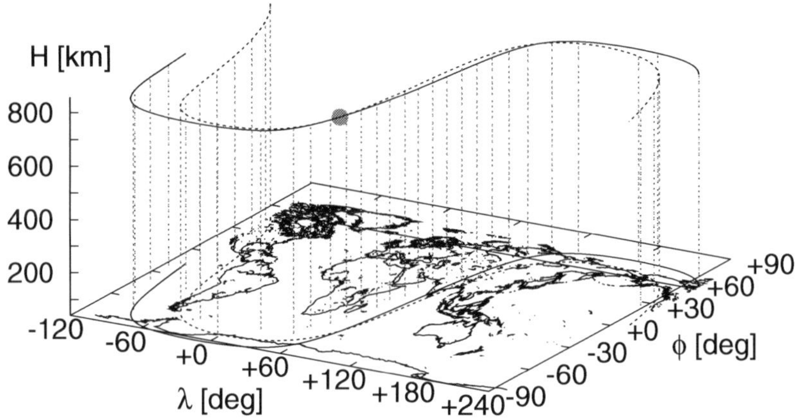

Fig. 8.6. Conjunction event scenario of Envisat (02-009A) with an explosion fragment of a Zenith 2 orbital stage (92-093EE) at epoch 2004/10/22 05:10:36 UTC (see case no.6 in Table 8.7).

Based on the consolidated information for the conjunction event the ESOC analysts and spacecraft operators prepared and up-linked a transversal avoidance maneuver of $\Delta V_V = -4.0$ cm/s (braking impulse) to be applied half a revolution (~50 minutes) before the conjunction, at 2004/10/22 04:20 UTC. The clearance distance in the radial direction (along the steepest gradient in the probability density) was thus increased from about 50 m to 316 m, with a corresponding reduction in the collision probability from $P_c \approx 5.5^{-4}$ to $P_c \approx 2.0^{-4}$, which was considered acceptable by the project team in the view of operational constraints and required continuity in the delivery of mission products. A restituting maneuver of $\Delta V_V = +3.9$ cm/s was performed half a revolution after the conjunction, at 2004/10/22 06:00 UTC, in order to recover the ground track repeat pattern.

Whenever an avoidance maneuver is planned, as part of this process the post-maneuver trajectory should be screened for new high-risk conjunction events resulting from the orbit change. This safety analysis is performed for 1.5 days ahead in the case of ISS, and for 7.0 days ahead for ESA spacecraft, prior to taking a final avoidance maneuver decision. For ESA's ERS-2 and Envisat satellites such a screening process is also applied prior to major orbit correction maneuvers (e.g. major adjustments in the ground track repeat pattern, or inclination control maneuvers). The routine ground track maintenance maneuvers for Envisat in 2004 required a total orbit velocity change of $\Delta v \approx 6.5$ m/s, corresponding to ~23 kg of hydrazine. The 2 avoidance maneuvers in the same timespan contributed to this consumption with only 2%.

Due to the low separation velocities during satellite releases by their upper stages, there is a risk of re-visits. For instance, Envisat passed its own upper stage at a distance of $\rho_{tca} \approx 2.0$ km at epoch 2004/09/15 01:23 UTC, and ERS-2 missed the same Ariane 5 orbital stage at epoch 2004/03/22 15:30 UTC, by $\rho_{tca} \approx 4.3$ km, and again at epoch 2005/01/16 00:47 UTC, by $\rho_{tca} \approx 0.5$ km.

Apart from the well-documented Cerise collision event (see page 233), two more unintentional on-orbit collisions could be identified by analysts of the US Space Surveillance Network (anon., 2005). In December 1991 the abandonned Russian navigation satellite Cosmos 1934 (1988-023A) collided with a fragment of Cosmos 926 (1977-062A), a predecessor sister spacecraft, at an altitude of 980 km, generating 2 additional catalog objects. On January 17, 2005, a 31-year-old US Thor Burner IIA final stage (1974-015B) collided with an explosion fragment (1999-057CV) of a Chinese CZ-4B third stage (1999-057C, see Table 2.2), at an altitude of 885 km, generating 3 additional, trackable objects which entered the US SSN catalog. The observed mean time between these collisions is consistent with predictions as illustrated in Fig. 6.7 on page 180.

8.6 REFERENCES

Alarcón, J. (2002). Development of a Collision Risk Assessment Tool. Technical Report (final report) ESA contract 14801/00/D/HK, GMV S.A.

Alby, F., Lansard, E., and Michal, T. (1997). Collision of Cerise with Space Debris. In *Proceedings of the Second European Conference on Space Debris*, ESA SP-393, pages 589–596.

Alfriend, K., Akella, M., Lee, D., Frisbee, J., and Foster, J. (1999). Probability of Collision Error Analysis. *Space Debris*, vol.1, no.1:21–35.

anon. (2005). Accidential Collisions of Cataloged Satellites Identified. *The Orbital Debris Quarterly News*, 9, Issue 2:1–2.

Bendisch, J., Bunte, K., Sdunnus, H., Wegener, P., Walker, R., and Wiedemann, C. (2002). Upgrade of ESA's MASTER Model. Technical Report (final report) ESA contract 14710/00/D/HK, ILR/TUBS, Braunschweig.

Bérend, N. (1997). Étude de la Probabilité de Collision Entre un Satellite et des Débris Spatiaux. Technical Report no. RT 38/3605 SY, ONERA.

Foster, J. (1992). A Parametric Analysis of Orbital Debris Collision Probability and Maneuver Rate for Space Vehicles. Technical Report NASA JSC-25898, NASA.

Foster, J. (2001a). International Space Station Debris Avoidance Operations. *The Orbital Debris Quarterly News*, 6, Issue 2.

Foster, J. (2001b). The Analytical Basis for Debris Avoidance Operations of the International Space Station. In *Proceedings of the Third European Conference on Space Debris*, ESA SP-473, pages 441–445.

Hoots, F., Crawford, L., and Roehrich, R. (1984). An Analytical Method to Determine Future Close Approaches Between Satellites. *Celestial Mechanics*, vol.33.

Hoots, F. and Roehrich, R. (1980). Spacetrack Report No.3 – Models for the Propagation of NORAD Element Sets. Technical report, Office of Astrodynamics, Aerospace Defense Center, Peterson AFB, CO.

Khutorovsky, Z., Boikov, V., and Kamensky, S. (1993). Direct Method for the Analysis of Collision Probability of Artificial Space Objects in LEO – Techniques, Results, and Applications. In *Proceedings of the First European Conference on Space Debris*, ESA SD-01, pages 491–508.

Klinkrad, H. (1997). One Year of Conjunction Events of ERS-1 and ERS-2 with Objects of the USSPACECOM Catalog. In *Proceedings of the Second European Conference on Space Debris*, ESA SP-393, pages 601–611.

Matney, N., Anz-Meador, P., and Foster, J. (2002). Covariance Correlations in Collision Avoidance Probability Calculations. *COSPAR Congress*, PEDAS1-B1.4-0038-02.

Montenbruck, O. and Gill, E. (2000). *Satellite Orbits – Models, Methods, and Applications*. Springer, Berlin, Heidelberg, New York.

Patera, R. (2001). General Method for Calculating Satellite Collision Probability. *Journal of Guidance, Control & Dynamics*, vol.24:716–722.

Sánchez-Ortiz, N., Belló-Mora, M., and Klinkrad, H. (2004). Collision Avoidance Maneuvers During Spacecraft Lifetime – Analysis of Operational Missions. *55th IAC Congress*, IAC-04-IAA.5.12.5.10.

9

Re-Entry Prediction and On-Ground Risk Estimation

H. Klinkrad, B. Fritsche, T. Lips and G. Koppenwallner

In Chapter 4 and 8 the statistical and quasi-deterministic collision risk was analyzed for modeled and trackable objects of the on-orbit population. When focusing on the trackable catalog objects in 2002, this on-orbit population represented only 33.3% of all launched objects since Sputnik 1, with only 16% of their total mass, and 33.0% of their cross-sectional area. The present chapter will hence be devoted to an analysis of the risk potential due to the large number of objects which re-enter into the Earth atmosphere.

9.1 HISTORY OF HAZARDOUS RE-ENTRY EVENTS

Between the first space launch on October 4, 1957, and January 2002, more than 18,000 trackable objects (66.7%) re-entered into the Earth atmosphere, with a total mass of ~27,000 tons (84%) and a total cross-sectional area of ~85,000 m^2 (67%). An analysis of re-entry structures suggests that 10% to 40% of the mass of larger objects may survive the severe structural and thermal loads during atmospheric descent to ground impact (Ailor et al., 2005). During a re-entry, the kinetic energy of an object is largely converted into heat energy by reducing the orbital velocity from typically 7.7 km/s to subsonic speeds within less than 5 minutes. For the Mir space station with a mass of 135 tons this corresponds to an energy dissipation rate of $\dot{E} \approx -13.0 \times 10^9$ W, which exceeds the output of a typical 1 GW nuclear power plant by more than one order of magnitude.

Table 9.1 lists historic re-entry events, which posed a risk either due to their large masses, or due to radioactive payloads. The latter ones were two singular events, both related to the reactor-powered RORSAT satellite class, which was operated until 1989 (see also Table 2.3). Details on the decays of Cosmos 954 and 1402 are provided in Section 9.7. The focus hereafter will be on large-size re-entry

Fig. 9.1. Delta II 2nd stage (top), with typical re-entry survivor objects, retrieved in South Africa on April 27, 2000 (center row), and retrieved in the central USA on January 22, 1997 (bottom row). Survivor objects (left to right): stainless steel tank of 250 kg, titanium pressurant tank of 30 kg, and thrust chamber of 45 kg (photos are courtesy of NASA, Aerojet, SAAO, and CSIR).

objects and on their associated risk to the on-ground population.

Despite the large number of catalog object re-entries, in more than 40 years of space flight, no more than ~250 major objects could be retrieved on ground, prior to the *Columbia* accident on February 1, 2003 (Ailor et al., 2005). As an example, Fig. 9.1 shows typical survivor objects of a Delta II second stage for two different re-entry events. In both cases the heat-resistant stainless steel propellant tank, a titanium pressurant tank, and the thrust chamber could be retrieved (the reason for the survival of particular shapes and materials will be explained in Section 9.5). The first event took place on January 22, 1997, at 09:30 UTC, when a Delta II second stage (Aerojet AJ10-118K) re-entered and broke up above Oklahoma and Texas. The 250-kg stainless steel tank impacted in Texas, just 50 m away from a farmhouse. The stage previously delivered the MSX satellite into orbit on April 24, 1996. The same kind of upper stage re-entered above the Cape Province of South Africa on April 27, 2000, at 13:30 UTC. Previously, it inserted Navstar 37 (GPS 2A-16) into orbit on March 28, 1996. Based on the recovered AJ10-118K second stage

Table 9.1. Re-entry objects with large masses or hazardous payloads. A single Buran flight, and some 100 controlled re-entries of the Space Shuttle (STS), with 78,000 to 82,000 kg dry mass each, are not included (source: Aerospace Corp.; with adaptations).

Object	Origin	Mass [kg]	Re-entry date	Re-entry type
Mir	CIS	135,000	2001/03/23	Controlled
Columbia (STS-107)	USA	82,000	2003/02/01	Uncontrolled
Skylab	USA	74,000	1979/07/11	Uncontrolled
Salyut 7/Cosmos 1686	USSR	40,000	1991/02/07	Uncontrolled
Salyut 6/Cosmos 1267	USSR	34,000	1982/07/29	Controlled
Cosmos 929	USSR	19,800	1978/02/02	Controlled
Cosmos 1443	USSR	19,800	1983/09/19	Controlled
Salyut 5	USSR	18,800	1977/08/08	Controlled
Salyut 4	USSR	18,700	1977/02/02	Controlled
Salyut 1	USSR	18,300	1971/10/11	Controlled
Salyut 2	USSR	18,300	1973/05/28	Uncontrolled
Salyut 3	USSR	18,300	1975/01/24	Controlled
Cosmos 557	USSR	18,300	1973/05/22	Uncontrolled
Apollo 5 Nose Cone	USA	17,100	1966/04/30	Uncontrolled
Apollo 6 CSM BP-13	USA	16,900	1964/07/01	Uncontrolled
Apollo 7 CSM BP-15	USA	16,700	1964/09/22	Uncontrolled
Apollo 10 CSM BP-9	USA	16,700	1975/11/22	Uncontrolled
Apollo 9 CSM BP-16	USA	16,700	1985/07/10	Uncontrolled
Apollo 8 CSM BP-26	USA	16,700	1989/07/08	Uncontrolled
Compton GRO	USA	14,910	2000/07/04	Controlled
Cosmos 954	USSR	4,500	1978/01/24	Radioactive
Cosmos 1402	USSR	990	1983/02/07	Radioactive

remnants one may safely assume that for all ~280 Delta launches through 2002 the same objects have probably survived to ground impact.

The statistics of retrieved re-entry fragments dramatically changed in the aftermath of the *Columbia* STS-107 break-up, which occurred on February 1, 2003, at about 14:00 UTC. A piece of insulation foam had apparently breached the left wing of the Shuttle 82 s after lift-off on January 16, 2003, leading to damage of the thermal protection system, and to an aerothermally induced disintegration on the re-entry approach to Kennedy Space Center. The break-up, which was observed by ground-based optical sensors, led to 84,000 recovered fragments, totaling 38% of *Columbia's* entry mass. The impact locations were mainly concentrated within a ground swath of 1,000 km $\times$ 40 km across Texas, covering a population of 216,000, with a corresponding risk of casualties on the order of 9% to 24%. This was by far the most critical re-entry event to date. If the same event had occurred one orbit later, for instance due to weather-induced delays, then the debris impact swath would have extended across the suburbs of Dallas and Fort Worth, affecting 632,000 citizens, and leading to a 3-fold increase of the risk level.

With a mass of 82 tons *Columbia* was the second largest re-entry risk object in space history. Only the Mir space station with 135 tons was more massive. However, due to its controlled entry mode, its incurred risk to the ground population was close to zero. Mir and the top-ranking entry events of Skylab and Salyut 7/Cosmos 1686 will be discussed further in this chapter. The remaining events listed in Table 9.1 mainly belong to the Cosmos 929 class (19.8 tons), to the Salyut 1 to 5 series of space stations (18.3 tons), and to the class of Apollo Command and Service Modules (CSM, 16.7 tons). Salyut stations were nominally de-orbited in a controlled fashion. This could not be accomplished for Salyut 2 and Cosmos 557[9.1] due to attitude acquisition problems, and for Salyut 7/Cosmos 1686 due to exhausted propellant reserves. Also the Compton Gamma-Ray Observatory (GRO), with a mass of 14.9 tons, was de-orbited in a controlled manner, into an uninhabited Pacific Ocean area, prior to a loss of attitude control capabilities due to failing gyros.

9.2 LONG- AND MEDIUM-TERM RE-ENTRY PREDICTIONS

To monitor and forecast the long- to medium-term evolution of an uncontrolled re-entry over periods of several years to a few weeks of the remaining lifetime, computationally efficient yet sufficiently accurate methods must be applied. For this purpose ESA uses a program called FOCUS-2, which is based on similar principles to the FOCUS propagator used for the MASTER model (see Section 3.1). FOCUS-2 integrates the combined time rates of change of singly averaged perturbation equations (following curve [2] in Fig. A.5), taking into account a non-spherical Earth gravity potential, a dynamic Earth atmosphere, luni-solar gravity perturbations, and solar radiation pressure in combination with an oblate, cylindrical Earth shadow. The integration is performed by a robust fourth-order Adams–Bashforth/Adams–Moulton predictor/corrector method, which is initiated by a self-starting fourth-order Runge–Kutta–Fehlberg method, using fixed time steps of typically 0.1 to 5 orbits, depending on the time to go until re-entry.

Perturbations from the Earth gravity potential are expressed in terms of the Lagrange equations for a truncated EGM-96 gravity model up to degree $n \leq 23$ and order $m \leq n$ (see Section C.2). To compute the derivatives of the gravity potential on the right-hand sides of Eq. A.40 to A.45, the spherical harmonics in Eq. C.3 are developed in terms of recursively computed expansions, using inclination functions and eccentricity functions according to (Kaula, 1966). The infinite expansions in eccentricities reduce to finite summations after the analytical averaging operation with respect to mean anomaly M according to Eq. A.46. A second-order Earth oblateness perturbation term $(J_2)^2$, adopted from (Liu and Alford, 1979), is superimposed on the results[9.2].

Third body effects are computed for point masses of the Sun and Moon, ap-

[9.1]Cosmos 557 was also considered to be a Salyut station

[9.2]the term $(J_2)^2$ is due to a series expansion of the geopotential; it has no physical meaning plying analytical ephemerides. Again, the Lagrange equations are used, with recursive expansions of the perturbing third body potential in terms of inclination and eccentricity functions according to (Cook, 1972). As for the geopotential, the perturbation equations are averaged with respect to the mean anomaly, maintaining the positions of Sun and Moon constant over the averaging time interval of one orbit.

The direct solar radiation pressure perturbations are determined with Aksnes' theory (Aksnes, 1976) for a given area-to-mass ratio A/m and momentum exchange coefficient c_R. The perturbation equations are averaged in a closed form over the illuminated orbit arc, taking into account an oblate, cylindrical Earth shadow.

The main improvement of the FOCUS-2 perturbation model, as compared to the streamlined FOCUS software, lies in the aerodynamic drag effects. They are computed from the Gauss equations (Eq. A.33 to A.38), which are expressed in terms of first-order (J_2) osculating elements, with the perturbing acceleration $\underline{p}_a$ and its components $p_{a,U}$, $p_{a,V}$, and $p_{a,W}$ in the radial, transversal, and out-of-plane direction.

$$p_{a,U} = -\frac{1}{2} c_D \frac{A}{m} \rho \; v_a \, v_{a,U} \qquad p_{a,U} \mapsto p_{a,V}, p_{a,W} \tag{9.1}$$

$$p_a = |\underline{p}_a| = \sqrt{(p_{a,U})^2 + (p_{a,V})^2 + (p_{a,W})^2} \tag{9.2}$$

Here, c_D is a free-molecular drag coefficient, A/m is the area-to-mass ratio, ρ is the local air density (varying with time, location, and activity conditions), and $v_a = |\underline{v}_a|$ is the aerodynamic velocity of the spacecraft with the radial, transversal, and out-of-plane components $v_{a,U}$, $v_{a,V}$, and $v_{a,W}$.

$$v_{a,U} = \underline{v}_a \underline{U} = \frac{\mu}{a(1-e^2)} + \Delta v_U\,(J_2) \tag{9.3}$$

$$\begin{aligned} v_{a,V} &= \underline{v}_a \underline{V} = \frac{\mu}{a(1-e^2)}\,(1 + e\cos f) - \omega_e\, r \cos i + \Delta v_V\,(J_2) \\ &\quad + \cos\tilde{A}\, v_{w,N} + \sin\tilde{A}\, v_{w,E} \end{aligned} \tag{9.4}$$

$$\begin{aligned} v_{a,W} &= \underline{v}_a \underline{W} = -\,\omega_e\, r \sin i \cos(\omega + f) + \Delta v_W\,(J_2) \\ &\quad + \sin\tilde{A}\, v_{w,N} - \cos\tilde{A}\, v_{w,E} \end{aligned} \tag{9.5}$$

$$v_a = |\underline{v}_a| = \sqrt{(v_{a,U})^2 + (v_{a,V})^2 + (v_{a,W})^2} \tag{9.6}$$

In these equations $\Delta v_{U,V,W}(J_2)$ are the first-order velocity corrections due to short-periodic Earth oblateness effects. The optional east and north wind components $v_{w,E}$ and $v_{w,N}$ are provided by a horizontal wind model (Hedin, 1991), which is superimposed on the Earth rotation rate ω_e (the azimuth angle $\tilde{A}$ is defined in Eq. 4.24 and 4.25). Local air densities in Eq. 9.1 are determined with the MSISe-90 model (Hedin, 1991), based on observed and predicted solar and geomagnetic activity proxies $\overline{F}_{10.7}$, $F_{10.7}$, and A_p (see Annex B).

To arrive at mean rates of change, which can be included in the propagation of the averaged equations of motion, the drag perturbations, expressed in osculating elements, are numerically averaged with respect to the mean anomaly M. The computation of the integral according to Eq. A.46 is performed by means of a 21-point Gauss–Legendre quadrature.

The procedure for long-, medium-, and short-term re-entry predictions shall be illustrated at the example of the Mir space station (see Fig. 9.2). The subsequent discussion is solely based on data which were available at the time of the Mir decay and re-entry.

Fig. 9.2. The Mir space station on orbit (X, Y, Z are body-fixed coordinate axes).

The assembly of the Mir space station started with the launch of its core module on February 19, 1986, followed by the launches of Kvant (March 1987), Kvant-2 (December 1989), Crystal (June 1990), Spectr (June 1995), and Priroda (April 1996). As part of the ISS Phase 1 mission (STS-74), also a docking module for STS was installed on Crystal (November 1995), which was subsequently used during 10 visits by the Space Shuttle. Mir reached its final configuration after the launch and docking of a modified Progress M1 vehicle at the end of January 2001. With this fuel supply and propulsion unit, especially adapted for the planned de-orbit operations, the Mir assembly had a mass of $\sim$135 tons, a volume of $\sim$400 m^3, and dimensions of 30 to 33 m along each of the main axes. The planned activities for the Mir disposal consisted of a monitored, natural orbit decay to an apogee altitude of 220 to 215 km, followed by 3 braking maneuvers on 4 consecutive orbits, which would de-orbit the space station in an uninhabited Pacific Ocean area.

In order to predict a natural orbit decay, it is common practice to retro-fit a

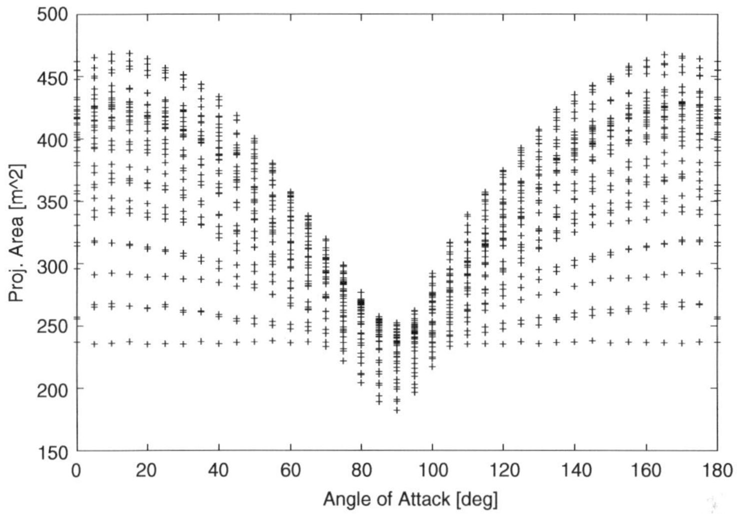

Fig. 9.3. Possible range of aerodynamic cross-section variations of Mir as a function of angle-of-attack $\alpha_a \in [0°, +180°]$ for different side-slip angles $\beta_a \in [-90°, +90°]$ (α_a is defined in the $\underline{X}, \underline{Y}$ plane; β_a is defined perpendicular to it; see Fig. 9.2).

given time history of the semimajor axis $\bar{a}_{TLE}(t)$ of one to two weeks, derived from TLE datasets, to a modeled history $\bar{a}(t)$, by an iterative adjustment of the ballistic parameter $B = c_D\, A/m$. The semimajor axis is selected in this $\bar{a}(t)$ fitting process, because it uniquely defines the total energy of the orbit.

$$E = E_{pot} + E_{kin} = -\frac{\mu}{2\,a} \tag{9.7}$$

The energy loss during re-entries is almost exclusively due to airdrag, and it only depends on its tangential component (see Eq. A.39). With $p_{a,T} = \sqrt{p_{a,U}^2 + p_{a,V}^2}$ (see Eq. 9.3 and 9.4) one obtains the following expression.

$$\frac{\mathrm{d}a}{\mathrm{d}t} = -c_D \frac{A}{m} \rho\, a \sqrt{\frac{a}{\mu} \frac{1 + e^2 + 2e\cos f}{1 - e^2}} \sqrt{(v_{a,U})^2 + (v_{a,V})^2}\; v_a \tag{9.8}$$

For the dominant class of near-circular re-entry orbits, this equation can be further simplified.

$$\frac{\mathrm{d}a}{\mathrm{d}t} \approx -c_D \frac{A}{m} \rho\, a\, v_a \tag{9.9}$$

The ballistic parameter $B = c_D\, A/m$, which governs the decay rate, can undergo significant changes, mainly due to aerodynamic cross-section variations. In the case of Mir, the cross-sectional area A ranged from 170 to 470 m^2, depending on the angle-of-attack α_a and side-slip angle β_a (see Fig. 9.3). Assuming a constant,

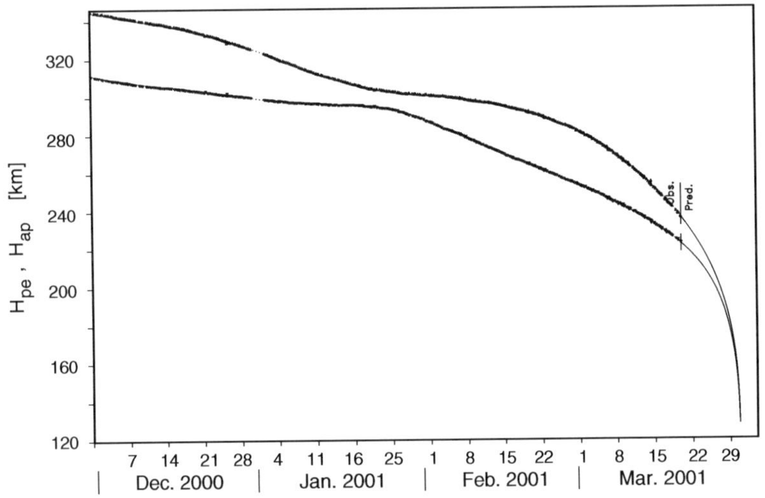

Fig. 9.4. Observed and predicted perigee and apogee altitude decay of the Mir space station, prior to its induced re-entry on March 23, 2001.

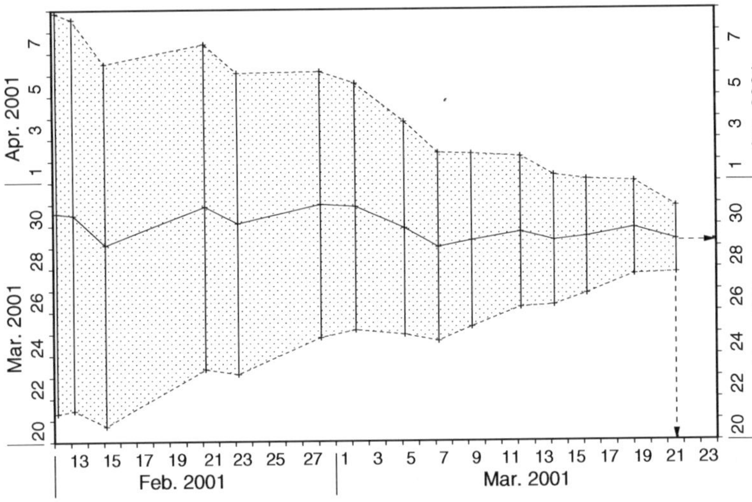

Fig. 9.5. Predicted re-entry time and uncertainty time window for the Mir space station, as a function of the prediction epoch, prior to its induced re-entry on March 23, 2001.

mean aerodynamic cross-section of $\overline{A} = 385\,\text{m}^2$, and a mass of $m = 135$ tons, the fitted drag coefficients maintained values of $c_D = 2.214 \pm 0.043$ during the last month of natural orbit decay, before the start of de-orbiting operations on March 23. These stable results for the c_D fit are a consequence of short-periodic attitude variations of Mir. After March 6, 2001, the FGAN radar noted that the space station started rotating about the Y-axis (Kvant-2/Spectr) at a period of $T_Y \approx 6\,\text{min}$, and about the X-axis (Kvant/core-module) at a period of $T_X \approx 15\,\text{min}$.

The RMS residuals of the fitted $\bar{a}_{TLE}(t)$ histories were on the order of $\Delta\bar{a}_{rms} =$ 70 to 135 m, with corresponding decays over the fitted arcs of $\Delta\bar{a} = -13$ to -20 km at the start and end of the 1-month analysis interval. Fig. 9.4 shows the observed (TLE-based) and predicted evolution of the perigee and apogee altitudes of the Mir orbit. The predicted, hypothetical re-entry for a natural decay converged towards 2001/03/30 $\pm$ 1 day since early February 2001. Fig. 9.5 shows the related nominal center of impact window (COIW) as a function of the prediction epoch, with an uncertainty time window of $\pm$20% of the remaining orbital lifetime (this uncertainty, combining errors from B, ρ, and the orbit state, is known to be a good estimate for nominal conditions). A final forecast of a natural re-entry of Mir indicated a COIW epoch of 2001/03/29 07:24 UTC $\pm$ 37 h. In the early morning of March 23, however, Mir had reached an orbit of $\overline{H}_{ap} \approx 217$ km (decaying at -5.7 km/d), and $\overline{H}_{pe} \approx 208$ km (decaying at -4.5 km/d). This marked the pass of the previously defined decision altitude for triggering the de-orbit operations.

The final three months of the Mir orbital lifetime fell into a period of moderate solar activity with $\overline{F}_{10.7} < 175$ and $F_{10.7} < 185$, in between two maxima of solar cycle 23, which occurred in mid 2000 (with $\overline{F}_{10.7} = 196$ and $F_{10.7} = 313$) and at the beginning of 2002 (with $\overline{F}_{10.7} = 225$ and $F_{10.7} = 274$). It is interesting to note that within 9 days of the Mir de-orbit the solar activity index $F_{10.7}$ increased from 179 on March 23 to 257 on April 1. The Sun spots which emitted this radio flux and the correlated extreme ultra-violet (EUV) radiation, also ejected a high-energy H^+ and He^{++} plasma, which reached the Earth with some delay, and caused a geomagnetic storm on March 31, with a daily mean level of $A_p = 192$, and a 3-hourly peak of $k_p = 9-$ (equivalent to $a_p = 300$, see Table B.4). If this peak had fallen into the final phase of the Mir descent, the strong increases in air density would have adversely affected the de-orbit operations. Section B.3 explains in more detail the cause of solar and geomagnetic activity, and their effect on the Earth atmosphere.

9.3 SHORT-TERM RE-ENTRY PREDICTIONS

Analytical and semi-analytical orbit prediction methods assume that the perturbing environment can be regarded as frozen during the averaging time interval, with perturbation magnitudes which are sufficiently small, and with perturbation effects which can be separated, with no cross-coupling effects. These assumptions lose their validity during the last day(s) of a re-entry, at orbit altitudes below $\sim$200 km. At that time a strictly numerical solution of the perturbed Newton equations is required, with an integration of the time evolution of the osculating position and velocity vector, considering perturbing accelerations and possible orbit maneuvers (see Eq. A.30 and A.31).

The perturbation model used for the numerical integration of re-entry trajectories at ESOC includes an Earth gravity model according to EGM-96 (see Section C.2), truncated at degree and order 7, luni-solar attraction, solar radiation pressure with umbra and penumbra shadow transits, and airdrag, with air densities computed from a hybrid MSISe-90/CIRA-72 model (see Eq. B.12), and with drag coefficients as a function of the flow regime. For the computation of air densities measurements and 27-day forecasts of the solar and geomagnetic activity proxies $\overline{F}_{10.7}$, $F_{10.7}$ and A_p are applied (see Section B.3). The integration is performed in a mean equatorial coordinate system of 2000.0 (J2000), using a Runge–Kutta/Shanks method of order 8, with step size control. Orbit maneuvers are considered as impulsive velocity changes.

Below re-entry altitudes of $\sim$120 km, airdrag becomes the dominant perturbation, increasing by 1 order of magnitude for every 30 km of altitude descent (see Fig. B.2). At the same time the drag coefficient c_D changes. For Mir the c_D coefficient was adjusted with the change in aerodynamic flow regime, from free-molecular, via transitional, to continuum conditions, using analytical solutions for simple shapes, as a function of the Knudsen number Kn_∞ in free-molecular flow, as a function of the Mach number Ma_∞ in continuum flow, and using Kn_∞-dependent bridging functions for the transitional regime. During the final descent of Mir (starting at $\sim$215 km altitude), the initial drag coefficient of $c_D = 2.21$ lost 10% of its starting value at each of the altitude marks 110 km, 30 km, 26 km, and 20 km. It finally attained a value of $c_D = 1.18$ in the sub-sonic regime.

The short-term re-entry prediction starts at approximately 200 km perigee altitude ($\sim$207 km for Mir). At this time the mean Kepler state of the long-term prediction is converted into an osculating cartesian state by means of a first-order (J_2) mapping function according to Eq. A.47 (Liu and Alford, 1979). This osculating state is then propagated to the time of the next (impulsive) maneuver, which is added to the right side of Eq. A.31, before re-starting the integrator. Though the integration is performed in an inertial frame, it is advantageous to express some results in an Earth-fixed frame (e.g. the impact velocity and the impact angle).

For the controlled re-entry of Mir the Russian Mission Control Center (MCC) had envisaged 3 de-orbit maneuvers on 4 consecutive tracks[9.3] prior to a splash-down in a designated, unpopulated area in the South Pacific. The final orbits were designated as #15, #16, #01, and #02, following an MCC convention, where ground track #01 is the first one in a given 1-day timespan with a nodal longitude λ_Ω west of 20°E (see Fig. 9.7). The de-orbit constellation was recurring on a daily basis, with a shift in nodal longitude by $\Delta\lambda_\Omega \approx -1.5°/\mathrm{d}$. Depending on the evolution of the natural orbit decay, three alternative scenarios were investigated for March 21, 22, and 23 of 2001. The start of the de-orbit sequence was finally initiated for March 23, after Mir had passed an orbit altitude of $\overline{H}_{ap} \approx 217$ km and $\overline{H}_{pe} \approx 208$ km. At this time Mir was in an uncontrolled rotation of $> 1°/\mathrm{s}$, which was stopped in the morning of March 22, on orbit #15, one day before the de-orbit. Thereafter, the Progress M1-5 thrusters were aligned with the transversal flight direction during the foreseen burn arcs on orbits #15, #16, and #01.

[9.3]a back-up maneuver on orbit #03 was included in the analysis, but not used

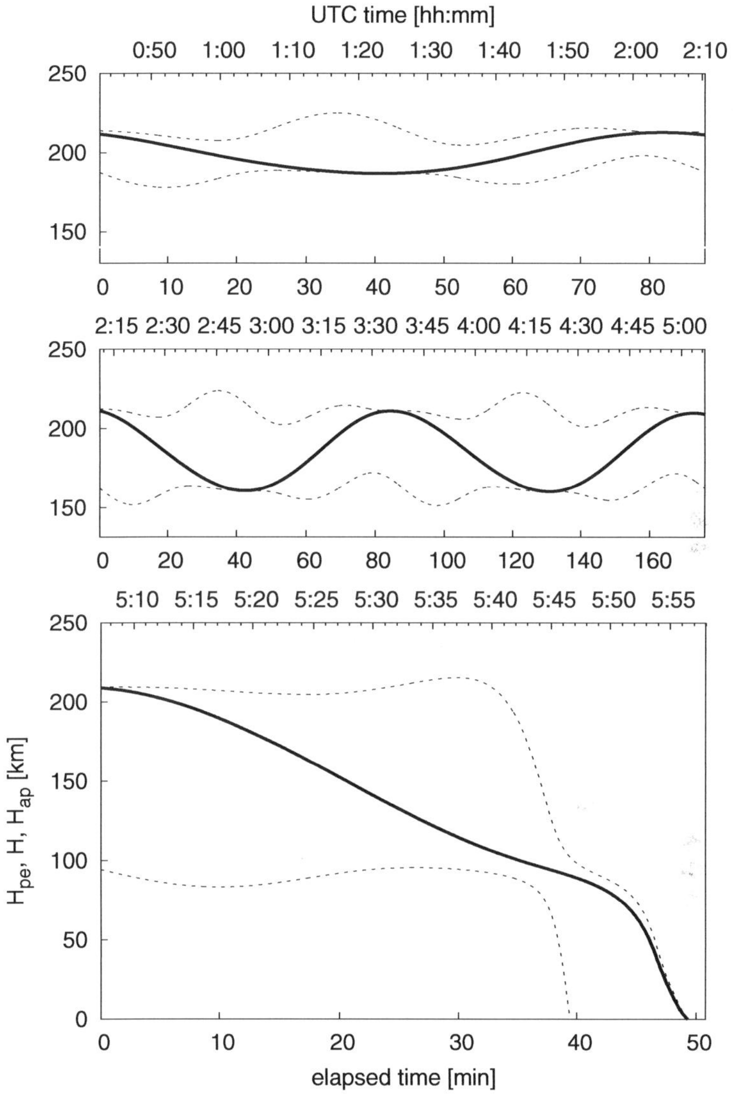

Fig. 9.6. Evolution of the Mir orbit altitude on March 21, 2001, following a sequence of 3 disposal maneuvers of $\Delta v_V = -9.23\,(-7.70)\,\mathrm{m/s}$ on orbit #15 (top), $\Delta v_V = -10.4\,(-7.70)\,\mathrm{m/s}$ on orbit #16 (center), and $\Delta v_V = -28.0\,(-20.30)\,\mathrm{m/s}$ on orbit #02 (bottom). Results are shown for the a-priori predicted maneuvers (in brackets; see Fig. 9.7 and Table 9.2 for maneuver details). The dashed lines show the envelopes of perigee altitude H_{pe} and apogee altitude H_{ap} of the osculating orbit. The solid lines indicate Mir's true altitude $H(t)$. UTC times refer to March 23, 2001 (continuous scale), while the lower time scale is counted from the epoch of the preceding maneuver.

Table 9.2. Predicted and actual de-orbiting procedure of Mir on March 23, 2001.

Modeled events	Maneuver #1	Maneuver #2	Maneuver #3
man. epoch [UTC]	00:42:00	02:10:25	05:06:45
man. position u [°]	40°	40°	40°
man. size ΔV [m/s]	−7.70	−7.70	−20.30
pre-man. H_{ap}/H_{pe} [km]	221.12/206.96	213.21/187.97	209.93/162.24
post-man. H_{ap}/H_{pe} [km]	213.81/188.22	212.24/163.01	209.43/95.01
Actual events	**Maneuver #1**	**Maneuver #2**	**Maneuver #3**
man. start [UTC]	00:31:59	02:00:24	05:07:36
man. end [UTC]	00:53:32	02:24:28	05:27:02
man. size ΔV [m/s]	−9.27	−10.40	−28.00
post-man. H_{ap}/H_{pe} [km]	213.50/188.00	213.20/156.10	213.0/~80.00
propulsion system	control thrusters	control thrusters	main engine
orbit no./node pos. λ_Ω [°]	#15/59.54°E	#16/37.04°E	#02/7.85°W

Table 9.2 summarizes the modeled and actual de-orbiting maneuver procedures, and their predicted and actual outcome. For the modeled maneuvers Fig. 9.6 shows the altitude evolution of the Mir orbit between successive braking impulses.

The first de-orbit burn (March 23, 00:31:50 to 00:53:32 UTC, on orbit #15) had a total Δv_V of -9.27 m/s. It moved the perigee location by almost 180° towards the southern hemisphere, and reduced the perigee altitude to 188.2 km. The second burn on the subsequent orbit #16 (02:00:24 to 02:24:28 UTC) generated a Δv_V of -10.4 m/s, and further reduced the perigee altitude to 156.1 km. Both of these maneuvers only used the combined Progress M1-5 attitude control thrusters. Orbit #01 was used as a maneuver-free coast trajectory to do orbit determination, and to align the Progress main engine for the final burn arc, which should be centered at $u \approx 90°$ from the ascending node. The final de-orbit burn with the Progress M1-5 main engine started at 05:07:26 UTC on orbit #02, well within coverage of the Russian surveillance network. The burn was completed successfully, close to the time of coverage loss, at 05:27:02 UTC, generating a total Δv_V of -28.0 to -28.5 m/s, which placed the perigee altitude close to 80 km, and led to a nominal re-entry location at $\phi \approx 40°$S and $\lambda \approx 160°$W. This outcome closely matches ESOC's predictions for the March 23 scenario as summarized in Table 9.2 and in Fig. 9.6.

The sensitivity of the Mir re-entry control can be illustrated by looking at the consequences of an under-performance of the final de-orbit maneuver. Based on the a-priori model of the disposal sequence, as given in Table 9.2, an analysis shows that a final Δv_V of -17 m/s would still result in a splash-down in the designated target area, $-17\,\text{m/s} \leq \Delta v_V \leq -15.2\,\text{m/s}$ would cause an impact prior to the next descending node, and $\Delta v_V = -10.0$ m/s, -5.0 m/s, and 0.0 m/s would entail an orbit lifetime extension by about 5, 20, and 40 hours, with no control of the impact footprint. Due to a thorough preparation and flawless execution of the Mir disposal sequence by the MCC, these contingencies remained hypothetical.

As compared with earlier MCC plans, the actual center of the thrust arc of the final de-orbit maneuver was shifted to $u \approx 90°$, towards the northernmost point of orbit #02. Moreover, the accomplished performance of the maneuver was better than anticipated. These two effects combined resulted in a steeper descent trajectory towards the planned splash-down zone in the South Pacific, with a safe altitude clearance during the pass over China, Korea and Japan. The associated on-ground risk is illustrated in Fig. 9.8. The chosen re-entry ground track has a maximum possible land coverage for the given orbit inclination of $51.6°$. Also the mean and maximum population densities of this orbit are at about 50% of the possible maxima. The reason to select this re-entry trajectory lies in its good coverage by the Russian ground stations, and in the extended ocean area around the targeted entry splash-down zone. In the lower part of Fig. 9.8 the land coverage, population density, and normalized casualty risk is shown over a time interval of ± 62.5 minutes, centered on the predicted impact location (COIW). The largest risk contribution in this timespan results from a pass over northern China and Korea (risk increase by $+30\%$). The delayed de-orbit burn at $\Delta u \approx +90°$, with a large Δv_V of $-28.5\,\mathrm{m/s}$, maximized the altitude clearance above the most densely populated areas below the entry trajectory, and minimized the associated risk on ground. Assuming a 2σ dispersion of the Mir impact footprint by $\pm 2,500$ km along-track and by ± 100 km cross-track, centered at 40°S and 160°W, the assessed risk of on ground casualties was virtually zero.

In an a-priori analysis of the Mir re-entry, the final de-orbit maneuver (see Table 9.2, and the lower part of Fig. 9.6) resulted in a predicted passage of the geodetic altitude $z = 120\,\mathrm{km}$ at 05:36 UTC, with a flight path angle of $\gamma_{a,120} = -0.44°$ and a velocity of $v_{a,120} = 7.567\,\mathrm{km/s}$ (in an aerodynamic coordinate system, relative to a co-rotating atmosphere). The intact Mir, with an area-to-mass ratio of $m/A \approx 350\,\mathrm{kg/m^2}$, would have reached a maximum heat flux of $\dot{q} \approx 10^6\,\mathrm{W/m^2}$ at $z \approx 59\,\mathrm{km}$ (05:53 UTC), and a maximum deceleration of $\dot{v} \approx -7.7\,g_\circ$ at $z \approx 46.5\,\mathrm{km}$ (05:36 UTC). The final impact was predicted to be at $\phi_i = 39.6°$S and $\lambda_i = 158.9°$E, at 05:58 UTC, on March 23, 2001, with a subsonic velocity of $v_{a,i} \approx 62\,\mathrm{m/s}$ at an impact angle of $\gamma_{a,i} = -89.8°$. The response of the entry orbit to its energy dissipation can be well monitored in Fig. 9.6. Outside the dense atmosphere the orbit altitude performs a harmonic oscillation at the orbital period. Below ~120 km, in the phase of atmospheric capture, the flight path angle steepens, with a corresponding increase in the orbit eccentricity. With the continuing dissipation of orbital energy the spacecraft starts to move towards the apogee location and maintains this position until the laws of orbit mechanics fail. At this point the spacecraft enters into a regime of equilibrium free fall, with balancing airdrag and gravitational attraction leading to the following impact velocity $v_{a,i}$ (with the gravity acceleration $g_\circ$ and density $\rho_\circ$ at ground level).

$$v_{a,i} \approx \sqrt{\frac{2\,m\,g_\circ}{c_D\,A\,\rho_\circ}} \tag{9.10}$$

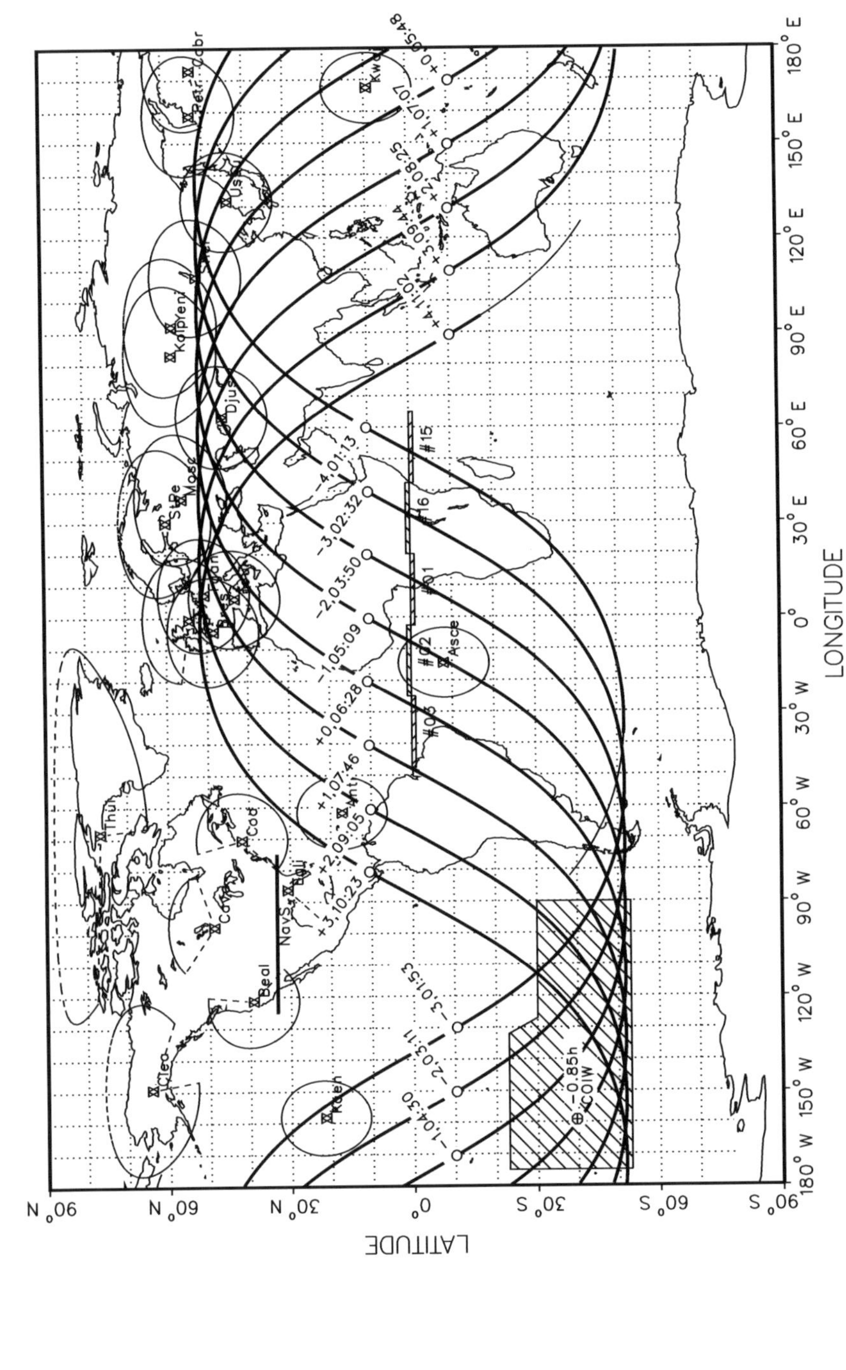

Fig. 9.7. The Mir de-orbit scenario. COIW marks the center of the impact window for a de-orbit strategy with 3 maneuvers on orbit #16, #15, and #02. Ground track markers (○) are labeled with the orbit number, relative to the COIW, and with the UTC time (hh:mm) of passage, for a hypothetical orbit of 120 km altitude. The hatched area marks the impact target zone in the Pacific Ocean.

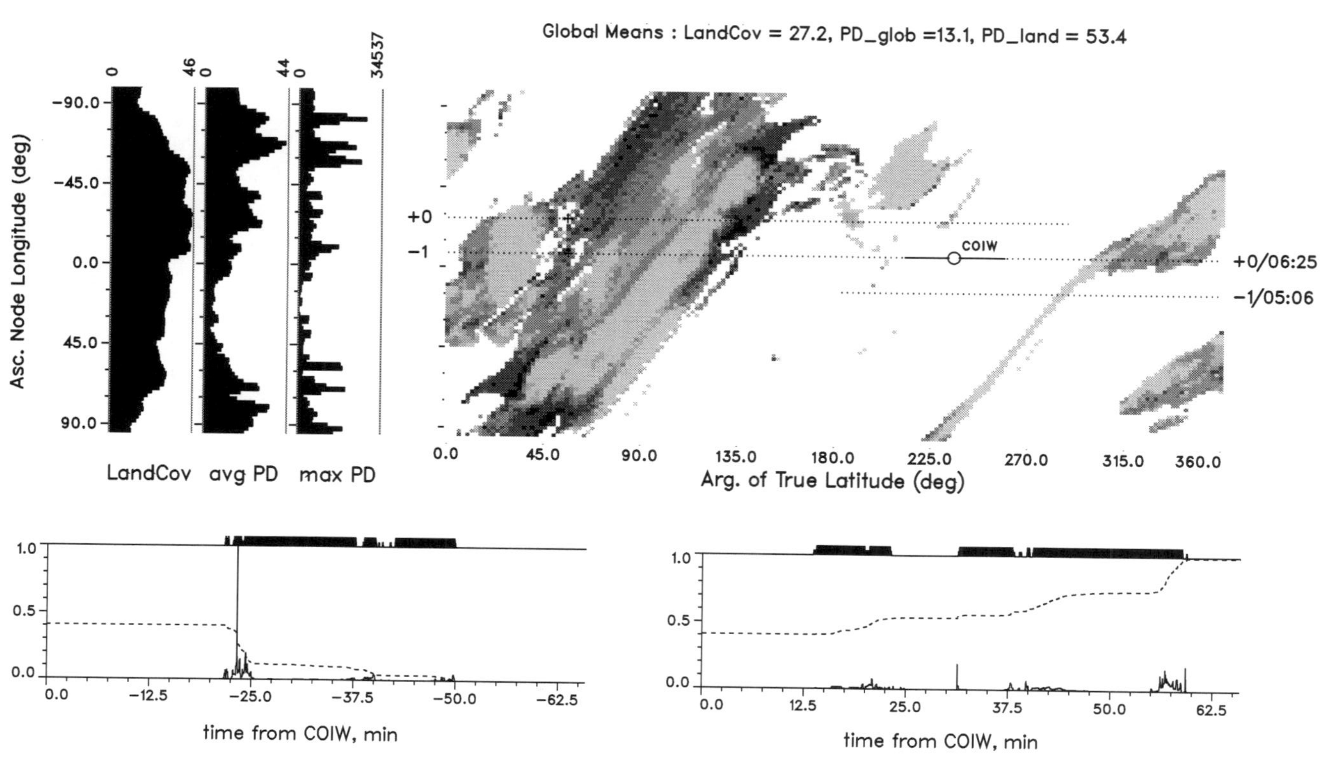

Fig. 9.8. Top chart: World map with population densities and overlaid ground tracks (±1 orbit around the COIW), as a function of the longitude of the ascending node λ_Ω, and of the orbit position $u = \omega + f$ of Mir, for an inclination of $i = 51.5°$. Mean land coverage, average and maximum population density are shown versus λ_Ω. The solid ground track marks a conservative dispersion area of $\pm 2,500$ km extension. Bottom charts: Land coverage (black silhouette), normalized local population density (solid line), and resulting cumulated risk (dashed line) in up-track (left) and down-track direction (right).

9.4 PREDICTION OF BREAK-UP AND SURVIVAL OF ENTRY OBJECTS

So far, in Sections 9.2 and 9.3, a re-entry object has been regarded as a point-mass, which passes the phase of aerodynamic deceleration and aerothermal heating in an intact manner, with no mass loss. The current Section 9.4 will highlight aspects of disintegration during re-entry, mass loss, and total demise.

An elegant, analytically tractable way of addressing the re-entry problem is provided by (Allen and Eggers, 1953), who analyze a 2-dimensional, ballistic (non-lifting) entry trajectory in an exponential atmosphere above a planar Earth surface. In this case the instantaneous velocity versus geodetic altitude $v(z)$ can be expressed as

$$v(z) = v_{a,120} \exp\left(-\frac{\rho_\circ H_{\rho,\circ} B}{2\,|\sin\gamma_{a,120}|} \exp\left(-\frac{z}{H_{\rho,\circ}}\right)\right) \tag{9.11}$$

where a constant ballistic parameter $B = c_D \cdot A/m$ is assumed, an exponential atmosphere $\rho_a(z) = \rho_\circ \cdot \exp(-z/H_{\rho,\circ})$, and an atmospheric entry velocity $v_{a,120}$ at the altitude threshold of $z = 120\,\text{km}$ with a constant flight path angle $\gamma_a = \gamma_{a,120} < 0°$. Velocity ratios of $v/v_{a,120} \approx 0.99$ and $v/v_{a,120} \approx 0.1$ mark the start and the end of the main re-entry phase. The aerodynamic deceleration $\dot{v}(z)$ is given by Eq. 9.12, with a peak of $\dot{v}_{max}(z)$ at $z_{\dot{v},max}$.

$$\dot{v}(z) = -\frac{1}{2} B\,\rho_\circ\, v_{a,120}^2 \exp\left(-\frac{z}{H_{\rho,\circ}} - \frac{\rho_\circ H_{\rho,\circ} B}{|\sin\gamma_{a,120}|} \exp\left(-\frac{z}{H_{\rho,\circ}}\right)\right) \tag{9.12}$$

$$\dot{v}_{max} = -\frac{1}{2}\frac{|\sin\gamma_{a,120}|}{e\,H_{\rho,\circ}}\, v_{a,120}^2 \quad \text{for} \quad \begin{cases} z_{\dot{v},max} = H_{\rho,\circ} \ln\left(\dfrac{\rho_\circ H_{\rho,\circ} B}{|\sin\gamma_{a,120}|}\right) \\ \dfrac{v(z_{\dot{v},max})}{v_{a,120}} = \dfrac{1}{\sqrt{e}} \end{cases}$$

In these equations and in the following ones $e = \exp(1) = 2.71828...$ is the base of the natural logarithm.

Using the analytical approximations in Eq. 9.11, the free-molecular heat flux $\dot{q}_f(z)$ for a spherical re-entry object of radius R (with a maximum of $\dot{q}_{f,max}$ at $z_{\dot{q}_{f,max}}$) can be described by Eq. 9.13 (Koppenwallner et al., 2001).

$$\dot{q}_f(z) = \frac{1}{2}\rho_\circ\, v_{a,120}^3 \exp\left(-\frac{z}{H_{\rho,\circ}} - \frac{3\,\rho_\circ H_{\rho,\circ} B}{2\,|\sin\gamma_{a,120}|} \exp\left(-\frac{z}{H_{\rho,\circ}}\right)\right) \tag{9.13}$$

$$\dot{q}_{f,max} = \frac{|\sin\gamma_{a,120}|}{3\,B\,e\,H_{\rho,\circ}}\, v_{a,120}^3 \quad \text{for} \quad \begin{cases} z_{\dot{q}_{f,max}} = H_{\rho,\circ} \ln\left(\dfrac{3\,\rho_\circ H_{\rho,\circ} B}{2\,|\sin\gamma_{a,120}|}\right) \\ \dfrac{v(z_{\dot{q},max})}{v_{a,120}} = \dfrac{1}{\sqrt[3]{e}} \end{cases}$$

The equivalent laminar, hypersonic continuum heat flux can be determined from Lees' theory, according to Eq. 9.14 (Koppenwallner et al., 2001).

$$\dot{q}_c(z) = c_{lam} \sqrt{\frac{\rho_\circ}{R}}\, v_{a,120}^3 \exp\left(-\frac{z}{H_{\rho,\circ}} - \frac{3\,\rho_\circ H_{\rho,\circ} B}{2\,|\sin\gamma_{a,120}|} \exp\left(-\frac{z}{H_{\rho,\circ}}\right)\right) \tag{9.14}$$

$$\dot{q}_{c,max} = c_{\text{lam}} \sqrt{\frac{|\sin\gamma_{a,120}|}{3\,B\,e\,H_{\rho,\circ}}} \sqrt{\frac{1}{R}}\, v_{a,120}^3 \quad \text{for} \quad \begin{cases} z_{\dot{q}_{c,max}} = H_{\rho,\circ} \ln\left(\dfrac{3\,\rho_\circ\,H_{\rho,\circ}\,B}{|\sin\gamma_{a,120}|}\right) \\ \dfrac{v(z_{\dot{q},max})}{v_{a,120}} = \dfrac{1}{\sqrt[6]{e}} \end{cases}$$

Here, $c_{\text{lam}} = 1.23 \times 10^{-4}\,\text{kg}^{1/2}\,\text{m}^{-1}$ is a dimensional constant for a sphere in laminar flow. In all flight regimes the heat flux is proportional to the Stanton number St, for which limiting cases are defined in free-molecular flow ($\text{Kn}_\infty > 10$) and in hypersonic continuum flow ($\text{Kn}_\infty < 0.01$). Intermediate results can be interpolated for the rarefied transitional flow regime ($0.01 \le \text{Kn}_\infty \le 10$).

$$\text{St} = \frac{\dot{q}}{\frac{1}{2}\,\rho\,v^3} \begin{cases} \text{St}_\text{f} = 1.0 & \text{Kn}_\infty > 10 \\ \text{St}_\text{c} = \dfrac{2.1}{\sqrt{\text{Re}_2}} & \text{Kn}_\infty < 0.01 \\ \text{St}_\text{t} = \dfrac{\text{St}_\text{c}}{\sqrt{1 + (\text{St}_\text{c}/\text{St}_\text{f})^2}} & 0.01 \le \text{Kn}_\infty \le 10 \end{cases} \tag{9.15}$$

where Re_2 is the Reynolds number behind the shock, and $\text{Kn}_\infty = \lambda/R$ is the ratio of mean free path length and object diameter in the undisturbed flow field.

For solid spheres of mass $m(R,\rho) = (4/3)\,\rho\,\pi\,R^3$ and cross-section $A(R) = \pi\,R^2$, Eq. 9.11 to 9.15 can be expressed as a function of the object diameter. Fig. 9.9 shows the corresponding size-dependent altitudes $z(v/v_{a,120}) \approx 0.99$ for the start of re-entry, $z(v/v_{a,120}) = 1/\sqrt[6]{e}$ for the maximum heat flux, and $z(v/v_{a,120}) \approx 0.1$ for the end of the main re-entry phase.

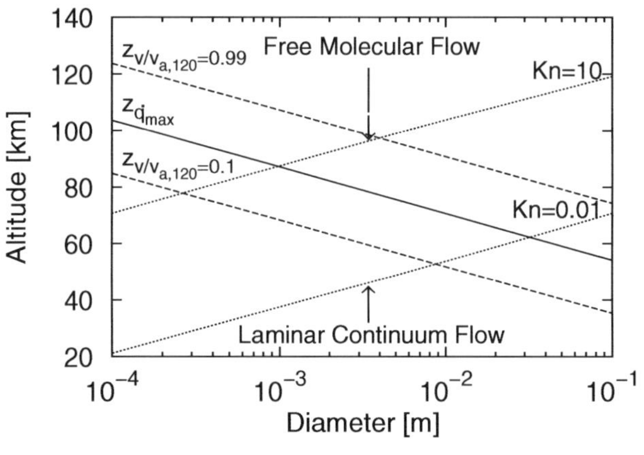

Fig. 9.9. Re-entry altitude regimes for free-molecular flow ($\text{Kn}_\infty > 10$), laminar continuum flow ($\text{Kn}_\infty < 0.01$), start of re-entry ($v/v_{a,120} \approx 0.99$), peak heat flux ($\dot{q}_{max}$), and start of equilibrium free fall ($v/v_{a,120} \approx 0.1$) as a function of the diameter of a spherical, stainless steel object, for an entry angle of $\gamma_{120} = -2.5°$.

If a re-entering body is of uniform composition, with infinite heat conduction, then an equilibrium temperature $T(t)$ will be established, resulting from the

Table 9.3. Properties of typical materials which are relevant for the survivability analysis of re-entry objects (density ρ, specific heat c_p, radiation emission coefficient ϵ, melting temperature T_m, and melting enthalpy h_m).

Material	ρ [g/cm^3]	c_p [J/kg K]	ϵ [–]	T_m [K]	h_m [J/g]
Titanium (TiAl6V4)	4.420	750.0	0.302	1900	400
Stainless steel (A316)	8.030	611.5	0.350	1650	274
Inconel	8.190	417.1	0.122	1570	309
Zerodur	2.530	1265.6	0.622	1424	250
Copper	8.960	434.1	0.216*	1356	243
CFRP	1.700	1100.0	0.850	1160	6650
Aluminum (AA7075)	2.800	751.1	0.141	870	385

* ϵ for an oxidized surface ($\epsilon = 0.012$ for a polished surface)

incoming aerothermal heat flux $\dot{q}$, and from the stored and re-radiated heat. Initially, the heat flux is largely converted into a temperature rise in the entry object according to Eq. 9.16. Once the melting temperature is reached ($T \geq T_m$), then the incoming energy is used to ablate the material at a rate of $\dot{m} = \rho \dot{V} < 0$ according to Eq. 9.17.

$$\dot{Q} = \rho c_p V \dot{T} + \epsilon \sigma A T^4 \quad \text{if} \quad T < T_m \tag{9.16}$$

$$\dot{Q} = -\rho h_m \dot{V} + \epsilon \sigma A T_m^4 \quad \text{if} \quad T = T_m \tag{9.17}$$

In these equations c_p is the specific heat, h_m is the melting enthalpy, ϵ is the emission coefficient, $\sigma = 5.67 \times 10^{-8}\,\mathrm{W\,m^{-2}\,K^{-4}}$ is the Stefan–Boltzmann constant, and $\dot{Q} = A\,\dot{q}$ is the instantaneous energy input, which results in a cumulated heat Q over the re-entry time span from $z(t_{120}) = 120$ km to ground impact at $t_\circ$.

$$Q = \int_{t_{120}}^{t_\circ} \dot{Q}\,dt \tag{9.18}$$

There are two ways to survive a re-entry. The first way is by effective re-radiation, such that the melting temperature is never reached, due to a dominance of the second term on the right-hand side of Eq. 9.16. This leads to the following conditions for re-entry survival in the presence of free-molecular heating (Eq. 9.19) or hypersonic continuum heating (Eq. 9.20).

$$T_m^4 > \frac{1}{\epsilon\sigma} \frac{\rho R\,|\sin\gamma_{a,120}|}{9\,c_D\,e\,H_{\rho,\circ}} v_{a,120}^3 \quad \text{if} \quad \mathrm{Kn}_\infty > 10 \tag{9.19}$$

$$T_m^4 > \frac{c_{\mathrm{rad}}\,c_{\mathrm{lam}}}{\epsilon\sigma} \sqrt{\frac{\rho\,|\sin\gamma_{a,120}|}{3\,c_D\,e\,H_{\rho,\circ}}}\, v_{a,120}^3 \quad \text{if} \quad \mathrm{Kn}_\infty < 0.01 \tag{9.20}$$

where $c_{\rm rad} = 1/\sqrt{3}$ for a non-rotating sphere, and $c_{\rm rad} = 1/\sqrt{12}$ for a tumbling sphere. According to Eq. 9.19 the demise condition in free-molecular flow is dependent on the object diameter, whereas, as a result of Eq. 9.20, in hypersonic continuum flow it is not.

A second, alternative way to survive re-entry is that due to a possible dominance of the first term on the right-hand side of Eq. 9.16. In this case heat storage causes an increase of the body temperature until the melting temperature T_m is reached. This leads to a minimum radius which is necessary to store the incoming energy within a sphere of given material.

$$R \geq \frac{1.5767 \times 10^6}{h_m^\star \cos^2 \gamma_{a,120}} \sqrt{\frac{1}{\rho \, c_D \, |\sin \gamma_{a,120}|}} \tag{9.21}$$

For smaller radii the specific demise enthalpy $h_m^\star$ according to Eq. 9.22 is reached, leading to a complete burn-up of the object.

$$h_m^\star = c_p \, (T_m - T_{120}) + h_m \tag{9.22}$$

Table 9.3 lists the density ρ, emission coefficient ϵ, melting temperature T_m, specific heat coefficient c_p, and melting enthalpy h_m for typical materials used in spacecraft manufacturing.

Fig. 9.10 shows results of a numerical integration of Eq. 9.13 and 9.14 in the free-molecular, transitional, and laminar continuum flow regimes for a US Standard Atmosphere (anon., 1976). The output is provided in terms of demise altitudes of solid spheres as a function of object diameter and material (Fig. 9.10, top), and as a function of the object shape (Fig. 9.10, center). Both diagrams indicate that below a certain minimum diameter $d < d_{min}$ and above a certain maximum diameter $d > d_{max}$ the objects tend to survive. For $d < d_{min}$ this is due to the low area-to-mass ratio (since $A/m \propto 1/d$), and due to the resulting early deceleration at high altitudes, with free-molecular heating, and with good re-radiation capabilities (since [re-radiation]/[heat storage] $\propto A/m \propto 1/d$). For large objects, where $d > d_{max}$, the altitude of peak heat flux is lowered into the continuum flow regime. In this case, the increased heat storage capacity (with [heat storage]/[re-radiation] $\propto m/A \propto d$) and the reduced integrated heat flux causes the object to survive.

For spherical shapes Fig. 9.10 (top) indicates that titanium has the best survival potential, followed by stainless steel, Inconel, and copper. Due to its low melting temperature and poor re-radiation capability aluminum has the lowest probability of survival. For titanium Fig. 9.10 (center) shows the influence of the body shape on demise altitude and survival. Disks, due to their large A/m ratio, have the best survival potential, while spheres and cylinders span a demise zone of more than a decade in diameters, with spheres more likely to reach ground at large sizes. The surviving mass fraction at ground impact is shown in Fig. 9.10 (bottom) for titanium. In a worst case, for titanium spheres, most objects in the size regime between 5 mm $\leq d_{\rm Ti} \leq$ 20 mm will completely burn up, and they will hence not contribute to the effective casualty cross-section at ground impact. For other materials the small survivor objects are limited to diameter ranges from $d_{\rm Fe} <$ 1 mm

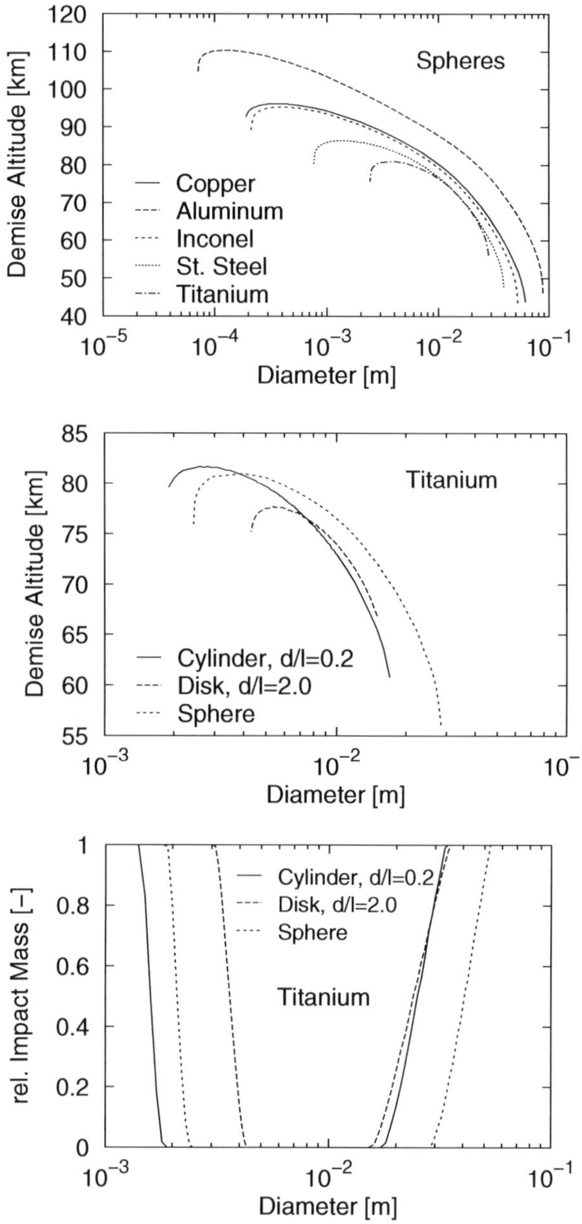

Fig. 9.10. Demise altitude of solid spheres as a function of material and diameter (top), demise altitude for titanium objects as a function of shape and diameter (center), and fractional impact mass of titanium objects as a function of shape and diameter (bottom). Entry conditions at $z = 120$ km: $v_{a,120} = 7.8\,\text{km/s}$, $\gamma_{a,120} = -2.5°$, and $T_{120} = 300\,\text{K}$.

for stainless steel to $d_{\mathrm{Al}} < 0.1$ mm for aluminum. Above these lower thresholds, and below an upper size threshold of a few cm (which is less material-dependent), all solid objects tend to burn up. This is an important finding, since most screws, nuts, and bolts used in spacecraft and rocket manufacturing fall into this domain. For the terminal velocity of solid spheres at ground impact the proportionality $v_i \propto \sqrt{m/A} \propto \sqrt{d}$ according to Eq. 9.10 holds. A solid, stainless steel sphere of diameter $d = 1$ cm, with $m/A \approx 100\,\mathrm{kg/m^2}$, would have a terminal velocity of $v_i \approx 40$ m/s, if it were to reach the ground (which is not the case according to Fig. 9.10, top). If the diameter is reduced to $d = 1$ mm, v_i reduces to 12.6 m/s, and if the diameter is further reduced to $d = 0.1$ mm, then v_i is merely 4.0 m/s. Hence, small survivor objects tend to carry a low risk potential due to their very low kinetic energies at impact ("rain down" of small particles).

The preceding analysis demonstrated that below a certain size threshold re-entry objects either burn up, or decelerate to equilibrium, free-fall velocities with kinetic energies which do not pose a risk to the population on ground. For non-spherical, simple-shaped hollow or solid objects of uniform material, with extensions larger than a few centimeters, Eq. 9.11 to 9.22 must be adapted. Assuming randomly tumbling entry objects, mean values of the ballistic parameter and of the radiation exchange properties must be established. Moreover, drag coefficients and heat exchange parameters (e.g. Stanton number) must be determined as a function of the changing flow regimes. This has for instance been accomplished in ESA's DRAMA[9.4] software (Klinkrad, 2004) and in NASA's ORSAT[9.5] program (Rochelle et al., 2004; Bouslog et al., 1994). These extensions of the previously described theory still have limitations in the case of spacecraft or orbital stages with complex geometries, non-uniform material properties, and non-random attitudes. In order to forecast the survival potential of such objects, the SCARAB[9.6] tool was developed (Fritsche et al., 1999).

SCARAB is a multi-disciplinary analysis tool which incorporates a CAD-like user interface to define the geometry, mass, and material properties of a complex spacecraft, a 6 degrees-of-freedom (6 DoF) flight dynamics analysis to predict the trajectory and attitude, an aerodynamic analysis to compute perturbing forces and torques, an aerothermal analysis to determine heat flux, a thermal analysis to determine the heat balance in each part of the spacecraft, and a structural analysis to monitor local stress levels. A break-up is initiated, if local stress limits are exceeded, or if load-bearing joints are molten.

The 6 DoF propagation of the dynamic state consists of an integration of the 3 DoF for the 3 components of the orbit position vector $\underline{r} = (r_X, r_Y, r_Z)$ according to Eq. 9.23, and of the 3 DoF for the 3 components of the spacecraft rotation vector $\underline{\omega} = (\omega_X, \omega_Y, \omega_Z)$ according to Eq. 9.24 ($\underline{\phi}$ is a resulting vector of attitude angles). In these state equations m is the instantaneous mass, and J is the tensor of the moments of inertia in the inertial reference frame X, Y, Z. $\underline{p}_a$ and $\underline{p}_g$ are perturbing

[9.4]Debris Risk Assessment and Mitigation Analysis

[9.5]Orbital Re-Entry Survival Analysis Tool

[9.6]Spacecraft Atmospheric Re-Entry and Aerothermal Break-Up translational accelerations, and $\underline{m}_a$ and $\underline{m}_g$ are perturbing rotational accelerations due to aerodynamic and Earth gravity effects.

$$\frac{1}{m}\frac{\mathrm{d}}{\mathrm{d}t}(m\,\dot{\underline{r}}) = \underline{p}_a + \underline{p}_g \quad\longrightarrow\quad \dot{\underline{r}}(t), \underline{r}(t) \tag{9.23}$$

$$\mathsf{J}^{-1}\left(\frac{\mathrm{d}}{\mathrm{d}t}(\mathsf{J}\,\underline{\omega}) + \underline{\omega}\times\mathsf{J}\,\underline{\omega}\right) = \underline{m}_a + \underline{m}_g \quad\longrightarrow\quad \underline{\omega}(t), \underline{\phi}(t) \tag{9.24}$$

The perturbations on the right-hand sides should be computed with models of similar resolution as in Section 9.3. However, for the perturbing moments due to the gravity gradient (as a result of the radial offset of the center of mass CoM relative to the center of gravity) it is sufficient to use the central attraction and the J_2 contribution.

Within SCARAB a spacecraft is composed of a large number of elementary geometric shapes, each with uniform material properties. All elementary shapes are discretized into volume elements (voxels) with planar surface facets which are adjacent to a neighboring voxel, or form a part of the outside or inside surface of the spacecraft. At every other integration step the mass properties are re-evaluated, and the aerodynamic, aerothermal, and thermal view factors of each voxel are re-determined to account for attitude changes, break-ups, or melting. The perturbing aerodynamic forces and moments (translational and rotational accelerations) are determined by a surface integral over all voxel surfaces which are exposed to the flow field of density ρ and aerodynamic velocity v_a.

$$m\,\underline{p}_a = \frac{1}{2}\rho v_a^2 \int_{A_a} (c_\sigma\,\underline{u}_\sigma + c_\tau\,\underline{u}_\tau)\,\mathrm{d}A \tag{9.25}$$

$$\mathsf{J}\,\underline{m}_a = \frac{1}{2}\rho v_a^2 \int_{A_a} (\underline{r}_A - \underline{r}_{CoM}) \times (c_\sigma\,\underline{u}_\sigma + c_\tau\,\underline{u}_\tau)\,\mathrm{d}A \tag{9.26}$$

The local pressure on a surface element has a normal pressure component of magnitude c_σ, positive along the outward unit vector $\underline{u}_\sigma$, and a shear-stress component of magnitude c_τ, along $\underline{u}_\tau$ within the tangent plane, positive in the stream line direction. The vector $\Delta\underline{r}_A = \underline{r}_A - \underline{r}_{CoM}$ marks the offset of the local surface facet relative to the center of mass. The integration is performed over the entire surface A_a which is exposed to the parallel flow field, taking into account aerodynamic shadowing by upstream components.

Let $S_\infty = v_a/\sqrt{2\,R\,T_\infty}$, Kn_∞, Ma_∞ and T_∞ be the free-stream molecular speed ratio, the Knudsen number, the Mach number and the temperature, let T_w and θ_w be the local wall temperature and incidence angle, and let κ be the ratio of specific heats of the ambient atmosphere, then the local pressure and shear stress coefficients c_σ and c_τ can be determined for each of the re-entry flow regimes according to Table 9.4. In the free-molecular regime SCARAB determines normal and tangential accommodation coefficients using a Nocilla or Schaaf–Chambré gas–surface interaction model. Effects due to dissociation and re-combination at catalytic surfaces are not considered. Further details are provided in (Fritsche et al., 2000).

Table 9.4. Coefficients of aerodynamic pressure c_σ, aerodynamic shear stress c_τ, and aerothermal heating c_h in free-molecular, laminar continuum, and transitional flow (where $\text{fct}(c_{\sigma,f}, c_{\sigma,c}, \text{Kn}_\infty)$, $\sigma \mapsto \tau, h$, denotes empirical bridging functions).

Free-molecular flow	Laminar continuum flow	Transitional flow
$\text{Kn}_\infty > 10$	$\text{Kn}_\infty < 0.01$	$0.01 \leq \text{Kn}_\infty \leq 10$
$c_{\sigma,f} = \text{fct}(S_\infty, \theta_w, T_w/T_\infty)$	$c_{\sigma,c} = \text{fct}(\text{Ma}_\infty, \theta_w, \kappa)$	$c_{\sigma,t} = \text{fct}(c_{\sigma,f}, c_{\sigma,c}, \text{Kn}_\infty)$
$c_{\tau,f} = \text{fct}(S_\infty, \theta_w)$	$c_{\tau,c} = 0$	$c_{\tau,t} = \text{fct}(c_{\tau,f}, c_{\tau,c}, \text{Kn}_\infty)$
$c_{h,f} = \text{fct}(S_\infty, \theta_w, T_w/T_\infty)$	$c_{h,c} = \text{fct}(\text{Re}_{\infty,0}, \theta_w)$	$c_{h,t} = \text{fct}(c_{h,f}, c_{h,c}, \text{Kn}_\infty)$

When substituting the dynamic pressure $\frac{1}{2}\,\rho\, v_a^2$ in Eq. 9.25 and 9.26 by the solar pressure constant at 1 A.U. ($p_\odot \approx 4 \times 10^{-6}\,\text{N}\,\text{m}^{-2}$), and when substituting the molecule–surface interaction parameters by corresponding photon–surface interaction parameters, then the same algorithms can be used to determine the perturbing solar radiation force and torque.

The calculation of local aerothermal heat flux onto the spacecraft surface has many similarities with the aerodynamic analysis, since it is governed by the same flow regimes, and it relates to the same exposed surface area. However, in contrast with the aerodynamics, the aerothermal analysis provides an important input to the local, voxel-based heat balance within the thermal analysis. The heat flux $\dot{q}$ into an external voxel surface is determined from

$$\dot{q} = \frac{1}{2}\, c_h\, \rho\, v_a^3 \left(1 - \frac{c_{p,0}\,(T_v - T_\infty)}{v_a^2/2}\right) \tag{9.27}$$

where $c_{p,0}$ is the heat capacity of the gas in the stagnation region, T_v is the voxel temperature, and c_h is a local heat transfer coefficient, which varies with the flight regime and with the spacecraft attitude as shown in Table 9.4 (where $\text{Re}_{\infty,0}$ is the local Reynolds number based on the free-stream density and the stagnation point temperature).

Aerothermal heating according to Eq. 9.27 can be regarded as a convective input into the exposed voxel surface, with a heating rate of $\dot{Q}_{\text{conv},\downarrow}$. Other heat sources are due to incoming radiation $\dot{Q}_{\text{rad},\downarrow}$ from parts of the spacecraft which are inside the field of view of a given voxel surface, and conductive heat flux $\dot{Q}_{\text{cond},\downarrow}$ from adjacent voxels. The latter two mechanisms can also release heat as $\dot{Q}_{\text{rad},\uparrow}$ and $\dot{Q}_{\text{cond},\uparrow}$. The resulting heating rate $\dot{Q}_v$ of a given voxel is governed by

$$\dot{Q}_v = \dot{Q}_{\text{conv},\downarrow} + \dot{Q}_{\text{rad},\downarrow} - \dot{Q}_{\text{rad},\uparrow} + \dot{Q}_{\text{cond},\downarrow} - \dot{Q}_{\text{cond},\uparrow} \tag{9.28}$$

If $\dot{Q}_v < 0$ (i.e. more heat is rejected than absorbed), then the voxel temperature decreases. Otherwise, if $\dot{Q} > 0$, the voxel temperature T_v increases inversely proportional to the heat capacity $c_{p,v}$ of the material and the voxel mass $m_v = \rho_v\, V_v$.

$$\dot{Q}_v = c_{p,v}\, \rho_v\, V_v\, \dot{T}_v \quad \text{if} \quad T_v < T_{m,v} \quad \longrightarrow \quad \text{heating} \tag{9.29}$$

Once the melting temperature $T_{m,v}$ of the voxel material is reached, the incoming heat is used to ablate mass (reducing the voxel volume) at a rate $\dot{V}_v$ which is determined by the melting enthalpy $h_{m,v}$ (see Table 9.3).

$$\dot{Q}_v = -\rho_v h_{m,v} \dot{V}_v \quad \text{if} \quad T_v = T_{m,v} \qquad \longrightarrow \qquad \text{melting} \tag{9.30}$$

In SCARAB, the change of temperature within the entire spacecraft is monitored down to the voxel level. This information is used to adjust local, temperature-dependent material properties for the thermal analysis itself, and for the structural analysis. The loss of mass according to Eq. 9.30 is used to update the overall spacecraft mass and its moments of inertia, and to adjust the external geometry for a refined aerodynamic and aerothermal analysis.

The structural integrity of the re-entry object is monitored at user-defined cut planes, where instantaneous bending moments, torques, normal and shear stress forces are combined into a resulting normal stress σ_v and shear stress τ_v at voxel level. These are used to determine an equivalent load $\tilde{\sigma}_v$.

$$\tilde{\sigma}_v = \sqrt{\sigma_v^2 + 3\tau_v^2} \tag{9.31}$$

If a material- and temperature-dependent ultimate tensile strength $\sigma_{max,v}$ is exceeded, then a local structural failure is logged. Break-up events can be triggered either by melting or by structural failure. In either case, SCARAB automatically installs an additional branch of a fragmentation tree, following all fragments to their demise altitude or ground impact. For the latter ones a log with mass, size, impact location, and terminal velocity is generated. This information can then be used to perform an event-related casualty risk assessment.

Fig. 9.11 shows a test satellite which was modeled with SCARAB, and which is released into a near-circular, retrograde re-entry trajectory with $i = 94.0°$, at a geodetic altitude of $z = 126.9$ km, with a velocity of $v = 7.875$ km/s, a flight path angle of $\gamma \approx 0°$, and an inertially fixed attitude. At the time of release the spacecraft has a mass of $m_\circ = 1{,}076.73$ kg, dimensions of 4.51 m $\times$ 2.99 m $\times$ 2.48 m, and a uniform temperature of $T_\circ = 300$ K. During the re-entry a total of 32 break-ups occur, each generating between 1 and 3 fragments. In the first sequence of release events, at altitudes between 109 km and 102 km, the solar arrays are completely removed in less than 3 minutes. In the following 9 minutes, within altitudes of 86.0 km and 50.5 km, 24 break-up events are recorded for the main spacecraft body. Most of the initial mass burns up between 103.4 km and 62.2 km, in 33 demise events. Only 10 objects survive to ground impact (see Fig. 9.12), with a total mass of 450.26 kg (41.82%) and a casualty cross-section of $A_c = 10.6 \pm 1.2\,\text{m}^2$ (depending on the attitude). 89.6% of the impacting mass is concentrated in the 403.37 kg remnant of the main spacecraft body, while 3 of the survivor objects have masses below 10 g. In total, 4 objects have kinetic energies below a fatality threshold of ~30 J, and one object has only a minor fatality potential of ~1%. The impact velocities vary between 6.6 m/s and 81.4 m/s, with impact locations dispersed within ~500 km along track, and impact times spread over ~30 minutes. An overview of the survivor object characteristics is provided in Table 9.6.

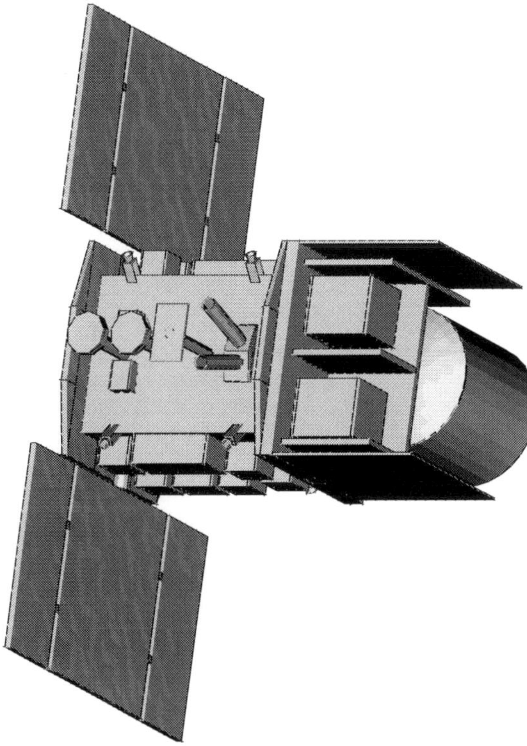

Fig. 9.11. Test satellite used for a detailed re-entry survival analysis (the satellite shown is a fictitious object; any resemblance to existing spacecraft is unintentional).

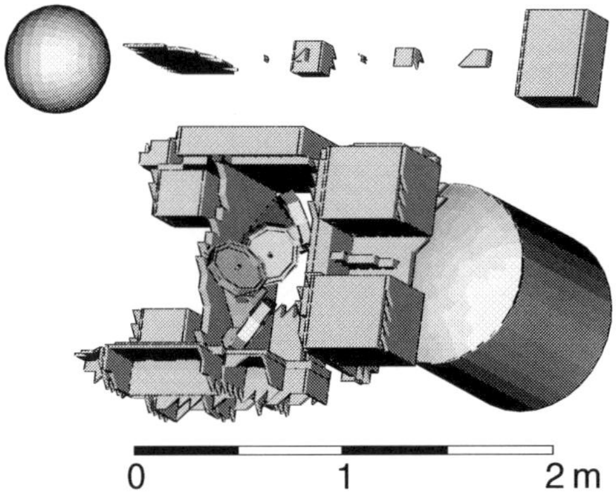

Fig. 9.12. Surviving ground impact fragments of the test satellite defined in Fig. 9.11.

9.5 ESTIMATION OF ON-GROUND RISK DUE TO RE-ENTRIES

To assess the risk which is imposed by a surviving re-entry object, it is first necessary to determine where it will impact on the Earth surface. Due to the stochastic nature of the re-entry dynamics in an equally stochastic ambient atmosphere, such an impact location can only by defined in a statistical manner. If one assumes a static, exponential atmosphere with density scale height $H_{\rho,o}$, then one can show that aerodynamic lift and side forces cause an impact dispersion in along-track direction Δx and cross-track direction Δy which can be expressed as follows.

$$\Delta x = -\frac{H_{\rho,o}}{2\,\gamma_{a,120}} \frac{\tilde{\nu}}{1+\tilde{\nu}\,(\gamma_r/\gamma_{a,120})^2} \left(\frac{c_L}{c_D} \cos\psi_{L,S} + \frac{c_S}{c_D} \sin\psi_{L,S} \right) \tag{9.32}$$

$$\begin{aligned} \Delta y &= -\frac{H_{\rho,o}}{\gamma_{a,120}} \left(\frac{\gamma_r}{\gamma_{a,120}}\right)^2 \ln\left(1+\tilde{\nu}\,(\gamma_r/\gamma_{a,120})^2\right) \\ &\quad \cdot \left(\frac{c_L}{c_D} \sin\psi_{L,S} - \frac{c_S}{c_D} \cos\psi_{L,S} \right) \end{aligned} \tag{9.33}$$

In these equations c_D, c_L, and c_S are the aerodynamic coefficients of the drag, lift, and side force, the angle $\psi_{L,S}$ describes the partitioning of the non-drag forces ($\psi_{L,S} = 0°, 180° \rightarrow$ maximum up/down lift, $\psi_{L,S} = \pm 90° \rightarrow$ maximum left/right side force), $\gamma_r = -\sqrt{H_{\rho,o}/a_e}$ is a reference angle ($-2°$ for the Earth), and $\nu = -\ln(v_a/\sqrt{\mu/r})$ is a velocity ratio, with a value of $\tilde{\nu} \approx 3$ in the relevant transonic region, prior to impact. The case $c_L = c_S = 0$ marks a ballistic re-entry.

Eq. 9.32 and 9.33 describe an elliptical impact dispersion area. Such an outcome can also be verified by numerical results of re-entry predictions with randomized c_L/c_D and c_S/c_D. When assuming an entry angle of $\gamma_{a,120} = -2°$, then the ratio of the maximum along-track to cross-track extension of the dispersion ellipse is on the order of $(\Delta x)_{max}/(\Delta y)_{max} \approx 2{,}400\,\text{km}/300\,\text{km} = 8$. One can show that under the influence of aerodynamically induced rotations of the entry object particularly the cross-track range is strongly affected. It reduces by a factor $f_{\Delta y} \propto (\gamma_{a,120}\sqrt{\mu/r})/(\dot{\psi}\,H_{\rho,o})$, which leads to $(\Delta y)_{max} \approx 1\,\text{km}$ for a rotation rate of $\dot{\psi} \approx 2°/\text{s}$ and an extreme ratio $c_S/c_D = 1$.

Table 9.5. Probability integrals of a 2D Gaussian distribution for elliptical and rectangular integration areas of 1σ to 4σ extensions.

Integration bounds	1σ	2σ	3σ	4σ
Elliptic area	0.39347	0.86467	0.98889	0.99966
Rectangular area	0.46606	0.91107	0.99461	0.99987

In reality the statistical dispersion of the impact location of an entry survivor object is driven by uncertainties in the effective area-to-mass ratio, in the aerodynamic coefficients, in the air density, and in the initial orbit and attitude state. A Monte Carlo analysis of these effects would provide an impact probability density

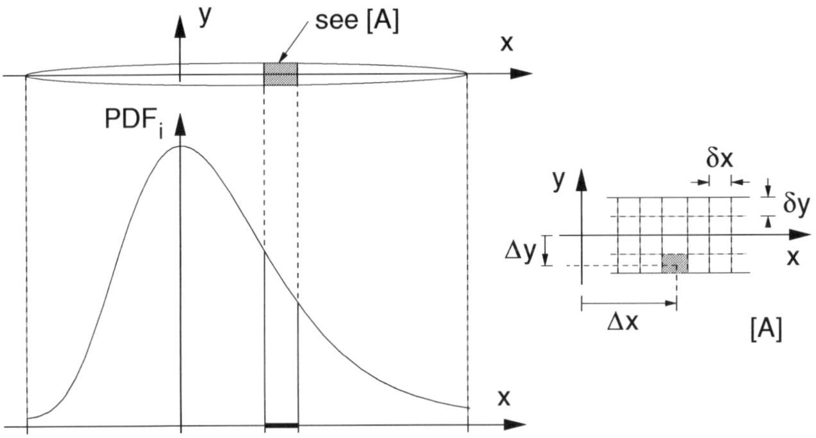

Fig. 9.13. Discretized impact probability density distribution PDF_i over a given impact swath, with resolution δx in along-track direction, and δy in cross-track direction.

function (PDF_i) which follows a distorted 2D Gaussian distribution as shown in Fig. 9.13. If a sufficiently large number of Monte Carlo runs is performed, then discrete, local PDF_i results can be obtained by determining the fraction of impacts in a bin of size $\delta x \times \delta y$ at a location x, y, in relation to the total number of runs. The integration bounds of the 2D dispersion area may then be derived by a summation over all top-ranking PDF_i results, until a user-defined threshold level is reached (e.g. an impact probability of $P_i = 0.86467$, equivalent to a 2σ probability for a 2D Gaussian distribution as listed in Table 9.5).

The outlined Monte Carlo approach can be prohibitive in computer run-time, particularly if meaningful statistics are to be obtained for a large number of randomized parameters. A standard 2D Gaussian distribution can in many cases be a viable alternative. Its standard deviation along-track $\Delta x_{1\sigma}$ can be determined by varying the drag term $\rho \cdot B$, and its cross-track component $\Delta y_{1\sigma}$ may be fixed to an empirical value (e.g. $\Delta y_{1\sigma} = 20\,\text{km}$). The analysis can be further simplified, if a rectangular area of integration is chosen for the distribution PDF_i. Table 9.5 shows how the resulting probability integrals deviate from those of elliptical integration areas within 1σ to 4σ limits. An integration over the rectangular area of extensions $\pm 2\sigma$ along-track and cross-track captures more than 91% of all impact events, which is a good trade-off of computing time versus statistical representation, and which shall be used hereafter. As illustrated in Fig. 9.13, a ground-swath related coordinate system in along-track direction x and cross-track direction y shall be centered at the most probable impact location, for which PDF_i has its maximum. The swath of 2σ extensions shall then be partitioned in rectangular bins of equal areas $\delta x \times \delta y$, centered at the bin coordinates Δx_n and Δy_m, where $-N_{2\sigma} \leq n \leq +N_{2\sigma}$ and $-M_{2\sigma} \leq m \leq +M_{2\sigma}$ are the bin numbers, with $n = m = 0$ for $PDF_{i,max}$. In the case of a more realistic non-Gaussian along-track dispersion, the bin indices $N_{2\sigma}$ at the heel and toe position of the 2σ footprint would be asymmetric, but the following algorithms would still apply. The local impact probability $(P_i)_{n,m}$ can be computed from the probability density and the bin area.

$$(P_i)_{n,m} = (\mathrm{PDF}_i)_{n,m} \cdot \delta x \cdot \delta y \tag{9.34}$$

The required, discrete probability density distribution $(\mathrm{PDF}_i)_{n,m}$ can be produced from a sufficiently large number of j_{MC} Monte Carlo runs.

$$(\mathrm{PDF}_i)_{n,m} = j_{n,m}/j_{MC} \tag{9.35}$$

where $j_{n,m}$ is the number of impacts within a given area bin. There may be cases where the same re-entry object completely survives in $j_i \leq j_{MC}$ cases, and burns up in $j_d = j_{MC} - j_i$ cases, depending on the randomized initial entry conditions. In such events, the 2σ bounds of a rectangular integration area are determined as outlined above, for a threshold probability of $\tilde{P}_{i,2\sigma} = P_{i,2\sigma} \cdot j_i/j_{MC}$ (with $P_{i,2\sigma} \approx 0.911$ according to Table 9.5).

For a 2D Gaussian distribution the probability density function $(\mathrm{PDF}_i)_{n,m}$ can be expressed analytically (where $\Delta x_n = n\,\delta x$ and $\Delta y_m = m\,\delta y$).

$$(\mathrm{PDF}_i)_{n,m} = \frac{1}{2\pi\,\Delta x_{1\sigma}\,\Delta y_{1\sigma}} \exp\left[-\frac{1}{2}\left(\left(\frac{\Delta x_n}{\Delta x_{1\sigma}}\right)^2 + \left(\frac{\Delta y_m}{\Delta y_{1\sigma}}\right)^2\right)\right] \tag{9.36}$$

This form of $(\mathrm{PDF}_i)_{n,m}$, leading to symmetric summation limits $-N_{2\sigma} \leq n \leq +N_{2\sigma}$ and $-M_{2\sigma} \leq m \leq +M_{2\sigma}$, shall be assumed hereafter (without compromising the general validity of the solutions). The dimensions δx, δy of the local sampling bins should be compatible with spatial resolutions $\delta\lambda$, $\delta\phi$ of discretized maps of land masses $L(\lambda, \phi)$, and population densities $\rho_p(\lambda, \phi)$, which are on the order of $\delta\lambda = \delta\phi = 5' = 9.25$ km at the equator (Tobler et al., 1995).

The impact probability $P_{i,2\sigma}$ for a given entry object can be determined by a summation over the rectangular swath area of 2σ extensions.

$$P_{i,2\sigma} = \sum_{n=-N_{2\sigma}}^{+N_{2\sigma}} \sum_{m=-M_{2\sigma}}^{+M_{2\sigma}} (P_i)_{n,m} \approx 0.91107\,(1 - P_d) \tag{9.37}$$

where $0 \leq P_d < 1$ is the overall probability of demise of the object for randomized initial entry conditions (with $P_d = j_i/j_{MC}$). For $P_d > 0$ the summation limits, mainly in along-track direction, are adjusted accordingly.

Assuming a local land mass fraction of $0 \leq L_{n,m} \leq 1$, the probability $P_{L,2\sigma}$ of a land impact can be determined from

$$P_{L,2\sigma} = \sum_{n=-N_{2\sigma}}^{+N_{2\sigma}} \sum_{m=-M_{2\sigma}}^{+M_{2\sigma}} (P_i)_{n,m} \cdot L_{n,m} \tag{9.38}$$

The casualty probability[9.7] to humans within a re-entry ground swath depends on the size of the impacting object and on the population density at the impact

[9.7]the term "casualty" shall comprise severe injury or death region. Let A_i be the geometric cross-section of the impactor, and let $A_h = 0.36\,\mathrm{m}^2$ be the vertical projection area of a human, then a casualty cross-section A_c can be defined according to the NASA Safety Standard NSS 1740.14 (anon., 1995b).

$$A_c = \left(\sqrt{A_h} + \sqrt{A_i}\right)^2 \tag{9.39}$$

Assuming circular cross-sections for the human and the impact object, A_c is the equally circular area within which A_h and A_i must touch or intersect each other (see illustration of A_c in Fig. 8.1). The casualty cross-section obtained from Eq. 9.39 can increase by a factor of up to 2 or 4 (depending on the surface), if the impact is not vertical, and if sliding or re-bounding occurs (Smith, 1999). An analysis of 17,620 re-entry objects between 1957 and 2001 shows that the mean geometric cross-section was on the order of $\overline{A}_i \approx 5.0\,\mathrm{m}^2$ (Klinkrad et al., 2003). This translates into a mean casualty cross-section of $\overline{A}_c \approx 8.0\,\mathrm{m}^2$.

Table 9.6. Risk-relevant characteristics of re-entry survivor objects of a test satellite (see Fig. 9.11 for the satellite and Fig. 9.12 for its survivor objects).

Impact object	Material	m_i [kg]	$\overline{A}_i$ [m^2]	$\overline{v}_{a,i}$ [m/s]	$\overline{E}_{kin}$ [J]	η_f [-]
Spherical tank	HC-AA7075	10.213	0.195	50.2	12,864.8	1.00
Radiator part	AA7075	1.880	0.117	50.9	2,435.4	1.00
Antenna part	HC-AA7075	0.006	0.001	17.4	0.9	0.00
Box #07	HC-AA7075	1.847	0.045	40.2	1,494.8	1.00
Antenna part	HC-AA7075	0.003	0.001	21.1	0.7	0.00
Box #08	AA7075	0.117	0.016	19.2	21.6	0.00
Structure part	HC-AA7075	0.008	0.010	6.6	0.2	0.00
Box #12	HC-AA7075	32.483	0.156	81.5	1.08×10^5	1.00
Main body	A316, AA7075	403.367	2.440	76.4	1.18×10^6	1.00
Box #10	HC-AA7075	0.339	0.013	36.4	38.0	0.01

A model of the world population density, suitable for re-entry risk analyses, is provided by the Global Demography Project 1994 (Tobler et al., 1995). This dataset for the year 1994, which is discussed in detail in Annex D, provides population densities $\rho_{p,1994}(\lambda, \phi)$ with a spatial resolution better than 9.25 km $\times$ 9.25 km (see Fig. D.1). The estimated world population in 1994 was $5.63 \times 10^{+9}$, distributed over a total land surface of $1.48 \times 10^{+8}\,\mathrm{km}^2$, which accounts for 28.97% of the surface of the Earth reference ellipsoid. This corresponds to a global mean population density of $\overline{\rho} = 11.0\,\mathrm{km}^{-2}$. When dropping a mean casualty cross-section of $\overline{A}_c \approx 8.0\,\mathrm{m}^2$ on a mean population density of $\overline{\rho} = 11.0\,\mathrm{km}^{-2}$, then the resulting casualty probability is $\overline{P}_c \approx 1/11,000$. This value is close to a commonly accepted risk level for a single re-entry event of $P_c \leq 1/10,000$, which has been adopted by several space agencies.

Assuming a discretized, epoch-adjusted world population density distribution $(\rho_p)_{n,m} = \rho_p(\lambda_{n,m}, \phi_{n,m}, t)$, and a bin-wise, ablation-adjusted casualty cross-

Table 9.7. Fatality index $\eta_f(E_{i,kin})$ of a falling re-entry object. The data, which are averaged over different exposure scenerios of a human body, are based on (anon., 2000).

Fatality index η_f [−]	0.01	0.10	0.50	0.90	0.99
Kinetic impact energy $E_{i,kin}$ [J]	37.0	52.0	104.0	205.0	292.0

section $(A_c)_{n,m}$, then the impact casualty probability $P_{c,2\sigma}$ for the rectangular integration swath of 2σ extensions is defined as

$$P_{c,2\sigma} = \sum_{n=-N_{2\sigma}}^{+N_{2\sigma}} \sum_{m=-M_{2\sigma}}^{+M_{2\sigma}} (P_i)_{n,m} \cdot (\rho_p)_{n,m} \cdot (A_c)_{n,m} \tag{9.40}$$

If there is more than one survivor object, then the $k = 1, \ldots, K$ individual results of $P_{c,2\sigma}$ can be combined into an overall casualty probability.

$$P_{c,2\sigma} = 1 - \prod_{k=1}^{K} \left(1 - (P_{c,2\sigma})_k\right) \tag{9.41}$$

Some authors tend to prefer the term casualty expectancy $E_{c,2\sigma}$ in this context, because one can think of extreme cases where Eq. 9.40 can yield a value which is larger than 1. This is acceptable for an expectancy, but not for a probability (Montgomery and Ward, 1995).

The casualty probability from Eq. 9.40 can be translated into a corresponding fatality probability $P_{f,2\sigma}$ by applying a fatality index $\eta_f \in [0, 1]$, which is commonly defined as a function of the kinetic energy $E_{i,kin}$ of the impact object.

$$P_{f,2\sigma} = \sum_{n=-N_{2\sigma}}^{+N_{2\sigma}} \sum_{m=-M_{2\sigma}}^{+M_{2\sigma}} (P_i)_{n,m} \cdot (\rho_p)_{n,m} \cdot (A_c)_{n,m} \cdot (\eta_f)_{n,m} \tag{9.42}$$

The fatality index $\eta_f(E_{i,kin})$, as defined in (anon., 2000), and as listed in Table 9.7, can be approximated to within $\pm 0.3\%$ by the following fitting function[9.8].

$$\eta_f(E_{i,kin}) = \begin{cases} 0 & \text{for} \quad E_{i,kin} < 32\,\mathrm{J} \\[2ex] \dfrac{1}{2} - \dfrac{1}{2}\cos\left(\pi\,\dfrac{\ln(E_{i,kin}/[\mathrm{J}]) - c_0}{c_1 - c_0}\right) & \text{for} \quad 32\,\mathrm{J} \le E_{i,kin} \le 338\,\mathrm{J} \\[2ex] 1 & \text{for} \quad E_{i,kin} > 338\,\mathrm{J} \end{cases} \tag{9.43}$$

with $c_0 = \ln(32)$ and $c_1 = \ln(338)$. The overall fatality risk caused by a single re-entry event with K survivor objects can be determined from Eq. 9.42 on the basis of the individual results.

$$P_{f,2\sigma} = 1 - \prod_{k=1}^{K} \left(1 - (P_{f,2\sigma})_k\right) \tag{9.44}$$

[9.8]NASA defines $\eta_f = 0$ for $E_{i,kin} < 15\,\mathrm{J}$ (anon., 1995b)

The fatality risk potential of individual survivor objects of the test satellite shown in Fig. 9.11 is summarized in Table 9.6. Out of 10 impacting fragments, only 5 would be likely to cause death, based on their kinetic energy. Hence, the effective mean casualty cross-section of $\overline{A}_c = 10.598\,\mathrm{m}^2$ for 10 survivor objects of the event is reduced to an effective fatality cross-section of $\overline{A}_\mathrm{f} = 8.305\,\mathrm{m}^2$, when only considering the 5 risk objects with $\eta_\mathrm{f} > 0$.

Table 9.8. Annual, personal risks of fatality due to common causes and activites, as compared to the individual risk of a fatal hit by re-entering space objects (anon., 1995a).

Activity or cause	Personal risk of fatality $[\mathrm{y}^{-1}]$	
Coal mining	9.3×10^{-3}	$= 1/107$
Cancer	1.8×10^{-3}	$= 1/545$
Fire fighting	8.0×10^{-4}	$= 1/1,250$
Motor vehicle operation	2.2×10^{-4}	$= 1/4,500$
Home accidents	1.2×10^{-5}	$= 1/83,000$
Air travel	2.0×10^{-6}	$= 1/500,000$
Lightning	5.0×10^{-7}	$= 1/2,000,000$
Space object re-entry	8.0×10^{-12}	$= 1/1.2 \times 10^{11}$

There may be occasions when it is important to put the re-entry risk into perspective with commonly accepted or tolerated risks in everyday life. To make a conservative assessment it shall be assumed that cataloged space objects re-enter at a mean rate of 450 per year, with each of these leading to a fatality cross-section of $A_\mathrm{f} \approx 10\,\mathrm{m}^2$. Furthermore, let the mean population density be $\rho_p \approx 11\,\mathrm{km}^{-2}$, for a total world population of 6×10^9. The fatality rate due to re-entry debris for a single person would then be $\sim 8.0 \times 10^{-12}\,\mathrm{y}^{-1}$. This compares with normal risks in day-to-day life as shown in Table 9.8. Fatality rates of different causes can be directly related by comparing the exposure timespans which would result in the same fatality probability. The annual risk of a person being killed by a re-entering man-made space object is equivalent to the risk of being killed while traveling 1 meter in a car, doing 10 seconds of skiing, working 1 second as a fire fighter, or spending 5 minutes of your life at the age of 60 (anon., 1995a).

Why do we care about re-entry risk assessments, if the individual risk is extremely low? In contrast to other causes listed in Table 9.8, a launch or re-entry related risk is not commonly accepted by the population as being inevitable. Moreover, a launching nation is liable for any launch-related consequences, such as damage to property or casualties, on ground or in air space, from falling space objects (anon., 2002). This is why launch and spacecraft operators request that the risk of *any* personal casualty due to a single re-entry event must be less than a given threshold value, such as $P_c < P_c^\star = 10^{-4}$, corresponding to 1 in 10,000. A mean, uncontrolled re-entry event with $A_c = 8\,\mathrm{m}^2$ will result in a casualty probability which is of this magnitude. Hence, risk management concepts and strategies are required to ascertain that safety requirements are met.

9.6 LONG- AND SHORT-TERM RE-ENTRY RISK MANAGEMENT

There are different ways to manage the risk imposed by re-entry events. All of these approaches aim for a reduction of contributing terms on the right-hand side of Eq. 9.40 or 9.42. In the early mission planning one can try to design a spacecraft or orbital stage in such a way that the total casualty cross-section of potential re-entry survivor objects with critical kinetic energies is minimized, by minimizing contributions due to $A_c \cdot \eta_f$. This can for instance be accomplished through the use of materials with low melting temperatures (see Table 9.3). Very often, however, functional requirements in context with structural or thermal design aspects pose limitations to such a mitigation concept.

The most common and most practicable concept of re-entry risk reduction is through a minimization of contributions from the term $P_i \cdot \rho_p$ for all ground-impacting objects. In the course of planning and execution of a space mission this can be accomplished in different ways. During the conceptual mission design phase an adaptation of the orbit inclination (if this is an option) can significantly affect the overall risk results. For frequently used orbit inclinations Table 9.9 summarizes global mean results of population density $\overline{\rho}_p$, land impact probability $\overline{P}_L$, and specific casualty probability $\overline{P}_c/A_c$, averaged over uniformly distributed nodal longitudes λ_Ω and orbit positions $u = \omega + f$ within a latitude band of $-i \leq \phi \leq +i$. Assuming discrete latitude bins of width $\Delta\phi$ the impact probability density $\Delta P_i(\phi)/\Delta\phi$ can be expressed analytically (Klinkrad et al., 2003).

$$\frac{\Delta P_i(\phi)}{\Delta\phi} = \Phi(\phi, \Delta\phi) - \frac{1}{\pi} \arcsin\left(\frac{\sin(\phi - \Delta\phi/2)}{\sin i}\right) \tag{9.45}$$

$$\Phi(\phi, \Delta\phi) = \begin{cases} \dfrac{1}{\pi} \arcsin\left(\dfrac{\sin(\phi + \Delta\phi/2)}{\sin i}\right) & \text{for} \quad \phi \leq i - \Delta\phi/2 \\ \dfrac{1}{2} & \text{for} \quad i - \Delta\phi/2 < \phi \leq i + \Delta\phi/2 \end{cases}$$

For common orbit inclinations between 7.0° and 98.5° the resulting probability density functions are plotted in Fig. D.2, D.4 and D.6, together with the resulting land impact probability densities $\Delta P_L(\phi)/\Delta\phi$ and the specific casualty probability densities $\Delta P_c(\phi)/(\Delta\phi\, A_c)$. Due to peaks in the population densities $\rho_p(\phi)$ at $+25° \leq \phi \leq +35°$ (see Fig. D.1), in combination with extended resident times in this latitude band for moderate inclinations, Table 9.9 shows the highest casualty probabilities for $i = 28.5°$ and 51.6° (~70% above the lowest value). Unfortunately, these orbit inclinations are of main interest for launches from Kennedy Space Center and from Baikonur. Especially $i \approx 51.6°$ is used by the ISS and its servicing missions. Along the ground tracks of such orbits the mean population density on land masses reaches a maximum of 60.1 km^{-2}, which is to some extent balanced by a lower land impact probability of $P_L = 27.2\%$ (as compared to a maximum of 33.4% on polar orbits).

Fig. D.2, D.4 and D.6 show that the mean results listed in Table 9.9 are hiding a strong latitude-dependence of P_i, P_L and P_c/A_c for different inclinations.

Table 9.9. On-ground risk characteristics for the year 2001, as a function of the inclination of a near-circular re-entry orbit, with mean land impact probability $\overline{P}_L$, mean population density $\overline{\rho}_{p,L}$ of overflown land masses, mean population density $\overline{\rho}_p$ along the entire ground tracks, and mean casualty probablity $\overline{P}_c$ per m^2 of casualty cross-section A_c. Results are averaged with respect to nodal longitudes $\lambda_\Omega \in [-\pi, +\pi]$ and orbit positions $u = \omega + f \in [0, 2\pi]$ (see Fig. D.8 to D.14).

i [°]	$\overline{\rho}_p$ [km^{-2}]	$\overline{\rho}_{p,L}$ [km^{-2}]	$\overline{P}_L$ [-]	$\overline{P}_c/A_c$ [1/m^2]
7.0	9.7	42.1	0.231	$9.7 \times 10^{-6} = 1/103{,}093$
28.5	15.9	55.2	0.271	$15.9 \times 10^{-6} = 1/62{,}893$
51.6	14.9	60.1	0.272	$14.9 \times 10^{-6} = 1/67{,}114$
65.0	11.1	36.6	0.284	$11.1 \times 10^{-6} = 1/90{,}090$
80.0	9.5	27.7	0.331	$9.5 \times 10^{-6} = 1/105{,}263$
89.9	9.3	26.8	0.334	$9.3 \times 10^{-6} = 1/107{,}527$
98.5	9.5	27.0	0.332	$9.5 \times 10^{-6} = 1/105{,}263$

Moreover, P_L and P_c/A_c are significantly different in the southern and northern hemisphere, since only 33.3% of the land masses and 11.7% of the world population are located south of the equator. The land impact probability densities $\Delta P_L(\phi)/\Delta\phi$ follow the trends of the much more pointed impact probability densities $\Delta P_i(\phi)/\Delta\phi$ in latitude regions with large land masses, where $-30° \leq \phi \leq +80°$. In contrast with the $\Delta P_i(\phi)/\Delta\phi$ distributions the $\Delta P_L(\phi)/\Delta\phi$ patterns show similar peak densities for $\phi \approx i$ at $\pm 7.0°$, $+28.5°$, $+51.6°$ and $+65.0°$. They drop off at higher and lower inclinations (except for the Antarctic region). The specific casualty probability density $\Delta P_c(\phi)/(\Delta\phi\, A_c)$ is obtained by weighting the mean population density $\rho_p(\phi)$ with the impact probability. Hence, due to a sparse population density near polar latitudes, the casualty risk is very low for $\phi < -40°$ and for $\phi > +60°$. As a direct consequence, all target impact areas for controlled re-entries are located in the southern hemisphere, at $-65° < \phi < -40°$. Equal ranking maxima of $\Delta P_c(\phi)/(\Delta\phi\, A_c)$ are obtained in the northern hemisphere for $i = 7.0°$ and $28.5°$, with a 50% lower peak for $i = 51.6°$. At higher inclinations the maximum resident probabilities at $\phi \approx \pm i$ are compensated by near-zero weights of population densities.

For the subsequent analysis it is important to note that "re-entry risk management" may include both controlled de-orbits and uncontrolled re-entries. In the first instance the procedures described hereafter can help optimizing the disposal maneuver location for a de-orbit into a low-risk impact zone, and in the second instance they can assist in minimizing the consequences of a re-entry to the endangered population by reliable forecasts of the impact ground swath, and by a timely issue of warnings to trigger alert procedures (taking shelter in buildings can already significantly reduce the casualty risk).

When a LEO satellite is nearing its end of mission, one should start to monitor

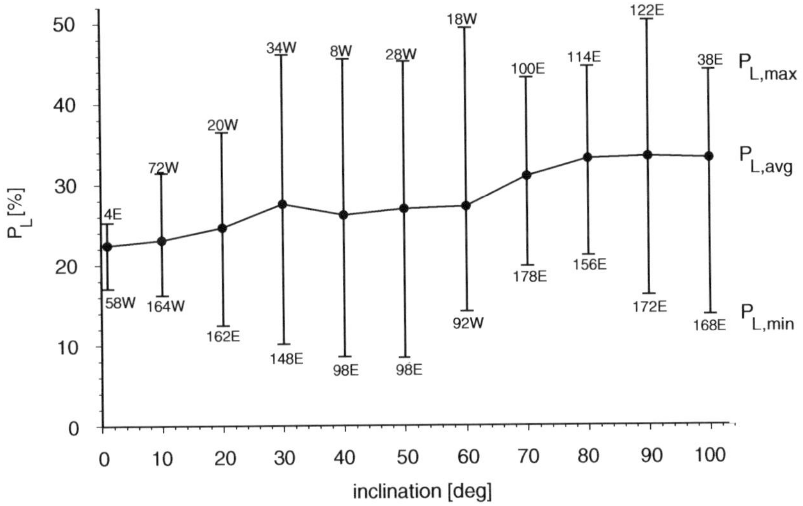

Fig. 9.14. Inclination-dependent, single-orbit means of land impact probability $P_L(i)$, averaged over all longitudes of ascending node λ_Ω (curve connecting dots), with local minima and maxima indicated by range bars, and labeled with the corresponding λ_Ω.

and control the final re-entry and its associated risk level. This can be accomplished by forecasting and ultimately targeting the final orbit to a nodal longitude λ_Ω with a low average population density, and with a correspondingly low casualty probability. For the adopted set of orbit inclinations between $i = 7.0°$ and $98.5°$ Fig. D.3, D.5 and D.7 show the dependence of single-orbit averages of land impact probability P_L and specific casualty probability P_c/A_c as a function of the nodal longitude λ_Ω. Almost regardless of the orbit inclination, the lowest risk level can be obtained for re-entries on orbits with $\lambda_\Omega \approx 175°\mathrm{E} \pm 5°$. For the important inclinations between $51.6°$ and $80.0°$ a secondary minimum exists at $\lambda_\Omega \approx 30°\mathrm{E} \pm 10°$.

For a family of inclinations from near-equatorial to $100°$, uniformly spaced at $\Delta i = 10°$, Fig. 9.14 and 9.15 depict the land impact probability P_L and population density ρ_p, averaged over one nodal revolution. The curves show the inclination-dependent global mean values over $\lambda_\Omega \in [-180°, +180°]$ (compare Table 9.9), with range bars indicating extremes, which are labeled with corresponding values of λ_Ω. The maximum range in land coverage is observed for $i \approx 40°$ and $50°$, with $8\% \leq P_L \leq 48\%$, while the largest mean values are obtained for near-polar orbits.[9.9]. Population densities along the inclination-dependent ground tracks have their largest mean value and widest variation for $i \approx 40°$, spanning a range of $2\,\mathrm{km}^{-2} \leq \rho_p \leq 62\,\mathrm{km}^{-2}$, with a mean of $\overline{\rho}_p = 16\,\mathrm{km}^{-2}$.

[9.9]the land impact probability may be important for radioactive re-entry fragments

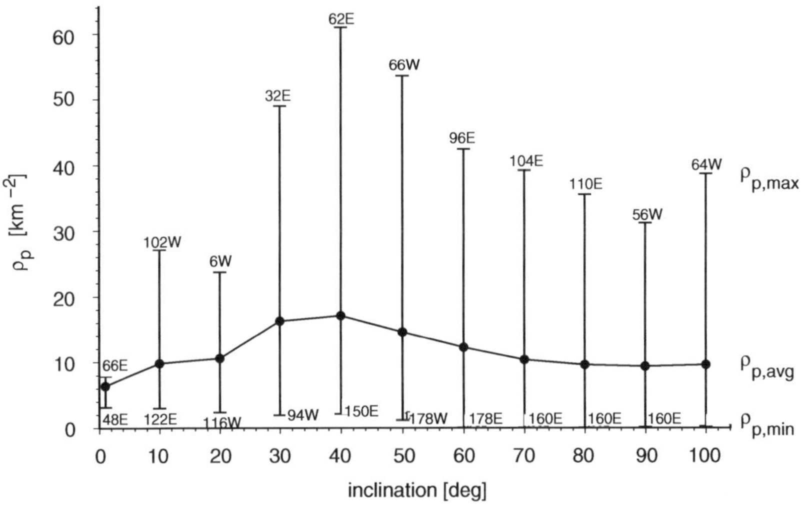

Fig. 9.15. Inclination-dependent, single-orbit means of population density $\rho_p(i)$ in 2001, averaged over all longitudes of ascending node λ_Ω (curve connecting dots), with local minima and maxima indicated by range bars, and labeled with the corresponding λ_Ω.

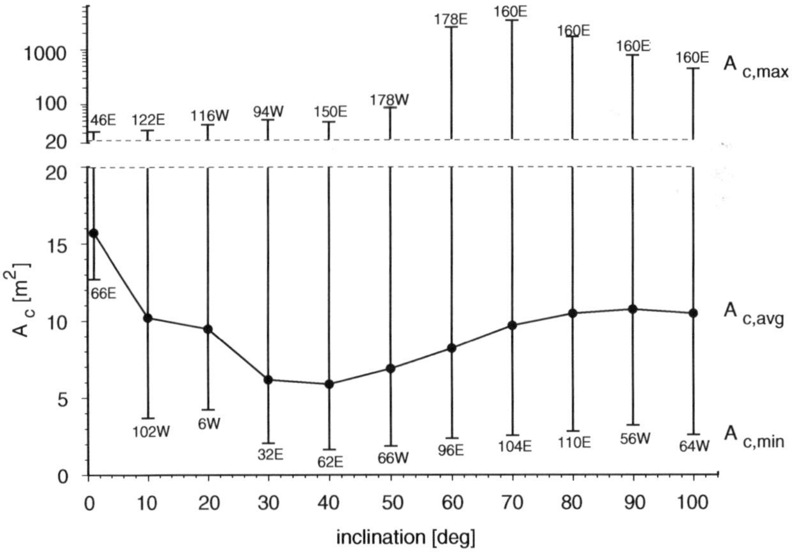

Fig. 9.16. Inclination-dependent, single-orbit means of permitted casualty cross-section $A_c(i)$ for an accepted casualty probability of $P_c \leq 10^{-4}$. Results are averaged over all longitudes of ascending node λ_Ω (curve connecting dots), with local minima and maxima indicated by range bars, and labeled with the corresponding λ_Ω.

For the assumed uniform impact probability along the orbit, and for a given casualty cross-section A_c the population densities can be directly translated into corresponding casualty probabilities P_c. Likewise, for a tolerated event-wise casualty probability (for instance $P_c \leq 10^{-4}$) a maximum casualty cross-section can be determined, which exactly meets the permitted risk threshold for a given population density. This dependency is plotted in Fig. 9.16. It shows that, averaged over all longitudes, the largest mean cross-section of $\overline{A}_{c,max} \approx 16\,\text{m}^2$ can be de-orbited from near-equatorial orbits, while at $i \approx 40°$ this value drops to a minimum of $\overline{A}_{c,min} \approx 6\,\text{m}^2$. However, with a proper phasing of the orbit node, the maximum allowed A_c can reach values from ~30 m^2 at low inclinations to ~3,000 m^2 at $i \approx 70°$. At $i \approx 51.6°$, which is the inclination of large space structures such as the ISS, Mir, Skylab and the Salyut space stations (see Table 9.1), a maximum casualty cross-section of $\overline{A}_{c,max} \approx 100\,\text{m}^2$ can be de-orbited, if the nodal longitude of the final ground track is close to $\lambda_\Omega \approx 178°$E. This ground track phasing was followed for the re-entry risk management of Skylab, in the absence of a de-orbiting capability (see Fig. 9.22 and 9.24). For the Mir and Salyut space stations, which were designed for a controlled re-entry, a low-risk impact zone was targeted by proper phasing of the de-orbit maneuver position on orbits with $\lambda_\Omega \approx 10°$E, which have the maximum land impact probability for this inclination of $P_L \approx 45\%$, and a mean population density which is ~20 times larger than at 178°E (see Fig. 9.23 and 9.24).

So far, all analyses were performed for uniform impact probability distributions. A statistical impact probability density function for a confined impact footprint, and the phasing of the re-entry position on the orbit may, however, play an important role in the re-entry risk management. Depending on the location of the impact zone on an orbit of given inclination i and nodal longitude λ_Ω, the casualty probability can be considerably reduced (as will be shown for Skylab), or increased (as will be shown in the following example) as compared to a single-orbit average.

Let a sample spacecraft consist of a box-shape satellite body of dimensions 2.5 m × 2.5 m × 2.0 m and mass 1,400 kg, with two solar arrays attached, each of size 3.0 m × 3.8 m with 80 kg mass. The spacecraft shall start its natural descent from a near-circular orbit of inclination 51.6° at a geodetic altitude of 133 km, west of New Zealand, on April 1, 2004, at 12:00 UTC. 46.7 minutes later the spacecraft passes an altitude of 95 km, north-west of Spain, when the solar arrays are torn off. 200 seconds further down-track, above the English Channel, the altitude mark of 78 km is trespassed at a flight path angle of $\gamma_a = -1.0°$ (see Fig. 9.17). At this point, the protective spacecraft shell is aerodynamically removed, and all objects listed in Table 9.10 are released into the ambient atmosphere with the same initial orbit state. Thereafter, each unique object is propagated with its own ballistic parameter $B = c_D A/m$, adjusted for flight regime and for mass loss due to aerothermal heating and ablation (Klinkrad, 2004).

Results of the re-entry survival analysis of the individual objects are summarized in Table 9.10. The survival potential largely depends on the object material, and on the ballistic coefficient during peak heating. Object #1, made of an aluminum alloy (AA7075), burns up for all atmospheric conditions. Object #4, made of the same material, always survives to ground impact, though with severe mass

Table 9.10. Characteristics of fragments of a re-entry test satellite at their time of release at 78 km altitude, and at ground impact (#2 to #5) or demise (#1). Data for objects #1 to #5 are given per item. The sum includes multiple items of the same object type.

	Object(s)	Dimensions [m] × [m] × [m]	Material	$m_\circ$ [kg]	$m_i/m_\circ$ [%]	A_c [m^2]	P_c [-]
#1	2 × Box	0.54 × 0.40 × 0.21	AA7075	50.0	0.00	0.00	0.00e-0
#2	4 × Cyl	0.31 × 0.05 × 0.05	A316	6.1	21.21	0.437	0.93e-4
#3	1 × Sph	0.45 × 0.45 × 0.45	TiAl6V4	5.5	95.77	0.955	2.06e-4
#4	1 × Cyl	0.50 × 0.66 × 0.66	AA7075	94.0	4.77	0.627	1.29e-4
#5	1 × Box	0.50 × 0.50 × 0.50	AA7075	160.0	29.92	0.845	1.61e-4
sum	—	—	—	383.9	16.36	4.193	8.66e-4

losses between 99% at the "heel" and 92% at the "toe" of the impact footprint. While object #5 is similar in size, its considerably larger initial mass (160 kg versus 94 kg) leads to a reduced integrated heat load and ablation. This goes along with reduced mass losses between 74% and 65%. The object with the best survival capability is #3, since it is made of a high-melting titanium alloy with a large ablation enthalpy. Its survives to 100%, which is reflected by 95.8% of the mass, when integrating over the $2\,\sigma$ impact swath. Object #2 represents 4 identical parts, all of which are made of a high-melting stainless steel alloy (A316). During its entry it loses between 82% and 74% of its original mass of 6.1 kg. This relatively high loss rate is due to the low melting enthalpy of A316 (see Table 9.3). Of the initial mass of 383.9 kg released at 78 km altitude only 62.8 kg (16.3%) reached the surface. Since object #2 with 3 replicates represents the largest contribution to the overall casualty risk, the underlying analysis shall be explained in some more depth hereafter.

For object #2 Fig. 9.18 shows the impact probability density $\Delta P_{i,2\sigma}/\Delta x$, as a function of the distance Δx relative to the center of impact window (COIW, see Fig. 9.13). This curve is interpolated[9.10] from impact locations generated for different atmospheric density biases $\Delta\rho = -20\%, \ldots, \pm0\%, \ldots, +20\%$, with each of these weighted by a Gaussian probability density function of 2σ bounds for $\Delta\rho = \pm20\%$. The COIW location is obtained for $\Delta\rho = 0\%$. The normally distributed density variations map onto an asymmetric along-track dispersion, with a 2σ "toe" point which is about 25% (510 km) further away from the COIW location than the "heel" of the footprint. The $2\,\sigma$ 1D distribution (along-track only) captures 95.8%, and the 2σ 2D distribution covers 91.5% of all possible impact locations (assuming a cross-track normal distribution with 2σ extensions of ±40 km).

Fig. 9.19 shows the casualty probability density distribution $\Delta P_c/\Delta x$ for object #2, as a function of the COIW distance Δx, for a 2D impact corridor, and for a 2D population density model $\rho_p(\lambda, \phi, t)$. Though the $2\,\sigma$ impact ground swath starts at $\Delta x \approx -1{,}500$ km, the first contribution to the ground risk only emerges at

[9.10]the visible discontinuities in Fig. 9.18 are discretization artefacts

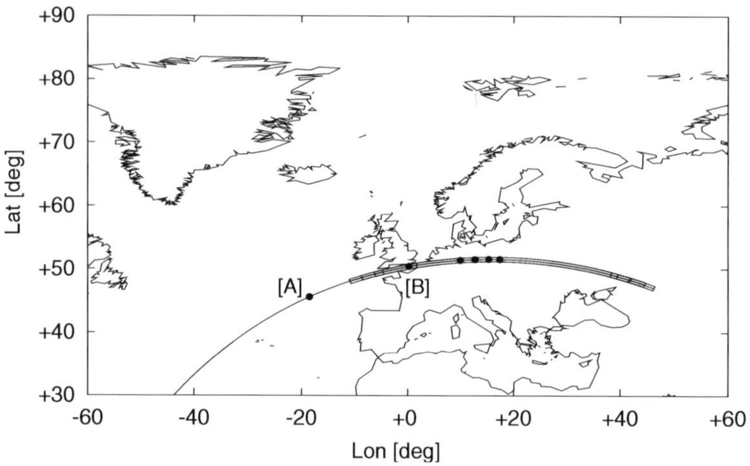

Fig. 9.17. Re-entry ground track and impact swath of a sample satellite defined in Table 9.10. [A] marks the break-off of the solar arrays at 95 km, and [B] marks the release of 4 survivor objects at 78 km altitude.

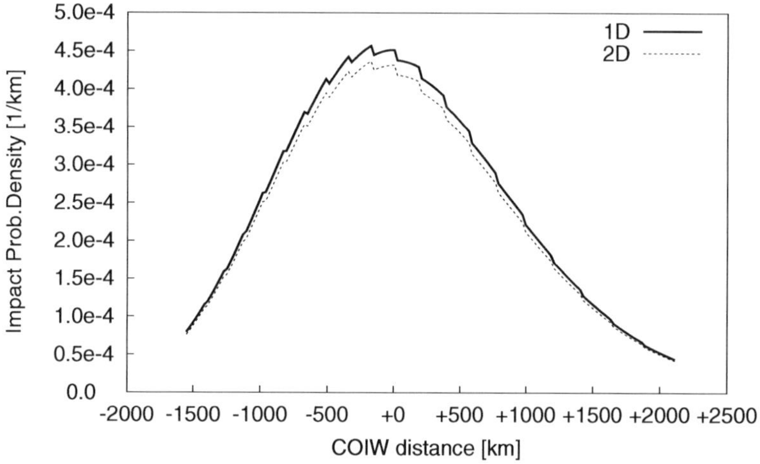

Fig. 9.18. Impact probability density distribution $\Delta P_i/\Delta x$ [km^{-1}], as a function of the COIW ground track distance, for a 2D and 1D impact swath of object #2 (see Table 9.10).

$\Delta x \approx -800$ km. This is due to the ground track reaching the shoreline in Belgium, after running up the English Channel, and before passing over Antwerp and the densely populated Ruhr area in Germany (see first and second peak). The COIW location is near Leipzig, where another peak in the risk distribution is noted. The minimum, maximum, and average casualty risk curves relate to the possible range

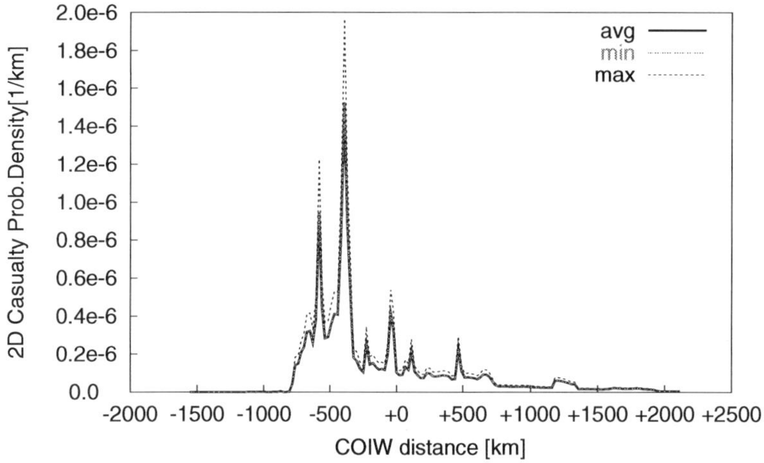

Fig. 9.19. Casualty probability density distribution $\Delta P_c/\Delta x$ [km^{-1}], as a function of the COIW ground track distance, for a 2D impact swath of object #2 (see Table 9.10).

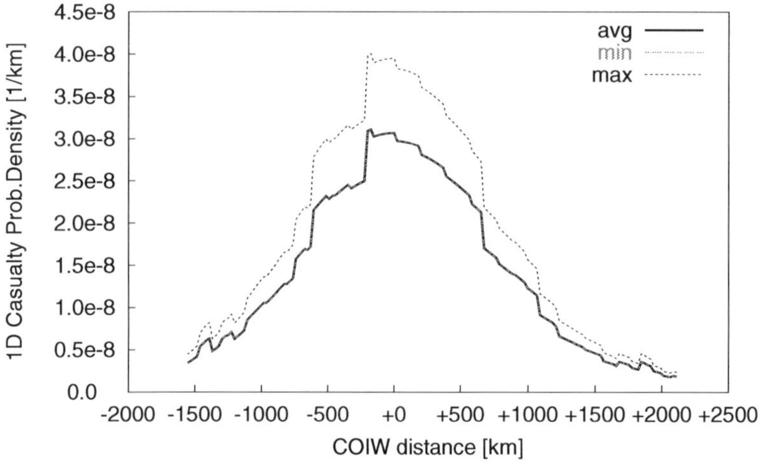

Fig. 9.20. Casualty probability density distribution $\Delta P_c/\Delta x$ [km^{-1}], as a function of the COIW ground track distance, for a 1D impact swath of object #2 (see Table 9.10).

of impact cross-sections for different attitudes of the survivor objects.

To illustrate the sensitivity of the solution with respect to model assumptions Fig. 9.20 shows the distribution of $\Delta P_c/\Delta x$ for a 1D (along-track only) impact probability distribution, and for a 1D world population density model $\overline{\rho}_p(\phi, t)$ (longitude averaged). The casualty probability distribution in this case closely follows the impact probability density distribution shown in Fig. 9.18. The aggregate casualty probability from all survivor objects for this case is $P_c = 1.25 \times 10^{-4}$, as compared with $P_c = 8.66 \times 10^{-4}$ when using a full-scale population density model $\rho_P(\lambda, \phi, t)$. For the combined casualty cross-section of $4.19\,\mathrm{m}^2$, these results compare with a global mean for this orbit inclination of $\overline{P}_c \approx 0.6 \times 10^{-4}$, and with a single-orbit mean for $\lambda_\Omega = 68.5°\mathrm{W}$ of $P_c \approx 2.1 \times 10^{-4}$. This illustrates that a detailed world population model and a good prediction of the impact ground swath is a prerequisite for a reliable casualty risk assessment of a re-entry event.

In Section 9.3 the Mir disposal operations were described as a good example of re-entry risk management, if an adequately sized and fully functional de-orbiting capability is available. In the case of Skylab, which had a similar orbit inclination as Mir, such a capability was not available, and the only means of managing the re-entry risk was through monitoring and controlling of the decay rate.

Fig. 9.21. The Skylab space station on orbit.

The Skylab space station (see Fig. 9.21) was launched on May 14, 1973, into a near-circular orbit of 50.0° inclination, with an initial altitude of 434 km, which was raised to 441 km in February 1974, after the last of three crews had left. Thereafter, no further orbit maneuvers were performed. Due to an unexpectedly high level of solar activity with the approach of the maximum of solar cycle 21 (see Fig. B.7), the 10-year lifetime predicted in early 1974, was significantly shortened, and Skylab re-entered on July 11, 1979, at 16:37 UTC above the Indian Ocean and Australia (Dreher et al., 1980). Prior to the re-entry, Skylab was re-activated on March 6, 1978. Subsequently, residual attitude control capabilities were used to perform an orbit energy management by changing the effective aerodynamic cross-section of the 74 ton compound of length 25.6 m and diameter 6.6 m. The mean area-to-mass ratio A/m could thus be altered by a factor of 2 between a Sun-inertial (SI) high drag, and an end-on-velocity-vector (EOVV) low drag attitude. On January 25, 1979, the attempt to prolong the orbital lifetime (in an EOVV configuration) was abandoned, and Skylab was left in an SI attitude, which was followed by a torque-equilibrium attitude (TEA) to control the longitude position λ_Ω of the ascending node of the final re-entry ground track (Klinkrad, 1999).

All re-entry predictions for the last 24 hours of Skylab's orbital lifetime were located on a single ground track which has its ascending node at $\lambda_\Omega \approx 174°$W (see Fig. 9.22). According to Fig. D.3 and 9.24, this longitude, attained by an orbit energy management strategy, results in a re-entry ground swath with a global minimum in mean population density, and hence in mean casualty probability (averaged over one orbit). During the last 24 hours, predictions indicated that the COIW location walked backwards towards the densely populated North America. At t_{COIW} − 13h the land impact probability was $P_L \approx 19.3\%$, and the specific casualty probability was $P_c/A_c \approx 1/316,400\,\mathrm{m}^{-2}$. In order to reduce the risk, Skylab was taken out of its torque-equilibrium attitude (TEA) and put into a tumbling motion at 07:45 UTC on July 11. This led to a drag reduction of about 20%, causing a shift of the impact location down-track by about 1/2 orbit. Due to the performed attitude maneuver, the probability of land impact was reduced to 16.5%, and the specific casualty risk was reduced by 22% to $1/405,100\,\mathrm{m}^{-2}$. This risk figure corresponds to only 16% of the global mean risk of $1/67,114\,\mathrm{m}^{-2}$ for this orbit inclination (see Table 9.9).

The COIW of Skylab was estimated to be on July 11, 1979, at 16:37 UTC, northeast of the Australian city of Esperance, at 32°S and 124°E. About 500 objects with a total mass of 20 tons (∼25%) were retrieved from an impact swath of 1000 km × 200 km, including a water tank, a heat exchanger, an airlock shroud, oxygen bottles, and a film vault (Dreher et al., 1980; Portree and Loftus, 1993).

Finally, the decay of Salyut-7 shall be reviewed at as an example of an uncontrolled re-entry. Salyut-7, a precursor of the Mir space station, was launched on August 19, 1982. A week later, its initial near-circular orbit of mean altitude 475 km and inclination 51.6° was reached. Following a series of crew visits (Sojus T-5 to T-14) and dockings with supply spacecraft (Progress 13 to 24), the 20-ton Salyut-7 station, with Sojus T-14 attached, was complemented by an unmanned Cosmos-1686 module of the same mass on October 2, 1985. After separation and return to Earth of Sojus T-14 on November 21, 1985, the Salyut-7/Cosmos-1686 compound of 40 tons mass and 26 m length was left mothballed at 475 km altitude in August 1986. From there it started its descent into the atmosphere, which led to a final re-entry above South America on February 7, 1991, at 03:45 UTC. The final ground track of Salyut-7 was effectively phased towards a standard longitude of $\lambda_\Omega \approx 13°$W for Russian entry missions (compare Fig. 9.7 and 9.23). Maneuvering capabilities, however, were insufficient to target the space station towards a low-risk impact location on that orbit.

The descent of Salyut-7/Cosmos-1686 occurred at the peak of solar cycle 22 (see Fig. B.7), 11 years after Skylab, in a stable, high-drag gravity gradient orientation. With increasing airdrag, a precession with a 10° to 20° coning angle was superimposed. Two days before the predicted re-entry, an attempt was made to re-orient the space station into a low-drag attitude in order to extend the lifetime and shift the probable impact location towards regions with minimized casualty risk. Due to an insufficient propellant reserve this strategy could not be realized.

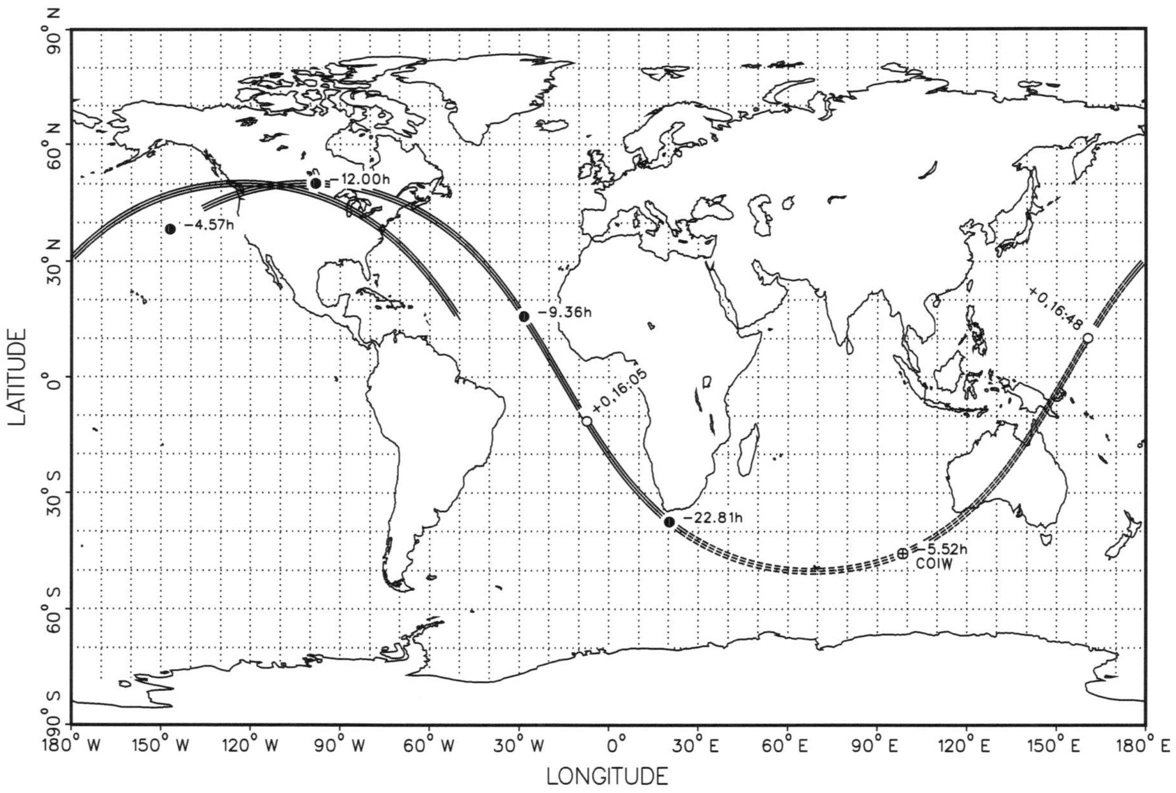

Fig. 9.22. The Skylab re-entry scenario. COIW marks the predicted center of impact window (July 11, 1979, 16:28 UTC). Ground track markers (○) are labeled with the orbit number, relative to the COIW, and with UTC times (hh:mm) of passage, for a hypothetical orbit of 120 km altitude. Impact predictions (●) are labeled with the epoch time-offset Δt_{TLE} (h) of the last available TLE dataset, relative to the COIW time of the final prediction.

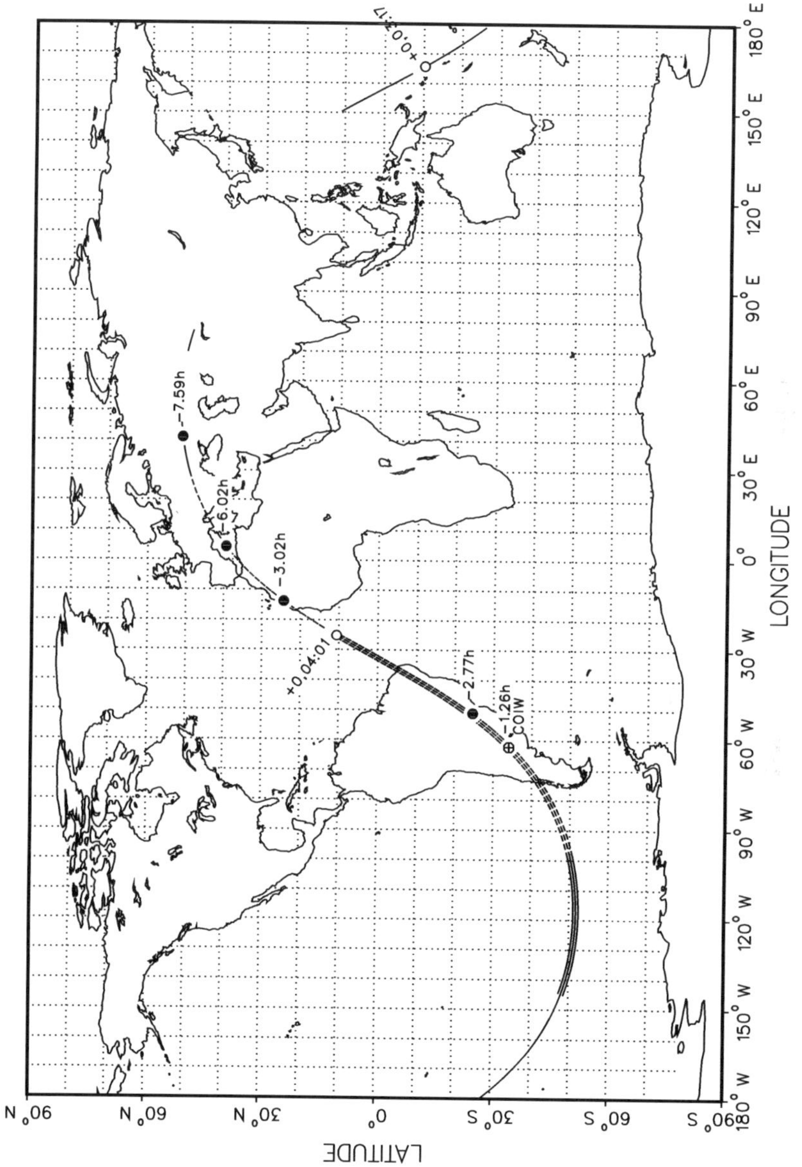

Fig. 9.23. The Salyut-7 re-entry scenario. COIW marks the predicted center of impact window (February 7, 1991, 03:50 UTC). Ground track markers (○) are labeled with the orbit number, relative to the COIW, and with UTC times (hh:mm) of passage, for a hypothetical orbit of 120 km altitude. Impact predictions (●) are labeled with the epoch time-offset Δt_{TLE} (h) of the last available TLE dataset, relative to the COIW time of the final prediction.

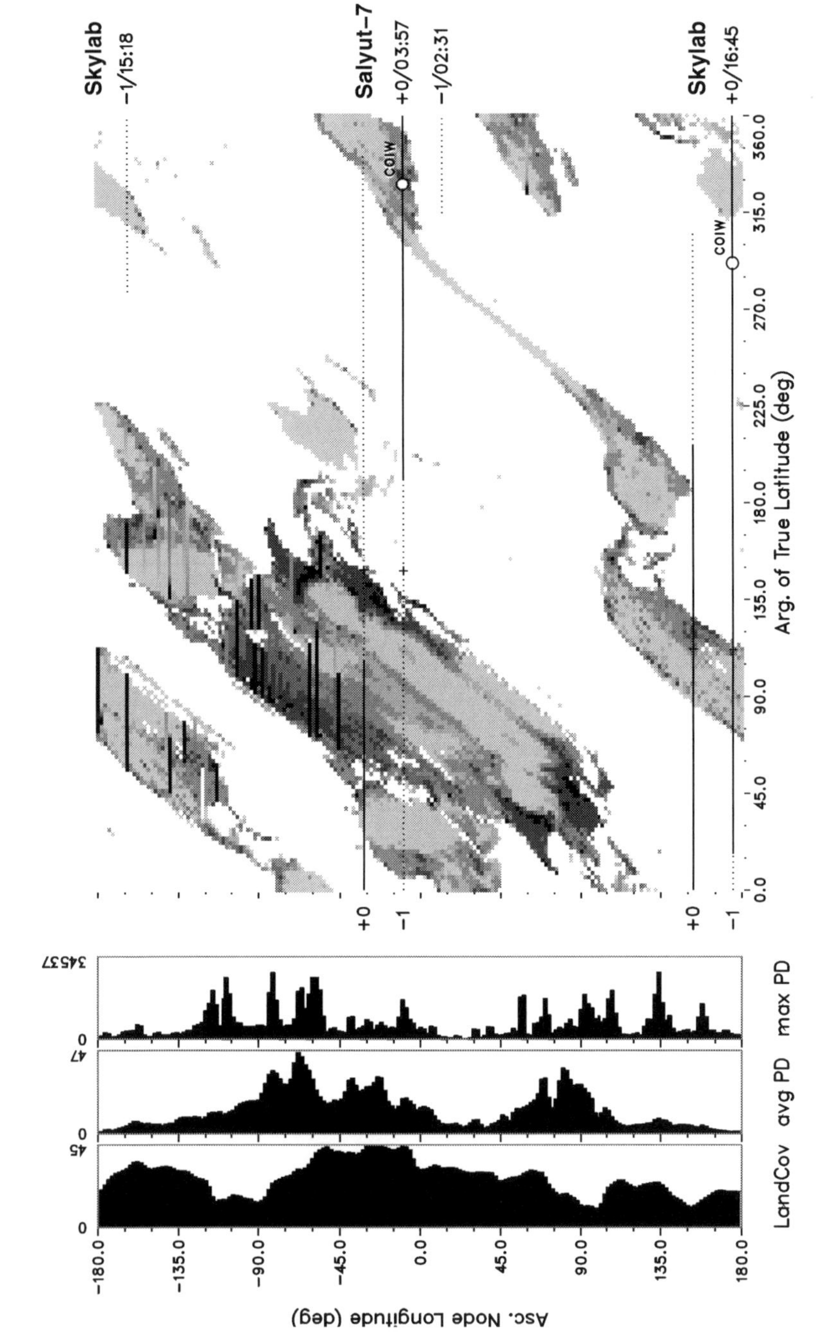

Fig. 9.24. World map with population densities and overlaid Skylab and Salyut-7 ground tracks (± 1 orbit around the COIW), as a function of the longitude of the ascending node λ_Ω, and of the orbit position $u = \omega + f$, for an inclination of $i = 51.6°$. Mean land coverage, average and maximum population densities versus λ_Ω are shown to the left.

Fig. 9.23 shows the final 2σ re-entry ground swath of Salyut-7 (thick curve), superimposed on a extended ground track of ± 40 min (thin curve) with respect to the last predicted center of impact window (COIW) on February 7, 1991, at 03:45 UTC, near $\lambda = 63.8°$W and $\phi = 34.9°$S. This final prediction is based on TLE data from $t_{COIW} - 1.26$ h, which only became available after the event. Four more predictions are marked on the ground track for prediction epochs between $t_{COIW} - 7.59$ h and $t_{COIW} - 2.77$ h. They indicate a backward walk of the impact location with progressing time (as was the case for Skylab). During the last 3 days of the orbital lifetime of Salyut-7, all predictions were within ± 1 revolution of the final COIW location. The geographic longitudes of the ascending nodes of the endangered orbits were in the vicinity of $\lambda_\Omega \approx 13°$W. According to Fig. D.3 and 9.24, this λ_Ω may lead to a global maximum of land coverage, with a risk of casualties close to the global mean. Due to its uncontrolled descent, the predicted population casualty risk was quite stable during the last 3 days of the Salyut-7 orbital lifetime. The final specific casualty probability was estimated to be $P_c/A_c \approx 1/70,057\,\mathrm{m}^{-2}$, based on an orbit determination from $t_{COIW} - 2.77$ h. This value is close to the global mean risk of $1/67,114\,\mathrm{m}^{-2}$ for this orbit inclination of 51.5°. The corresponding land impact probability was about 42.0%, which is close to the global maximum of 45.0% over a single orbit. In the final real-time analysis 60% of the casualty risk was attributed to the pass over South America and its densely populated coastline, while another 30% were due to the downrange possibility of a pass across North Africa and Europe (Klinkrad, 1991; Klinkrad, 1999).

According to the Russian Mission Control Center, the reconstructed COIW of Salyut-7/Cosmos-1686 was on February 7, 1991, at 03:47 UTC, over Argentina. At least 3 major fragments could be retrieved after ground impact.

9.7 HAZARDOUS RE-ENTRY MATERIALS

Apart from the primary risk of being hit by fragments which have survived aero-thermal heating during re-entry, some spacecraft can pose a secondary risk due to a possible dispersion of radioactive and/or poisonous material on ground and/or in the lower atmosphere.

Two different types of nuclear power sources (NPS) have been employed in space missions since 1959: radio-isotope thermo-electric generators (RTGs), using the heat produced during the natural decay of ^{238}Pu, ^{210}Po, ^{144}Ce, or ^{90}Sr to drive a thermopile, and nuclear reactors, using the heat produced during the fission of enriched ^{235}U to drive thermionic converters.

Radio-isotope power sources, also denoted as radio-thermal generators (RTGs), have been used in LEO orbits between 1961 and 1972 (6 missions), in GEO orbits up to 1976 (2 missions), and in Earth escape missions (14 lunar and interplanetary missions) since 1969 (Johnson and McKnight, 1991). The total mass of the RTG nuclear fuel in Earth orbit today is on the order of 150 kg, mainly resulting from early US Transit missions using SNAP RTGs. RTGs in LEO orbits are ^{238}Pu fuelled, and they are predominantly at altitudes around 1,000 km, with orbital lifetimes of several hundred years, as compared with the ^{238}Pu half-life time of 24,400 years. Following an accident in April 1964, when Transit 5BN-3 with a SNAP-9A RTG failed to reach orbit, and 1 kg of ^{238}Pu were dispersed in the atmosphere, the RTG design was improved to withstand re-entry conditions from Earth orbits. In another launch accident in 1968 the RTG could demonstrate its robustness and was recovered intact (Scala, 1998). However, due to the risk of accidental re-entries or on-orbit collisions with space debris, the use of RTGs on terrestrial orbits should be avoided. For interplanetary missions, due to the reduced efficiency of solar cells at large solar distances, RTGs are still a justifiable choice. They should, however, be designed to withstand launch accidents, and accidental atmospheric captures when using Earth gravity assists on interplanetary trajectories (e.g. Cassini's Earth swing-by in August 1999, with a total of 33 kg of ^{238}Pu nuclear fuel in 3 RTG units).

Another form of nuclear power sources used in space activities are nuclear reactors. Between 1965 and 1988, 36 space reactors using ^{235}U as nuclear fuel for the fission process had been deployed. Of these, 35 were of Russian origin (using Topaz and Buk reactors with about 30 kg ^{235}U each), and 1 was of US origin (a SNAP-10A reactor with 11 kg ^{235}U). In total, close to 1,000 kg of ^{235}U nuclear inventory is still in LEO orbits. Initially, most of these reactors were used on low-altitude, high-drag orbits in radar reconnaissance missions of Russian RORSATs (Radar Ocean Reconnaissance Satellites). After a mission duration of a few weeks to a few months, when the orbit control capacity was exhausted, the nominal RORSAT mission separated the reactor from the payload, and injected it into a near-circular disposal orbit at about 950 km altitude (see Table 2.3). The payload then re-entered, and the switched-off reactor remained in an orbit of several hundred years' lifetime.

The radiation power (radioactivity) of the decaying fission products after reactor shut-down decreases to about 1/1,000 of the original power within 1 week, assuming a preceding operational (switch-on) period of 1 week. The highest risk potential from nuclear reactors used on LEO missions is due to a possible uncontrolled re-entry into the Earth atmosphere directly after a preceding operational period of the reactor. Two such events have occurred in the past. On January 24, 1978, the Cosmos-954 RORSAT re-entered over the Northwest Territories of Canada after the attitude and orbit control of the combined payload–reactor compound failed. Numerous radioactive fragments were collected on the ground along a swath of more than 1,000 km length. In a second accident Cosmos-1402 failed to reach the intended reactor disposal orbit after its payload and its reactor had been separated. Thereafter, both objects re-entered. The payload burnt up over the Indian Ocean on January 23, 1983, and the reactor decayed over the South Atlantic on February 7, 1983. Following these accidents and a near-failure of Cosmos-1900, the former USSR ceased to use space reactors after 1988. The total count of reactors in Earth orbits today is 34 (all of them in LEO).

In the aftermath of the Cosmos-954 re-entry, the United Nations have adopted a set of principles as to the use of nuclear power sources in space (anon., 2002). Some of these principles also address the risk potential and liability issues in the case of a re-entry.

9.8 REFERENCES

Ailor, W., Hallman, W., Steckel, G., and Weaver, M. (2005). Analysis of Re-Entered Debris and Implications for Survivability Modeling. In *Proceedings of the Fourth European Conference on Space Debris*, ESA SP-587.

Aksnes, K. (1976). Short-Period and Long-Period Perturbations of a Spherical Satellite Due to Direct Solar Radiation Pressure. *Celestial Mechanics*, vol.13:89–104.

Allen, H. and Eggers, A. (1953). A Study of the Motion and Aerodynamic Heating of Missiles Entering the Earth's Atmosphere at High Supersonic Speeds. Technical Report NACA RM A53D28, National Advisory Committee for Aeronautics.

anon. (1976). U.S. Standard Atmosphere 1976. Technical Report NOAA-S/T 76-1562, NOAA/NASA/USAF, Washington, DC.

anon. (1995a). Hazard Analysis of Commercial Space Transportation. Technical Report DOT Office of Commercial Space Transportation.

anon. (1995b). *NASA Safety Standard – Guidelines and Assessment Procedures for Limiting Orbital Debris.* NSS 1740.14.

anon. (2000). Common Risk Criteria for National Test Ranges – Inert Debris. Technical Report MIL Standard 321-00.

anon. (2002). *United Nations Treaties and Principles on Outer Space.* United Nations publications, Vienna, Austria.

Bouslog, S., Ross, B., and Madden, C. (1994). Space Debris Re-Entry Risk Analysis. In *32nd Aerospace Sciences Meeting, Reno, USA*, AIAA Paper No.94-0591.

Cook, G. (1972). Basic Theory of PROD, a Program for Computing the Development of Satellite Orbits. *Celestial Mechanics*, Vol. 7:301–314.

Dreher, P., Little, R., and Wittenstein, G. (1980). Skylab Orbital Lifetime Prediction and Decay Analysis. Technical Report NASA TM 78308, NASA/MSFC.

Fritsche, B., Klinkrad, H., Kashkovsky, A., and Grinberg, E. (1999). Spacecraft Disintegration During Uncontrolled Atmospheric Re-Entry. *50th IAC Congress*, IAA-99-IAA.6.7.02.

Fritsche, B., Koppenwallner, G., and Kashkovsky, A. (2000). Advanced Model for Spacecraft Disintegration During Atmospheric Re-Entry. Technical Report ESA Contract 12804/98/D/IM, ESA.

Hedin, A. (1991). Extension of the MSIS Thermosphere Model into the Lower Atmosphere. *Journal of Geophysical Research*, vol.96, no.A2:1159–1172.

Johnson, N. and McKnight, D. (1991). *Artificial Space Debris.* Krieger Publishing Company, Malabar, FL, second edition.

Kaula, W. (1966). *Theory of Satellite Geodesy.* Blaisdell Publishing Company, Waltam, MA/Toronto/London.

Klinkrad, H. (1991). Salyut-7/Cosmos-1686 Re-Entry Prediction Activities at ESOC. In *Proceedings of an International Workshop on the Salyut-7/Cosmos-1686 Re-Entry*, ESA SP-345, pages 17–34.

Klinkrad, H. (1999). Evolution of the On-Ground Risk During Uncontrolled Re-Entries. *50th IAC Congress*, IAA-99-IAA.6.7.05.

Klinkrad, H. (2004). A Standardized Method for Re-Entry Risk Evaluation. *55th IAC Congress*, IAC-04-IAA.5.12.2.07.

Klinkrad, H. et al. (2003). *ESA Space Debris Mitigation Handbook.* ESA/ESOC, second edition. issue 1.0.

Koppenwallner, G., Fritsche, B., and Lips, T. (2001). Survivability and Ground Risk Potential of Screws and Bolts of Disintegrating Spacecraft During Uncontrolled Re-Entry. In *Proceedings of the Third European Conference on Space Debris*, ESA SP-473, pages 533–539.

Liu, J. and Alford, R. (1979). A Semi-Analytic Theory for the Motion of a Close-Earth Artificial Satellite with Drag. In *17th Aerospace Sciences Meeting, New Orleans*, AIAA Paper No. 79-0123.

Montgomery, R. and Ward, J. (1995). Casualty Areas from Impacting Inert Debris for People in the Open. Technical Report RTI/5180/60-31F, Research Triangle Institute.

Portree, D. and Loftus, J. (1993). Orbital Debris and Near-Earth Environmental Management: A Chronology. Technical Report 1320, NASA Reference Publication.

Rochelle, W., Marichalar, J., and Johnson, N. (2004). Analysis of Re-Entry Survivability of UARS Spacecraft. *Advances in Space Research*, vol.34, no.5:1049–1054.

Scala, K. (1998). A History of Nuclear Space Accidents. *Spaceflight*, vol.40.

Smith, P. (1999). Expected Casualty Calculations for Commercial Space Launch and Re-Entry Missions. Technical Report Advisory Circular No. 431.35-1, US Department of Transportation & Federal Aviation Administration.

Tobler, W., Deichmann, U., Gottsegen, J., and Maloy, K. (1995). The Global Demography Project. Technical Report Technical Report TR-95-6, National Center for Geographic Information and Analysis, Santa Barbara, CA.

10

Modeling of the Terrestrial Meteoroid Environment

H. Klinkrad and E. Grün

During almost 50 years of space activities approximately 27,000 tons of man-made material re-entered into the Earth atmosphere at a mean rate of ∼600 tons per year. This compares with an estimated 40,000 tons of natural meteoroid material which reaches the Earth atmosphere each year. If this is so, why are meteoroids only dealt with at the end of this book? The answer lies in the size spectrum of meteoroids, which is dominated by particles with diameters of ∼200 μm and corresponding masses of $\sim 1.5 \times 10^{-5}$ g. The resulting risk to operational spacecraft is generally low as compared with space debris, in spite of much higher impact velocities of up to 72 km/s (corresponding to a heliocentric escape orbit of 42 km/s intercepted by the Earth orbit at 30 km/s).

10.1 THE DIVINE–STAUBACH METEOROID MODEL

The meteoroid environment of the Earth is comparatively well known from ground-based photographic and radar observations of meteors, from space-borne detector experiments, from retrieved space surfaces, from meteoritic material captured by high-flying airplanes, and from micro-craters on returned lunar rocks. In contrast with the highly dynamic space debris environment, the dominant part of the meteoroid environment can be assumed to be sporadic and invariant with time. There are, however, seasonally recurring meteoroid stream events, associated with material in cometary orbits, which can give noticeable contributions, particularly in the case of meteoroid storms. The topic of cosmic dust and meteoroid research has many interesting facets, and it deserves more attention than can be given in this brief summary. A comprehensive overview of the current state of research and its historic roots is provided in (Grün et al., 2001).

During the Apollo Program, meteoroids were identified as a potential risk, and a first empirical flux model was developed, based on ground-based optical and radar observations of meteors, and in-situ impact detections on LEO spacecraft. Assuming a mean impact speed of $\overline{v} = 20\,\mathrm{km/s}$ and a uniform meteoroid density of $\overline{\rho} = 0.5\,\mathrm{g/cm}^3$, the unobscured impact flux F on a sphere of $1\,\mathrm{m}^2$ diameter can be estimated by an empirical fitting function (Cour-Palais, 1969).

$$\log_{10} \hat{F} = c_1 + c_2 \log_{10} \hat{m} + c_3 \left(\log_{10} \hat{m}\right)^2 \tag{10.1}$$

where $\hat{F} = F/[\mathrm{m}^{-2}\,\mathrm{s}^{-1}]$ and $\hat{m} = m/[\mathrm{g}]$ are dimensionless quantities. There are two different parameter sets defined for two individual size spectra.

$$\begin{aligned} 10^{-12}\,\mathrm{g} \leq m \leq 10^{-6}\,\mathrm{g} \quad &\rightarrow \quad c_1 = -14.339,\ c_2 = -1.548,\ c_3 = -0.063 \\ 10^{-6}\,\mathrm{g} \leq m \leq 10^{0}\,\mathrm{g} \quad &\rightarrow \quad c_1 = -14.370,\ c_2 = -1.213,\ c_3 = 0.0 \end{aligned}$$

Geometric shielding due to a field of view obscuration by the Earth disk can be accounted for through a correction factor $0 \leq \eta_S \leq 1$, which should be applied to the flux from Eq. 10.1, as a function of the altitude H above the Earth.

$$\eta_S = \frac{1}{2} + \frac{1}{2} \cos\left(\arcsin\left(\frac{a_e}{a_e + H}\right)\right) \tag{10.2}$$

With new data from zodiacal light observations, lunar micro-craters, and in-situ detectors on interplanetary spacecraft (e.g. HEOS and Pioneer 8 and 9), an updated flux model was presented by (Grün et al., 1985). It assumes a mean meteoroid mass of $\overline{\rho} = 2.5\,\mathrm{g/cm}^3$, a mean meteoroid speed of $\overline{v} = 20\,\mathrm{km/s}$, and a planar surface at 1 AU, rotating around a spin axis which is perpendicular to the ecliptic.

$$\hat{F} = \left(c_1\, \hat{m}^{\gamma_1} + c_2\right)^{\gamma_2} + c_3\, \left(\hat{m} + c_4\, \hat{m}^{\gamma_3} + c_5\, \hat{m}^{\gamma_4}\right)^{\gamma_5} + c_6\, \left(\hat{m} + c_7\, \hat{m}^{\gamma_6}\right)^{\gamma_7} \tag{10.3}$$

This equation is valid for the mass range $10^{-18}\,\mathrm{g} \leq m \leq 10^{+2}\,\mathrm{g}$, with the fitted coefficients $c_1 = 2.2 \times 10^{+3}$, $c_2 = 15.0$, $c_3 = 1.3 \times 10^{-9}$, $c_4 = 10^{+11}$, $c_5 = 10^{+27}$, $c_6 = 1.3 \times 10^{-16}$, $c_7 = 10^{+6}$, $\gamma_1 = 0.306$, $\gamma_2 = -4.38$, $\gamma_3 = 2.0$, $\gamma_4 = 4.0$, $\gamma_5 = -0.36$, $\gamma_6 = 2.0$, and $\gamma_7 = -0.85$.

The flux models according to Eq. 10.3 and 10.1 do not give indications on the flux direction and on associated velocities. This shortcoming was overcome by Divine (Divine, 1993), who employed principles outlined earlier by Kessler (Kessler, 1981) to determine meteoroid number densities N in a phase space of heliocentric orbits with given distributions p_r in perihelion radius r_p, p_e in eccentricity, and p_i in inclination. These distributions have the following properties.

$$\int_0^\infty p_r\, r^2\, dr = 1, \qquad \int_0^1 \frac{p_e}{(1-e)^{3/2}}\, de = 1, \qquad \int_0^\pi p_i\, \sin i\, di = 1 \tag{10.4}$$

With the auxiliary variables $\chi = \arcsin(r_p/r)$ and $e_\chi = (1 - \sin\chi)/(1 + \sin\chi)$, the number density D of meteoroids with masses larger than m can be expressed by integrals over the three probability densities (Staubach et al., 2001).

$$D = \frac{H}{\pi} \int_{0}^{\pi/2} \sin\chi \, p_r \, d\chi \int_{e_\chi}^{1} \frac{p_e}{\sqrt{e - e_\chi}} \, de \int_{|\phi|}^{\pi - |\phi|} \frac{\sin i \, p_i}{\sqrt{\cos^2\phi - \cos^2 i}} \, di \qquad (10.5)$$

In this equation p_r must be determined at $r_p = r \sin\chi$, and H represents the fraction of meteoroids with masses larger than m, defined as

$$H = \int_{m}^{\infty} H_m \, dm \qquad (10.6)$$

with H_m representing the differential number of meteoroids in a bin dm, at mass m, divided by the number of meteoroids of mass $m > 1$ g.

The velocity vector $\underline{v} = (v_x, v_y, v_z)$ of a meteoroid in the heliocentric coordinate system x, y, z can be derived as a function of the phase space parameters r_p, e, and i (Staubach et al., 2001). For the same parameter set, 4 different orientations of the velocity vector must be considered: inbound or outbound, and towards the north or south of the ecliptic. For each of the 4 constellations the velocity relative to the Earth moving with velocity $\underline{v}_\oplus$ can be determined.

$$\Delta\underline{v}_\oplus = \underline{v}(r_p, e, i) - \underline{v}_\oplus \qquad (10.7)$$

The corresponding position relative to the Earth center is

$$\Delta\underline{r}_\oplus = \underline{r}(r_p, e, i) - \underline{r}_\oplus \qquad (10.8)$$

with a projected, radial velocity component of $\Delta v_{\oplus,U}$. Divine shows that due to the presence of the Earth, gravitational focusing becomes effective, with a flux enhancement factor of $1.0 \leq \eta_F \leq 2.0$ (Divine, 1993).

$$\eta_F = \left| \frac{1}{2} - \left(\Delta v_\oplus - \Delta v_{\oplus,U} + \frac{2\,\mu_\oplus}{\Delta r_\oplus \, \Delta v_\oplus} \right) \left(\frac{\pm 1}{4\,B_F} \right) \right| \qquad (10.9)$$

with the auxiliary quantity B_F defined as

$$B_F = \frac{1}{2} \sqrt{(\Delta v_\oplus - \Delta v_{\oplus,U}) \left(\Delta v_\oplus - \Delta v_{\oplus,U} + \frac{4\,\mu_\oplus}{\Delta r_\oplus \, \Delta v_\oplus} \right)} \qquad (10.10)$$

An approximation of the gravitational focusing effect can be determined from a simple function of the orbital altitude H (with $H_{120} = 120$ km corresponding to the altitude of atmospheric capture).

$$\eta_F = 1 + \frac{a_e + H_{120}}{a_e + H} \qquad (10.11)$$

Also the geometric shielding by the Earth can be expressed in phase space parameters. Let B_S be another auxiliary quantity defined as

$$B_S = \sqrt{1 + \left(\frac{\Delta r_\oplus \, \Delta v_\oplus}{\mu_\oplus}\right)^2 \left((\Delta v_\oplus)^2 + \frac{2\,\mu_\oplus}{\Delta r_\oplus} - \left(\frac{\Delta v_\oplus + \Delta v_{\oplus,U}}{2} \pm B_F\right)^2\right)} \tag{10.12}$$

The perigee radius r_{pe} and velocity v_{pe} of the Earth passing meteoroid can then be expressed as

$$r_{pe} = \frac{\mu_\oplus}{(\Delta v_\oplus)^2} (B_S - 1) \tag{10.13}$$

$$v_{pe} = \frac{\Delta r_\oplus}{r_{pe}} \sqrt{(\Delta v_\oplus)^2 + \frac{2\,\mu_\oplus}{\Delta r_\oplus} - \left(\frac{\Delta v_\oplus + \Delta v_{\oplus,U}}{2} \pm B_F\right)^2} \tag{10.14}$$

Using these quantities, the true anomaly f at the position $\Delta \underline{r}_\oplus$ can be determined (with 2 possible approaches, depending on the sign of B_F).

$$f = \frac{1+e}{e\,v_{pe}} \left(\frac{\Delta v_\oplus + \Delta v_{\oplus,U}}{2} \pm B_F\right) \tag{10.15}$$

The condition for geometric shielding by a spherical Earth of radius a_e can be described by an on/off weighting factor $\eta_S = 1/0$ (for no shielding/shielding).

$$\eta_S = \begin{cases} 0 & \text{if } f > 0 \text{ and } a_e - r_{pe} > 0 \\ 1 & \text{otherwise} \end{cases} \tag{10.16}$$

Let the $\underline{v}_t$ be the velocity of a target satellite on an Earth orbit, and let $\underline{v}_m$ be the (hyperbolic) approach velocity $\Delta \underline{v}_\oplus$ of a meteoroid in the Earth-centered coordinate system X, Y, Z, then the relative velocity of an impact will be

$$\Delta \underline{v}_m = \underline{v}_m - \underline{v}_t \tag{10.17}$$

The total impact flux can now be determined from Eq. 10.5 by cumulating contributions from $J_p = 5$ meteoroid populations, $J_d = 4$ possible meteoroid velocity directions, and $J_a = 2$ possible hyperbolic Earth approach directions.

$$F = \sum_{j_p=1}^{J_p} \sum_{j_d=1}^{J_d} \sum_{j_a=1}^{J_a} D\, \eta_F\, \eta_S\, \eta_D\, |\Delta \underline{v}_m| \tag{10.18}$$

In this equation, which is analogous to Eq. 4.9, the term $\eta_D \in [0, 1]$ is a weighting factor, depending on the orientation of the target surface. It is 1 for a sphere, and between 0 and 1 for a planar surface. The integrals which need to be determined for the spatial density D in Eq. 10.5 are generally solved by means of numerical quadrature for discrete probability densities of the 3 phase space parameters r_p, e and i, and of the mass m.

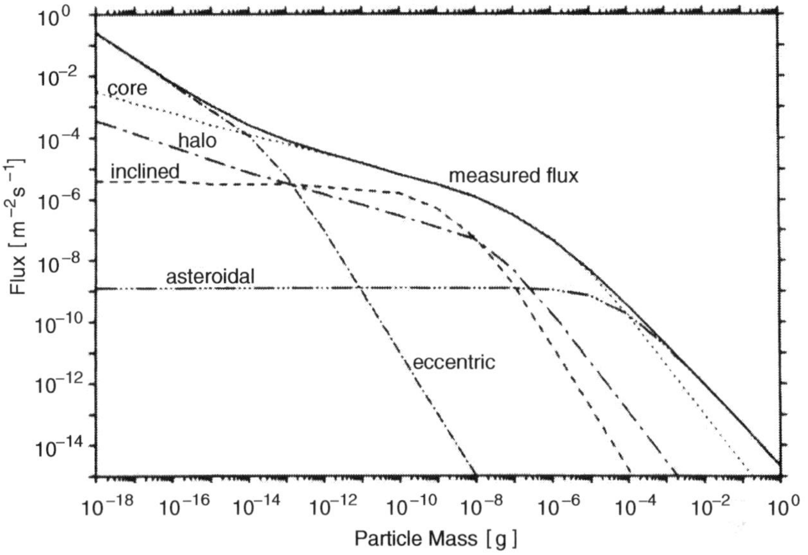

Fig. 10.1. Predicted meteoroid flux in a heliocentric system at 1 AU according to the Divine model (Divine, 1993; Staubach et al., 2001). The curves show cumulative fluxes $F(>m)$ due to meteoroids of masses larger than m.

Divine defined 5 individual meteoroid populations, with distinct properties in their parameter distributions p_r, p_e, p_i (see Eq. 10.4) and their mass spectra (Divine, 1993).

- Core Population: $\overline{m} \approx 10^{-5}$ g; near-circular orbits, close to the ecliptic; number densities increasing towards to Sun
- Asteroidal Population: $\overline{m} \approx 10^{-3}$ g; near-circular orbits, close to the ecliptic; number densities increasing away from the Sun; population exists outside of 1 AU radial distance
- Halo Population: $\overline{m} \approx 10^{-7}$ g; near-circular orbits at random inclinations; population exists outside of 2.5 AU radial distance
- Inclined Population: $\overline{m} \approx 10^{-8}$ g; near-circular orbits at moderate inclinations; population exists inside of 1 AU radial distance
- Eccentric Population: $\overline{m} \approx 10^{-12}$ g; eccentric orbits at moderate inclinations; population exists mainly inside of 1 AU radial distance

After a mathematically more rigorous re-definition of Divine's probability densities p_r, p_e and p_i by Kessler and Matney, in the light of the identities in Eq. 10.4, the original naming conventions for the halo, inclined, and eccentric populations lost their meaning. Hence, and due to the integration of new measurement data from Ulysses and Galileo, Staubach re-fitted the probability densities, and renamed them as A, B, and C populations (Staubach et al., 2001). Fig. 10.1 shows the cumulative meteoroid flux contributions for the Divine model. For LEO satellites the meteoroid flux matches or exceeds the space debris flux only in the size range $10\,\mu\text{m} \leq d \leq 1\,\text{mm}$ (compare Fig. 2.38).

The shape of meteoroids is mostly assumed to be spherical. The density of the meteoroid material is known to be source- and size-dependent. It can range from $\sim$0.5 g/cm^3 for cometary meteoroids to $\sim$3.5 g/cm^3 for meteoroids of asteroidal origin, and it varies with size from $\sim$3.0 g/cm^3 for micrometer-size dust particles to $\sim$1.5 g/cm^3 for millimeter-size visual meteors (Staubach et al., 2001).

10.2 METEOROID FLUX ASSESSMENT FOR TYPICAL TARGET ORBITS

The Divine–Staubach meteoroid model is part of ESA's MASTER-2001 program[10.1] (Bendisch et al., 2002). Its implementation follows the principles of a discretized control volume to determine local density contributions and transient velocities for a large, representative set of meteoroid orbits (see Section 3.1). The control volume for the meteoroids is defined in an Earth-centered, Sun-oriented coordinate system, with the $\underline{x}$ axis towards the Sun, the $\underline{z}$ axis towards the north pole of the ecliptic, and the $\underline{y}$ axis completing the right-hand system. The volume cells have resolutions of 22.5° in ecliptic latitude and ecliptic longitude. The radial resolution increases from 400 km in near-Earth orbits to 10,000 km in the GEO ring. Six mass classes are defined, covering the range $10^{-12}\,\text{g} \leq m \leq 10^{+3}\,\text{g}$, with logarithmic class widths of 2 up to 4 decades for the upper and lower mass thresholds.

Due to the selected reference frame all cell passage events have invariant properties throughout the year, and symmetries can be used to reduce the volume of data. Based on spatial density and velocity distributions across the control volume, collision flux contributions can be determined for a given target orbit, using the principles described in Section 4.1. Separate flux contributions can be discriminated according to their source population (core, asteroidal, A, B, or C). All meteoroid flux results are expressed in the same J2000.0 coordinate system as in the case of space debris.

Table 10.1 summarizes results of meteoroid flux on a spherical object of 1 m^2 cross-section, as a function of the target orbit, for different meteoroid size thresholds. The meteoroid fluxes according to the MASTER-2001 implementation of the Divine–Staubach model (Bendisch et al., 2002) are expressed in terms of mean time between collisions, to facilitate comparisons with corresponding debris results in Table 4.3. Particularly in the size range $0.1\,\text{mm} < d < 1.0\,\text{mm}$ the meteoroid flux is prevailing on Earth orbits, except for the most densely populated debris altitude bands around $\overline{H} = 900 \pm 200\,\text{km}$ (e.g. ERS), and at $\overline{H} = 1{,}400 \pm 100\,\text{km}$ (e.g. Globalstar). For a mean meteoroid density of $\overline{\rho} = 2.5\,\text{g/cm}^3$, as assumed in the Divine–Staubach model, the size range $0.1\,\text{mm} < d < 1.0\,\text{mm}$ translates into a mass range of $1.3 \times 10^{-6}\,\text{g} < m < 1.3 \times 10^{-3}\,\text{g}$ in Fig. 10.1. Outside these size and mass ranges debris tend to dominate in the LEO environment (compare Fig. 2.38).

Table 10.1 shows mean times between impacts as a function of meteoroid

[10.1]Meteoroid and Space Debris Terrestrial Environment Reference 2001

Table 10.1. Mean time between meteoroid impacts as a function of target orbit (see Table 4.2 and 4.3) and meteoroid size for a spherical object of 1 m^2 cross-section. Results are based on MASTER-2001, averaged over $0° \leq i \leq 120°$ ($0° \leq i \leq 30°$ for GEO).

Orbit	$d > 1\,\mu m$	$d > 10\,\mu m$	$d > 100\,\mu m$	$d > 1\,mm$
ISS	2.22 h	13.0 h	59 d	55.6 y
ERS	2.23 h	13.2 h	59 d	56.4 y
GStar	2.27 h	13.3 h	60 d	57.3 y
GTO	2.95 h	19.5 h	97 d	100.5 y
GEO	2.83 h	20.7 h	95 d	99.7 y

sizes, averaged over inclinations bands which are populated by Earth satellites. The inclination-dependent fluxes cover a range of approximately ±6% around the mean value, with peaks at $i \approx 23.5°$, near the ecliptic plane. The maximum possible impact velocity in this case is $\Delta v_m = \sqrt{(v_{\odot,esc} + v_{\oplus})^2 + v_{\oplus,esc}^2} \approx 72.8\,km/s$, with contributions from the heliocentric escape velocity at the Earth orbit radius ($v_{\odot,esc} \approx 42\,km/s$), the orbital velocity of the Earth ($v_{\oplus} \approx 30\,km/s$), and the escape velocity from a near-Earth orbit ($v_{\oplus,esc} \approx 11\,km/s$).

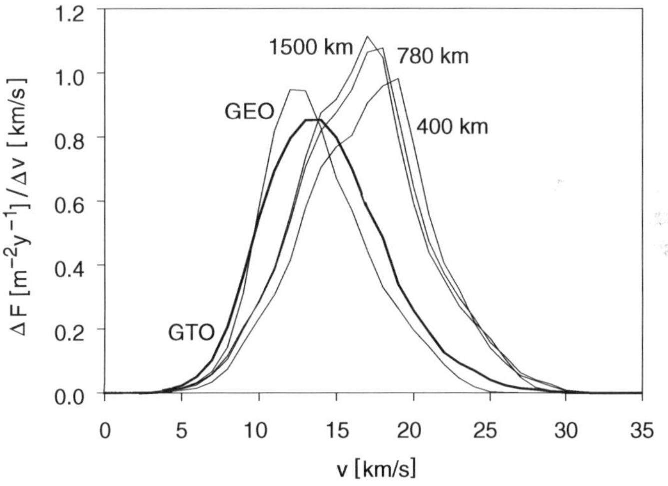

Fig. 10.2. Illustration of the effect of Earth gravitational focusing on the velocity profile of the meteoroid flux distribution $\Delta F/\Delta v$ for spherical targets on typical Earth orbits (velocity class width: $\Delta v = 0.5\,km/s$).

Apart from the inclination of the target orbit, the mean meteoroid impact velocity also depends on the target orbit altitude and the related effects due to gravitational focusing. Fig. 10.2 shows meteoroid flux distributions as a function of impact velocity for typical terrestrial target orbits. The most probable impact velocities range from $\Delta v_m \approx 13\,km/s$ in GEO to $\Delta v_m \approx 20\,km/s$ in a low-Earth orbit. The increasing velocities and raised flux levels due to gravitational focusing with decreasing orbit altitude are to some extent compensated through geometric shielding by the Earth, which can intercept up to 50% of the possible approach directions. Fig. 10.3, which shows the meteoroid flux distribution versus approach elevation, illustrates this effect. For a target object in GEO, only 2.3% of the approach directions are covered by the Earth disk, while it is 33.3% for an ISS orbit and 27.3% for an ERS orbit (see Eq. 10.2). Hence, meteoroid flux in high-altitude orbits follows a cosine distribution with respect to the approach elevation, while in low-Earth orbits contributions from $h < 0°$ are suppressed by Earth shielding.

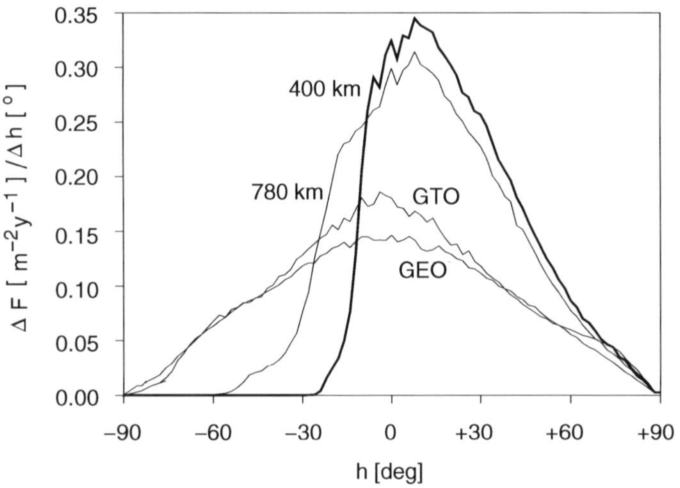

Fig. 10.3. Illustration of the effect of Earth shielding on the elevation profile of the meteoroid flux distribution $\Delta F/\Delta h$ for spherical targets on typical Earth orbits (elevation class width: $\Delta h = 1°$).

10.3 MODELING METEOROID STREAM EVENTS

The sporadic meteoroid flux, which is uncorrelated in time, with no pronounced directionality, is by far dominating the overall contribution from natural sources. The remainder, on the average less than 5%, is due to meteor streams. They can often be correlated with cometary orbits (e.g. Lyrids → Comet 1861 I, Perseids → Comet 109P Swift–Tuttle, Leonids → Comet 55P Tempel–Tuttle), or can be associated with asteroidal sources from within the ecliptic (e.g. Virginids, Aquarids, Geminids). In either case the Earth orbit intersects trails of meteoritic material at fixed time intervals within an annual repetition cycle, from apparently fixed directions (so-called radiants) on the celestial sphere. Meteors from such streams can become visible to observers on Earth as light-emitting trails of material burning up in the upper atmosphere. Such observations can be formally translated into zenith hourly rates (ZHR). Under normal conditions these peak ZHR counts do not exceed values of about 130. In extreme cases, however, when the perihelion pass of a source comet is close to the epoch of the stream event, the ZHR can reach values of several thousands, due to fresh, undispersed clouds of material produced during recent perihelion passes. Due to this loss of material, short-periodic comets can reduce in brightness by up to 5 magnitudes during 10 revolutions about the Sun.

As part of his meteoroid environment description, Cour-Palais devised a meteor stream model which consists of 18 distinct streams (Cour-Palais, 1969). For each of these stream events a mean velocity $\overline{v}$, and an activity profile $\tilde{F}(t)$ is defined, where $\tilde{F} = F_s/F$ is the cumulative stream flux for masses larger than m, related to the corresponding background flux from Eq. 10.1.

$$\log_{10} F_s = -14.41 - \log_{10} \hat{m} - 4.0 \log_{10} \hat{v} + \log_{10} \tilde{F}(t) \qquad (10.19)$$

$\hat{v} = \overline{v}/[20\,\text{km/s}]$ and $\hat{m} = m/[\text{g}]$ are dimensionless parameters of this equation, which is valid for the mass range $10^{-6}\,\text{g} \leq m \leq 1\,\text{g}$.

A more detailed meteor stream model was defined by Jenniskens and McBride (Jenniskens, 1994; McBride, 1997). It is based on world-wide meteor observations during a timespan from 1981 to 1991, totaling 4482 hours, with 110,538 meteor detections. The parameter set of this model is listed in Table 10.2 for a total of 50 meteor streams. For each of these the right ascension α and declination δ of the radiant is defined as a function of the instantaneous solar longitude $\lambda_\odot(t)$ relative to the vernal equinox, and of α_{max}, δ_{max}, and $\lambda_{\odot,max}$ at the time of the stream maximum (with all angles in degrees).

$$\alpha = \alpha_{max} + \Delta\alpha_\lambda \cdot (\lambda_\odot - \lambda_{\odot,max}) \qquad (10.20)$$

$$\delta = \delta_{max} + \Delta\delta_\lambda \cdot (\lambda_\odot - \lambda_{\odot,max}) \qquad (10.21)$$

The solar longitude $\lambda_\odot(t)$ can be translated into a corresponding day of the year $t(\lambda_\odot)/[\text{d}] \approx 80.0 + 0.9856 \cdot \lambda_\odot/[^\circ]$.

The zenith hourly rate ZHR(t) indicating the activity of a given meteor stream at time t can be expressed as a function of $\lambda_\odot(t)$, with stream-specific parameters from Table 10.2. In its most general formulation ZHR$(\lambda_\odot)$ is a superposition of a background profile (index "b") and of a peak profile (index "p"), with ascending (index "$+$") and descending slopes (index "$-$").

$$\text{ZHR}(\lambda_\odot) = Z_p \cdot 10^{-B_p\,|\lambda_\odot - \lambda_{\odot,max}|} + Z_b \cdot 10^{-B_b\,|\lambda_\odot - \lambda_{\odot,max}|} \qquad (10.22)$$

$$B_p = \begin{cases} B_p^- & \text{for } \lambda_\odot \geq \lambda_{\odot,max} \\ B_p^+ & \text{for } \lambda_\odot < \lambda_{\odot,max} \end{cases}$$

$$B_b = \begin{cases} B_b^- & \text{for } \lambda_\odot \geq \lambda_{\odot,max} \\ B_b^+ & \text{for } \lambda_\odot < \lambda_{\odot,max} \end{cases}$$

The maximum zenith hourly rate ZHR$_{max}$ is obtained for $\lambda_\odot = \lambda_{\odot,max}$.

$$\text{ZHR}_{max} = Z_p + Z_b \qquad (10.23)$$

Table 10.2. Parameters of 50 meteor streams used in the Jenniskens model (Jenniskens, 1994), with extensions by McBride (McBride, 1997). The stream parameters are explained in Eq. 10.20 to 10.24. B_p^- is not listed ($B_p^- = B_p^+$; except for the Geminids, where $B_p^- = 0.810$). v is the mean velocity of the meteor stream relative to the Earth.

	Stream name	$\lambda_{\odot,max}$	α_{max}	$\Delta\alpha_\lambda$	δ_{max}	$\Delta\delta_\lambda$	Z_p	B_p^+	Z_b	B_b^+	B_b^-	α_k	c_k	v
δ	Pavonids	10.4	310	1.6	-63.0	-0.2	5.3	0.075	0	0.00	0.00	0.67	6.1e-18	60
	Lyrids	34.7	273	1.2	33.0	0.2	12.8	0.220	0	0.00	0.00	0.69	2.1e-17	49
μ	Virginids	39.0	229	0.5	-8.0	-0.3	2.2	0.045	0	0.00	0.00	0.77	8.0e-18	30
ε	Aquarids	45.8	339	0.9	-1.0	0.3	36.7	0.080	0	0.00	0.00	0.69	2.1e-17	66
β	Corona Au.	55.0	283	1.3	-40.0	0.1	3.0	0.200	0	0.00	0.00	0.79	2.0e-18	45
α	Scorpiids	55.2	251	1.1	-27.0	-0.2	3.2	0.130	0	0.00	0.00	0.64	2.8e-17	35
ο	Scorpiids	71.9	240	1.0	-20.0	-0.1	5.2	0.150	0	0.00	0.00	0.77	6.3e-17	21
δα	Arietids	76.0	46	0.7	24.0	0.6	54.0	0.100	0	0.00	0.00	0.69	2.0e-16	38
γ	Sagitarids	88.5	285	1.1	-25.0	0.1	2.4	0.037	0	0.00	0.00	0.74	1.3e-17	29
τ	Cetids	95.0	23	0.9	-12.0	0.4	3.6	0.180	0	0.00	0.00	0.64	4.3e-18	66
τη	Ophiucids	97.0	291	1.1	-11.0	0.1	2.3	0.037	0	0.00	0.00	0.72	2.0e-17	27
τ	Aquarids	97.3	341	1.0	-12.0	0.4	7.1	0.240	0	0.00	0.00	0.64	9.9e-18	63
ν	Phoenicids	110.5	27	1.0	-40.0	0.5	5.0	0.250	0	0.00	0.00	0.77	3.5e-18	48
ο	Cygnids	116.0	305	0.6	47.0	0.2	2.5	0.130	0	0.00	0.00	0.69	1.0e-17	37
χ	Apricomids	121.7	301	0.9	-10.0	0.3	2.2	0.041	0	0.00	0.00	0.48	2.3e-16	25
δ	Aquarids N	123.4	323	1.0	-8.0	0.2	1.0	0.063	0	0.00	0.00	0.83	5.0e-19	42
	Pisces Aus.	123.7	338	1.0	-33.0	0.4	2.0	0.400	0	0.030	0.10	0.81	1.9e-18	42
δ	Aquarids S	124.9	339	0.8	-17.0	0.2	11.4	0.091	0	0.00	0.00	0.83	5.2e-18	43
λ	Aquarids S	131.0	334	1.0	-15.0	0.3	1.5	0.070	0	0.00	0.00	0.83	1.3e-18	36
	Perseids	139.5	47	1.3	58.0	0.1	70.0	0.350	23	0.050	0.092	0.64	1.3e-16	61
κ	Cygnids	146.0	290	0.6	52.0	0.3	2.3	0.069	0	0.00	0.00	0.55	1.0e-16	27
π	Eridanids	152.0	50	0.8	-16.0	0.3	40.0	0.200	0	0.00	0.00	0.72	2.5e-17	59
γ	Doradids	155.0	60	0.5	-50.0	0.2	4.8	0.180	0	0.00	0.00	0.72	1.0e-17	41
	Aurigids	157.5	72	1.0	43.0	0.2	9.0	0.190	0	0.00	0.00	0.69	4.5e-18	69
κ	Aurigids	176.5	338	0.9	-5.0	0.4	2.7	0.110	0	0.00	0.00	0.72	7.0e-17	19
ε	Geminids	206.0	103	0.7	28.0	0.1	2.9	0.082	0	0.00	0.00	0.77	4.7e-19	71
	Orionids	207.9	95	0.7	16.0	0.1	25.0	0.120	0	0.00	0.00	0.79	3.7e-18	67
	Leominorids	209.0	160	1.0	38.0	-0.4	1.9	0.140	0	0.00	0.00	0.69	1.5e-18	61
	Taurids	222.9	49	0.3	18.0	0.1	7.3	0.026	0	0.00	0.00	0.58	1.9e-16	30
δ	Eridanids	228.0	53	0.9	-2.0	0.2	0.9	0.200	0	0.00	0.00	0.72	4.9e-18	31
ζ	Puppids	231.5	117	0.7	-42.0	-0.2	3.2	0.130	0	0.00	0.00	0.86	1.4e-18	41
	Leonids	234.4	153	1.0	22.0	0.4	19.0	0.550	4	0.025	0.150	0.86	1.1e-18	71
	Puppids/Vel	251.0	128	0.8	-48.0	-0.4	4.5	0.034	0	0.00	0.00	0.74	7.9e-18	40
	Phoenicids	251.7	18	0.8	-58.0	0.4	2.8	0.300	0	0.00	0.00	0.72	8.6e-17	18
	Monodcerotids	260.2	99	1.0	14.0	-0.1	2.0	0.250	0	0.00	0.00	0.88	5.5e-19	43
	Geminids	261.4	112	1.0	32.0	0.1	74.0	0.810	18	0.090	0.310	0.67	5.3e-16	36
σ	Hydrusids	264.8	132	0.9	0.0	-0.3	2.5	0.100	0	0.00	0.00	0.77	8.1e-19	59
	Ursids	270.3	223	-0.2	78.0	-0.3	10.0	0.900	2	0.080	0.20	0.86	9.3e-18	35
	Bootids	282.6	232	0.6	45.0	-0.3	110.0	2.500	20	0.370	0.450	0.64	6.3e-16	43
γ	Velids	285.0	124	0.5	-47.0	-0.2	2.4	0.120	0	0.00	0.00	0.77	5.1e-18	35
α	Crucids	294.7	192	1.1	-63.0	-0.4	3.0	0.110	0	0.00	0.00	0.74	2.4e-18	50
α	Hydrusids	299.0	137	0.7	-13.0	-0.3	2.0	0.200	0	0.00	0.00	0.72	3.4e-18	44
α	Carinids	310.5	99	0.4	-54.0	0.0	2.3	0.160	0	0.00	0.00	0.64	5.5e-17	25
δ	Velids	317.0	127	0.5	-50.0	-0.3	1.3	0.200	0	0.00	0.00	0.77	2.8e-18	35
α	Centaurids	318.7	209	1.3	-58.0	-0.3	7.3	0.180	0	0.00	0.00	0.58	3.0e-17	57
ο	Centaurids	322.7	175	0.9	-55.0	-0.4	2.2	0.150	0	0.00	0.00	0.72	2.3e-18	51
θ	Centaurids	333.0	219	1.1	-44.0	-0.4	4.5	0.200	0	0.00	0.00	0.67	5.2e-18	60
δ	Leonids	334.0	168	1.0	17.0	-0.3	1.1	0.049	0	0.00	0.00	0.77	9.9e-18	23
	Virginids	339.0	164	0.9	9.0	-0.2	1.5	0.200	0	0.00	0.00	0.77	8.9e-18	26
γ	Normids	352.3	284	1.3	-56.0	-0.2	5.8	0.190	0	0.00	0.00	0.61	1.7e-17	56

The cumulative meteoroid flux F_s for masses larger than m can be determined as a function of m and $\lambda_\odot$ from the following equation (McBride, 1997).

$$\frac{F_s\left(m,\lambda_\odot\right)}{\left[\mathrm{m}^{-2}\,\mathrm{s}^{-1}\right]} = c_k \left(\frac{m}{[\mathrm{kg}]}\right)^{-\alpha_k} \frac{\mathrm{ZHR}(\lambda_\odot)}{\mathrm{ZHR}_{max}} \tag{10.24}$$

Finally, the total flux contribution from all $J_s = 50$ meteor streams at a given time of year is obtained from a superposition of the individual events.

$$F_s\left(m,\lambda_\odot\right) = \sum_{j=1}^{J_s} F_{s,j}\left(m,\lambda_\odot\right) \tag{10.25}$$

When seasonally resolved flux results from seasonal meteor streams are combined with results of sporadic meteoroid background flux, one must observe that some of the latter models, for instance (Grün et al., 1985) according to Eq. 10.3, already incorporate streams events as a uniformly spread contribution over the year. Hence, when applying time-dependent stream profiles, the annual mean flux $\overline{F}_s(m)$ from Eq. 10.26 must first be removed from the background flux.

$$\overline{F}_s(m) = \sum_{j=1}^{J_s} \left(\frac{1}{2\pi} \int_0^{2\pi} F_{s,j}\left(m,\lambda_\odot\right) \,\mathrm{d}\lambda_\odot \right) \tag{10.26}$$

The averaged annual stream flux $\overline{F}_s$ is a factor 2 higher for the Jenniskens/McBride model than for the Cour-Palais model, and it is smaller by a factor of 20 to 200 (depending on the particle mass) as compared with the background meteoroid flux. Meteor streams tend to have larger mean particle masses than the background meteoroid population.

Meteoroids, due to their prevailing sizes in the sub-millimeter domain, are generally not considered hazardous for the integrity of a spacecraft. They may, however, damage satellite subsystems or payloads, either directly, due to cratering of sensitive surfaces (e.g. CCDs, lenses, mirrors), or indirectly, due to discharges through an impact-induced plasma (e.g. on solar arrays). In both cases, the high impact velocities play an important role. For the background flux the most probable meteoroid impact velocities are on the order of 13 to 20 km/s (see Fig. 10.2). For stream events they are generally a factor 2 to 3 larger (see Table 10.2). The resulting impact crater diameters according to Eq. 7.7 are $d_c \propto (\Delta v_m)^{2/3}$. The induced plasma charges $Q \propto (\Delta v_m)^{3.48}$ and plasma currents $I \propto (\Delta v_m)^{4.48}$ are even more sensitive to the velocity. For the latter two quantities accelerator tests suggest the following dependencies (McDonnell et al., 1997).

$$\frac{Q}{[\mathrm{C}]} = 0.1\,\frac{m}{[\mathrm{g}]} \left(\frac{m}{[10^{-11}\,\mathrm{g}]}\right)^{0.02} \left(\frac{\Delta v_m}{[5\,\mathrm{km/s}]}\right)^{3.48} \tag{10.27}$$

$$\frac{I}{[\mathrm{A}]} = \frac{Q}{[\mathrm{C}]}\,\frac{\Delta v_m}{[\mathrm{km/s}]} \left(\frac{\lambda_P}{[\mathrm{km}]}\right)^{-1} \tag{10.28}$$

Here, λ_P is a mean plasma path length, which together with the meteoroid velocity defines a characteristic time $\Delta t_P = \lambda_P / \Delta v_m$. The impact plasma can trigger discharges of electrostatically charged dielectric spacecraft surfaces which may by far exceed the pure impact-induced charge according to Eq. 10.27, and which may reach levels of several amperes (Drolshagen, 2001). The failure of the Olympus satellite may have been caused by such an event.

Olympus, a geostationary communication satellite, lost its attitude control on August 11, 1993, at 23:23 UTC, 4 hours before the Perseid meteoroid stream reached its peak of $ZHR_{max} \approx 400$. This level is considerably above the mean Perseid stream activity, due to the perihelion passage of the source comet P109/Swift–Tuttle less than one year before, in December 1992 (see Table 10.3). During the Olympus anomaly an on-board micro-accelerometer was operating. However, it did not indicate any momentum change which could have been attributed to a meteoroid impact. A more likely explanation of the failure may be the triggering of an electrostatic discharge by an impact below the accelerometer detection threshold. This scenario is supported by an observed raise of electron flux in GEO during the week preceding the event (Drolshagen, 2001). Due to a sequence of anomalies following the initial failure, it was decided to remove the spacecraft from the GEO ring. Limitations in the remaining propellant forced the operators to maneuver Olympus on August 30, 1993, into a non-standard disposal orbit 213 km below GEO (see Table 6.4).

Table 10.3. Heliocentric orbits of the source comets of major meteoroid streams (the epoch of perihelion passage t_p is the one closest to the year 2000).

Event	Source comet	T [y]	r_p [AU]	e [–]	i [°]	t_p
Perseids	P109/Swift–Tuttle	135.3	0.959	0.964	113.4	Dec. 1992
Leonids	P055/Tempel–Tuttle	32.9	0.982	0.904	162.7	Feb. 1998
Draconids	P021/Giacobini–Zinner	6.6	1.028	0.708	31.9	Oct. 1998

In the light of the Olympus anomaly the Leonid stream became of interest to spacecraft operators. Its source comet P055/Tempel–Tuttle, with a period of 32.9 years, had its perihelion pass in February 1998. In 1966, during its previous approach of the Sun, zenith hourly rates of $ZHR_{max} \approx 15,000$ were observed, with the most spectacular meteor showers since 1866 (caused by the same comet). Models by several authors predicted Leonid meteoroid storm levels ($ZHR_{max} > 1,000$) for the years 1999, 2001 and 2002 (Jenniskens, 1991). The observed peak zenith hourly rates were on the order of 3,000 in 1999 (due to remnants of the 1899 perihelion pass), 430 in 2000, and 2,600 to 2,700 both in 2001 and 2002 (due to remnants of the 1866 perihelion pass). During these storm periods, which were limited to a few hours, several operators put their spacecraft in a safe mode and oriented them so as to minimize their cross-section with respect to the radiant direction of the Leonids, which was 2.8° away from the apex direction of the Earth orbit. The mitigation of this time-concentrated impact risk, however, is equalized by a few extra days of exposure to the unavoidable background meteoroid flux

(McDonnell et al., 1997).

Two more meteoroid streams reached storm level during the past 100 years. The Draconids, associated with the short periodic comet P021/Giacobini–Zinner ($T = 6.6\,\text{y}$, $\lambda_{\odot,max} = 196.2°$), reached peaks of $\text{ZHR}_{max} = 10,000$ to $12,000$ in 1933 and 1946. Otherwise, they did not notably contribute to the mean meteoroid stream flux, and they are hence not part of the model data in Table 10.2. In 1982 the Puppids ($\lambda_{\odot,max} = 231.5°$) reached peak counts of $\text{ZHR}_{max} = 2,500$.

A comprehensive overview of meteoroid stream events and associated risk mitigation techniques is provided in (Peterson, 1999).

10.4 NEAR-EARTH OBJECTS AND ASSOCIATED RISKS

During the preceding analysis of debris and meteoroid impact fluxes it turned out that the frequency of impacts above a certain size threshold follows an almost linear trend in a log-log scale of impact flux versus impactor size, with reduced impact frequencies when going to larger sizes (see Fig. 2.38). There is strong evidence that a similar trend can also be observed on a larger scale, where the Earth is the target, and celestial objects of different origins are the impactors. With increasing mass, the meteoroids entering the Earth atmosphere are likely to become meteorites, which impact on the Earth surface. During their pass through the atmosphere meteoroids may cause meteors. These are light and ionization trails which are observable in the optical and radio spectrum. The light trails can start at ~140 km and end at ~50 km altitude.

Meteors of mag $= 1^m0$ (corresponding to the brightest stars) have masses of typically $0.01\,\text{g} \leq m \leq 1.5\,\text{g}$. Large-size meteors are called fireballs. They may, or may not be accompanied by thundering sound. A spectacular example was a bright daylight fireball (mag $= -15^m0$ to -19^m0) which was observed in the USA and Canada on August 10, 1972. Between 20:28:29 and 20:30:10 UTC the object produced a trail of 1,500 km length which was followed by ground-based observers and by a US satellite. The flight path descended to about 58 km above Montana, and then left the atmosphere again (the last detection was at 102 km). Depending on its material composition, the size of the fireball was estimated to be between 3 and 14 m before, and between 2 and 10 m after the atmosphere pass, with an initial mass of up to several thousand tons[10.2]. Table 10.4 shows that events in this size class are not infrequent and can be expected every few years. In most of the observed cases, however, the objects expired in mid-air explosions at altitudes between 25 and 35 km due to aerothermal heating and aerodynamic deceleration. All of these objects belong to the dominant class of stony meteorites (92.8%), which is further subdivided into chondrites (85.7%) and achondrites (7.1%). Stony iron meteorites (1.5%) and iron meteorites (5.7%) are in the minority.

Stony meteorites of $d < 10\,\text{m}$ normally explode before reaching the Earth sur-

[10.2]source: G.W. Kronk, *Comets and Meteor Showers* (http://comets.amsmeteors.org)

Table 10.4. Recent meteorite impacts or atmospheric entries.

Year	Date	Location	Circumstances	d [m]
1908	Jun. 30	Russia/Siberia	Tunguska valley, expl. at $H \approx 8$ km	~60
1930	Aug. 13	Brazil/Mato Grosso	Tunguska-class, mid-air explosion	~50
1947	Feb. 12	Russia/Sikhote-Alin	break-up at 6 km, craters up to 14 m	5.5
1992	Oct. 09	USA/New York	iron meteorite punctured a car trunk	0.1
1994	Jul. 16	Jupiter	Shoemaker–Levy comet fragments	2,000
1996	Mar. 30	Chile	off-coast to the west (Pacific Ocean)	11
1997	Apr. 27	Australia	off-coast to the west (Indian Ocean)	27
1997	Sep. 05	Mauritius	off-coast to the south (Indian Ocean)	14
1997	Sep. 30	South Africa	off-coast	5
1997	Oct. 09	USA/Texas	near El Paso, explosion at $H \approx 36$ km	12
1998	Jan. 11	USA/Colorado	near Denver	2
1999	Jul. 07	New Zealand	North Island, explosion at $H \approx 29$ km	2
1999	Dec. 05	USA/Alabama	explosion at $H \approx 23$ km	2
2000	Jan. 18	Canada/Yukon	explosion at $H \approx 25$ km	5

face. On October 9, 1992, however, a fireball crossed Kentucky and released a small stony meteorite of 10 cm diameter and 12 kg mass which punched a hole into the trunk of a parked automobile in Peekskill, New York. Several similar events of damage to property by meteorites are documented worldwide.

Iron meteorites have a much better survival potential during atmospheric entry than their stony counterparts. This is due to their higher heat capacity and material strength. On February 12, 1947, an iron meteorite of ~5.5 m diameter and 136 tons mass fell into the Sikhote-Alin mountains in the far east of Russia. Its fragments of up to 1,741 kg covered an area of ~2 km^2, creating 102 craters of diameters $1\,\text{m} \leq d \leq 26.5\,\text{m}$. Meteorites of this size are likely to fall at annual rates, though mostly unnoticed.

With increasing mass of the meteorites, their collision frequency with the Earth decreases (see Fig. 10.5), but their probability of reaching the Earth surface increases. A limiting case was the so-called Tunguska event. At 07:17 local time on June 30, 1908, a stony meteorite of about 60 m diameter entered the Earth atmosphere as a giant fireball over Siberia and exploded at approximately 8 km above the Tunguska valley, north of Lake Baikal. The energy release was equivalent to 10 or 20 megatons TNT (1,000 times stronger than the Hiroshima bomb), causing a blast wave which burnt and flattened a forest area of 2,200 km^2, and caused an atmospheric pressure wave which traveled around the Earth twice before dissipating. First expeditions visiting the area in the 1930s found grounded trees, all pointing to a common detonation center. However, they did not find evidence of an impact. A similar type of event occurred on August 13, 1930, above the Brazilian Mato Grosso region. A stony meteorite of 50 m diameter also dissolved in a low-altitude explosion, with no retrievable fragments on ground (see Table 10.4).

Table 10.5. Major meteorite impact craters on the Earth.

Crater name	Location	λ_i [°]	ϕ_i [°]	d_c [km]	Age [10^6 y]	Rank
Vredefort	South Africa	27°00′S	27°30′E	300	$2,023 \pm 4$	1
Sudbury	Canada	46°36′N	81°11′W	250	$1,850 \pm 3$	2
Chicxulub	Mexico	21°20′N	89°30′W	170	64.98 ± 0.05	3
Popigai	Russia	71°39′N	111°11′E	100	35.7 ± 0.2	4
Manicouagan	Canada	51°23′N	68°42′W	100	214 ± 1	5
Acraman	Australia	32°01′S	135°27′E	90	~590	6
Chesapeake Bay	USA	37°17′N	76°01′W	90	35.5 ± 0.3	7
Puchezh-Katunki	Russia	56°58′N	43°43′E	80	167 ± 3	8
Morokweng	South Africa	26°28′S	23°32′E	70	145.0 ± 0.8	9
Kara	Russia	69°06′N	64°09′E	65	70.3 ± 2.2	10
Beaverhead	USA	44°36′N	113°00′W	60	~600	11
Tookoonooka	Australia	27°07′S	142°50′E	55	128 ± 5	12
Kara-Kul	Tajikistan	39°01′N	73°27′E	52	< 5	15
Ries	Germany	48°53′N	10°37′E	24	15.1 ± 0.1	36
Steinheim	Germany	48°41′N	10°04′E	3.8	15.1 ± 0.1	117
Barringer	USA	35°02′N	111°01′W	1.2	0.049 ± 0.003	143

Fig. 10.4. The Barringer meteorite impact crater in Arizona (see Table 10.5 for details).

As meteorites start to exceed sizes of 50 to 60 m, their survival potential increases. The Barringer crater in Arizona, often denoted as "Meteor Crater", was caused by such an object in the 50 m size category, which fell 49,000 years ago (see Fig. 10.4 and Table 10.5). In contrast with the similar-size Tunguska object, the Barringer meteorite was of nickel-iron composition. It hence survived to ground impact, creating a crater of 1.2 km diameter and 183 m depth, and releasing an energy of 6 to 7 $\times 10^{+16}$ J ($\approx$ 15 to 17 Mt TNT). As Table 10.6 indicates, such events occur about once every 1,000 years.

For those meteorites which cause ground impacts, the resulting crater diameters d_c can be estimated by an empirical formula proposed by Eugene (Gene) Shoemaker, a pioneer in the field of meteorite impact research.

$$d_c \approx d_\circ \left(\frac{E_i}{E_\circ} \right)^{0.294} \tag{10.29}$$

where $d_\circ = 15\,\text{km}$ and $E_\circ = 10^{+20}\,\text{J} \approx 25,000$ Mt TNT are calibration quantities, and E_i is the impact energy, depending on the mass and velocity of the meteorite. Eq. 10.29 corresponds well with results listed in Table 10.6.

Table 10.6. Frequency and consequences of meteorite impacts on Earth as a function of their size. A stony meteorite at $v_i = 20\,\text{km/s}$ with $\rho_i = 3.5\,\text{g/cm}^3$ was assumed for estimating impact energy E_i and impact crater size d_c (based on (anon., 2004)).

d_i [m]	Δt_i [y]	E_i [Mt TNT]	d_c [km]	Consequences
30	200	2	—	fireball, shock-wave, minor damage
50	1,000	10	≤ 1	Tunguska event, or small crater
100	5,000	80	2	equivalent to a large H-bomb
200	47,000	600	4	destruction on a national scale
500	200,000	10,000	10	destruction on a continental scale
1,000	600,000	80,000	20	global effects, with millions dead
5,000	20×10^6	10×10^6	100	global change, with billions dead
10,000	100×10^6	80×10^6	200	global mass extinction (K/T event)

A meteorite of the kilometer-size class, with an estimated diameter of 1.5 km, impacted in the south of Germany some 15.1 millions years ago. The object of asteroidal or cometary origin released an energy of about 100,000 Mt TNT and excavated a crater of 24 km diameter, which has eroded to a remaining depth of 200 m today. This site is called "Nördlinger Ries", with the city of Nördlingen right in its center. A smaller crater of 3.8 km diameter and 100 m depth, created by a fragment of the same meteorite, is located 40 km south-west of the Ries crater, hosting the city of Steinheim. The Ries impact spread an estimated 10^{12} tons of material across central Europe. Today, such an event, which is likely to occur once in a million years, could entail a death toll of hundreds of millions and cause long-lasting consequences at a continental level, and climate changes on the hemisphere where the impact occurred.

Consequences are escalating to a world-wide level once a meteorite reaches sizes of $d > 10$ km. For a long time paleontologists had speculated that the extinction of the dinosaurs within a geologically short timespan (so-called Cretaceous-Tertiary or K/T event) might have been caused by a meteorite impact. Only in the 1980s did scientists find the first concrete pieces of evidence. Distributed across

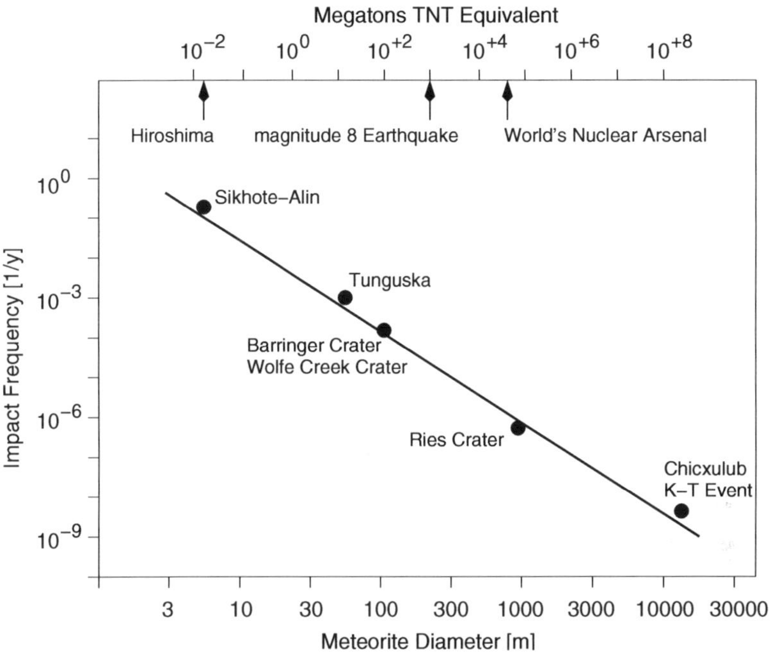

Fig. 10.5. Event frequency and size-dependent risk due to meteorite impacts or mid-air explosions (see also Table 10.6).

the entire Earth, sediment layers dating back to the end of the Cretaceous epoch (65 million years ago) showed unusually large enrichments of iridium. This metal is known to be contained in larger concentrations in the primordial material of asteroids, in an almost unchanged state since the formation of the solar system 4.5 billion years ago. The iridium content in a globally distributed sediment layer of ~1 cm thickness could be interpreted as residues of a 10-km chondritic asteroid. It was another 10 years before in 1990 geological data from the Yucatan peninsula (which were collected during oil exploration in the 1980s) provided proof that the missing impact location was the Chicxulub crater, a ring structure of 170 km diameter, centered on the northern coastline of Yucatan. The crater size and its age, determined from an $^{40}Ar/^{39}Ar$ isotope analysis, exactly matched the previous forecasts by scientists. The Chicxulub event caused a catastrophe on a global scale, with seismic shock waves, atmospheric shock waves, and tsunamis traveling around the Earth. The local heat input and the long-lasting cooling due to high-altitude dust clouds led to a mass extinction of about 70% of all species (including the dinosaurs). Such an event, with an energy release of about 10^{+8} Mt TNT, is estimated to happen every 50 to 100 million years (see Table 10.6). This is consistent with records of geochronology which indicate 7 mass extinctions over the past 438 million years.

The Chicxulub crater is the largest confirmed meteorite impact on the Earth.

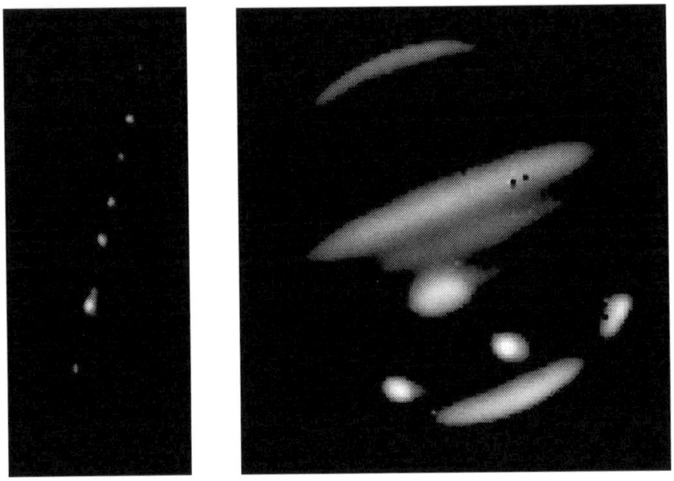

Fig. 10.6. Fragments of the comet Shoemaker–Levy 9 (left image) impacting on the southern hemisphere of Jupiter on June 16–22, 1994 (right image).

Table 10.5 list two ring structures of 300 km diameter in South Africa, and of 250 km diameter in Canada, which are larger than the Chicxulub crater, but which are still considered with some uncertainty, partly owing to their probable age of 2 billion years.

A more recent impact event could have caused global consequences, if it had happened on the Earth. Instead, mankind was able to witness it on the celestial stage. In March 1993 the Shoemaker–Levy team identified a trail of objects on an orbit around Jupiter. The observation data indicated that the 23 distinct objects must have been generated by the fragmentation of a comet about 1 year before. This common source object, which probably had been captured in a highly eccentric orbit by the gravity field of Jupiter several decades ago, was denoted as comet Shoemaker–Levy 9 (SL-9). The observation data suggested that the periapsis of the fragment orbits was descending, and that they would impact on Jupiter between July 16 and 22, 1994. These predictions were very well matched by observations at that time. Of the 23 identified comet fragments 20 could be correlated with an impact. All impacts occurred at $-43°$ Jovian latitude, under a flight path angle of $-45°$, with a velocity of $\sim$60 km/s (close the the Jovian escape velocity), well distributed in Jovian longitude due to time-delayed arrivals. Fig. 10.6 shows infrared images of the trail of comet fragments of up to 2 km in size and their impact features in the atmosphere of Jupiter. Some impact signatures are the size of the Earth. The whole sequence lasted 6.5 days, starting on July 16, 1994, at 20:11 UTC, and ending on July 22, at 08:00 UTC. Due to the high velocity of 60 km/s, the largest fragment alone (SL-9G) released an energy of about 6 million Mt TNT, corresponding to a terrestrial impact object of 3 km diameter according to Table 10.6. All 20 impacts combined would have led to a K/T-type event, if they had occurred on the Earth.

In order to predict potential collisions of celestial risk objects with the Earth, it is necessary to understand their sources, and to determine their orbits. Risk objects which can intersect the Earth orbit shall hereafter be called Near-Earth Objects (NEO). Most of the known ones belong to the asteroids, which form a belt of minor planets of sizes up to 913 km in diameter (Ceres), located between the orbits of Mars and Jupiter. The combined mass of the 15,000 asteroids detected by the year 2000 corresponds to a planet of approximately 1,500 km diameter (less than half the size of the Moon). Of all asteroids discovered by 2000, about 1,000 were classified as NEOs. After a NASA initiative in 1998 to discover 90% of the NEOs larger than 1 km within 10 years, the detection rates increased considerably, and the NEO count in 2005 has already exceeded 3,500. NEOs of diameters 1 km and more have visual magnitudes of 18^m0 or brighter. Based on the ever-increasing volume and quality of observation data, NEO orbit determinations can be steadily improved, and NEO sizes can be estimated more reliably. From the orbit data and their associated uncertainties long-term forecasts of close conjunctions with the Earth can be provided, including probabilities of collision (see algorithms described in Chapter 8). Based on these impact probabilities, often cumulated over several conjunction events, and based on the estimated NEO size, a hazard index can be determined. The most frequently used index is the so-called Torino scale. It classifies the collision risk with a given NEO in 10 categories, with 0 for "no risk", 1 for "dedicated observation advised", 2 to 4 for different levels of "alarming", 5 to 6 for different levels of "hazardous", and 7 to 10 for "certain collision" with different levels of consequences (Steel, 2000).

For the next 100 years ahead Table 10.7 shows the top-ranking NEO risk objects, sorted by their risk index. Only 3 of the objects reach class 1 of the Torino scale, mainly due to their large size. Object 2004 MN4, however, is also among the closest predicted fly-bys in Table 10.8 with a miss distance of 0.1 Moon orbit radii (~38,400 km), leading to a pericenter within the geostationary orbit.

Table 10.7. Top-ranking NEO collision risk scenarios in 2005 for the next 100 years. Impact probabilities P_i are cumulated over the number of events during the given timespan of close conjunctions. c_T shows the Torino scale of the risk level ($0 \leq c_T \leq 10$).

NEO name	Timespan	Events	P_i [–]	v_i [km/s]	mag$_i$ [–]	d_i [m]	c_T [–]
2004 VD17	2091–2104	5	1.1e-04	18.22	18.8	0.580	1
2004 MN4	2029–2055	9	1.5e-04	7.49	19.2	0.320	1
1997 XR2	2101–2101	2	9.7e-05	7.17	20.8	0.230	1
1994 WR12	2054–2102	134	1.0e-04	9.84	22.4	0.110	0
1979 XB	2056–2101	3	3.3e-07	24.54	18.5	0.685	0
2000 SG344	2068–2101	68	1.8e-03	1.37	24.8	0.040	0
2000 QS7	2053–2053	2	1.3e-06	12.32	19.6	0.420	0
1998 HJ3	2100–2104	3	2.1e-07	24.09	18.4	0.694	0
2004 XK3	2029–2104	66	1.7e-04	6.55	24.5	0.040	0
1994 GK	2051–2071	7	6.1e-05	14.87	24.2	0.050	0

Table 10.8. Top-ranking NEO conjunction scenarios in 2005 for the next 100 years. The miss distance $\Delta\hat{r}$ is given in multiples of the Moon orbit radius (384,400 km). Visual magnitudes mag_i can be converted to approximate object diameters d_i ($19^m0 \rightarrow 500$ m, $25^m0 \rightarrow 40$ m, $30^m0 \rightarrow 4$ m).

NEO name	Close approach time (UTC)	$\Delta\hat{r}$ [–]	v_i [km/s]	mag_i [–]
2004 FU162	2004-Mar-31 15:34 ± 00:28	0.03	13.39	28.7
2004 YD5	2004-Dec-19 20:37 ± 00:01	0.10	25.37	29.3
2004 MN4	2029-Apr-13 21:45 ± 00:01	0.10	7.49	19.2
2004 FH	2004-Mar-18 22:08 ± 00:01	0.10	8.00	26.4
2003 SQ222	2003-Sep-27 22:56 ± 00:01	0.20	15.44	30.0
1994 XM1	1994-Dec-09 18:54 ± 00:01	0.30	11.31	28.2
2002 XV90	2002-Dec-11 08:23 ± 00:01	0.30	8.02	25.2
2002 MN	2002-Jun-14 02:02 ± 00:01	0.30	10.57	23.3
2005 FN	2005-Mar-18 21:43 ± 00:01	0.40	10.62	26.9
2003 XJ7	2003-Dec-06 19:04 ± 00:01	0.40	16.93	26.3

Once in a while the possible risk of a NEO collision with the Earth receives public attention. This was for instance the case when the 1 km-class Hermes asteroid (1937 UB) passed the Earth at less than 2 lunar orbit radii on January 18, 1937, and when in 1998 the asteroid 1997 XF11 was predicted to collide with the Earth in 2028. In 1992, it was suspected that the comet Swift–Tuttle would hit the Earth during its next perihelion pass in 2126 (Steel, 2000). In all these cases, a more detailed analysis reduced the risk to an insignificant level, but a general concern with this problem still remained. As a consequence, an international group of astronomers founded the Spacewatch Project in 1981, and formed the Spaceguard Foundation in 1992 to improve our knowledge of the NEOs. Later, several space agencies sent space missions to asteroids and comets to better understand their dynamics and physical characteristics. This combined knowledge will be a prerequisite to engage in visions of a planetary defense, with the objective to deflect potential Earth colliders. Such a deflection maneuver is most efficient, if it is carried out well in advance. It may be achieved by a sufficiently large impulsive change of the NEO orbit velocity (e.g. through the detonation of an H-bomb on the NEO surface), or by a long-duration, low-level acceleration (e.g. through the deployment of large solar sails on the NEO surface). If a tangential avoidance maneuver is considered, then the integrated velocity change up to the predicted impact epoch should at least result in a shift of the arrival time by 425 s ($\Delta t \geq 2 \cdot a_e \cdot v_\oplus = 2 \cdot 6778\,\text{km} \cdot 30\,\text{km/s}$), to let the Earth move on its orbit by an angular segment exceeding its diameter.

The number of asteroidal NEOs of $d > 1$ km is estimated to be 500 to 1,000. At sizes of $d > 0.1$ km the count may be 30,000 to 300,000 (anon., 2000). Long-periodic comets from the Oort Cloud, located at 40,000 to 100,000 AU, and short-periodic comets from the Edgeworth–Kuiper belt, just outside the Neptune orbit, are expected to influence the NEO population only by a few percent. Comets, however, carry a large risk potential due to high relative velocities, and short warning times. Moreover, comets which have consumed their volatile material during prior perihelion passes are difficult to observe. Some astronomers assume that a sizeable fraction of the known NEO population may be extinct comets, captured by the gravity field of Jupiter. For such objects, due to their amorphous structure, a deflection attempt by an H-bomb explosion could result in a break-up into a large number of hazardous fragments, with no overall risk reduction.

The topic of NEO detection, the assessment of their collision risk, and the planning of planetary defense strategies has received increasing public and governmental interest since the early 1990s. This is for instance manifested by several hearings and reports at the US Congress, by discussions at the Technical Sub-Committee of the United Nations Committee on the Peaceful Uses of Outer Space (UNCOPUOS), by debates at the House of Commons in the UK, and by initiatives which were started at NASA/JPL and ESA.

10.5 REFERENCES

anon. (2000). Potentially Hazardous Near-Earth Objects. Technical Report by the NEO Task Force, BNSC, London, UK.

anon. (2004). Space Mission Priorities for Near-Earth Object Risk Assessment and Reduction. Technical Report by the NEOMAP Panel, ESA.

Bendisch, J., Bunte, K., Sdunnus, H., Wegener, P., Walker, R., and Wiedemann, C. (2002). Upgrade of ESA's MASTER Model. Technical Report (final report) ESA contract 14710/00/D/HK, ILR/TUBS, Braunschweig.

Cour-Palais, B. (1969). Meteoroid Environment Model 1969 – Near Earth to Lunar Surface. Technical Report SP-8013, NASA.

Divine, N. (1993). Five Populations of Interplanetary Meteoroids. *Journal of Geophysical Research*, vol. 98:17,029–17,048.

Drolshagen, G. (2001). Hypervelocity Impact Effects on Spacecraft. In *Proceedings of the Meteoroids Conference 2001, Kiruna, Sweden*, ESA SP-495, pages 533–541.

Grün, E., Gustafson, B., Dermott, S., and Fechtig, H. (2001). *Interplanetary Dust*. Springer Verlag, Berlin/Heidelberg/New York.

Grün, E., Zook., H., Fechtig, H., and Giese, R. (1985). Collisional Balance in the Meteoritic Complex. *Icarus*, vol.62:244–272.

Jenniskens, P. (1991). *Discoveries from Observations and Modeling of the 1998/99 Leonids*, pages 233–252. In (Grün et al., 2001).

Jenniskens, P. (1994). Meteor Stream Activity – I. The Annual Meteor Streams. *Journal of Astronomy and Astrophysics*, vol.287:990–1013.

Kessler, D. (1981). Derivation of Collision Probability Between Orbiting Objects: The Lifetimes of Jupiter's Outer Moons. *Icarus*, vol.48:39–48.

McBride, N. (1997). The Importance of the Annual Meteoroid Streams to Spacecraft and their Detectors. *Advances in Space Research*, vol.20, no.8:1513–1516.

McDonnell, J., McBride, N., and Gardner, D. (1997). The Leonid Meteoroid Stream: Spacecraft Interactions and Effects. In *Proceedings of the Second European Conference on Space Debris*, ESA SP-393, pages 391–396.

Peterson, G. (1999). *Dynamics of Meteor Outbursts and Satellite Mitigation Strategies*. The Aerospace Press, El Segundo, CA.

Staubach, P., Grün, E., and Matney, M. (2001). *Synthesis of Observations*, pages 345–384. In (Grün et al., 2001).

Steel, D. (2000). *Target Earth*. Quarto Publishing, London, UK.

11

Space Debris Activities in an International Context

H. Klinkrad

Space debris is a problem of the Earth environment with global dimensions, to which all spacefaring nations have contributed during half a century of space activities. As the space debris environment progressively evolved, it became evident that understanding its causes and controlling its sources is a prerequisite to ensure safe space flight also in the future. Space debris researchers and decision takers in space agencies and commercial space companies came to a consensus that this could only be achieved through international cooperation.

11.1 INTERNATIONAL FORUMS FOR INFORMATION EXCHANGE

Space debris activities started to collect momentum in the early 1960s, with initial research activities in the USA. As of the mid 1970s, first results were presented to the international community, mainly at the Safety & Rescue Symposia of the IAF[11.1] congresses. In July 1982 NASA JSC conducted the first dedicated conference on orbital debris, with limited international participation, mainly from Europe. In response to the decays of Skylab in July 1979 and Cosmos-1402 in February 1983, ESA organized a first workshop on the re-entry of space debris, in September 1985, with scientists from Europe. In July 1984 a first international workshop on space debris was organized during the XXV-th COSPAR[11.2] conference in Graz. Two years later, at the XXVI-th COSPAR conference in Toulouse, a first dedicated space debris session was held. In the aftermath of the Salyut-7 re-entry in February 1991, ESA conducted a workshop dedicated to this event, with participants from Europe and Russia. In April 1993, ESA organized the First European Conference on Space Debris, with participants from all major spacefar-

[11.1]Int. Astronautical Federation
[11.2]Committee on Space Research

ing nations. Later in 1993 the IAF congress had its first dedicated space debris sessions.

Since 1993 results on space debris research have been regularly presented to an international audience at dedicated sessions of IAF conferences[11.3] (every year), COSPAR congresses (every 2 years), and European Conferences on Space Debris (every 4 years). On a smaller scale, or with a reduced scope, space debris is for instance addressed at the International Symposium on Space Technology and Science (ISTS) in Japan, at various meetings of the American Institute of Aeronautics & Astronautics (AIAA), at AMOS[11.4] Conferences, and at Hyper-Velocity Impact Symposia (HVIS).

11.2 INTERNATIONAL COOPERATION AT A TECHNICAL LEVEL

From the mid 1960s onward the trackable on-orbit population of space objects increased at a rate of $\sim$300 y^{-1}, supported by annual launch rates of up to 110, and fueled by on-orbit fragmentation rates of up to 9 y^{-1}. In an attempt to control the growth of the space debris population, expertise from different disciplines was first coordinated at agency or national level, for instance in the USA to investigate the Delta second stage break-ups in the 1970s, and in ESA/France to understand the Ariane H-10 break-up in 1986. The latter event also led to the formation of an ESA Space Debris Working Group in 1987, which published a status report on space debris in 1988 (anon., 1988). In early 1989 a report with similar scope was produced by the US Inter-Agency Group (anon., 1989).

The global nature of the space debris problem and a lack in knowledge transfer on its possible causes led to first bilateral meetings on the initiative of NASA. These included regular NASA/ESA coordination meetings since October 1987, regular meetings between NASA and the Technical University of Braunschweig (TUBS) since October 1989, and regular NASA/Russian coordination meetings since December 1989. In September 1990 a first NASA/NASDA technical interchange meeting was held. ESA intensified their cooperation with Russia during bilateral workshops on the Salyut-7/Cosmos-1686 re-entry in April 1991, and on the GEO debris environment in August 1992. At this time the International Academy of Astronautics (IAA) also published their Position Paper on Orbital Debris, produced by an international ad-hoc group of experts (see (anon., 2001) for the third release).

In February 1993 the seventh NASA/ESA coordination meeting took place, with the participation of NASDA. This first multi-lateral meeting prepared the ground for the Inter-Agency Space Debris Coordination Committee (IADC), which was founded in April 1993 by ESA (Europe), NASA (USA), NASDA[11.5] (Japan),

[11.3] now: International Astronautical Congress (IAC)
[11.4] Air Force Maui Optical & Supercomputing Site
[11.5] later: JAXA

and RSA[11.6] (Russia). They were in later years joined by ASI (Italy), BNSC (United Kingdom), CNES (France), CNSA (China), DLR (Germany), ISRO (India) and NSAU (Ukraine). The IADC meets annually and consists of 4 working groups to channel the technical information exchange in the areas of debris measurements, debris environment modeling techniques, hypervelocity impact protection, and debris mitigation. The overall work program is coordinated by a steering group. IADC can today be regarded as the leading international body with technical expertise in the field of space debris. This is also manifested by its technical support role in the Scientific & Technical Subcommittee of the United Nations' Committee on the Peaceful Uses of Outer Space (UNCOPUOS), where *space debris* has been an agenda item since 1994.

11.3 INTERNATIONAL STANDARDS AND POLICIES

As the knowledge on space debris sources increased, and as the understanding of effective remedial actions improved, several space agencies, governmental bodies, and international space operators developed their own space debris mitigation standards, guidelines, codes of conduct, or policies (see Chapter 6, page 193ff).

A first step to a wider, international application of debris mitigation measures was taken by the IADC in 2002, with the publication of the *IADC Space Debris Mitigation Guidelines.* This document, which was presented at the UNCOPUOS Scientific & Technical Subcommittee in 2003 (anon., 2003), serves as a baseline for the development of space debris mitigation principles in two directions: towards a non-binding policy document, and towards applicable implementation standards. The former route is followed by a UNCOPUOS working group, while the latter direction is pursued by an Orbital Debris Coordination Working Group (ODCWG) within the Technical Committee 20 and Sub-Committee 14 of the International Organization for Standardization (ISO TC20/SC14). To a large extent these UN and ISO working groups recruit their experts from IADC member organizations.

International space debris mitigation policies and standards, based on the consensus of the IADC guidelines, could in the future facilitate and harmonize the implementation of space debris mitigation measures. Internationally agreed standards could enforce appropriate debris mitigation measures on spacecraft operators and launch service providers through the mechanisms of conditional launch license issuance and insurance coverage, depending on the acceptance of a space debris mitigation plan.

At the time of writing this book, the voluntary implementation of debris mitigation and disposal measures by many space operators has shown first indications of a changing trend towards a safer environment in the LEO and GEO region. Such measures, however, must be applied in a consistent manner, by all users of space, to preserve a safe orbit environment for future generations.

[11.6]later: ROSCOSMOS

11.4 REFERENCES

anon. (1988). Space Debris. Technical Report ESA SP-1109, ESA Space Debris Working Group, Darmstadt, Germany.

anon. (1989). Report on Orbital Debris. Technical report, Inter-Agency Group (Space) of the National Security Council, Washington, DC.

anon. (2001). *Position Paper on Orbital Debris.* International Academy of Astronautics, Paris, France.

anon. (2003). IADC Space Debris Mitigation Guidelines. Technical Report A/AC/105/C.1/L.2260, UNCOPUOS, Vienna, Austria.

Epilog

Space operations should comply with a general rule of the National Park Service: "What you take in you must take out".

Joseph P. Loftus, Jr.

A

Basics of Orbit Mechanics

H. Klinkrad

A.1 KEPLER ORBITS

According to Kepler, an object which orbits the Earth travels on an ellipse (or on a circle as special case), with the Earth's center of mass in one of the two foci (see Fig. A.1). Parabolic and hyperbolic orbits shall not be considered here.

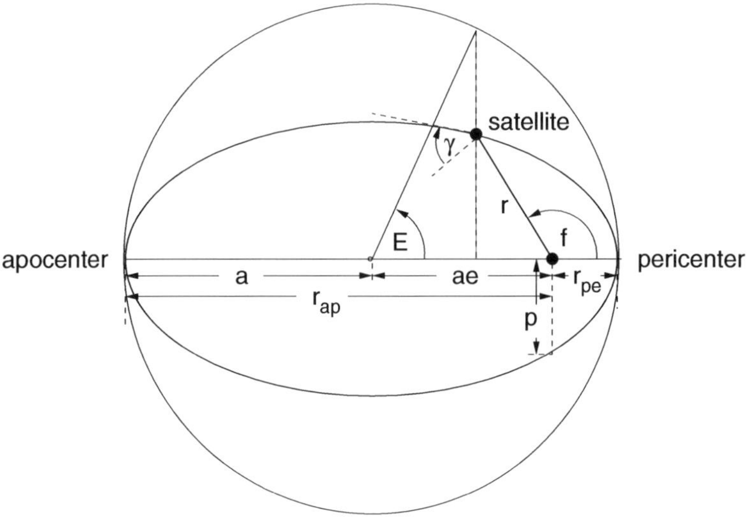

Fig. A.1. Definition of a Kepler orbit with semimajor axis a, semilatus rectum p, eccentricity e, true anomaly f, eccentric anomaly E, and flight path angle γ.

A planar, unperturbed 2-body Kepler motion can be uniquely described by the Kepler elements a (semimajor axis), e (eccentricity), and f (true anomaly). Alternatively, the eccentric anomaly E, or the mean anomaly M may be used as time-dependent parameters. They are defined as

$$\tan\left(\frac{E}{2}\right) = \sqrt{\frac{1-e}{1+e}}\tan\left(\frac{f}{2}\right) \tag{A.1}$$

$$M = E - e\sin E = n\,\Delta t_{pe} \tag{A.2}$$

with Δt_{pe} the elapsed time since pericenter passage, which can also be expressed as a function of the true anomaly f.

$$\Delta t_{pe} = \sqrt{\frac{a^3}{\mu}}\left[2\arctan\left(\sqrt{\frac{1-e}{1+e}}\tan\left(\frac{f}{2}\right)\right) - \frac{e\sqrt{1-e^2}\sin f}{1+e\cos f}\right] \tag{A.3}$$

The mean orbital motion n is directly related to the orbital period T.

$$n = \sqrt{\mu/a^3} \tag{A.4}$$

$$T = 2\pi\sqrt{a^3/\mu} = 2\pi/n \tag{A.5}$$

where $\mu = 3.986013 \times 10^{14}\,\mathrm{m^3\,s^{-2}}$ is the central attraction term of the Earth. The orbit radius r, and its perigee and apogee values r_{pe} and r_{ap} can be written as

$$r = \frac{a(1-e^2)}{1+e\cos f} = a(1-e\cos E) \tag{A.6}$$

$$r_{pe} = a(1-e) \qquad r_{ap} = a(1+e) \tag{A.7}$$

The magnitude of the orbit radius at $f = 90°$ is denoted as the parameter of the conical section (see Fig. A.1).

$$p = a\left(1-e^2\right) \tag{A.8}$$

For the orbit velocity v, and its perigee and apogee values v_{pe} and v_{ap} the following relationships hold.

$$v = \sqrt{\frac{\mu}{a}\frac{1+e^2+2e\cos f}{1-e^2}} = \sqrt{\frac{\mu}{a}\frac{1+e\cos E}{1-e\cos E}} \tag{A.9}$$

$$v_{pe} = \sqrt{\frac{\mu}{a}\frac{1+e}{1-e}} \qquad v_{ap} = \sqrt{\frac{\mu}{a}\frac{1-e}{1+e}} \tag{A.10}$$

Unless stated otherwise, all orbital elements and state vectors are defined in an inertial, Earth-centered, mean equatorial coordinate system of 2000.0 (so-called "J2000"), which corrects for precession and nutation of the CIO pole (Conventional International Origin) at this date. In this system, $\underline{X}$ is towards the mean equinox, within the mean equatorial plane, $\underline{Z}$ is perpendicular to the mean equator plane, towards the CIO north pole, and $\underline{Y} = \underline{Z} \times \underline{X}$ completes the right-hand system, as shown in Fig. A.2 (Montenbruck and Gill, 2000). All epochs shall be provided in coordinated universal time (UTC).

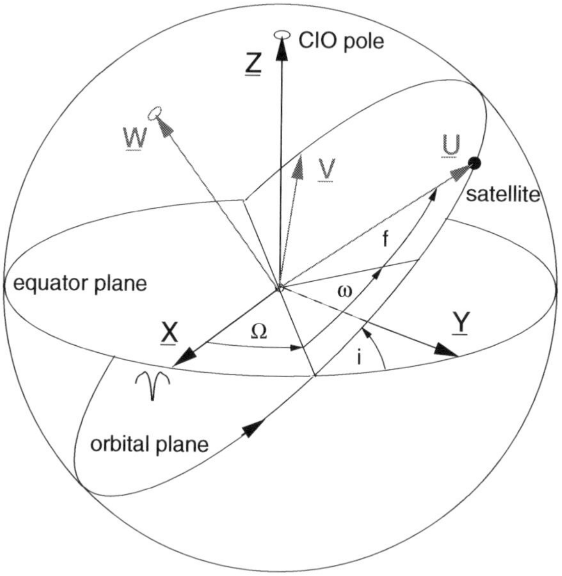

Fig. A.2. Illustration of the angular Kepler elements Ω, i, ω, and f, and their meaning for the rotation from the inertial system $\underline{X}$, $\underline{Y}$, $\underline{Z}$ to the orbit-related system $\underline{U}$, $\underline{V}$, $\underline{W}$ (radial, transversal, and out-of-plane). The $\underline{X}$ axis points to the vernal equinox ("first point of Aries"), the $\underline{Z}$ axis points to the Conventional International Origin (CIO pole), and the axis $\underline{Y} = \underline{Z} \times \underline{X}$ completes the right-hand system.

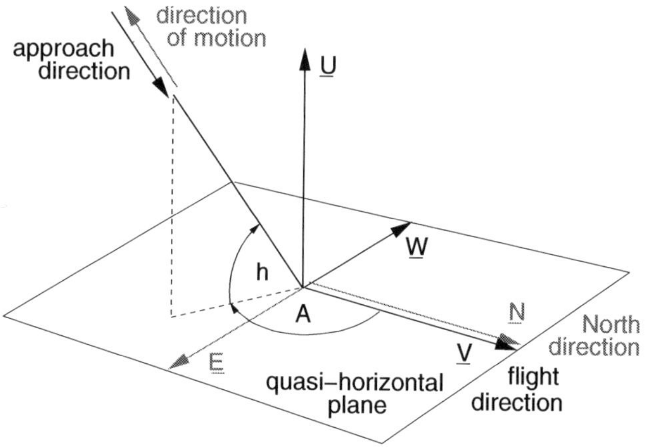

Fig. A.3. Definition of the azimuth A and elevation h of an approach direction in the orbit-related coordinate system $\underline{U}$, $\underline{V}$, $\underline{W}$ (radial, transversal, and out-of-plane). Definition of the azimuth A and elevation h of a direction of motion in the quasi-horizontal coordinate system $\underline{U}$, $\underline{E}$, $\underline{N}$ (radial, east, and north).

In the $\underline{X}$, $\underline{Y}$, $\underline{Z}$ inertial reference system the orientation of a satellite orbit is defined by Ω, the right ascension of the ascending orbit node (measured in the equatorial plane from the vernal equinox to the intersection of the ascending orbit with the equator plane), by i, the inclination of the orbit plane with respect to the equator, and by ω, the position of the pericenter (measured from the ascending node within the orbital plane, in the direction of motion). These orbital elements, together with the true anomaly f define the unit vectors in radial direction $\underline{U}$, in transversal in-flight direction $\underline{V}$, and in out-of-plane direction $\underline{W}$ (see Fig. A.2).

$$\underline{U} = \begin{pmatrix} U_X \\ U_Y \\ U_Z \end{pmatrix} = \begin{pmatrix} \cos u \cos\Omega - \sin u \sin\Omega \cos i \\ \cos u \sin\Omega + \sin u \cos\Omega \cos i \\ \sin u \sin i \end{pmatrix} \tag{A.11}$$

$$\underline{V} = \begin{pmatrix} V_X \\ V_Y \\ V_Z \end{pmatrix} = \begin{pmatrix} -\sin u \cos\Omega - \cos u \sin\Omega \cos i \\ -\sin u \sin\Omega + \cos u \cos\Omega \cos i \\ \cos u \sin i \end{pmatrix} \tag{A.12}$$

$$\underline{W} = \begin{pmatrix} W_X \\ W_Y \\ W_Z \end{pmatrix} = \begin{pmatrix} \sin\Omega \sin i \\ -\cos\Omega \sin i \\ \cos i \end{pmatrix} \tag{A.13}$$

where $u = \omega + f$ is the argument of true latitude. With the help of these unit vectors the satellite position and velocity vector, $\underline{r}$ and $\underline{v} = \dot{\underline{r}}$, can be written as

$$\underline{r} = r\underline{U} \tag{A.14}$$

$$\underline{v} = \dot{r}\underline{U} + r\dot{f}\,\underline{V} \tag{A.15}$$

with r from Eq. A.6, and the radial and transversal velocity magnitudes, $\dot{r}$ and $r\dot{f}$, according to Eq. A.16 and A.17.

$$\dot{r} = \sqrt{\frac{\mu}{a}}\,\frac{e \sin f}{\sqrt{1-e^2}} \tag{A.16}$$

$$r\dot{f} = \sqrt{\frac{\mu}{a}}\,\frac{1 + e \cos f}{\sqrt{1-e^2}} \tag{A.17}$$

The radial and transversal components of the velocity also determine the flight path angle $\gamma \in [-\pi/2, +\pi/2]$, which is measured from the quasi-horizontal plane (which has $\underline{U}$ as its normal vector) to the velocity vector, positive towards space.

$$\cos\gamma = \frac{r\dot{f}}{v} = \frac{1 + e \cos f}{\sqrt{1 + e^2 + 2e \cos f}} \tag{A.18}$$

$$\sin\gamma = \frac{\dot{r}}{v} = \frac{e \sin f}{\sqrt{1 + e^2 + 2e \cos f}} \tag{A.19}$$

$$\gamma = \arcsin\left(\frac{e \sin f}{\sqrt{1 + e^2 + 2e \cos f}}\right) \tag{A.20}$$

A.2 PLANAR ORBIT TRANSFER MANEUVERS

A satellite can be transferred from a circular orbit of radius r_1 to another, co-planar circular orbit of radius r_2 by 2 impulsive, tangential maneuvers, as illustrated in Fig. A.4. A first tangential impulse Δv_1 injects the spacecraft into an elliptic transfer trajectory with a pericenter at r_1, and an apocenter at r_2. A second tangential impulse Δv_2, half a revolution later, circularizes the target orbit.

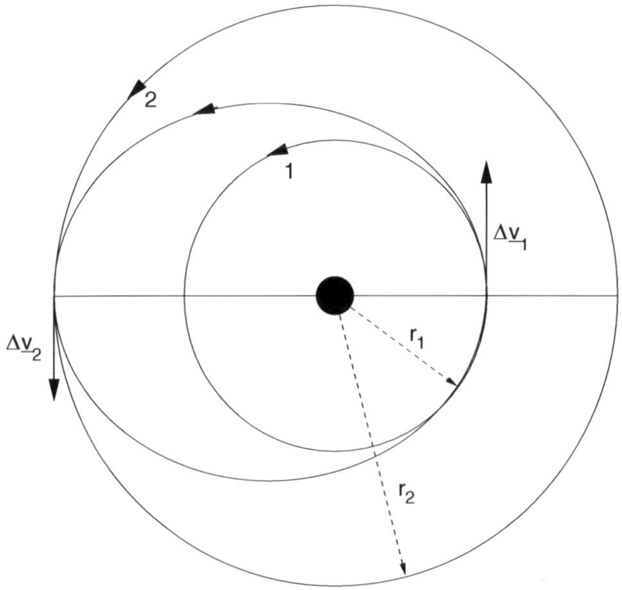

Fig. A.4. Two-impulse, planar Hohmann transfer from an inner circular orbit of radius r_1 to an outer, co-planar circular orbit of radius r_2.

This transfer scenario according to Hohmann can be shown to be optimal in terms of Δv, if the ratio $\beta = r_2/r_1$ is less than 11.939 (i.e. for all applications between LEO and GEO). For a given $\beta = r_2/r_1$, the required impulses are

$$\Delta v_1 = |\Delta \underline{v}_1| = \sqrt{\frac{\mu}{r_1}} \left(\sqrt{\frac{2\beta}{1+\beta}} - 1 \right) \tag{A.21}$$

$$\Delta v_2 = |\Delta \underline{v}_2| = \sqrt{\frac{\mu}{r_2}} \left(1 - \sqrt{\frac{2}{1+\beta}} \right) \tag{A.22}$$

The total Δv for the 2-impulse Hohmann transfer is hence

$$\Delta v_h = \Delta v_1 + \Delta v_2 = \sqrt{\frac{\mu}{r_1}} \left[\left(1 - \frac{1}{\beta} \right) \sqrt{\frac{2\beta}{1+\beta}} + \sqrt{\frac{1}{\beta}} - 1 \right] \tag{A.23}$$

A.3 SHAPE OF THE EARTH AND THE GEODETIC POSITION

The Earth surface, to first approximation, takes a hydrostatic equilibrium shape of an elastic, rotating body. This rotationally symmetric reference ellipsoid has an equatorial radius of $a_e = 6378.137\,\text{km}$, a polar radius of $b_e = 6356.752\,\text{km}$, an oblateness of $f_e = (a_e - b_e)/a_e = 1/298.257$, and a sidereal rotation rate of $\omega_e = 0.7292115 \times 10^{-4}\,\text{rad/s}$, according to WGS 84 (Seidelmann et al., 1992). In celestial mechanics, orbit altitudes $H(r)$ are generally referred to a spherical Earth of radius $r_e = a_e$. Geodetic altitudes $z(r,\delta)$, however, are vertical distances to the reference Earth ellipsoid of radius $r_e(\phi)$, connecting the satellite position and sub-satellite point of the same geodetic latitude ϕ. For a given orbit position $\underline{r}(r,\delta,\alpha)$ according to Eq. 3.4, the orbital altitude H, geodetic altitude z, and geodetic latitude ϕ can be expressed as

$$H(r) \quad = \quad r - a_e \tag{A.24}$$

$$z(r,\delta) \quad = \quad r - a_e \left(1 - f_e \sin^2\delta - \left(\frac{a_e}{r} - \frac{1}{4}\right)\frac{f_e^2}{2}\sin^2 2\delta + O(f_e^3)\right) \tag{A.25}$$

$$\phi(r,\delta) \quad = \quad \delta + \frac{a_e}{r}\left(f_e \sin 2\delta + \left(\frac{a_e}{r} - \frac{1}{4}\right) f_e^2 \sin 4\delta + O(f_e^3)\right) \tag{A.26}$$

where the declination δ can be determined from the cartesian coordinates, or from the Kepler elements of the orbit position. The maximum difference between geodetic latitude and declination is $\phi - \delta \approx 0.19°$ for $z = 0$ at $45°$ declination.

$$\delta \; = \; \arcsin\left(r_Z/r\right) \; = \; \arcsin\left(\sin i \, \sin(\omega + f)\right) \tag{A.27}$$

A.4 MAJOR PERTURBATIONS ON EARTH ORBITS

A satellite or debris object on an Earth orbit does not strictly follow a planar, circular or elliptic Kepler orbit of constant energy and orbit momentum. Real Earth orbits undergo several perturbations, which can be characterized as follows:

- Geopotential perturbations ($\underline{p}_g$) are due to asymmetries in the Earth's gravitational field which can be expressed in terms of harmonic coefficients $C_{n,m}$ and $S_{n,m}$ of an expansion in spherical harmonics. One can distinguish zonal harmonics $C_{n,0}$, leading to $\underline{p}_g = \text{fct}(C_{n,0}, \phi, r)$, sectorial harmonics $C_{n,n}$ and $S_{n,n}$, leading to $\underline{p}_g = \text{fct}(C_{n,n},\ S_{n,n}, \lambda, r)$, and tesseral harmonics $C_{n,m}$ and $S_{n,m}$, leading to $\underline{p}_g = \text{fct}(C_{n,m}, S_{n,m}, \phi, \lambda, r)$. Here, ϕ is the geodetic latitude, λ is the geodetic longitude, and r is the geocentric distance of the satellite. For space debris applications, normally a truncated geopotential model of degree $n \leq 5$ and order $m \leq n$ is sufficient. The harmonic coefficients $C_{n,m}$ and $S_{n,m}$ can be adopted from contemporary gravity models. As short-hand notation for zonal harmonics, often $J_n = -C_{n,0}$ is used. For most Earth orbits the Earth oblateness term $J_2 = 1.082 \times 10^{-3}$ ($\approx 1/550$ of the central attraction term) dominates all other perturbations by a factor of ~600. J_2 is hence the only first order perturbation. Geopotential perturbations are conservative (not changing orbital energy), with the exception of resonant cases (e.g. geosynchronous orbits). In (Montenbruck and Gill, 2000; Seidelmann et al., 1992) models and techniques are outlined to define the perturbing geopotential, and to derive perturbing accelerations.

- Third body perturbations ($\underline{p}_t$) are dominated by the Sun and Moon. They are a function of the geocentric positions of the satellite, the Sun, and the Moon at a given epoch, where the last two can be read from precise ephemerides files, or be approximated by analytical expressions. Third body perturbations are conservative. Their computation is explained in (Montenbruck and Gill, 2000), and required physical ephemerides of Sun and Moon can be obtained from (Seidelmann et al., 1992).
- Aerodynamic perturbations ($\underline{p}_a$) are produced by the motion of a satellite through a rotating Earth atmosphere with superimposed wind patterns. In general $\underline{p}_a = \mathrm{fct}(\rho, A/m, \underline{c}_a, \underline{v}_a)$, where A/m is the ratio of projected cross-sectional area to mass, $\underline{c}_a$ represents aerodynamic coefficients for lift (c_L) and drag (c_D), $\underline{v}_a$ is the aerodynamic velocity relative to the atmospheric medium, and ρ is the local air density, which depends on the geodetic satellite position, on the relative position to the Sun, and on solar and geomagnetic activity indicators. Aerodynamic perturbations, and in particular the dominant drag force, are non-conservative, energy-dissipating. During re-entries the drag force may become a first-order perturbation (equal to or exceeding the J_2 perturbation). Details on the aerodynamic model parameters, and on the formulation of the aerodynamic acceleration function are provided in (Montenbruck and Gill, 2000).
- Radiation pressure perturbations ($\underline{p}_r$) are generated by the interaction of emitted photons from the Sun, or of reflected photons from the Earth atmosphere or surface, with the satellite. For debris-related analyses, only the direct solar radiation pressure is important, with $\underline{p}_r = \mathrm{fct}(A/m, c_R, \underline{r}, \underline{r}_\odot)$, where A/m is the ratio of projected cross-sectional area to mass, $0 \le c_R \le 2$ is a radiation pressure coefficient (normally, $c_R \approx 1.3$), $\underline{r}$ is the satellite position, and $\underline{r}_\odot$ is the Sun position. $\underline{p}_r$ is a conservative perturbation, unless resonances with eclipse periods play a role. Due to its dependence on A/m, the solar radiation pressure (as the airdrag) is denoted as "skin force". Depending on the solar activity level it exceeds the aerodynamic perturbation level above altitudes of 500 km to 600 km. Perturbing accelerations due to direct solar radiation pressure are derived in (Montenbruck and Gill, 2000).

All listed perturbation effects can be expressed in terms of perturbing accelerations $\underline{p}$, or as perturbing potentials P. The former are used for integrating the perturbed Newton equations, Eq. A.29, or the Gauss perturbation equations, Eq. A.33–A.38. The latter are used in context with the Lagrange perturbation equations Eq. A.40–A.45.

A.5 THE PERTURBED NEWTON EQUATIONS

The motion of a satellite whose mass m is small as compared to the mass $m_\oplus$ of the Earth can be described by Newton's equation of the unperturbed 2-body problem.

$$\underline{\ddot{r}} = -\frac{\mu}{r^3}\underline{r} \tag{A.28}$$

where $\mu = m_\oplus G \approx 3.986013 \times 10^{14}\,\mathrm{m^3\,s^{-2}}$ is the Earth's central attraction term, and $G = 6.67 \times 10^{-11}\,\mathrm{m^3\,kg^{-1}\,s^{-2}}$ is the universal gravitational constant. $\underline{\ddot{r}} = (\ddot{r}_X, \ddot{r}_Y, \ddot{r}_Z)$ is the acting acceleration vector, and $\underline{r} = (r_X, r_Y, r_Z)$ is the geocentric radius vector to the satellite. In an inertial coordinate system the unperturbed 2-body problem results in a planar motion, on a conical section (circle, ellipse, parabola, or hyperbola) according to the Kepler laws.

In the presence of a perturbing acceleration $\underline{p}$ (i.e. other than the central attraction term), Eq. A.28 changes to

$$\underline{\ddot{r}} = -\frac{\mu}{r^3}\underline{r} + \underline{p} \tag{A.29}$$

which can be rewritten as

$$\underline{\dot{v}} = -\frac{\mu}{r^3}\underline{r} + \underline{p} \tag{A.30}$$

$$\underline{\dot{r}} = \underline{v} \tag{A.31}$$

This is a system of 6 first-order differential equations which can be integrated in time to propagate the orbit position vector $\underline{r}(t)$ and velocity vector $\underline{\dot{r}}(t) = \underline{v}(t)$. As in the unperturbed problem, these 6 parameters (3 position and 3 velocity components) at any time uniquely define the motion of a mass point on an orbit. The resulting trajectory is an infinite sequence of locally tangent (so-called "osculating") Kepler orbits. When the infinitesimal orbit segments are patched together, they generally do not result in a pure conical section, nor in a planar motion, nor in a trajectory with conserved orbital momentum. The perturbing acceleration $\underline{p}$ can be modeled as a sum of dominant individual perturbation sources as outlined in Section A.4.

The perturbed Newton equations as in Eq. A.30–A.31 are used by numerical integration schemes, if a good prediction accuracy is required (as in conjunction event analyses), or if perturbation levels violate underlying assumptions in alternative analytical theories (as in re-entry predictions). The orbit parameters $\underline{r}(t)$ and $\underline{v}(t)$ have the advantage of being non-singular. Depending on the context, the integration should be performed in a quasi-inertial frame (e.g. for conjunction event predictions), or in an Earth-fixed frame (e.g. for re-entry predictions). The perturbing accelerations $\underline{p} = (p_X, p_Y, p_Z)$ must be formulated in the same coordinate system. The integration of Eq. A.30–A.31 is often referred to as the "Cowell Method".

A.6 THE GAUSS PERTURBATION EQUATIONS

In Section A.5 the orbit state was expressed in terms of six cartesian coordinates r_X, r_Y, r_Z of the position vector $\underline{r}$, and $\dot{r}_X, \dot{r}_Y, \dot{r}_Z$ of the velocity vector $\dot{\underline{r}}$. There are several alternative sets of orbital parameters into which the cartesian state can be converted. One such set are the six Kepler orbital elements a (semimajor axis), e (eccentricity), i (inclination), Ω (right ascension of the ascending orbit node), ω (argument of pericenter), and M (mean anomaly of the orbit position), which are well suited to describe a conical section, oriented in inertial space. In the case of an unperturbed motion, only the mean anomaly M changes with time at the rate of the mean motion n.

$$\frac{dM}{dt} = n = \sqrt{\frac{a^3}{\mu}} \tag{A.32}$$

With a perturbing acceleration $\underline{p} = (p_U, p_V, p_W)$ expressed in terms of its radial (p_U), transversal (p_V), and out-of-plane component (p_W) in the orbit oriented coordinate system, the resulting rates of change of the osculating Kepler elements can be expressed by the Gauss formulation of the Lagrange perturbation equations.

$$\frac{da}{dt} = \sqrt{\frac{a}{\mu}} \frac{2a}{\sqrt{1-e^2}} \left(e \sin f \, p_U + (1 + e\cos f) \, p_V \right) \tag{A.33}$$

$$\frac{de}{dt} = \sqrt{\frac{a}{\mu}} \sqrt{1-e^2} \left(\sin f \, p_U + \frac{e + 2\cos f + e\cos^2 f}{1 + e\cos f} \, p_V \right) \tag{A.34}$$

$$\frac{di}{dt} = \sqrt{\frac{a}{\mu}} \frac{\sqrt{1-e^2}}{1+e\cos f} \cos(\omega + f) \, p_W \tag{A.35}$$

$$\frac{d\Omega}{dt} = \sqrt{\frac{a}{\mu}} \frac{\sqrt{1-e^2}}{1+e\cos f} \frac{\sin(\omega + f)}{\sin i} \, p_W \tag{A.36}$$

$$\frac{d\omega}{dt} = \sqrt{\frac{a}{\mu}} \frac{\sqrt{1-e^2}}{e} \left(-\cos f \, p_U + \frac{2 + e\cos f}{1 + e\cos f} \sin f \, p_V \right) - \cos i \frac{d\Omega}{dt} \tag{A.37}$$

$$\frac{dM}{dt} = n + \frac{1-e^2}{nae} \left[\left(\cos f - \frac{2e}{1+e\cos f} \right) p_U - \frac{2 + e\cos f}{1+e\cos f} \sin f \, p_V \right] \tag{A.38}$$

These equations are particularly well suited for analytical integrations of non-gravitational perturbations, such as airdrag and solar radiation pressure.

If instead of $\underline{p} = (p_U, p_V, p_W)$ the perturbing acceleration is expressed in tangential (p_T), normal (p_N), and out-of-plane direction (p_W), one can show that the rate of change of the semimajor axis only depends on the tangential component.

$$\frac{da}{dt} = 2a \sqrt{\frac{a}{\mu} \frac{1 + e^2 + 2e\cos f}{1-e^2}} \, p_T \tag{A.39}$$

Eq.A.33–A.38 are also useful to assess the effects of a small (!) maneuver with $\Delta \underline{v} = \underline{p} \, \Delta t$ (where $\underline{p}$ is the acceleration imposed by a thruster actuation). In this case, (p_U, p_V, p_W) become $(\Delta v_U, \Delta v_V, \Delta v_W)$, and $da/dt, ..., dM/dt$ become $\Delta a, ..., \Delta M$.

A.7 THE LAGRANGE PERTURBATION EQUATIONS

If a perturbation results from a perturbing potential P, then the Lagrange equations in their original form can be applied. They provide the rates of change of the osculating Kepler elements as a function of the derivatives $\partial P/\partial a$, $\partial P/\partial e$, $\partial P/\partial i$, $\partial P/\partial \Omega$, $\partial P/\partial \omega$, and $\partial P/\partial M$.

$$\frac{\mathrm{d}a}{\mathrm{d}t} = -\frac{2}{na}\frac{\partial P}{\partial M} \tag{A.40}$$

$$\frac{\mathrm{d}e}{\mathrm{d}t} = -\frac{1-e^2}{na^2e}\frac{\partial P}{\partial M} + \frac{\sqrt{1-e^2}}{na^2e}\frac{\partial P}{\partial \omega} \tag{A.41}$$

$$\frac{\mathrm{d}i}{\mathrm{d}t} = -\frac{\cos i}{na^2\sqrt{1-e^2}\sin i}\frac{\partial P}{\partial \omega} + \frac{1}{na^2\sqrt{1-e^2}\sin i}\frac{\partial P}{\partial \Omega} \tag{A.42}$$

$$\frac{\mathrm{d}\Omega}{\mathrm{d}t} = -\frac{1}{na^2\sqrt{1-e^2}\sin i}\frac{\partial P}{\partial i} \tag{A.43}$$

$$\frac{\mathrm{d}\omega}{\mathrm{d}t} = \frac{\cos i}{na^2\sqrt{1-e^2}\sin i}\frac{\partial P}{\partial i} - \frac{\sqrt{1-e^2}}{na^2e}\frac{\partial P}{\partial e} \tag{A.44}$$

$$\frac{\mathrm{d}M}{\mathrm{d}t} = n + \frac{1-e^2}{na^2e}\frac{\partial P}{\partial e} + \frac{2}{na}\frac{\partial P}{\partial a} \tag{A.45}$$

These equations are particularly well suited for analytical integrations of gravitational perturbations, such as from geopotential harmonics, and from Sun and Moon attraction.

A.8 INTEGRATING THE PERTURBATION EQUATIONS

All perturbations acting on a satellite orbit can be categorized according to the following classification scheme:

- secular perturbations: time-proportional, secular changes of orbital elements with propagation time
- long-periodic perturbations: harmonic changes in orbital elements with periods on the order of the rotation of the pericenter (for geopotential perturbations), and/or on the order of a month or year (for luni-solar perturbations)
- short-periodic perturbations: harmonic changes in orbital elements with periods on the order of the orbital period
- resonant changes: changes in orbital elements due to energy inputs into an otherwise conservative system, as a consequence of synchronism between the orbital motion and a perturbing environment (mostly geopotential)

A real satellite orbit, manifested by the time evolution of its 6 osculating elements $\underline{\psi}(t)$, is a superposition of all 4 types of perturbations. The resonant contributions, however, are only important in a few special cases (e.g. in GEO). The focus shall hence be on the first 3 categories.

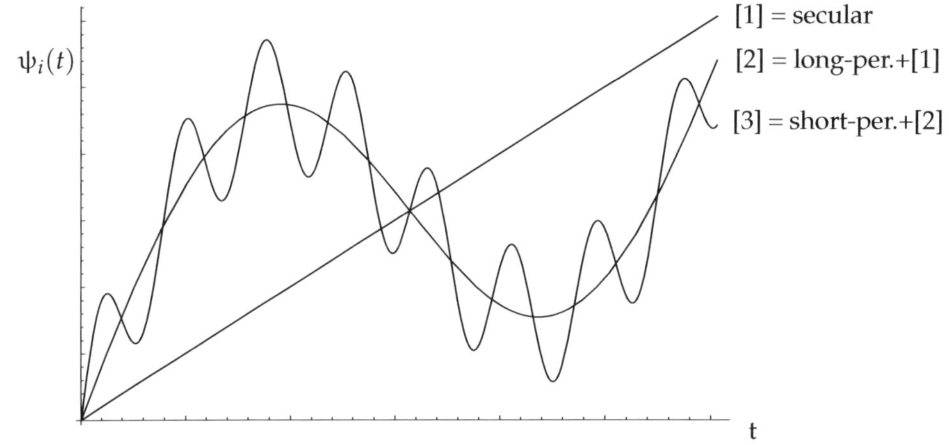

Fig. A.5. Definition of perturbation classes according to their periodicity, with time-proportional, secular changes [1], long-periodic changes [2], modulated on the secular slope, and short-periodic changes [3] modulated on the sum of [1] and [2] (perturbation amplitudes and frequencies at not to scale).

Fig. A.5 sketches the effects of secular, long-periodic, and short-periodic perturbations on an orbital element $\psi_i(t)$, which is part of a state vector $\underline{\psi}(t)$.

In numerical integrations the orbit state is propagated along the osculating trajectory (curve [3] in Fig. A.5), with sufficiently small time steps Δt of typically a few minutes, to follow the "high frequency" perturbation signal. For the propagation of cartesian state vectors, with $\underline{\psi} = (r_X, r_Y, r_Z, \dot{r}_X, \dot{r}_Y, \dot{r}_Z)$, Eq. A.30–A.31 should be solved. In case of Kepler state vectors, where $\underline{\psi} = (a, e, i, \Omega, \omega, M)$, Eq. A.33–A.38 or Eq. A.40–A.45 should be integrated. The integration is mostly performed with single-step, self-starting methods (e.g. Runge–Kutta), or with multi-step methods (e.g. Adams–Bashforth). The latter require an externally produced starting arc (e.g. from a Runge–Kutta method). They do, however, have advantages with respect to CPU time consumption, since (in contrast to single-step schemes) they perform only one evaluation of the perturbation function $\underline{p}(t)$ per time step. Single-step methods, on the other hand, have advantages concerning time step control, which can be important in rapidly changing perturbation environments (e.g. during re-entries).

In order to reduce CPU time consumption analytical orbit theories try to increase the size of the prediction time step. This can be achieved by propagating along the considerably smoother curve [2], with time steps on the order of hours or days. The propagated orbit state in this case is denoted as "singly averaged" (over short-periodic perturbations). The time rate of change of the singly averaged state can be determined by averaging with respect to the mean anomaly M.

$$\frac{\mathrm{d}\overline{\psi}_i}{\mathrm{d}t} = \frac{1}{2\pi} \int_0^{2\pi} \frac{\mathrm{d}}{\mathrm{d}t} \left(\psi_i(\underline{\psi}(t), \underline{p}(t)) \right) \mathrm{d}M \tag{A.46}$$

For Kepler elements $\mathrm{d}\psi_i/\mathrm{d}t$ can be obtained from Eq. A.33–A.38 or Eq. A.40–A.45. The averaging operation can be simplified by a prior separation of variables in the perturbing function $\underline{p}(t)$. For geopotential and luni-solar perturbations this can be accomplished by means of expansions in inclination functions $F(i)$ and eccentricity functions $G(e)$ (Brumberg, 1995). The resulting equations of the mean rates of change require the use of mean elements $\overline{\psi}_i$ in their right-hand sides (with $i = 1,...,6$ for $\overline{a}, \overline{e}, \overline{i}, \overline{\Omega}, \overline{\omega}$, and $\overline{M}$). Short-periodic transition functions $\Delta\psi_i$ enable the explicit translation from mean to osculating states.

$$\psi_i = \overline{\psi}_i + \Delta\overline{\psi}_i(\underline{\overline{\psi}}, \underline{p}) \tag{A.47}$$

The transition from osculating to mean states requires an iterative solution of Eq. A.47, starting with $\overline{\psi}_i \approx \psi_i$. The short-periodic perturbation function $\underline{\mathrm{p}}$ is in most cases only approximated to first order, including the zonal harmonic J_2.

A given, singly averaged orbit state is hence propagated by step-wise integration of Eq. A.46. Osculating states can then be recovered via Eq. A.47.

When propagating along curve [1] in Fig. A.5, the time step can be further increased. In this case it is mainly limited by the need to update the time-dependent perturbation function. The "doubly averaged" time rates of change (over short- and long-periodic perturbations) are determined by averaging the functions $\mathrm{d}\overline{\psi}_i/\mathrm{d}t$ with respect to $\overline{\omega}$.

$$\frac{\mathrm{d}\overline{\overline{\psi}}_i}{\mathrm{d}t} = \frac{1}{2\pi}\int_0^{2\pi} \frac{\mathrm{d}}{\mathrm{d}t}\left(\overline{\psi}_i(\underline{\overline{\psi}}(t), \underline{p}(t))\right)\, \mathrm{d}\overline{\omega} \tag{A.48}$$

In order to perform the transition from the propagated doubly averaged elements $\overline{\overline{\psi}}_i$ (curve [1]) to the singly averaged elements $\overline{\psi}_i$ (curve [2]), long-periodic correction functions must be applied as provided in Eq. A.49. In a second step, indicated by Eq. A.50, a conversion to the osculating state ψ_i of curve [3] in Fig. A.5 can be accomplished.

$$\overline{\psi}_i = \overline{\overline{\psi}}_i + \Delta\overline{\overline{\psi}}_i(\underline{\overline{\overline{\psi}}}, \underline{p}) \tag{A.49}$$

$$\psi_i = \overline{\overline{\psi}}_i + \Delta\overline{\overline{\psi}}_i(\underline{\overline{\overline{\psi}}}, \underline{p}) + \Delta\overline{\psi}_i(\underline{\overline{\psi}}, \underline{p}) \tag{A.50}$$

For Kepler elements, $\underline{\overline{\overline{\psi}}}$ consists of $(\overline{\overline{a}}, \overline{\overline{e}}, \overline{\overline{i}}, \overline{\overline{\Omega}}, \overline{\overline{\omega}}, 0)$. The concept of singly and doubly averaged elements, and corresponding conversions according to Eq. A.49–A.50 are explained in (Liu, 1974) for a perturbing geopotential up to J_4, with short-periodic perturbations considered up to first order (J_2), only.

A.9 RESULTING PERTURBATIONS ON EARTH ORBITS

An analysis of spectra of different perturbation source terms on terrestrial orbits shows the following dominant effects on the Kepler orbital elements:

- semimajor axis a: short-periodic variations due to the Earth oblateness term J_2, at periods of $T/2$ with amplitudes of up to ± 9 km; long-periodic variations due to luni-solar and solar radiation perturbations; secular decrease due to airdrag (see Eq. A.51)
- eccentricity e: short-periodic variations at periods of T and $T/3$ due to J_2; long-periodic variations due to odd zonal harmonics J_{2n+1}, luni-solar and solar radiation perturbations; secular decrease due to airdrag (see Eq. A.52)
- inclination i: short-periodic variations at periods of $T/2$ due to J_2; long-periodic variations due to odd zonal harmonics J_{2n+1}; small secular change due to airdrag and luni-solar perturbations
- right ascension of the ascending node Ω: short-periodic variations at periods of $T/2$ due to J_2; long-periodic variations due odd zonal harmonics J_{2n+1}; secular change due to even zonal harmonics J_{2n} (particularly J_2) at a rate of up to 10° per day (see Eq. A.54)
- argument of pericenter ω: short-periodic variations at periods of T and $T/3$ due to J_2; long-periodic variations due odd zonal harmonics J_{2n+1}; secular change due to even zonal harmonics J_{2n} (particularly J_2) at a rate of up to 20° per day (see Eq. A.55)
- mean anomaly M: short-periodic variations at periods of T and $T/3$ due to J_2; long-periodic variations due odd zonal harmonics J_{2n+1}; secular change due to the central attraction term μ and (to a lesser extent) J_2 at a rate of up to 5,900° per day (see Eq. A.56)

For an assessment of first-order effects on low-altitude orbits its is helpful to summarize the major secular perturbations due to Earth oblateness and due to airdrag.

$$\frac{da}{dt} \approx -c_D \frac{A}{m} \rho_{pe}\, a^2 \frac{n}{2\pi} \exp(-z) \Big(I_0(z) + 2e\, I_1(z) + \frac{3}{4} e^2 \left(I_0(z) + I_2(z) \right) \Big) \tag{A.51}$$

$$\frac{de}{dt} \approx -c_D \frac{A}{m} \rho_{pe}\, a(1-e^2) \frac{n}{2\pi} \exp(-z) \Big(I_1(z) + \frac{1}{2} e \left(I_0(z) + I_2(z) \right) + \frac{1}{8} e^2 \left(3I_1(z) + I_3(z) \right) \Big) \tag{A.52}$$

$$\frac{di}{dt} \approx 0 \tag{A.53}$$

$$\frac{d\Omega}{dt} \approx -\frac{3}{2} n\, J_2 \left(\frac{a_e}{p} \right)^2 \cos i \tag{A.54}$$

$$\frac{d\omega}{dt} \approx \frac{3}{4} n\, J_2 \left(\frac{a_e}{p} \right)^2 \left(4 - 5\sin^2 i \right) \tag{A.55}$$

$$\frac{dM}{dt} \approx n \left(1 + \frac{3}{4} J_2 \left(\frac{a_e}{p} \right)^2 \left(2 - 3\sin^2 i \right) \sqrt{1-e^2} \right) \tag{A.56}$$

In these equations all elements should be regarded as doubly averaged, leaving only secular perturbation terms. Eq. A.51 and A.52 are based on analytical solutions by (King-Hele, 1987), neglecting higher-order terms in eccentricity. The quantity $c_D A/m = B$ is denoted as the ballistic parameter, ρ_{pe} is the air density at pericenter, $H_{\rho,pe}$ is the density scale height at pericenter, and I_k are integer Bessel functions of order $k = 1$ to 3 and argument $z = ae/H_{\rho,pe}$. They can be computed by polynomial expansions in z^{-j} for large values of z, and in z^{+j} for small values of z, with $j = 0, 1, 2, ...$ (Abramowitz and Stegun, 1984).

In the case of near-circular orbits, $\exp(-z) \rightarrow 1$, $I_0 \rightarrow 1$, and $I_k \rightarrow 0$ for $k \geq 1$, leaving a non-vanishing drag contribution only for the decay of the semimajor axis. In the case of eccentric orbits, the drag concentrated near the pericenter causes the eccentricity and semimajor axis to decrease in such a way that the pericenter remains constant, and the apocenter decays secularly, until a near-circular orbit is reached, from which the object spirals down to re-entry.

Depending on the orbit altitude, a value of $i \in [96°, 100°]$ in Eq. A.54 yields a rotation of the line of nodes at a rate of $\dot{\Omega} \approx 0.9856°/\mathrm{d}$, which corresponds to the mean motion of the Sun. Hence, such orbits are denoted as "Sun-synchronous". For $i = \arcsin(\sqrt{4/5})$ the rotation rate of the pericenter according to Eq. A.55 becomes zero. The corresponding inclination of $i \approx 63.4°$ is known as the "critical inclination".

A.10 REFERENCES

Abramowitz, M. and Stegun, A. (1984). *Pocketbook of Mathematical Functions*. Verlag Harri Deutsch, Frankfurt a.M.

Brumberg, V. (1995). *Analytical Techniques of Celestial Mechanics*. Springer, Berlin, Heidelberg, New York.

King-Hele, D. (1987). *Satellite Orbits in an Atmosphere: Theory and Applications*. Blackie, Glasgow, UK.

Liu, J. (1974). Satellite Motion About an Oblate Earth. *AIAA Journal*, vol.12, no. 11:1511–1516.

Montenbruck, O. and Gill, E. (2000). *Satellite Orbits – Models, Methods, and Applications*. Springer, Berlin, Heidelberg, New York.

Seidelmann, P. et al., editors (1992). *Explanatory Supplement to the Astronomical Almanac*. University Science Books, Mill Valley, CA.

B

The Atmosphere of the Earth

H. Klinkrad

B.1 STRUCTURE OF THERMOSPHERE MODELS

The Earth atmosphere can be subdivided into three main altitude regions: the homosphere, the thermosphere, and the exosphere. The homosphere reaches from ground level to the homopause, at geodetic altitude $z \approx 120\,\text{km}$. In this region turbulent mixing, also denoted as "eddy diffusion", causes a uniform composition with similar fractions of atmospheric constituents as on the Earth surface. Above the homopause vertical diffusion prevails, and each of the main constituents (N_2, O, O_2, He, H, Ar, and N) establishes its own concentration profile with altitude under the influence of Earth gravity, and energy inputs from solar radiation and solar winds. This thermospheric region shows a strong increase in temperature, which reaches an asymptotic value near the thermopause, at an altitude $300\,\text{km} \leq z \leq 600\,\text{km}$ which depends on the current energy input into the atmosphere (see Fig. B.1). Above the thermopause the excitation of the light-weight constituents H and He by EUV radiation can cause thermal speeds which exceed the Earth escape velocity. Since the mean free paths at these altitudes are almost infinite, this phenomenon leads to a continuous escape flux of constituents at the upper rims of the atmosphere. Hence, this region is designated the exosphere.

The main energy sources for thermospheric processes are extreme ultra-violet radiation (EUV) and solar winds, both of which are emitted from active Sun spot regions. EUV radiation is traveling at the speed of light, reaching the Earth within a timespan of $8^{\text{m}}\,20^{\text{s}}$. Solar winds consist of a high-energy plasma, mainly composed of H^+ and He^{++} ions, traveling at velocities of $170\,\text{km/s} \leq v \leq 2,000\,\text{km/s}$, with mean values of $\overline{v} = 470\,\text{km/s}$. Because the solar wind only travels with 0.06% to 0.6% of the speed of light, its reaches the Earth with time delays of 1 to 10 days. The energy which the EUV radiation and the solar wind plasma transport into the upper atmosphere is dissipated by means of thermal excitation through photons, Joule heating by ions and electrons, ionization, dissociation, and chemical reactions.

If one assumes that the instantaneous energy balance in the thermosphere is fully described by the temperature profile $T(z)$, then the diffusive equilibrium concentration profile $n_i(z)$ for the i-th constituent can be expressed as

$$0 = \frac{1}{n_i}\frac{\mathrm{d}n_i}{\mathrm{d}z} + \frac{1}{H_{n_i}} + \frac{1+\alpha_i}{T}\frac{\mathrm{d}T}{\mathrm{d}z} \tag{B.1}$$

where $i = 1, \ldots, 7$ for N_2, O, O_2, He, H, Ar, and N. The thermal diffusion coefficient α_i is defined as $\alpha_i = 0.0$ for N_2, O, O_2, Ar and N, and as $\alpha_i \approx -0.4$ for He and H. The quantity H_{n_i} is denoted as concentration scale height, or number density scale height of the i-th constituent. It is defined as

$$H_{n_i} = \frac{R\,T}{M_i\,g} \tag{B.2}$$

with the mole masses $M_i \approx 28, 16, 32, 4, 1, 40$ and 14 kg/kmol, the universal gas constant $R = 8,314.41$ J kmol^{-1} K^{-1}, and the local gravitational acceleration $g = g(z, \phi)$, which is a function of geodetic altitude z and latitude ϕ.

On the right-hand side of Eq. B.1 the first two terms represent a hydrostatic equilibrium for an isothermal atmosphere, with an exponential concentration decrease with altitude. The third term reflects contributions due to thermal diffusion, where light-weight species tend to separate from the heavier species in the direction of higher temperatures.

The solution of Eq. B.1 may be considerably simplified, if $g(z, \phi)$ could be replaced by a constant value g_o. This can be accomplished by a change from geodetic altitude z to a fictitious geopotential altitude $\hat{z}$.

$$\hat{z} = \frac{(z - z_o)\,(a_e + z_o)}{a_e + z} \tag{B.3}$$

where $z_o = 120$ km marks the begin of the thermosphere (corresponding to $\hat{z}_o = 0$). Substitution of z by this new variable $\hat{z}$ leads to

$$g_o\,\mathrm{d}\hat{z} \approx g(z, \phi)\,\mathrm{d}z \tag{B.4}$$

With this substitution Eq. B.1 can be solved for $n_i(z)$.

$$n_i\,(z) = n_i\,(z_o)\left(\frac{T\,(\hat{z}_o)}{T\,(\hat{z})}\right)^{1+\alpha_i} \exp\left(-\int_{\hat{z}_o}^{\hat{z}} \frac{M_i\,g_o}{R\,T\,(\hat{z})}\,\mathrm{d}\hat{z}\right) \tag{B.5}$$

In this equation the temperature profile $T\,(\hat{z})$ is still undefined. Measurements by in-situ sensors and ground-based incoherent scatter radars have shown that a good fit to $T\,(\hat{z})$ can be obtained by a temperature function according to Bates.

$$T\,(\hat{z}) = T_\infty\,[1 - \tau\,\exp\,(-\sigma\,\hat{z})] \tag{B.6}$$

where $\tau = 1 - T(z_o)/T_\infty$, with the exospheric temperature T_∞ and the homopause temperature $T(z_o)$. σ is a fitted temperature gradient parameter. Substitution of

Eq. B.6 into Eq. B.5, and an expansion in a truncated series of the "small" parameter τ leads to an analytical expression for the $i = 1, \ldots, 7$ concentration profiles according to Walker.

$$n_i\,(z) = n_i\,(z_o)\,\exp\left(-\sigma\gamma_i\hat{z}\right)\left[\frac{1-\tau}{1-\tau\,\exp\left(-\sigma\hat{z}\right)}\right]^{1+\alpha_i+\gamma_i} \tag{B.7}$$

with $\gamma_i = M_i\,g_o/(R\,T_\infty\,\sigma)$.

Once the partial concentrations are known, then the total density $\rho(z)$ can be determined from a weighted sum of the contributions by the individual species.

$$\rho\,(z) = \frac{1}{A}\sum_{i=1}^{7} n_i\,(z)\,M_i \tag{B.8}$$

where $A = 6.022169 \times 10^{26}\,\mathrm{kmol}^{-1}$ is Avogadro's constant. In the vicinity of a given geodetic altitude z_o, the density profile can be approximated by an exponential function, with a constant density scale height $H_{\rho,o} = H_\rho(z_o)$, representing a barometric equilibrium condition for an isothermal atmosphere.

$$\rho\,(z) \approx \rho\,(z_o)\,\exp\left(-\frac{z-z_o}{H_{\rho,o}}\right) \tag{B.9}$$

Appropriate values of $\rho(z_o)$ and $H_\rho(z_o)$ are listed in Tables B.1 to B.3 for LEO altitudes, and for low, medium, and high activity conditions.

From Eq. B.7 it becomes evident that the decrease in the logarithmic number density n_i for a given species i can be approximated by a linear function of the geopotential altitude $\hat{z}$, with a slope which is proportional to the inverse of the concentration scale height H_{n_i}.

$$\frac{d\left[\ln\left(n_i\right)\right]}{d\hat{z}} \propto -\frac{M_i}{T_\infty} \propto \frac{1}{H_{n_i}} \tag{B.10}$$

Hence, the heavier the species, and the smaller the exospheric temperature, the stronger the decrease in concentration with increasing altitude. Consequently, for given initial conditions at $z = 120$ km, the overlay of the individual concentration profiles leads to a dominance of N_2, followed by O, H, and He (see Tables B.1 to B.3 for low, mean, and high activities). The possible range of variations in local temperature $T(z)$ and local air density $\rho(z)$ is shown in Fig. B.1 and B.2 for CIRA-72, the COSPAR International Reference Atmosphere of 1972 (Stickland, 1972).

The thermodynamic state of the thermosphere is known to be a function of geodetic altitude z, local solar time $t_{LST} = (12^h/\pi)\,(\alpha - \alpha_\odot + \pi)$, universal time t_{UT}, day of the year t_d, geodetic latitude ϕ, mean solar activity $\overline{F}_{10.7}$, current solar activity $F_{10.7}$, and geomagnetic activity A_p (these activity parameters are discussed in more detail in Section B.3).

At a given altitude level the number densities n_i, total densities ρ, and temperatures T undergo periodic changes which are dominated by diurnal periods (fct($t_{LST}, \phi, ...$), see Fig. B.3 and B.4), and semi-annual periods (fct($t_d, \phi, ...$), see Fig. B.5 and B.6). Both of these are strongly latitude-dependent. The diurnal variations have maxima at $t_{LST} \approx 15^h$, following the sub-solar point with a delay of typically 3 hours. The semi-annual variations have peaks in early April and late October, and minima at the end of January and in late July. The mean values of n_i, ρ, and T at a given altitude, and the variation amplitudes across that level are largely driven by the solar activity (with proxies $\overline{F}_{10.7}$ and $F_{10.7}$), and by the geomagnetic activity (with proxies A_p, K_p, or a_p), as explained in Section B.3.

B.2 IMPLEMENTATIONS OF THERMOSPHERE MODELS

The history of models of the upper atmosphere reaches back to the first years of spaceflight. Early models were entirely based on total density data, derived from drag information which could be deduced from observed orbit decay rates (mainly via $\dot{a}$ and $\dot{T}$). The data were processed to establish mean, static density profiles with altitude (e.g. the US Standard Atmosphere 1962 and 1976 (anon., 1976)), and more sophisticated models with spatial and temporal variations as a function of solar and geomagnetic activity (e.g. the series of Jacchia models of 1960, 1965, 1971, and 1977). The latter models integrate the diffusion equations for the main thermospheric species according to Eq. B.1. The integration is performed by numerical quadrature for empirical temperature profiles. In the Jacchia models the density variations are mainly driven by T_∞, the limiting value of the temperature profile at exospheric altitudes, with T_∞ varying with z, t_{LST}, t_d, ϕ, $\overline{F}_{10.7}$, $F_{10.7}$, and K_p (Jacchia, 1971; Jacchia, 1977). The Jacchia-71 model was adopted as COSPAR International Reference Atmosphere, CIRA-72, in 1972 (Stickland, 1972).

In the USSR drag data from Cosmos satellites were processed to derive a series of Russian reference atmosphere models (anon., 1985). These GOST models use a strict separation of atmosphere parameters to determine total density as a product of individual effects due to diurnal variations (K_1), semi-annual variations (K_2), solar activity variations (K_3), and geomagnetic activity variations (K_4), which modulate a night-time reference density profile with altitude (ρ_n).

$$\rho = \rho_n\left(z, \overline{F}_{10.7}\right) \cdot K_1\left(t_{LST}, \phi\right) \cdot K_2\left(t_d, \phi\right) \cdot K_3\left(F_{10.7}\right) \cdot K_4\left(K_p\right) \qquad \text{(B.11)}$$

All of the functional correction terms on the right-hand side are provided as empirical expressions, with no direct correspondence to the physical principles reflected in Eq. B.1 to B.7.

Starting in the early 1970s, in-situ mass spectrometer data from atmospheric research satellites (e.g. AE-B, OGO 6, San Marco 3, Aeros, and AE-C) led to an improved understanding of the spatial and temporal variations of number densities of thermospheric gas species. Ground-based incoherent scatter data of atmospheric research radars (e.g. Millstone Hill, St. Santin, Arecibo, and Jicamarca) furthermore increased the knowledge of temperature distributions in the thermosphere. Based on such extended data the MSIS model series was developed (e.g. the Mass Spectrometer and Incoherent Scatter atmosphere models of 1977, 1983,

1986, and 1990). They use species-wise analytical temperature and concentration profiles according to Eq. B.6 and B.7 to derive total densities via Eq. B.8. Early versions of the MSIS model, and also the Density and Temperature Model, DTM-77 (Barlier et al., 1978), used the same set of model parameters as the Jacchia models, i.e. z, t_{LST}, t_d, ϕ, $\overline{F}_{10.7}$, $F_{10.7}$, and a daily A_p or 6-hourly k_p index. As of MSIS-86, the diurnal variability $\Delta\rho = \text{fct}(t_{LST}, \phi)$ was extended by a longitude dependency $\Delta\rho = \text{fct}(t_{UT}, \phi)$, as a function of universal time t_{UT}, to reflect the Earth-fixed magnetic field component of the variation. The thermospheric part of MSIS-86 (Hedin, 1987) was later adopted as CIRA-90 reference atmosphere. In 1990 the MSISe-90 model was extended into the homosphere, down to ground level, with smooth transitions at the turbopause (Hedin, 1991).

In spite of several decades of research effort, the model accuracy for total densities is in general on the order of $\Delta\rho_{1\sigma} \approx 15\%$, even for perfectly known input parameters. In phases of extreme solar or geomagnetic activities local density errors can temporarily exceed 100%. Even at mean activity levels differences between individual thermosphere models can be appreciable. This is particularly so at low altitudes of $120\,\text{km} \le z \le 200\,\text{km}$, when comparing drag-based models (e.g. CIRA-72), which integrate the diffusion equations (Eq. B.1) for empirical temperature profiles, with mass-spectrometer based models (e.g. MSIS-86), which use analytical approximations of the diffusion and temperature profiles (Eq. B.7). In the course of the MSIS model development the analytical profiles were adjusted to provide best fits at altitudes where most of the in-situ measurement data are concentrated. Since the shapes of the density profiles are fixed according to analytical expressions, the good fit at higher altitudes is achieved at the expense of larger model errors at low altitudes, where in-situ measurement data are sparse. As a consequence, the differences between CIRA-72 and MSIS-86 at low altitudes can reach 30% to 40%. From monitoring re-entry events one can conclude that drag-based models (e.g. CIRA-72 or GOST-85) are mostly superior in accuracy at low thermospheric altitudes, and should get preference for final orbital decay predictions. The advantages of both model types (drag based and mass-spectrometry based) can be combined in a hybrid model, for instance by an overlay of MSISe-90 and CIRA-72 density profiles, with altitude dependent weighting.

$$\rho = w_{\text{msis}}\,\rho_{\text{msis}} + (1 - w_{\text{msis}})\,\rho_{\text{cira}} \tag{B.12}$$

$$w_{\text{msis}}(z) = \begin{cases} \dfrac{1}{2}\cos\left(\pi\,\dfrac{z_{200} - z}{z_{200} - z_{120}}\right) + \dfrac{1}{2} & \text{for} \quad 120\,\text{km} \le z \le 200\,\text{km} \\[2ex] \dfrac{1}{2}\cos\left(\pi\,\dfrac{z - z_{90}}{z_{120} - z_{90}}\right) + \dfrac{1}{2} & \text{for} \quad 90\,\text{km} \le z < 120\,\text{km} \\[2ex] 1 & \text{for} \quad z < 90\,\text{km} \quad \text{and} \quad z > 200\,\text{km} \end{cases}$$

According to Eq. B.12 the MSISe-90 model is used above 200 km and below 90 km, the CIRA-72 model is used near 120 km (homopause), and a smooth fairing between the two density profiles is used at intermediate altitudes.

Based on the same principles as the MSIS models, a series of horizontal wind models (HWM-87, HWM-91, and HWM-93) was developed (Hedin et al., 1991). They provide the north and east components of horizontal wind vectors in the thermosphere for the same eight input parameters as used in MSIS-86 and CIRA-90. Fig. B.9 and B.10 show the diurnal and semi-annual variation of the wind pattern. For mean activity conditions, at 400 km altitude, the magnitude of the horizontal wind can reach 250 m/s. During high solar or geomagnetic activity thermospheric wind speeds can exceed 600 m/s, with maximum velocities across the poles, towards the night side. Due to increased viscosity with higher altitude, wind models are mainly important below the thermopause.

A comprehensive overview of thermospheric models is provided by (Marcos, 1993; Vallado, 1997; Montenbruck and Gill, 2000).

B.3 SOLAR AND GEOMAGNETIC ACTIVITIES

Variations in temperature, density, and composition of the Earth thermosphere are mainly driven by changes in solar activity. Active regions on the Sun are manifested by Sun spots. These are holes in the photosphere, up to 10,000 km deep, with temperatures of about 4,000 K (~2,000 K lower than in the photosphere), and with sizes of up to 35 times the Earth surface. The formation of Sun spots can be correlated with solar cycles of ~11 years length. At the beginning of a cycle Sun spots start forming near ±35° solar latitude, symmetric to the Sun's equator, which is inclined to the ecliptic by 7°15′. Due to varying rotation periods between 30 days at high latitudes and 25 days near the equator, the Sun's magnetic field lines tend to wrap around small circles of latitude. Sun spots generally occur in pairs when magnetic field lines leave the photosphere in a loop, with surface entry and exit points of clearly discernible polarities, which are opposite on the north and south hemisphere, and which are also alternating between successive solar cycles on the same hemisphere (see Fig. B.8, left and top). Hence, one often refers to a 22-year solar cycle. The lifetime of Sun spot pairs can range from a few days to a few weeks. During this timespan they drift apart by up to 150,000 km before the magnetic loop submerges again in the photosphere. With progressing time in the solar cycle the number of Sun spots increases as they migrate towards the equator. The activity reaches a maximum after ~4 years, when Sun spots concentrate around ±25° latitude. The generation of Sun spots, and hence the solar cycle ends when reaching latitudes of about ±10°. At this time a new cycle starts, with activity regions developing again near ±35°, with an inverse polarity. When plotting the Sun spot coverage of the solar disk versus latitude and time, so-called butterfly diagrams are produced (see Fig. B.8, right bottom). At peak activities up to 1% of the solar disk are covered with Sun spots.

Sun spot areas emit large amounts of energy in the EUV and X-ray frequency bands. When reaching the Earth, particularly the EUV radiation is absorbed by different physical and chemical processes in the thermosphere, causing strong temperature variations and corresponding density changes. A proxy for the absorbed EUV radiation is the solar flux index $F_{10.7}$ for emissions from the Sun spots in the 10.7 cm wavelength. These are well correlated with the EUV radiation, and can be measured on the ground through the radio frequency window of the Earth atmosphere. $F_{10.7}$ is measured in solar flux units (SFU) or Jansky units (Jy), where $1\,\mathrm{SFU} = 10^4\,\mathrm{Jy} = 10^{-22}\,\mathrm{W/(m^2\,Hz)}$. Since the atmosphere reacts to the incoming energy with some delay, most thermosphere models use $F_{10.7}$ data from $t-1\,\mathrm{d}$ for predictions at time t. Moreover, all models use a mean value $\overline{F}_{10.7}$, which is either a moving average over the preceeding 3 solar rotations (81 days), or a centered average over 6 solar rotations (162 days). Fig. B.7 (top) shows $F_{10.7}$ and $\overline{F}_{10.7}$ for solar cycles 21 and 22. The center diagram shows the corresponding Zürich Sun spot numbers R which are determined on the basis of the observable number of spots f and groups g, with a correction factor k to account for observation conditions.

$$R = k\,(f + 10\,g) \tag{B.13}$$

The proxies R and $F_{10.7}$ can be cross-converted with good accuracy by a linear relationship ($\Delta F_{10.7,1\sigma} \approx 5$ for solar cycles 20 and 21).

$$F_{10.7} \approx 0.9\,R + 59.6 \tag{B.14}$$

This conversion allows one to use measurements of Sun spot numbers which reach back some 250 years. Using this extended database together with genuine $F_{10.7}$ data (which are available since 1948) improves the reliability of long-term forecasts. ESA produces long-term forecasts of monthly mean $\overline{F}_{10.7}$ values on the basis of the McNish & Lincoln technique (McNish and Lincoln, 1949; Kerridge et al., 1989), where a mean solar cycle is established from historic data, and modulations for predicted cycles are forecast from offsets of recent data with respect to the mean cycle. This technique, which is also adopted by NOAA and NASA, is mainly applicable to the next 11 or 22 years. Further into the future, a replicate of the mean cycle is reproduced. For short-term forecasts, spanning the period of about one solar rotation, ESA uses an ARIMA[B.1] time-series model to predict daily $F_{10.7}$ values for the next 27 days, with model coefficients derived from solar flux data of the last 24 months (Thomson et al., 1992).

Sun spot regions not only emit electromagnetic radiation, but also particles in the form of solar wind (~96% H^+ and ~4% He^{++}). When reaching the Earth, this high-energy plasma interacts with the magnetic field, causing increases in the horizontal component of the magnetic induction $B_{m,\ominus}$ (see Eq. 6.12) by up to 400 nT ($400 \times 10^{-9}\,\mathrm{V\,s/m^2}$). This peak excursion marks the range of possible values for the linear geomagnetic index $A_p \in [0, 400]$, which maps onto equivalent quasi-logarithmic indices $K_p \in [0, 9]$ according to Table B.4. Apart from the daily mean indices A_p and K_p, there are corresponding 3-hourly data a_p and k_p. Fig. B.7 (bottom) shows the A_p history for solar cycles 20 and 21. Due to the highly stochastic nature of geomagnetic activities, their prediction is only possible with some confidence for a single solar rotation period of 27 days (Thomson et al., 1992).

[B.1]auto-regressive integrated moving average

Table B.1. Atmospheric properties versus geodetic altitude z according to MSISe–90 for low solar and geomagnetic activities ($\overline{F}_{10.7} = F_{10.7} = 70$, $A_p = 0$), at reference conditions with $t_d = 152$ d, $t_{UT} = 12$ h, $t_{LST} = 11$ h, and $\phi = 0°$.

z [km]	T [K]	ρ [kg/m³]	n_{N_2} [1/m³]	n_O [1/m³]	n_{O_2} [1/m³]	n_{He} [1/m³]	n_{Ar} [1/m³]	n_H [1/m³]	n_N [1/m³]	$\overline{M}_{mol}$ [kg/kmol]	H_ρ [km]
100.0	174.5	6.57e–07	1.08e+19	4.28e+17	2.55e+18	1.32e+14	1.00e+17	3.01e+13	3.87e+11	28.45	5.32
125.0	413.3	9.79e–09	1.62e+17	4.64e+16	1.89e+16	2.32e+13	5.06e+14	4.19e+12	1.79e+12	25.91	9.78
150.0	606.0	1.57e–09	2.40e+16	1.31e+16	2.03e+15	1.13e+13	3.52e+13	1.22e+12	1.01e+13	24.19	17.88
175.0	689.0	4.68e–10	6.34e+15	5.59e+15	4.52e+14	8.74e+12	5.31e+12	6.26e+11	2.12e+13	22.71	23.31
200.0	724.9	1.74e–10	2.02e+15	2.76e+15	1.23e+14	7.25e+12	1.04e+12	4.61e+11	2.19e+13	21.29	27.73
225.0	740.6	7.29e–11	6.90e+14	1.45e+15	3.62e+13	6.15e+12	2.26e+11	4.01e+11	1.58e+13	19.99	30.27
250.0	747.4	3.31e–11	2.45e+14	7.86e+14	1.11e+13	5.28e+12	5.16e+10	3.73e+11	9.83e+12	18.86	33.11
275.0	750.4	1.60e–11	8.85e+13	4.34e+14	3.47e+12	4.56e+12	1.21e+10	3.55e+11	5.81e+12	17.95	35.68
300.0	751.8	8.10e–12	3.24e+13	2.42e+14	1.10e+12	3.95e+12	2.88e+09	3.41e+11	3.39e+12	17.23	38.01
325.0	752.4	4.27e–12	1.20e+13	1.37e+14	3.53e+11	3.42e+12	6.95e+08	3.28e+11	2.01e+12	16.64	40.09
350.0	752.7	2.32e–12	4.46e+12	7.75e+13	1.14e+11	2.97e+12	1.70e+08	3.17e+11	1.20e+12	16.15	41.76
375.0	752.8	1.29e–12	1.68e+12	4.43e+13	3.74e+10	2.58e+12	4.19e+07	3.06e+11	7.22e+11	15.67	43.22
400.0	752.8	7.30e–13	6.35e+11	2.54e+13	1.23e+10	2.25e+12	1.05e+07	2.95e+11	4.39e+11	15.16	44.58
425.0	752.9	4.19e–13	2.42e+11	1.46e+13	4.09e+09	1.96e+12	2.64e+06	2.85e+11	2.69e+11	14.54	45.58
450.0	752.9	2.45e–13	9.30e+10	8.46e+12	1.37e+09	1.71e+12	6.73e+05	2.76e+11	1.66e+11	13.77	27.98
475.0	752.9	1.45e–13	3.60e+10	4.91e+12	4.63e+08	1.49e+12	1.73e+05	2.67e+11	1.02e+11	12.82	48.42
500.0	752.9	8.74e–14	1.40e+10	2.87e+12	1.57e+08	1.30e+12	4.50e+04	2.58e+11	6.38e+10	11.68	50.22
525.0	752.9	5.38e–14	5.49e+09	1.68e+12	5.40e+07	1.14e+12	1.18e+04	2.49e+11	3.99e+10	10.40	52.82
550.0	752.9	3.39e–14	2.17e+09	9.87e+11	1.87e+07	9.99e+11	3.13e+03	2.41e+11	2.51e+10	9.07	56.09
575.0	752.9	2.21e–14	8.62e+08	5.83e+11	6.50e+06	8.76e+11	8.38e+02	2.33e+11	1.58e+10	7.79	60.58
600.0	752.9	1.49e–14	3.45e+08	3.45e+11	2.28e+06	7.69e+11	2.27e+02	2.26e+11	1.00e+10	6.65	66.46
625.0	752.9	1.05e–14	1.39e+08	2.05e+11	8.08e+05	6.75e+11	6.18e+01	2.19e+11	6.35e+09	5.70	74.86
650.0	752.9	7.64e–15	5.63e+07	1.23e+11	2.88e+05	5.93e+11	1.70e+01	2.12e+11	4.04e+09	4.94	85.39
675.0	752.9	5.82e–15	2.30e+07	7.35e+10	1.03e+05	5.22e+11	4.73e+00	2.05e+11	2.58e+09	4.36	98.43
700.0	752.9	4.59e–15	9.43e+06	4.42e+10	3.74e+04	4.60e+11	1.33e+00	1.99e+11	1.66e+09	3.93	113.79

Table B.2. Atmospheric properties versus geodetic altitude z according to MSISe–90 for mean solar and geomagnetic activities ($\overline{F}_{10.7} = F_{10.7} = 140$, $A_p = 15$), at reference conditions with $t_d = 152$ d, $t_{UT} = 12$ h, $t_{LST} = 11$ h, and $\phi = 0°$.

z [km]	T [K]	ρ [kg/m^3]	n_{N_2} [1/m^3]	n_O [1/m^3]	n_{O_2} [1/m^3]	n_{He} [1/m^3]	n_{Ar} [1/m^3]	n_H [1/m^3]	n_N [1/m^3]	$\overline{M}_{mol}$ [kg/kmol]	H_ρ [km]
100.0	190.9	5.80e–07	9.59e+18	4.24e+17	2.21e+18	1.17e+14	9.02e+16	2.01e+13	5.31e+11	28.39	5.74
125.0	449.2	1.01e–08	1.67e+17	5.33e+16	1.79e+16	2.42e+13	5.72e+14	2.22e+12	3.03e+12	25.64	9.95
150.0	725.5	1.79e–09	2.70e+16	1.57e+16	2.12e+15	1.44e+13	4.89e+13	4.96e+11	1.86e+13	24.00	19.69
175.0	880.6	6.20e–10	8.45e+15	7.33e+15	5.73e+14	1.11e+13	9.75e+12	2.17e+11	4.43e+13	22.74	27.65
200.0	968.1	2.75e–10	3.33e+15	4.09e+15	2.00e+14	9.30e+12	2.66e+12	1.49e+11	5.27e+13	21.60	36.08
225.0	1017.7	1.39e–10	1.46e+15	2.48e+15	7.86e+13	8.08e+12	8.34e+11	1.26e+11	4.43e+13	20.57	39.09
250.0	1045.9	7.58e–11	6.77e+14	1.57e+15	3.28e+13	7.15e+12	2.82e+11	1.17e+11	3.21e+13	19.66	43.35
275.0	1062.0	4.36e–11	3.24e+14	1.02e+15	1.42e+13	6.39e+12	9.92e+10	1.11e+11	2.21e+13	18.88	46.97
300.0	1071.3	2.60e–11	1.59e+14	6.75e+14	6.28e+12	5.75e+12	3.59e+10	1.07e+11	1.51e+13	18.21	50.26
325.0	1076.6	1.61e–11	7.87e+13	4.51e+14	2.82e+12	5.19e+12	1.32e+10	1.04e+11	1.04e+13	17.65	53.22
350.0	1079.7	1.02e–11	3.94e+13	3.03e+14	1.28e+12	4.69e+12	4.92e+09	1.01e+11	7.23e+12	17.18	55.82
375.0	1081.5	6.55e–12	1.99e+13	2.05e+14	5.86e+11	4.25e+12	1.86e+09	9.88e+10	5.06e+12	16.79	58.12
400.0	1082.5	4.29e–12	1.01e+13	1.39e+14	2.71e+11	3.86e+12	7.06e+08	9.64e+10	3.58e+12	16.45	60.08
425.0	1083.2	2.85e–12	5.18e+12	9.49e+13	1.26e+11	3.51e+12	2.71e+08	9.41e+10	2.54e+12	16.15	61.66
450.0	1083.5	1.91e–12	2.66e+12	6.49e+13	5.88e+10	3.19e+12	1.05e+08	9.19e+10	1.81e+12	15.86	62.98
475.0	1083.7	1.30e–12	1.37e+12	4.45e+13	2.76e+10	2.90e+12	4.08e+07	8.97e+10	1.29e+12	15.57	64.78
500.0	1083.9	8.86e–13	7.14e+11	3.06e+13	1.31e+10	2.64e+12	1.60e+07	8.76e+10	9.32e+11	15.25	66.05
525.0	1083.9	6.10e–13	3.73e+11	2.11e+13	6.22e+09	2.41e+12	6.32e+06	8.56e+10	6.73e+11	14.91	67.67
550.0	1084.0	4.23e–13	1.95e+11	1.46e+13	2.97e+09	2.19e+12	2.51e+06	8.37e+10	4.88e+11	14.51	68.96
575.0	1084.0	2.95e–13	1.03e+11	1.01e+13	1.43e+09	2.00e+12	1.01e+06	8.18e+10	3.54e+11	14.05	70.22
600.0	1084.0	2.08e–13	5.45e+10	7.03e+12	6.91e+08	1.83e+12	4.05e+05	7.99e+10	2.57e+11	13.52	71.49
625.0	1084.0	1.47e–13	2.90e+10	4.90e+12	3.36e+08	1.67e+12	1.64e+05	7.81e+10	1.88e+11	12.91	73.53
650.0	1084.0	1.05e–13	1.55e+10	3.43e+12	1.64e+08	1.53e+12	6.71e+04	7.64e+10	1.37e+11	12.23	75.43
675.0	1084.0	7.59e–14	8.30e+09	2.40e+12	8.05e+07	1.40e+12	2.76e+04	7.47e+10	1.00e+11	11.48	77.79
700.0	1084.0	5.53e–14	4.48e+09	1.69e+12	3.97e+07	1.28e+12	1.14e+04	7.31e+10	7.38e+10	10.69	80.53

Table B.3. Atmospheric properties versus geodetic altitude z according to MSISe–90 for high solar and geomagnetic activities ($\overline{F}_{10.7} = F_{10.7} = 380$, $A_p = 300$), at reference conditions with $t_d = 152$ d, $t_{UT} = 12$ h, $t_{LST} = 11$ h, and $\phi = 0°$.

z [km]	T [K]	ρ [kg/m^3]	n_{N_2} [1/m^3]	n_O [1/m^3]	n_{O_2} [1/m^3]	n_{He} [1/m^3]	n_{Ar} [1/m^3]	n_H [1/m^3]	n_N [1/m^3]	$\overline{M}_{mol}$ [kg/kmol]	H_ρ [km]
100.0	315.7	3.02e–07	5.00e+18	3.22e+17	1.08e+18	6.12e+13	5.10e+16	4.01e+12	1.33e+12	28.17	8.17
125.0	572.5	1.18e–08	1.87e+17	8.50e+16	1.50e+16	2.62e+13	9.35e+14	2.36e+11	2.01e+13	24.70	10.83
150.0	1033.5	2.72e–09	3.84e+16	3.01e+16	2.20e+15	3.37e+13	1.37e+14	2.39e+10	1.81e+14	23.02	25.31
175.0	1312.9	1.24e–09	1.56e+16	1.73e+16	8.04e+14	2.66e+13	4.33e+13	6.72e+09	6.24e+14	21.79	39.93
200.0	1482.9	7.27e–10	7.92e+15	1.18e+16	3.79e+14	2.28e+13	1.76e+13	3.81e+09	1.02e+15	20.69	67.05
225.0	1586.8	4.75e–10	4.49e+15	8.62e+15	2.00e+14	2.03e+13	8.05e+12	3.01e+09	1.12e+15	19.79	63.83
250.0	1650.4	3.28e–10	2.69e+15	6.52e+15	1.12e+14	1.86e+13	3.96e+12	2.72e+09	1.02e+15	19.08	71.18
275.0	1689.6	2.34e–10	1.67e+15	5.01e+15	6.51e+13	1.71e+13	2.02e+12	2.58e+09	8.50e+14	18.52	76.72
300.0	1713.7	1.70e–10	1.06e+15	3.87e+15	3.87e+13	1.60e+13	1.06e+12	2.50e+09	6.89e+14	18.07	81.21
325.0	1728.7	1.26e–10	6.78e+14	3.02e+15	2.34e+13	1.49e+13	5.64e+11	2.44e+09	5.55e+14	17.69	85.04
350.0	1738.0	9.46e–11	4.40e+14	2.37e+15	1.42e+13	1.40e+13	3.04e+11	2.39e+09	4.46e+14	17.36	88.31
375.0	1743.9	7.16e–11	2.87e+14	1.86e+15	8.75e+12	1.31e+13	1.66e+11	2.35e+09	3.59e+14	17.07	91.45
400.0	1747.5	5.48e–11	1.88e+14	1.46e+15	5.41e+12	1.24e+13	9.09e+10	2.31e+09	2.90e+14	16.83	93.93
425.0	1749.8	4.22e–11	1.24e+14	1.16e+15	3.36e+12	1.16e+13	5.02e+10	2.28e+09	2.35e+14	16.61	96.33
450.0	1751.2	3.27e–11	8.23e+13	9.14e+14	2.10e+12	1.10e+13	2.79e+10	2.24e+09	1.91e+14	16.42	99.22
475.0	1752.1	2.55e–11	5.47e+13	7.23e+14	1.32e+12	1.03e+13	1.55e+10	2.21e+09	1.55e+14	16.26	101.24
500.0	1752.7	2.00e–11	3.64e+13	5.74e+14	8.28e+11	9.76e+12	8.71e+09	2.18e+09	1.27e+14	16.11	103.31
525.0	1753.1	1.57e–11	2.44e+13	4.56e+14	5.23e+11	9.22e+12	4.90e+09	2.15e+09	1.04e+14	15.97	105.81
550.0	1753.3	1.24e–11	1.64e+13	3.63e+14	3.31e+11	8.71e+12	2.77e+09	2.12e+09	8.50e+13	15.85	107.48
575.0	1753.5	9.88e–12	1.10e+13	2.89e+14	2.11e+11	8.23e+12	1.57e+09	2.09e+09	6.97e+13	15.73	109.01
600.0	1753.6	7.87e–12	7.42e+12	2.31e+14	1.34e+11	7.78e+12	8.97e+08	2.06e+09	5.73e+13	15.62	110.35
625.0	1753.6	6.29e–12	5.02e+12	1.85e+14	8.60e+10	7.35e+12	5.14e+08	2.03e+09	4.71e+13	15.51	112.27
650.0	1753.7	5.04e–12	3.41e+12	1.48e+14	5.52e+10	6.96e+12	2.95e+08	2.00e+09	3.88e+13	15.40	113.72
675.0	1753.7	4.05e–12	2.32e+12	1.19e+14	3.56e+10	6.59e+12	1.70e+08	1.97e+09	3.20e+13	15.28	115.19
700.0	1753.7	3.27e–12	1.58e+12	9.55e+13	2.30e+10	6.24e+12	9.87e+07	1.95e+09	2.64e+13	15.17	116.75

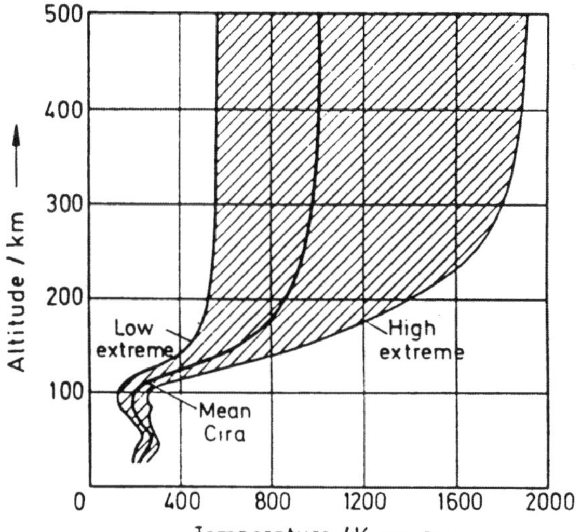

Fig. B.1. Range of CIRA-72 temperature variations with geodetic altitude for low, mean, and high solar and geomagnetic activities (Stickland, 1972).

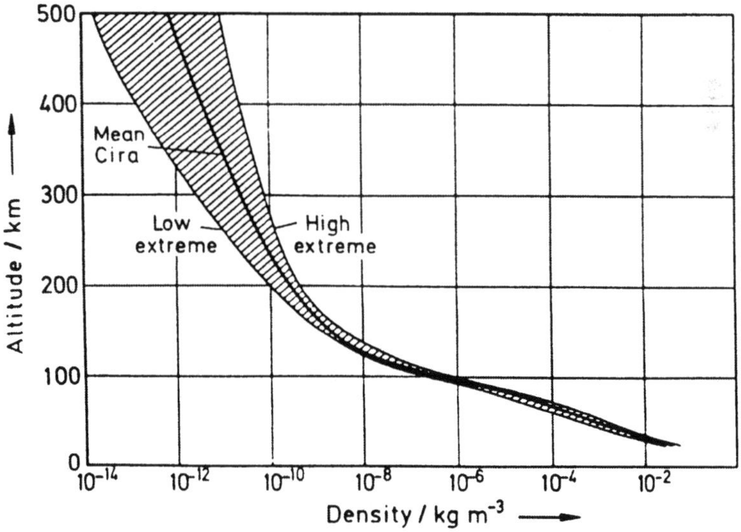

Fig. B.2. Range of CIRA-72 air density variations with geodetic altitude for low, mean, and high solar and geomagnetic activities (Stickland, 1972).

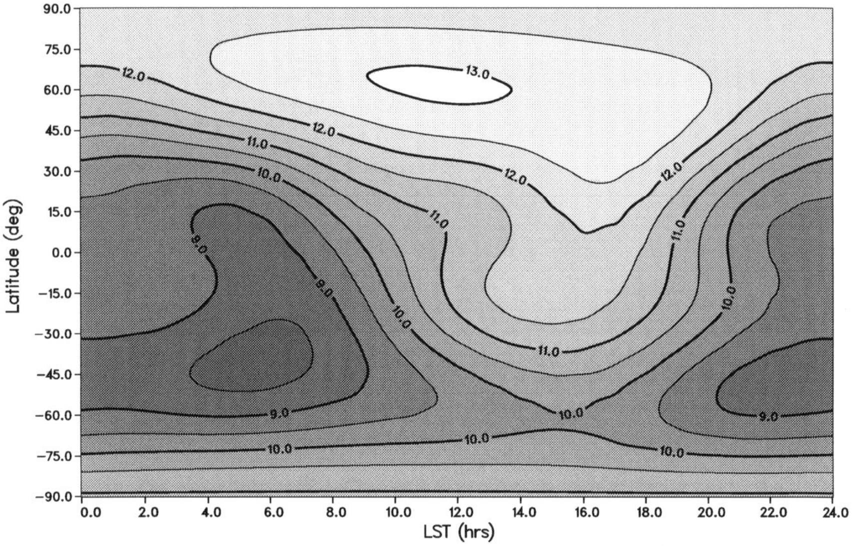

Fig. B.3. Diurnal temperature variation $T(t_{LST}, \phi)$ according to MSISe-90 for mean activity conditions ($\overline{F}_{10.7} = F_{10.7} = 140$, $A_p = 15$) and reference conditions ($z = 400$ km, $t_d = 152$ d, $t_{UT} = 12$ h). Level lines are in units of 100 K.

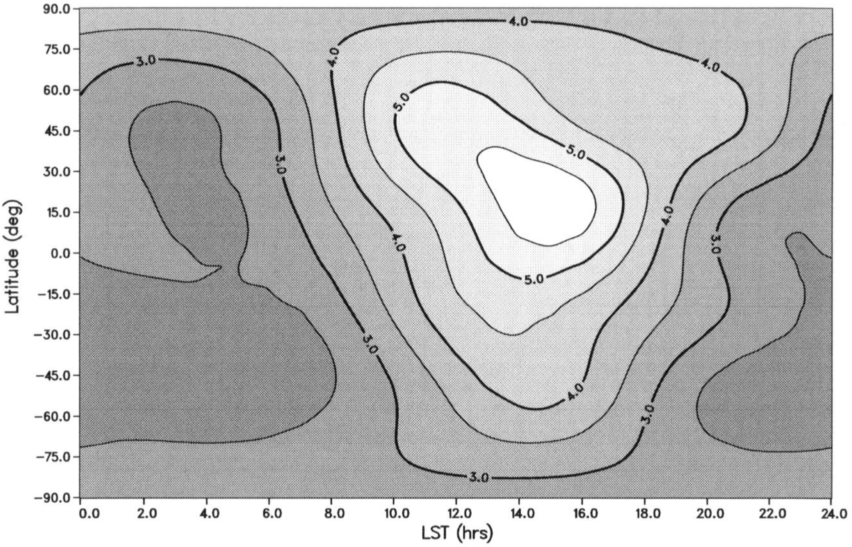

Fig. B.4. Diurnal air density variation $\rho(t_{LST}, \phi)$ according to MSISe-90 for mean activity conditions ($\overline{F}_{10.7} = F_{10.7} = 140$, $A_p = 15$) and reference conditions ($z = 400$ km, $t_d = 152$ d, $t_{UT} = 12$ h). Level lines are in units of 10^{-12} kg/m^3.

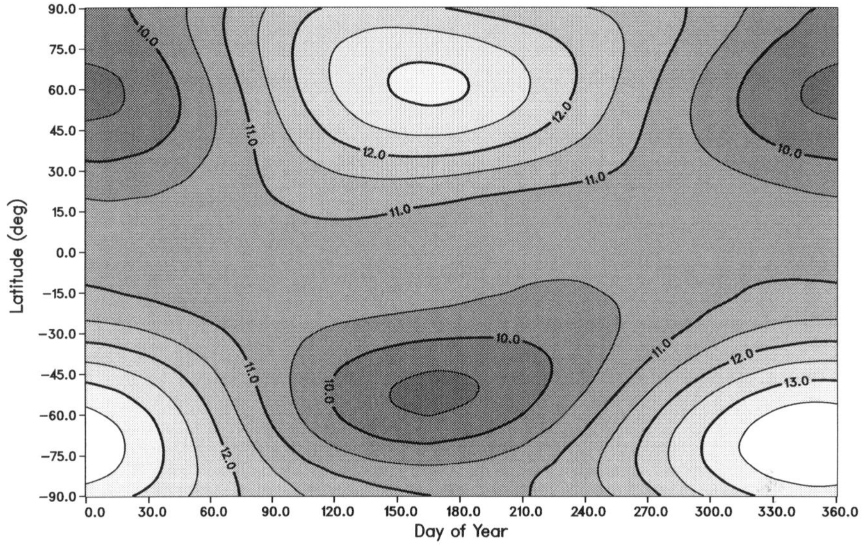

Fig. B.5. Semi-annual temperature variation $T(t_d, \phi)$ according to MSISe-90 for mean activity conditions ($\overline{F}_{10.7} = F_{10.7} = 140$, $A_p = 15$) and reference conditions ($z = 400$ km, $t_{LST} = 11$ h, $t_{UT} = 12$ h). Level lines are in units of 100 K.

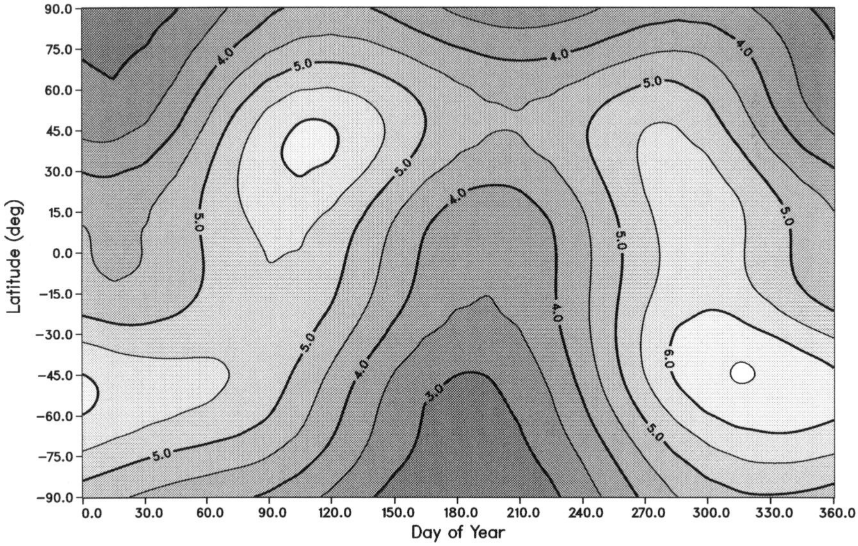

Fig. B.6. Semi-annual air density variation $\rho(t_d, \phi)$ according to MSISe-90 for mean activity conditions ($\overline{F}_{10.7} = F_{10.7} = 140$, $A_p = 15$) and reference conditions ($z = 400$ km, $t_{LST} = 11$ h, $t_{UT} = 12$ h). Level lines are in units of 10^{-12} kg/m^3.

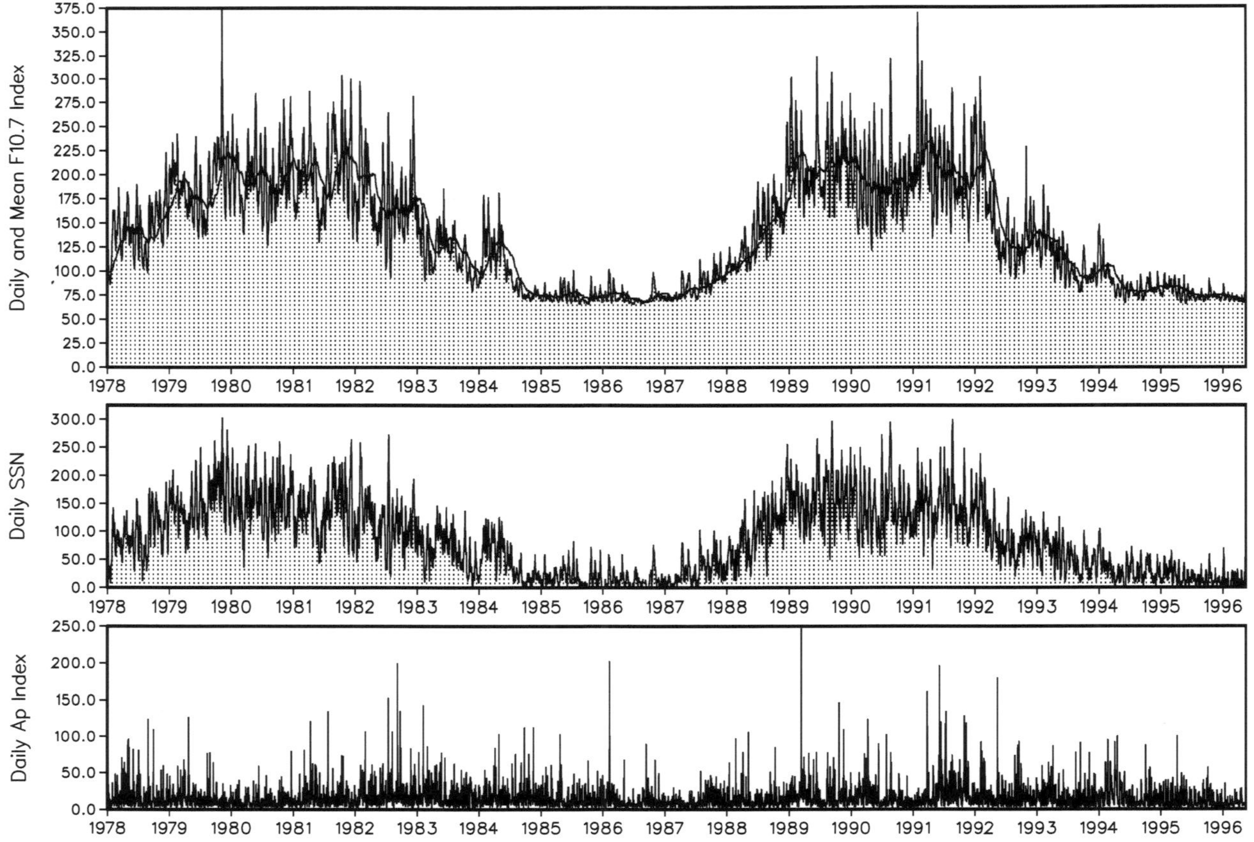

Fig. B.7. Solar and geomagnetic activity measurement data of solar cycle 21 and 22 (top: daily $F_{10.7}$ value in units of 10^{-22} W m^{-2} Hz^{-1}, and its moving average $\overline{F}_{10.7}$ over 81 days; center: daily Sun spot number SSN; bottom: planetary geomagnetic index A_p).

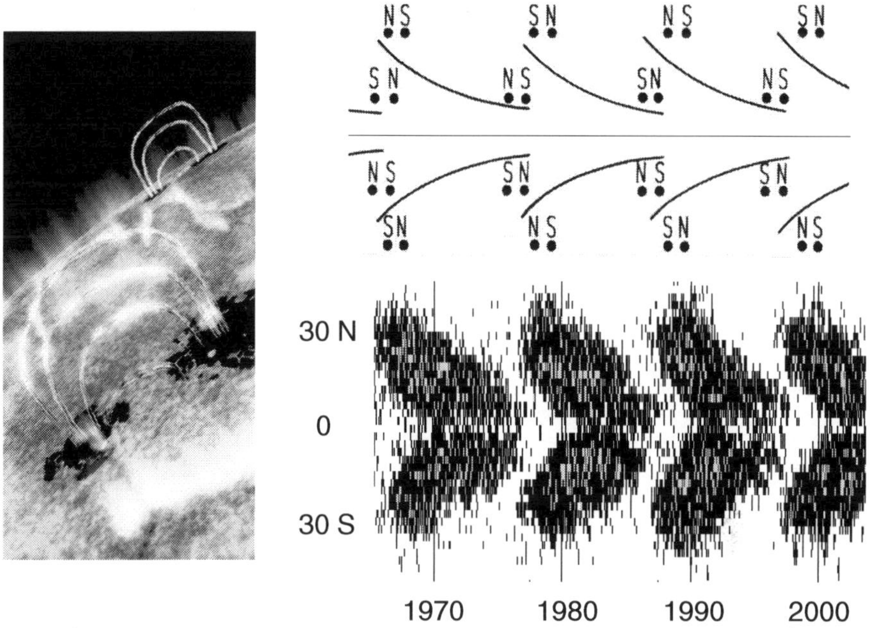

Fig. B.8. Evolution of the Sun spot distributions during solar cycles 20 to 23 (left: evolution of bi-polar Sun spots where magnetic field line loops penetrate the photosphere and chromosphere; right/bottom: butterfly diagrams of Sun spot regions during solar cycles 20 to 23; right/top: change of the polarization of Sun spot groups during successive solar cycles; see color plate on page 385).

Table B.4. Correspondence between the quasi-logarithmic geomagnetic index K_p and the linear geomagnetic index A_p (Rawer, 1971).

K_p	0o	0+	1−	1o	1+	2−	2o	2+	3−	3o
A_p	0	2	3	4	5	6	7	9	12	15
K_p	3+	4−	4o	4+	5−	5o	5+	6−	6o	6+
A_p	18	22	27	32	39	48	56	67	80	94
K_p	7−	7o	7+	8−	8o	8+	9−	9o		
A_p	111	132	154	179	207	236	300	400		

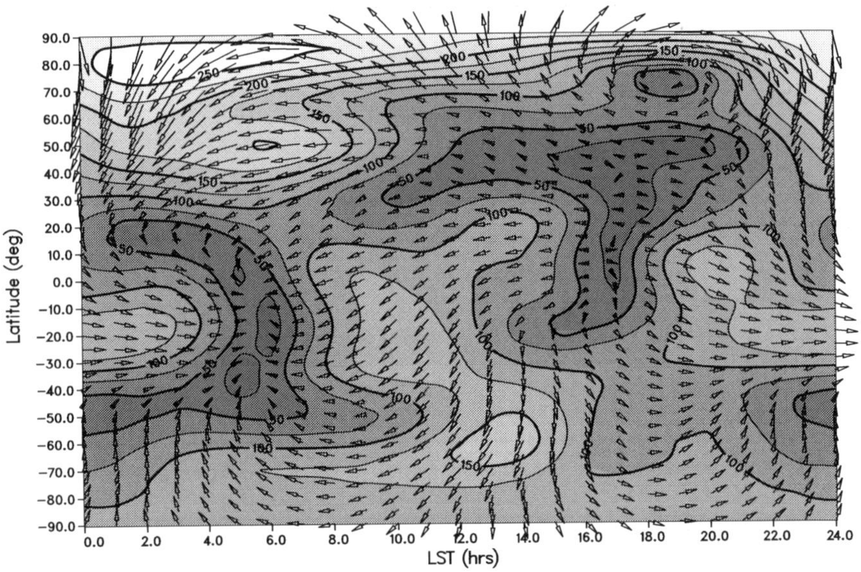

Fig. B.9. Diurnal horizontal wind variation $\underline{v}_w(t_{LST}, \phi)$ according to the HWM-93 model for mean activity conditions ($\overline{F}_{10.7} = F_{10.7} = 140$, $A_p = 15$) and reference conditions ($z = 400$ km, $t_d = 152$ d, $t_{UT} = 12$ h). Level lines are in units of 25 m/s.

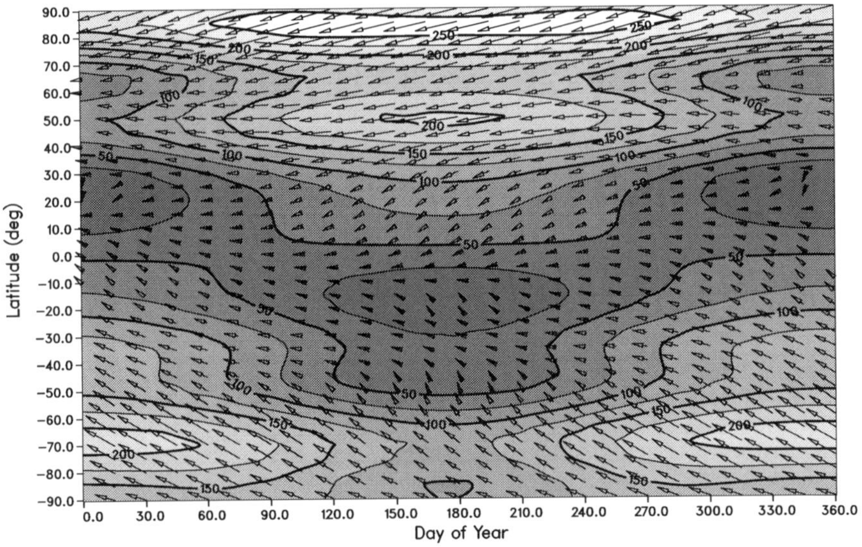

Fig. B.10. Semi-annual horizontal wind variation $\underline{v}_w(t_d, \phi)$ according to the HWM-93 model for mean activity conditions ($\overline{F}_{10.7} = F_{10.7} = 140$, $A_p = 15$) and reference conditions ($z = 400$ km, $t_{LST} = 6$ h, $t_{UT} = 12$ h). Level lines are in units of 25 m/s.

B.4 REFERENCES

anon. (1976). U.S. Standard Atmosphere 1976. Technical Report NOAA-S/T 76-1562, NOAA/NASA/USAF, Washington, DC.

anon. (1985). Earth Upper Atmosphere – Density Model for Ballistic Support of Flights of Artificial Satellites. Technical Report GOST 25645.115-84, USSR State Standard, Moscow.

Barlier, F., Berger, C., Falin, L., Kockarts, G., and Tuillier, G. (1978). A Thermospheric Model Based on Satellite Drag Data. *Annales de Géophysique*, 34(1):9–24.

Hedin, A. (1987). MSIS-86 Thermospheric Model. *Journal of Geophysical Research*, vol.92, no.A5:4649–4662.

Hedin, A. (1991). Extension of the MSIS Thermosphere Model into the Lower Atmosphere. *Journal of Geophysical Research*, vol.96, no.A2:1159–1172.

Hedin, A., Bondi, M., Burnside, R., Hernandez, G., Johnson, R., Killeen, T., Mazaudier, C., Meriwether, J., Salah, J., Sica, R., Smith, R., Spencer, N., Wickwar, V., and Virdi, T. (1991). Revised Model of the Thermosphere Winds Using Satellite and Ground-Based Observations. *Journal of Geophysical Research*, vol.96, no.A5:7657–7688.

Jacchia, L. (1971). Revised Static Models of the Thermosphere and Exosphere with Empirical Temperature Profiles. Technical Report SAO Special Report 332, Smithsonian Astrophysical Observatory, Cambridge, MA.

Jacchia, L. (1977). Thermospheric Temperature, Density and Composition. Technical Report SAO Special Report 375, Smithsonian Astrophysical Observatory, Cambridge, MA.

Kerridge, D., Carlaw, V., and Beamish, D. (1989). Development and Testing of Computer Algorithms for Solar and Geomagnetic Activity Forecasting. Technical Report WM/89/22C [ESA CR(P) 3039], British Geological Survey.

Marcos, F. (1993). Satellite Drag Models – Current Status and Prospects. *AAS/AIAA Conference*, paper AAS-93-621.

McNish, A. and Lincoln, J. (1949). Prediction of Sunspot Numbers. *Transactions of the American Geophysical Union*, vol.30:673–685.

Montenbruck, O. and Gill, E. (2000). *Satellite Orbits – Models, Methods, and Applications*. Springer, Berlin, Heidelberg, New York.

Rawer, K., editor (1971). *Encyclopedia of Physics*, volume XLIX/3 of *Geophysics III*. Springer Verlag, Berlin, Heidelberg, New York.

Stickland, A., editor (1972). *COSPAR International Reference Atmosphere 1972.* Akademie Verlag, Berlin.

Thomson, A., Clark, T., and Kerridge, D. (1992). Computer Algorithms and FORTRAN Programs for Forecasting Solar and Geomagnetic Activity in the Short Term. Technical Report WM/92/19C, British Geological Survey.

Vallado, D. (1997). *Fundamentals of Astrodynamics and Applications.* McGraw-Hill, New York, NY.

C

The Gravitational Potential of the Earth

H. Klinkrad

C.1 MATHEMATICAL FORMULATION OF THE GEOPOTENTIAL

The Earth can be considered as an elastic body, with a surface which takes the shape of an oblate ellipsoid, under the balancing forces of the central attraction and centrifugal acceleration, as a function of the geodetic latitude (see, for instance, (Beutler, 2005b)). Let a_e and b_e be the semimajor and semiminor axis of the rotationally symmetric Earth ellipsoid which has a sidereal rotation rate of ω_e. Furthermore, let M_e be the mass and μ be the central attraction term of an elastic Earth, then the shape of the equilibrium surface which is tangent to the mean equatorial radius $r_e(\phi = 0°) = a_e$ can be described as

$$\frac{a_e - b_e}{a_e} = \frac{3}{2}\frac{I_{ZZ} - I_{XX}}{M_e\, a_e^2} + \frac{1}{2}\frac{\omega_e^2\, a_e^3}{\mu} \approx \frac{1}{297.6} \tag{C.1}$$

where I_{ZZ} and $I_{XX} \stackrel{!}{=} I_{YY}$ are the moments of inertia with respect to the axes $\underline{Z}$ and $\underline{X}$ in Fig. A.2. This mathematically derived oblateness of $1/297.6$ matches the best estimate of $f_e = 298.257$ to within 0.22% (see Section A.3). With a similar accuracy the first term on the right-hand side of Eq. C.1 defines the second zonal harmonic J_2 of the geopotential.

$$\frac{I_{ZZ} - I_{XX}}{M_e\, a_e^2} \approx J_2 \tag{C.2}$$

This oblateness term J_2 (see Fig. C.1(a)) is by far dominating the perturbations on Earth orbits. Its perturbing force is about a factor of 550 smaller than the central attraction, but a factor of 600 larger than all other perturbations in a LEO environment.

In a more general form the gravity potential $U(r, \phi, \lambda)$ of the Earth can be expressed by means spherical harmonic expansions, as a function of spherical position coordinates r, ϕ, and λ in an Earth-fixed system (Seidelmann et al., 1992;

Montenbruck and Gill, 2000).

$$U(r,\phi,\lambda) = \frac{\mu}{r} \sum_{n=0}^{\infty} \left(\frac{a_e}{r}\right)^n \sum_{m=0}^{n} P_{n,m}(\sin\phi) \left[C_{n,m} \cos(m\lambda) + S_{n,m} \sin(m\lambda)\right] \tag{C.3}$$

By convention, the central attraction term for a spherical Earth of uniform mass distribution is $C_{0,0} = 1$. If the center of mass coincides with the origin of the Earth-fixed coordinate system, then $C_{1,0} = C_{1,1} = S_{1,1} = 0$. For the stronger assumption that the Earth-fixed coordinate axes coincide with the main axes of inertia, one would also get $C_{2,1} = S_{2,1} = S_{2,2} = 0$ (Beutler, 2005a).

In Eq. C.3, $P_{n,m}(\sin\phi)$ are associated Legendre functions of degree n and order m, and $C_{n,m}$ and $S_{n,m}$ are spherical harmonic coefficients of the geopotential. In particular, coefficients with $n \geq 2$ and $m = 0$, which are only latitude-dependent, are denoted as zonal harmonics. They describe the rotationally symmetric contributions (see Fig.C.1(a–d)). Coefficients with $n \geq 2$ and $m = n$ are called sectorial harmonics. They describe the purely longitude-dependent part of the potential (see Fig. C.1(e–f)). In the most general case, for tesseral harmonics with $n \geq 2$, $n \geq m \geq 1$ and $n \neq m$, there is no strict symmetry (see Fig. C.1(g–i)).

Eq. C.3 can be rewritten in an alternative way, with a single cosine term, a phase shift of $-\lambda_{n,m}$, and an adjusted amplitude $J_{n,m}$ of the combined harmonic function.

$$U(r,\phi,\lambda) = \frac{\mu}{r} - \frac{\mu}{r} \sum_{n=2}^{\infty} \left(\frac{a_e}{r}\right)^n \sum_{m=0}^{n} P_{n,m}(\sin\phi)\, J_{n,m} \cos\left[m\left(\lambda - \lambda_{n,m}\right)\right] \tag{C.4}$$

In this equation the following conventions were used for $n \geq m > 0$.

$$C_{n,m} \stackrel{\text{def}}{=} -J_{n,m} \cos\left(m\,\lambda_{n,m}\right) \tag{C.5}$$

$$S_{n,m} \stackrel{\text{def}}{=} -J_{n,m} \sin\left(m\,\lambda_{n,m}\right) \tag{C.6}$$

$$J_{n,m} = \sqrt{C_{n,m}^2 + S_{n,m}^2} \tag{C.7}$$

$$\lambda_{n,m} = \frac{2}{m} \arctan\left(\frac{S_{n,m}}{J_{n,m} + C_{n,m}}\right) \tag{C.8}$$

For the special case $m = 0$ (zonal harmonics), Eq. C.5 is often written in a shorthand notation as $J_n = J_{n,0} \stackrel{\text{def}}{=} -C_{n,0}$, with $\lambda_{n,0} \stackrel{\text{def}}{=} 0°$.

C.2 HARMONIC COEFFICIENTS OF THE GEOPOTENTIAL

For the geopotential model EGM-96 (Lemoine et al., 1998) the coefficients up to degree and order $n \times m = 8 \times 8$ are listed in Table C.1. The corresponding geoid of the full EGM-96 model is shown in Fig. C.2.

Table C.1. Coefficients of the geopotential model EGM-96 up to degree and order 8. The complete EGM-96 model extends up to $n \times m = 360 \times 360$ (Lemoine et al., 1998).

n	m	$C_{n,m}$	$S_{n,m}$	$J_{n,m}$	$\lambda_{n,m}$
2	0	−1.082627e−03	0.000000e+00	1.082627e−03	0.000°
2	1	−2.414000e−10	1.543100e−09	1.561868e−09	98.891°
2	2	1.574460e−06	−9.038038e−07	1.815430e−06	−14.929°
3	0	2.532656e−06	0.000000e+00	−2.532656e−06	0.000°
3	1	2.192639e−06	2.684249e−07	2.209008e−06	6.979°
3	2	3.089892e−07	−2.114376e−07	3.744065e−07	−17.192°
3	3	1.005488e−07	1.972226e−07	2.213748e−07	20.995°
4	0	1.619622e−06	0.000000e+00	−1.619622e−06	0.000°
4	1	−5.087994e−07	−4.491449e−07	6.786810e−07	−138.563°
4	2	7.841759e−08	1.481779e−07	1.676484e−07	31.056°
4	3	5.920994e−08	−1.200777e−08	6.041526e−08	−3.821°
4	4	−3.984074e−09	6.525714e−09	7.645770e−09	30.351°
5	0	2.272961e−07	0.000000e+00	−2.272961e−07	0.000°
5	1	−5.318030e−08	−8.085869e−08	9.677951e−08	-123.333°
5	2	1.055872e−07	−5.232919e−08	1.178431e−07	−13.182°
5	3	−1.493006e−08	−7.097342e−09	1.653115e−08	−51.525°
5	4	−2.299300e−09	3.867123e−10	2.331594e−09	42.613°
5	5	4.308225e−10	−1.648183e−09	1.703559e−09	−15.070°
6	0	−5.406812e−07	0.000000e+00	5.406812e−07	0.000°
6	1	−5.986567e−08	2.068412e−08	6.333823e−08	160.940°
6	2	5.992912e−09	−4.649304e−08	4.687769e−08	−41.328°
6	3	1.185421e−09	1.871637e−10	1.200105e−09	2.991°
6	4	−3.263618e−10	−1.784503e−09	1.814101e−09	−25.091°
6	5	−2.155941e−10	−4.329813e−10	4.836876e−10	−23.294°
6	6	2.254350e−12	−5.526094e−11	5.530690e−11	−14.611°
7	0	3.523599e−07	0.000000e+00	−3.523599e−07	0.000°
7	1	2.048460e−07	6.985031e−08	2.164277e−07	18.829°
7	2	3.284327e−08	9.269671e−09	3.412634e−08	7.881°
7	3	3.527092e−09	−3.059439e−09	4.669105e−09	−13.646°
7	4	−5.842136e−10	−2.628938e−10	6.406393e−10	−38.943°
7	5	6.857781e−13	6.277782e−12	6.315127e−12	16.753°
7	6	−2.490816e−11	1.053569e−11	2.704473e−11	26.179°
7	7	2.025443e−14	4.534042e−13	4.538564e−13	12.492°
8	0	2.047995e−07	0.000000e+00	−2.047995e−07	0.000°
8	1	1.604040e−08	4.054803e−08	4.360547e−08	68.417°
8	2	6.595204e−09	5.373019e−09	8.506824e−09	19.585°
8	3	−1.939889e−10	−8.729546e−10	8.942491e−10	−34.176°
8	4	−3.192517e−10	9.139431e−11	3.320762e−10	41.006°
8	5	−4.621840e−12	1.613533e−11	1.678423e−11	21.197°
8	6	−1.835924e−12	8.636619e−12	8.829599e−12	17.000°
8	7	3.430703e−13	3.811240e−13	5.127891e−13	6.858°
8	8	−1.581886e−13	1.536513e−13	2.205274e−13	16.979°

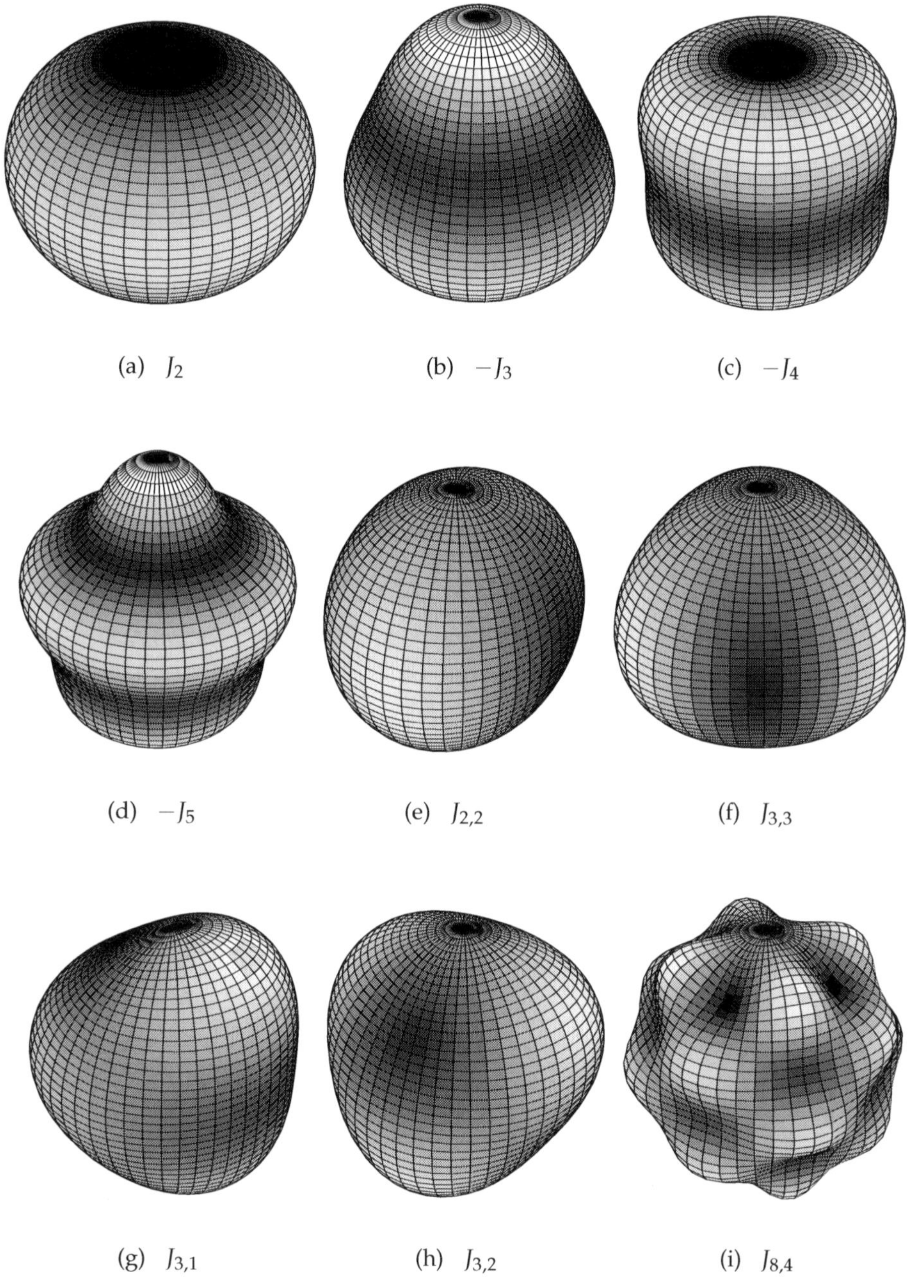

Fig. C.1. Illustration of the spherical harmonics $J_{n,m}$ of the geopotential for low degrees n and orders $m \leq n$ (see Table C.1). Bright regions indicate mass accumulations, while dark regions indicate mass deficiencies (note that for J_3 to J_5 the real mass distribution is inverse to the displayed spherical harmonic function). The grid lines show small circles of latitude and meridians of longitude in an Earth-fixed system, viewed from the northern hemisphere.

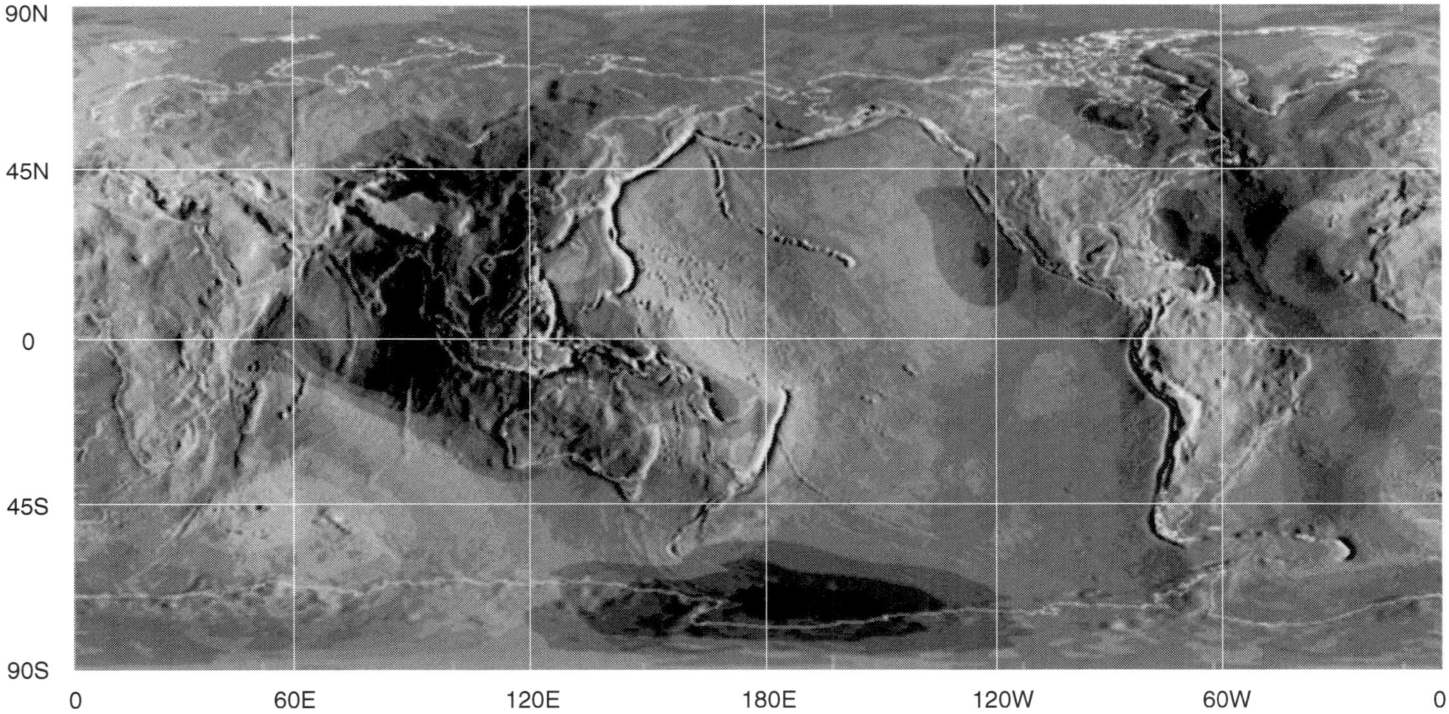

Fig. C.2. Geoid of the geopotential model EGM-96 (Lemoine et al., 1998). The geoid is an equipotential surface which approximates the mean sea level in the presence of a non-symmetric geopotential field. Level contours of geoid altitudes ΔH_g are provided relative to an Earth ellipsoid of $a_e \approx 6378.136\,\mathrm{km}$ and $f_e = 298.257$. Bright (red) regions indicate mass concentrations leading to a global maximum of $\Delta H_{g,max} \approx +85\,\mathrm{m}$, north of New Guinea, while dark (blue) regions indicate mass deficits leading to a global minimum of $\Delta H_{g,min} \approx -105\,\mathrm{m}$, south of India (see color plate on page 386).

C.3 REFERENCES

Beutler, G. (2005a). *Methods of Celestial Mechanics – Volume I: Physical, Mathematical, and Numerical Principles*. Springer Verlag, Berlin, Heidelberg, New York.

Beutler, G. (2005b). *Methods of Celestial Mechanics – Volume II: Application to Planetary System, Geodynamics, and Satellite Geodesy*. Springer Verlag, Berlin, Heidelberg, New York.

Lemoine, F., Kenyon, S., Factor, J., Trimmer, R., Pavlis, N., Chinn, D., Cox, C., Klosko, S., Luthcke, S., Torrence, M., Wang, Y., Williamson, R., Pavlis, E., Rapp, R., and Olson, T. (1998). The Development of the Joint NASA GSFC and NIMA Geopotential Model EGM96. Technical Report NASA TP-1998-206891, NASA/GSFC.

Montenbruck, O. and Gill, E. (2000). *Satellite Orbits – Models, Methods, and Applications*. Springer, Berlin, Heidelberg, New York.

Seidelmann, P. et al., editors (1992). *Explanatory Supplement to the Astronomical Almanac*. University Science Books, Mill Valley, CA.

D

The World Population Viewed from Orbit

H. Klinkrad

D.1 MODEL OF THE WORLD POPULATION DENSITY DISTRIBUTION

Population density maps with sufficient spatial resolution are necessary to associate the impact of re-entry survivor objects with a casualty risk in the affected ground track swath. Fig. D.1 shows a world map of $1^\circ \times 1^\circ$ mean population densities, derived from $5' \times 5'$ (9.25 km × 9.25 km) high resolution data of the Global Demography Project (Tobler et al., 1995). In the subsequent analysis these population data for the year 1994 are augmented by discretized maps of land masses.

The total world population in 1994 was $5.63 \times 10^{+9}$, distributed over a total land surface of $1.48 \times 10^{+8}\,\text{km}^2$, which accounts for 28.97% of the surface of the Earth reference ellipsoid. This corresponds to a global mean population density of $11.0\,\text{km}^{-2}$, and to a mean land population density of $38.1\,\text{km}^{-2}$. The global maximum of the population density in 1994 (averaged over the $5' \times 5'$ bin size) was $34{,}537\,\text{km}^{-2}$, in the area of Bombay, India. By the year 2000, the world population had reached $6.23 \times 10^{+9}$. Based on an analysis of historic population data, this value is predicted to increase two-fold within ~40 years, following an exponential law according to Eq. D.1 (see also Table D.1). For simplicity it is assumed that the 2D population density distribution $\rho_{p,1994}(\lambda, \phi)$ can be linearly scaled by a uniform, epoch-dependent growth factor.

$$\rho_p(\lambda, \phi, t) = \rho_{p,1994}(\lambda, \phi) \exp\left(\frac{t - 1994.5}{59.63}\right) \tag{D.1}$$

Histograms on the right-hand side of Fig. D.1 give the small circle of latitude mean values of land cover (in %) and average population density (km^{-2}), plus the maximum population density in each latitude band of 1° width. An imbalance between the northern and southern hemisphere becomes evident from these charts. In fact, only 11.7% of the world population, and only 33.3% of the land masses are located south of the equator. This has a direct impact on an equally imbalanced

Table D.1. Estimated evolution of the world population, extrapolated from 1994 census data of the Global Demography Project (Tobler et al., 1995). Results are provided for an exponential growth model according to Eq. D.1, for mid-year conditions.

Year	Growth factor	Population forecast
1994	1.000	$5.580 \times 10^{+9}$
2000	1.106	$6.172 \times 10^{+9}$
2005	1.203	$6.714 \times 10^{+9}$
2010	1.309	$7.303 \times 10^{+9}$
2015	1.424	$7.945 \times 10^{+9}$
2020	1.549	$8.642 \times 10^{+9}$
2025	1.685	$9.400 \times 10^{+9}$
2030	1.833	$10.226 \times 10^{+9}$
2035	1.993	$11.123 \times 10^{+9}$
2040	2.168	$12.100 \times 10^{+9}$
2045	2.359	$13.162 \times 10^{+9}$
2050	2.566	$14.317 \times 10^{+9}$

risk distribution between the hemispheres, with a much higher risk of re-entry casualties at northern latitudes. A re-entry which occurs on the northern hemisphere carries a 1.75 times larger risk, and a re-entry in the southern hemisphere carries a 4.30 times lower risk than a global mean event. This result is directly related to the corresponding mean population densities for the year 2000, which are 12.3 km^{-2} for the whole world, 21.6 km^{-2} for the northern hemisphere, and 2.9 km^{-2} for the southern hemisphere (Klinkrad et al., 2003).

D.2 SAMPLING THE WORLD POPULATION ALONG GROUND TRACKS

A re-entry ground track samples the underlying world population in a way which is mainly determined by the orbit inclination. A typical, near-circular re-entry orbit of mean altitude $\overline{H} \approx 120$ km has a nodal orbit period of $T_\Omega \approx 87^m16^s$. Within this orbit period the geographic longitude of the ascending node λ_Ω is displaced towards the west by an interval $\Delta S_\Omega = \lambda_{\Omega,n+1} - \lambda_{\Omega,n}$, which results from the Earth rotation ω_e, and from a precession $\dot{\Omega}$ of the orbital plane due to the Earth oblateness term J_2.

$$T_\Omega = T\left(1 - \frac{3}{2} J_2 \left(\frac{a_e}{a}\right)^2 \left(3 - 4\sin^2 i\right)\right) \quad \text{(D.2)}$$

$$\dot{\Omega} = -\frac{3\pi}{T} J_2 \left(\frac{a_e}{a}\right)^2 \cos i \quad \text{(D.3)}$$

$$\Delta S_\Omega = \lambda_{\Omega,n+1} - \lambda_{\Omega,n} = -T_\Omega\left(\omega_e - \dot{\Omega}\right) \quad \text{(D.4)}$$

where T is the unperturbed orbital period according to Eq. A.5. Close to a re-entry the following approximation can be used.

$$\Delta S_\Omega \approx -21.8^\circ - 0.6^\circ \cos i \qquad \text{(D.5)}$$

Using a drag-free, first-order perturbation model (J_2 only), single orbit arcs can be generated for a near-circular re-entry orbit of a given inclination, assuming a uniform impact probability density $\Delta P_i/\Delta u$ for an argument of true latitude of $0^\circ \leq u \leq 360^\circ$. With every new orbit the longitude of ascending node λ_Ω is shifted by 0.5°, until the full range of $-180^\circ \leq \lambda_\Omega \leq +180^\circ$ is covered. Below each of the 720 sample orbits the Earth surface is scanned for land masses and local population densities within a swath of ±25 km cross-track. These data are then processed to determine global statistics for the scanned part of the Earth surface (see Table 9.9), to determine averages for certain latitudes ϕ (see Fig. D.2, D.4 and D.6), and to determine averages for given nodal longitudes λ_Ω (see Fig. D.3, D.5 and D.7). Results are provided for $i = 7^\circ, 28.5^\circ, 51.6^\circ, 65^\circ, 80^\circ, 89.9^\circ$, and 98.5°, which are the most commonly used inclination bands (see Fig. 2.12). The greatest detail is obtained when plotting the underlying land masses and population density distribution against the argument of true latitude $u = \omega + f$ of the orbit position, and against the longitude of ascending node λ_Ω of the sampling orbits. Such plots are provided in Fig. D.8 to D.14, with population densities superimposed in color code (grey scales) on an Earth map which has a resolution of 1° both in λ_Ω and $u = \omega + f$. Due to the ground sampling technique the "pixels" of the map can be considerably distorted, particularly for low inclinations. The mapping principle can be best followed in Fig. D.13, for a near-polar orbit, which shows a proper Earth map for the ascending part of the orbit (where $-90^\circ \leq u \leq +90^\circ$), and a mirror image on the descending arc (where $+90^\circ \leq u \leq +270^\circ$).

Fig. D.2, D.4 and D.6 can be helpful for re-entry risk assessment in the early phase of mission planning (e.g. choice of orbit inclination)[D.1]. Fig. D.3, D.5 and D.7 can be used for medium-term risk management by phasing of the ground track node (λ_Ω), and Fig. D.8 to D.14 can assist in the short-term selection of an optimal orbit position $u = \omega + f$ for a low-risk impact zone.

Due to the chosen mapping technique in Fig. D.8 to D.14, an analyst can "propagate" orbits for a given inclination by drawing straight, horizontal lines, starting at $u = 0^\circ$, and at an initial nodal longitude $\lambda_{\Omega,n}$ (compare Fig. 9.8). The end of the line at $u = 360^\circ$ marks the new ascending node $\lambda_{\Omega,n+1}$, which is displaced by ΔS_Ω to the west (see Eq. D.4 and D.5). The next orbit can be produced by starting a new line at $u_{n+1} = 0^\circ$ and $\lambda_{\Omega,n+1} = \lambda_{\Omega,n} - \Delta S_\Omega$, or by glueing a copy of the map to the left edge of the original, displaced by ΔS_Ω to the top (towards the west), and continuing the existing straight line at $u_n = 360^\circ$ (corresponding to $u_{n+1} = 0^\circ$).

[D.1] $i = 51.6^\circ$ is used as common reference in Fig. D.2 to D.7 to facilitate comparisons

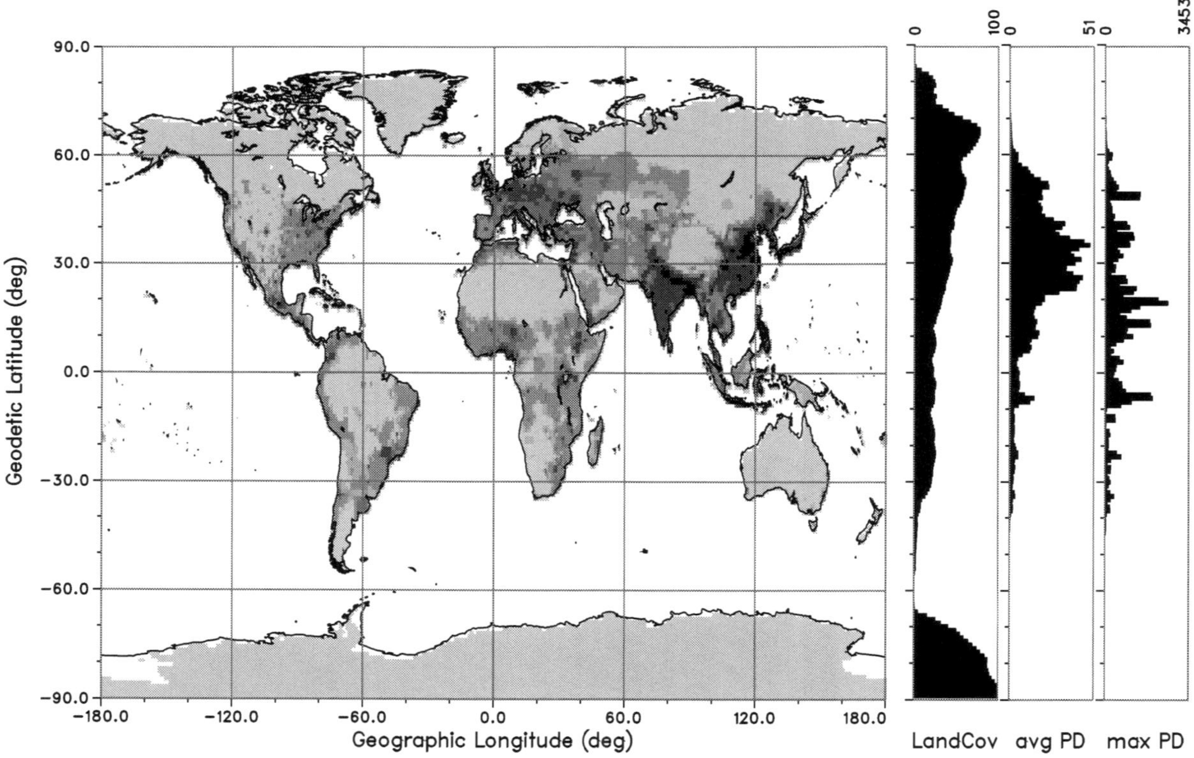

Fig. D.1. World population density distribution $\rho_p(\lambda, \phi)$ in 1994, according to the Global Demography Project (Tobler et al., 1995), with resolution $\Delta\lambda \times \Delta\phi \leq 9.25\,\text{km} \times 9.25\,\text{km}$. Small circle of latitude averages of land coverage [%], mean population density [1/km^2], and maximum population density [1/km^2] are provided as bar charts to the right.

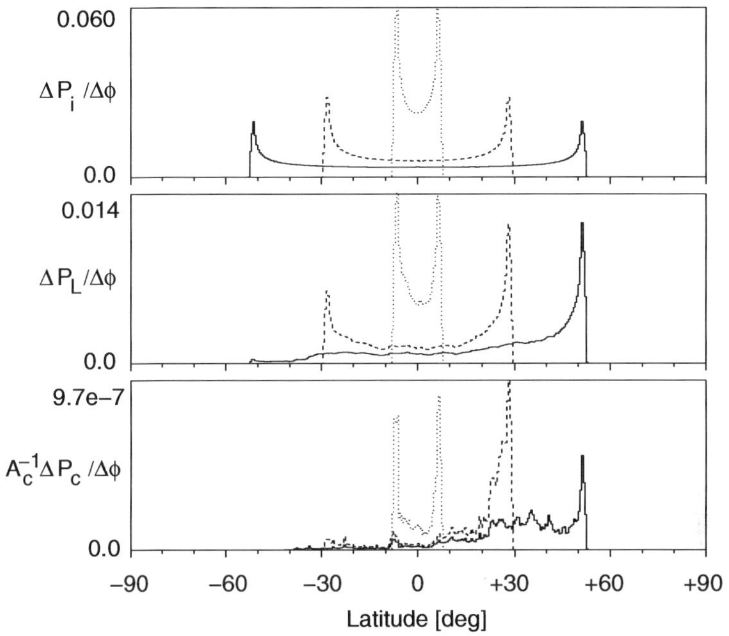

Fig. D.2. Latitude-dependent distributions of impact probability density $\Delta P_i(\phi)/\Delta\phi$, land impact probability density $P_L(\phi)/\Delta\phi$, and specific casualty probability density $P_c(\phi)/(\Delta\phi\, A_c)$ for orbits with $i = 7.0°$ (dotted line), 28.5° (dashed line) and 51.6° (solid line), assuming a uniform impact probability density $\mathrm{PDF}_i(u) = \mathrm{const.}$ (latitude class width: $\Delta\phi = 0.5°$).

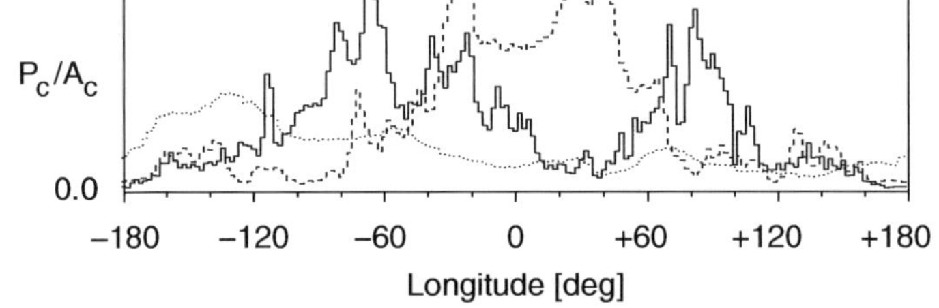

Fig. D.3. Longitude-dependent distributions of land impact probability $P_L(\lambda_\Omega)$ and specific casualty probability $P_c(\lambda_\Omega)/A_c$ for single orbit arcs with $i = 7.0°$ (dotted line), 28.5° (dashed line) and 51.6° (solid line), assuming a uniform impact probability density $\mathrm{PDF}_i(u) = \mathrm{const.}$ (longitude class width: $\Delta\lambda_\Omega = 2.0°$).

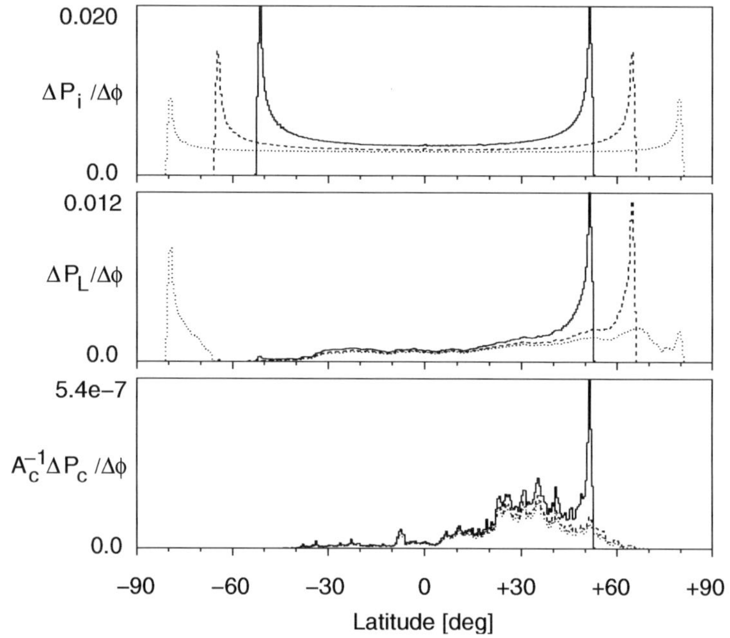

Fig. D.4. Latitude-dependent distributions of impact probability density $\Delta P_i(\phi)/\Delta\phi$, land impact probability density $P_L(\phi)/\Delta\phi$, and specific casualty probability density $P_c(\phi)/(\Delta\phi\, A_c)$ for orbits with $i = 80.0°$ (dotted line), 65.0° (dashed line) and 51.6° (solid line), assuming a uniform impact probability density $\mathrm{PDF}_i(u) = \mathrm{const.}$ (latitude class width: $\Delta\phi = 0.5°$).

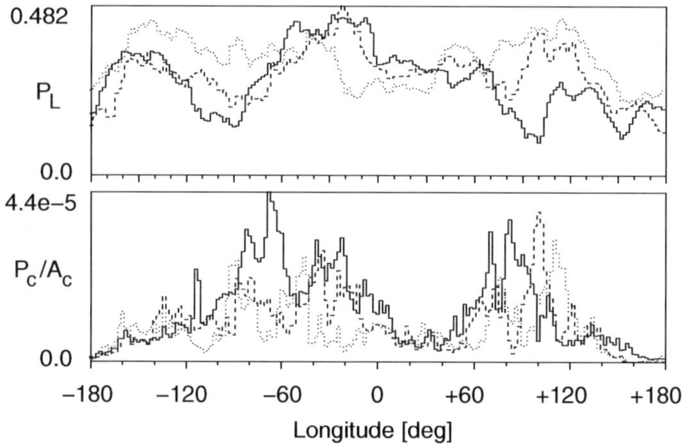

Fig. D.5. Longitude-dependent distributions of land impact probability $P_L(\lambda_\Omega)$ and specific casualty probability $P_c(\lambda_\Omega)/A_c$ for single orbit arcs with $i = 80.0°$ (dotted line), 65.0° (dashed line) and 51.6° (solid line), assuming a uniform impact probability density $\mathrm{PDF}_i(u) = \mathrm{const.}$ (longitude class width: $\Delta\lambda_\Omega = 2.0°$).

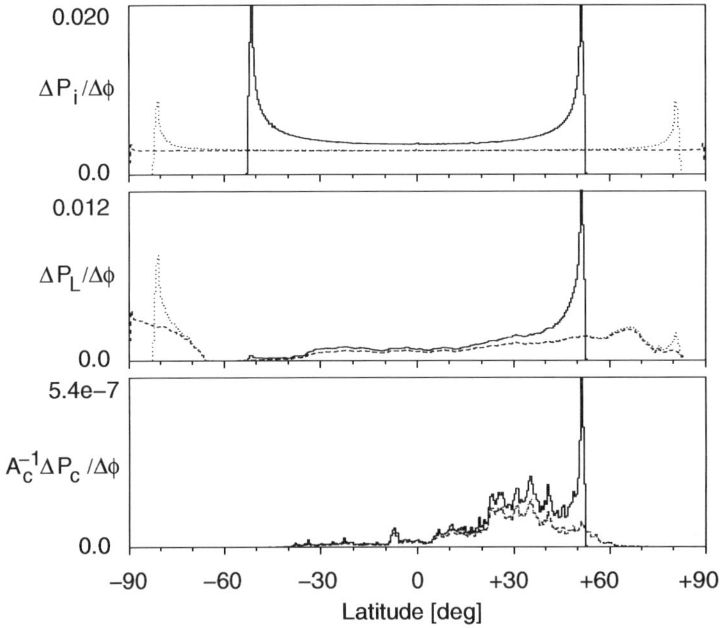

Fig. D.6. Latitude-dependent distributions of impact probability density $\Delta P_i(\phi)/\Delta\phi$, land impact probability density $P_L(\phi)/\Delta\phi$, and specific casualty probability density $P_c(\phi)/(\Delta\phi\, A_c)$ for orbits with $i = 98.5°$ (dotted line), 89.9° (dashed line) and 51.6° (solid line), assuming a uniform impact probability density $\mathrm{PDF}_i(u) = $ const. (latitude class width: $\Delta\phi = 0.5°$).

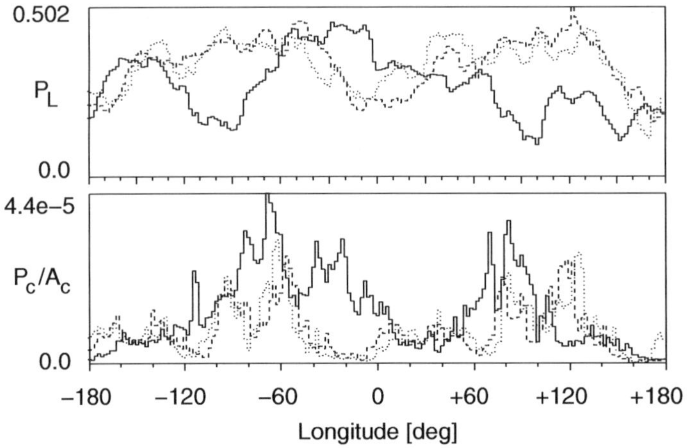

Fig. D.7. Longitude-dependent distributions of land impact probability $P_L(\lambda_\Omega)$ and specific casualty probability $P_c(\lambda_\Omega)/A_c$ for single orbit arcs with $i = 98.5°$ (dotted line), 89.9° (dashed line) and 51.6° (solid line), assuming a uniform impact probability density $\mathrm{PDF}_i(u) = $ const. (longitude class width: $\Delta\lambda_\Omega = 2.0°$).

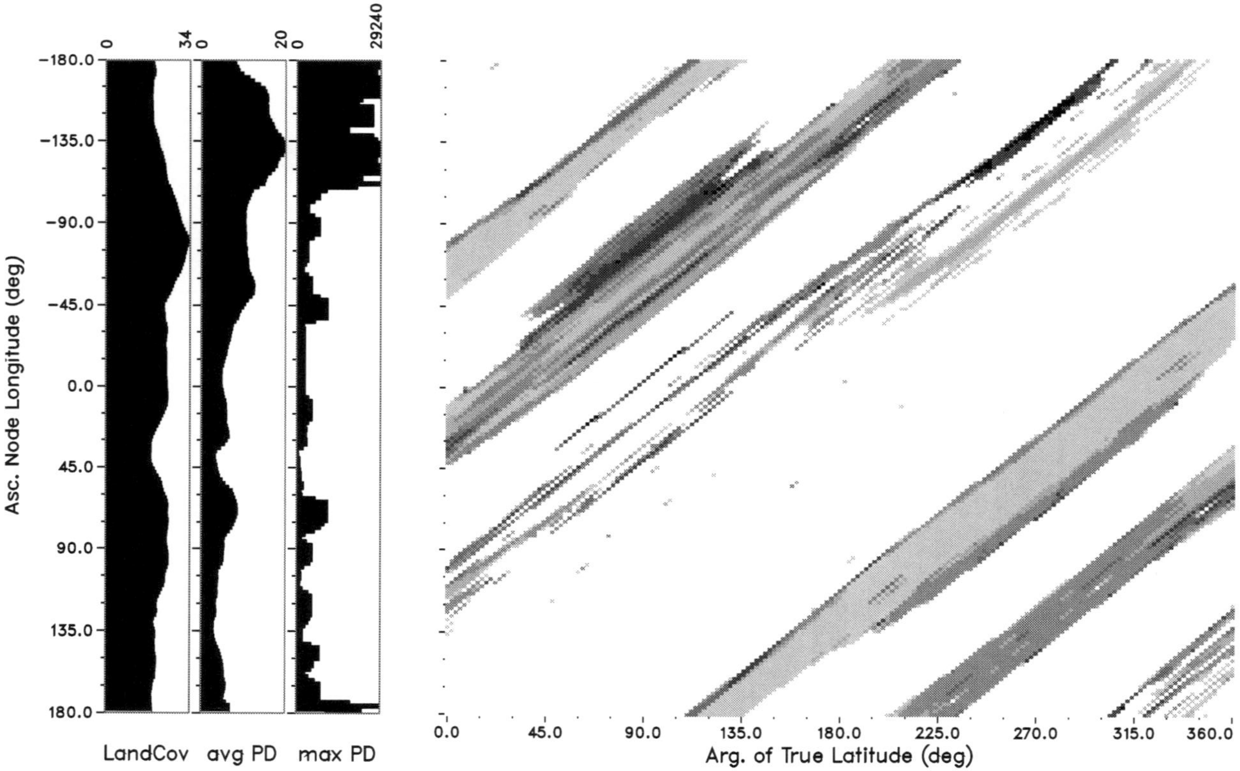

Fig. D.8. World population density distribution in 1994, as sampled by ground tracks of a near-circular orbit of $H = 120\,\text{km}$ and $i = 7.0°$, as a function of longitude of ascending node λ_Ω and orbit position $u = \omega + f$. Bar charts to the left show mean land coverage [%], average population density [$1/\text{km}^2$], and maximum population density [$1/\text{km}^2$] per orbit, as fct(λ_Ω).

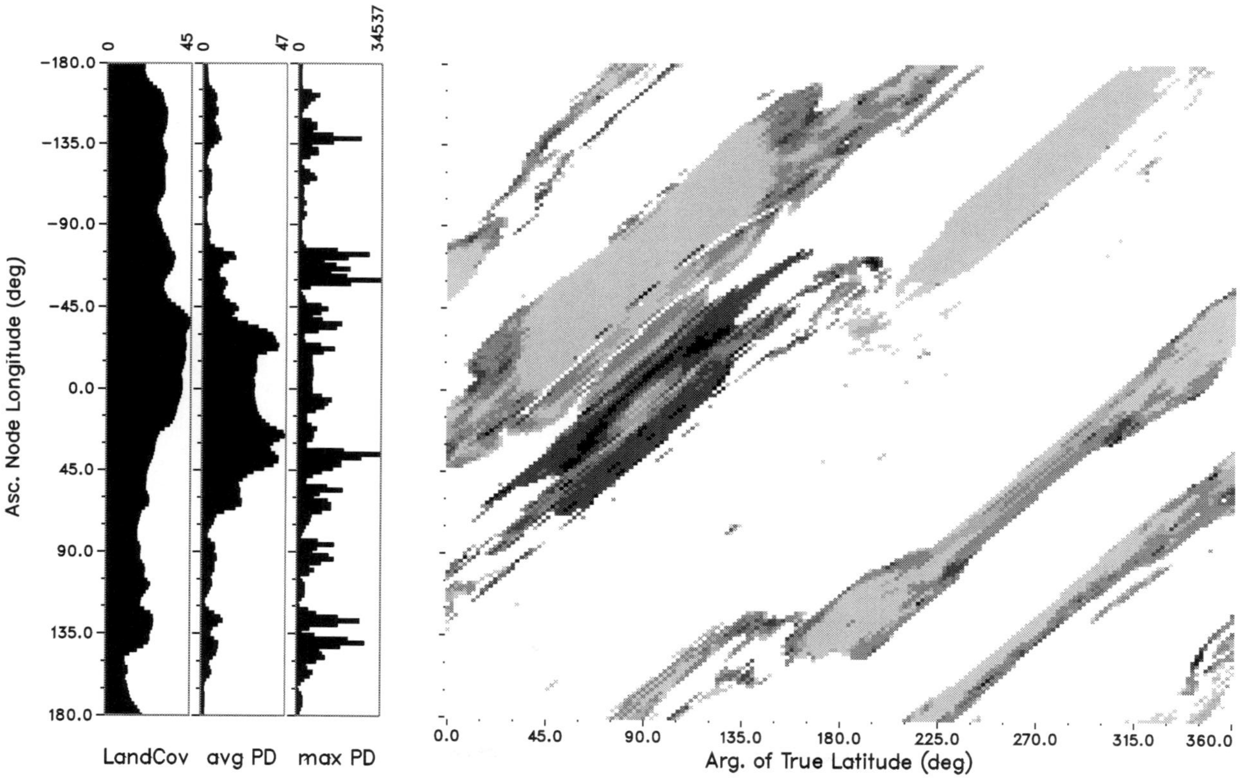

Fig. D.9. World population density distribution in 1994, as sampled by ground tracks of a near-circular orbit of $H = 120\,\text{km}$ and $i = 28.5^\circ$, as a function of longitude of ascending node λ_Ω and orbit position $u = \omega + f$. Bar charts to the left show mean land coverage [%], average population density [$1/\text{km}^2$], and maximum population density [$1/\text{km}^2$] per orbit, as fct(λ_Ω).

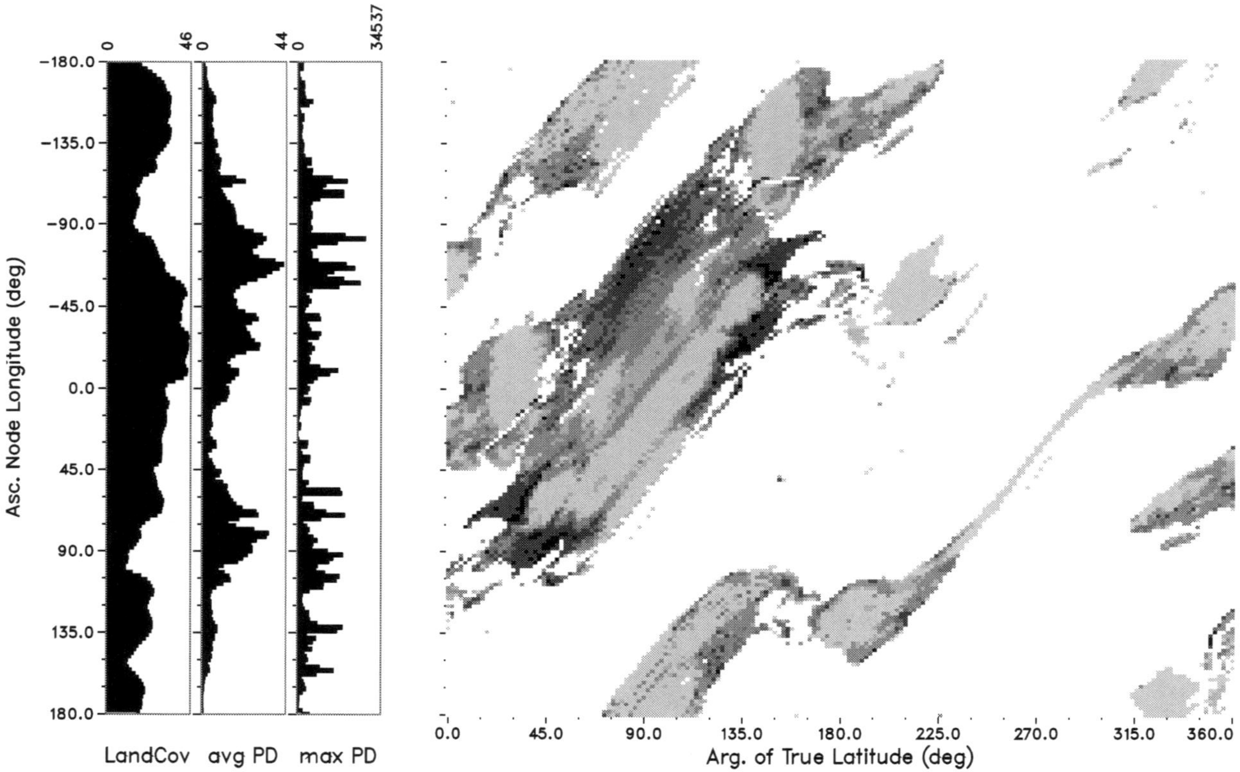

Fig. D.10. World population density distribution in 1994, as sampled by ground tracks of a near-circular orbit of $H = 120\,\mathrm{km}$ and $i = 51.6°$, as a function of longitude of ascending node λ_Ω and orbit position $u = \omega + f$. Bar charts to the left show mean land coverage [%], average population density $[1/\mathrm{km}^2]$, and maximum population density $[1/\mathrm{km}^2]$ per orbit, as $\mathrm{fct}(\lambda_\Omega)$.

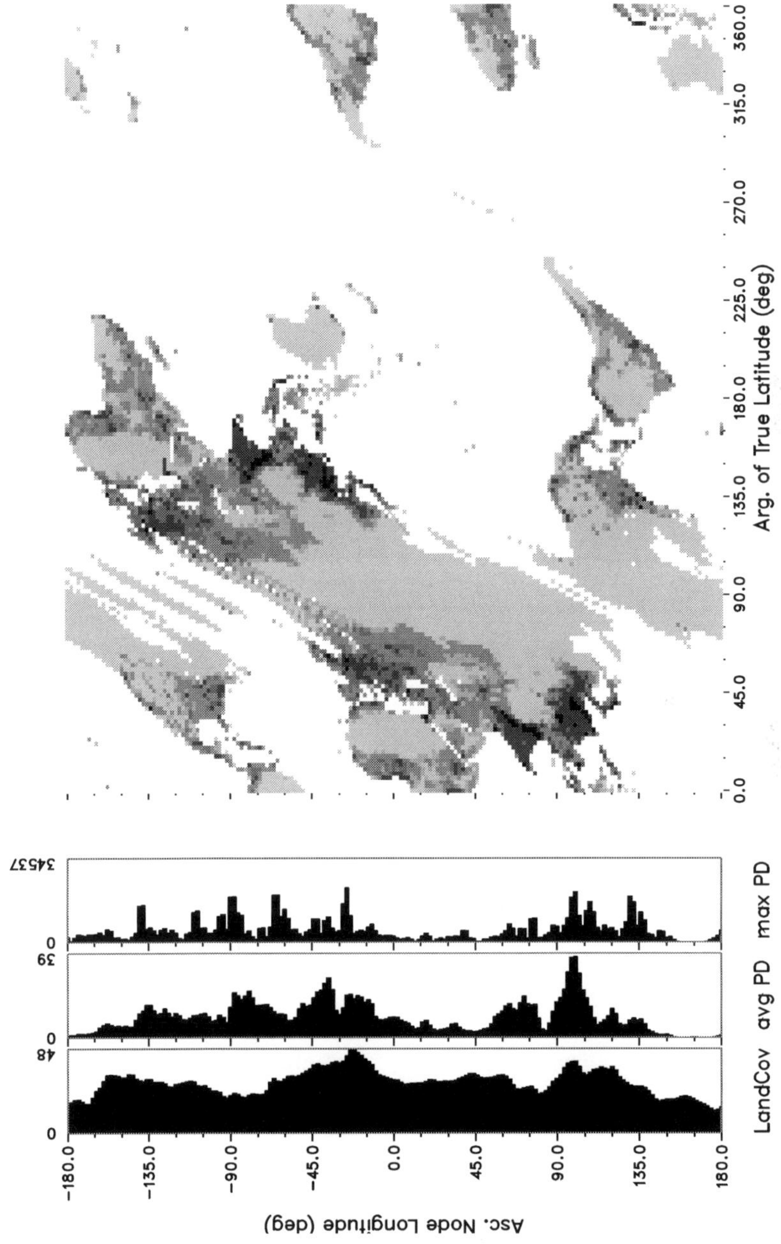

Fig. D.11. World population density distribution in 1994, as sampled by ground tracks of a near-circular orbit of $H = 120\,\mathrm{km}$ and $i = 65.0^\circ$, as a function of longitude of ascending node λ_Ω and orbit position $u = \omega + f$. Bar charts to the left show mean land coverage [%], average population density [$1/\mathrm{km}^2$], and maximum population density [$1/\mathrm{km}^2$] per orbit, as $\mathrm{fct}(\lambda_\Omega)$.

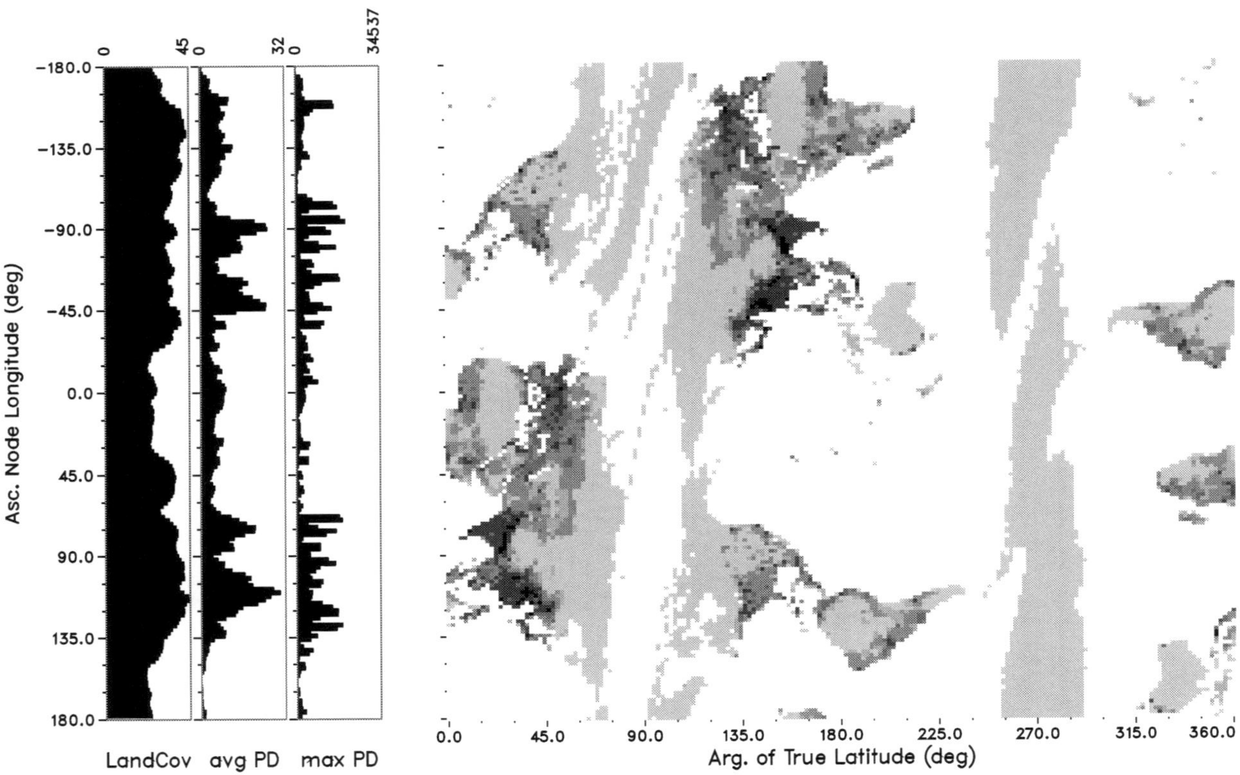

Fig. D.12. World population density distribution in 1994, as sampled by ground tracks of a near-circular orbit of $H = 120\,\text{km}$ and $i = 80.0°$, as a function of longitude of ascending node λ_Ω and orbit position $u = \omega + f$. Bar charts to the left show mean land coverage [%], average population density [1/km^2], and maximum population density [1/km^2] per orbit, as fct(λ_Ω).

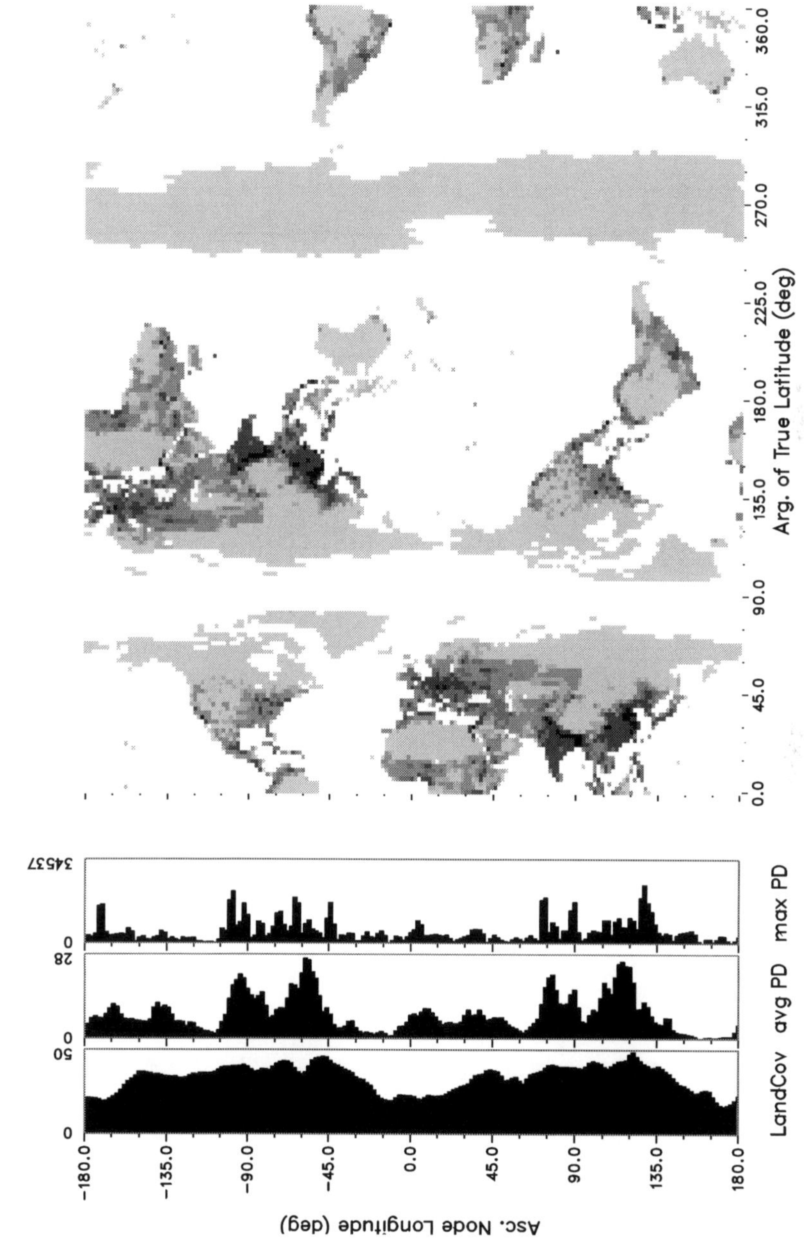

Fig. D.13. World population density distribution in 1994, as sampled by ground tracks of a near-circular orbit of $H = 120\,\mathrm{km}$ and $i = 89.9^\circ$, as a function of longitude of ascending node λ_Ω and orbit position $u = \omega + f$. Bar charts to the left show mean land coverage [%], average population density [$1/\mathrm{km}^2$], and maximum population density [$1/\mathrm{km}^2$] per orbit, as $\mathrm{fct}(\lambda_\Omega)$.

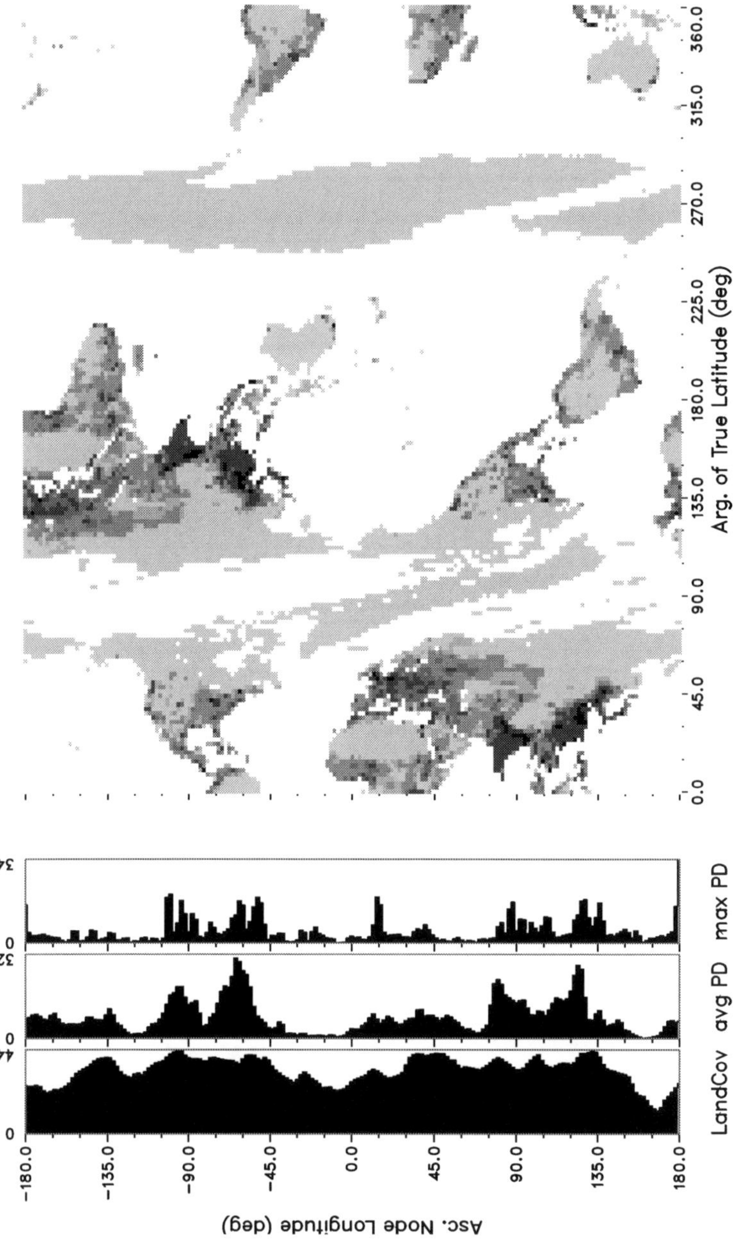

Fig. D.14. World population density distribution in 1994, as sampled by ground tracks of a near-circular orbit of $H = 120$ km and $i = 98.5^{\circ}$, as a function of longitude of ascending node λ_{Ω} and orbit position $u = \omega + f$. Bar charts to the left show mean land coverage [%], average population density [1/km^2], and maximum population density [1/km^2] per orbit, as fct(λ_{Ω}).

D.3 REFERENCES

Klinkrad, H. et al. (2003). *ESA Space Debris Mitigation Handbook.* ESA/ESOC, second edition. issue 1.0.

Tobler, W., Deichmann, U., Gottsegen, J., and Maloy, K. (1995). The Global Demography Project. Technical Report Technical Report TR-95-6, National Center for Geographic Information and Analysis, Santa Barbara, CA.

E

Color Supplement

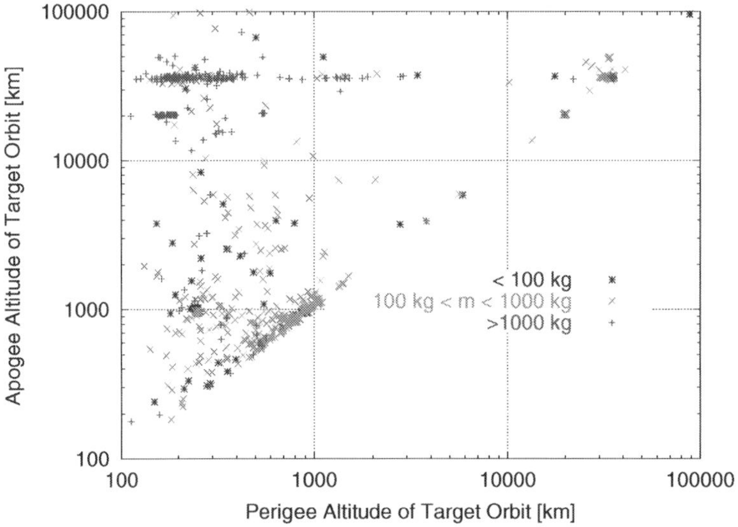

Fig. E.1. Perigee and apogee altitudes of the delivery orbits for SRM supported injection maneuvers. The SRM propellant masses are indicated by color codes (see page 81).

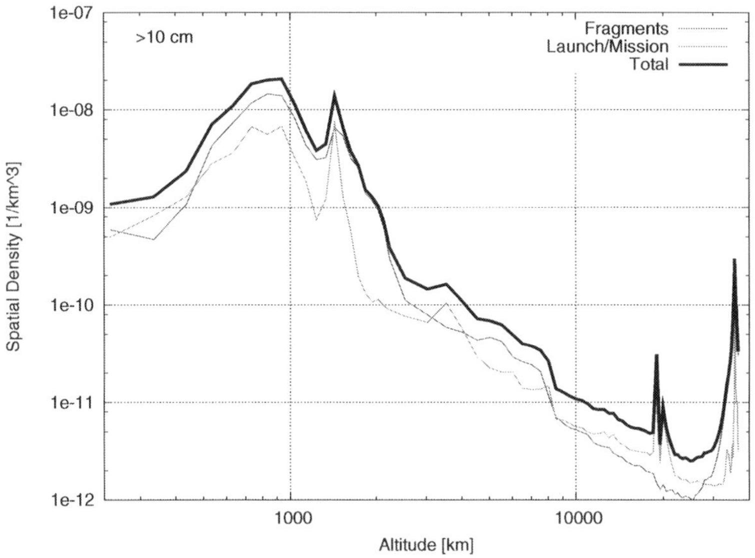

Fig. E.2. Spatial density versus altitude for objects of diameters $d > 10$ cm according to the MASTER-2001 model, for May 2001. Different sources are color coded. The thick line shows the envelope of total spatial density (see page 97).

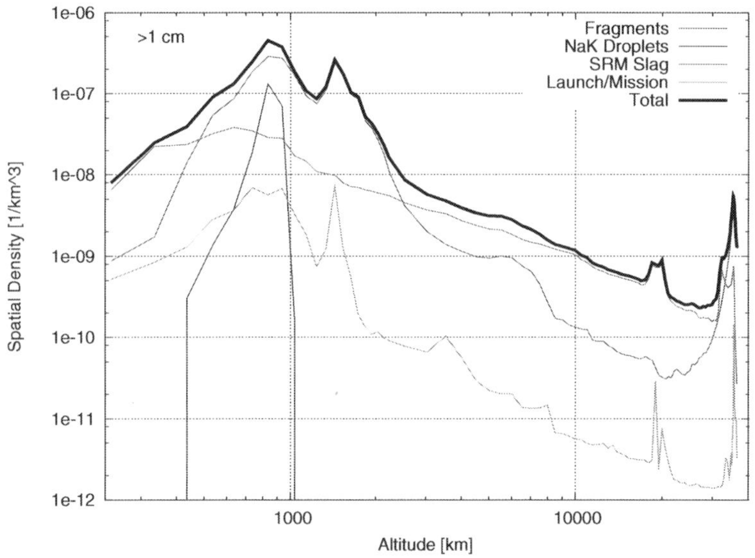

Fig. E.3. Spatial density versus altitude for objects of diameters $d > 1$ cm according to the MASTER-2001 model, for May 2001. Different sources are color coded. The thick line shows the envelope of total spatial density (see page 97).

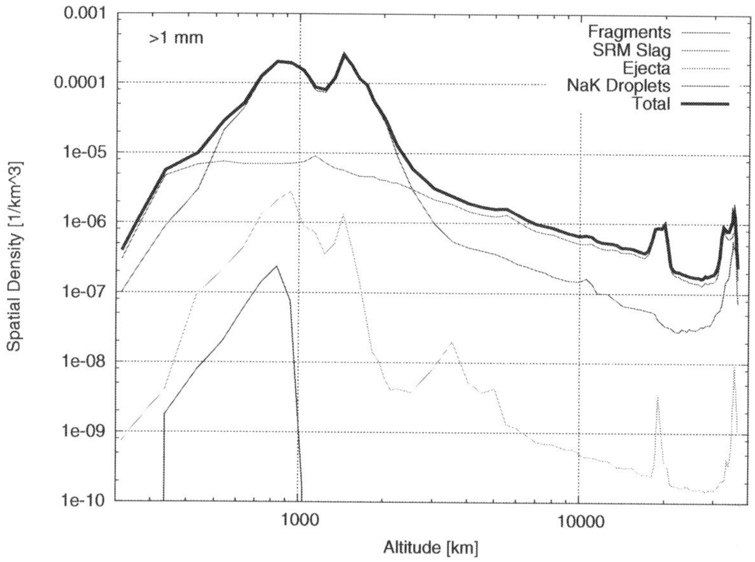

Fig. E.4. Spatial density versus altitude for objects of diameters $d > 1$ mm according to the MASTER-2001 model, for May 2001. Different sources are color coded. The thick line shows the envelope of total spatial density (see page 98).

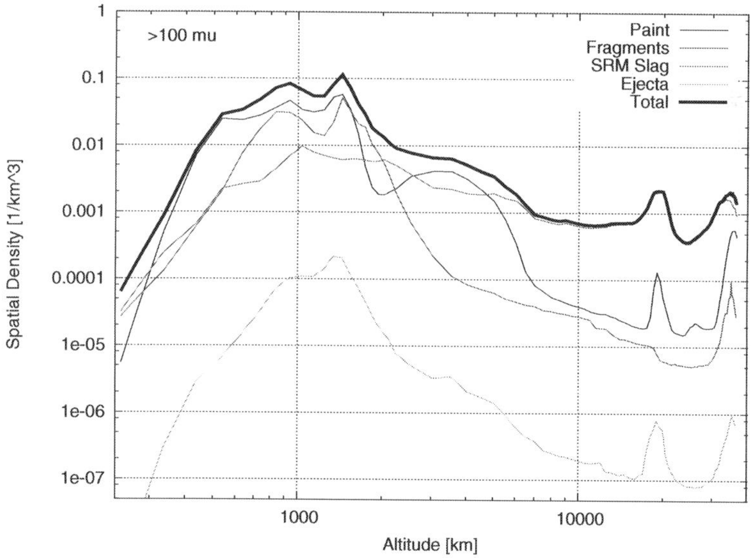

Fig. E.5. Spatial density versus altitude for objects of diameters $d > 0.1$ mm according to the MASTER-2001 model, for May 2001. Different sources are color coded. The thick line shows the envelope of total spatial density (see page 98).

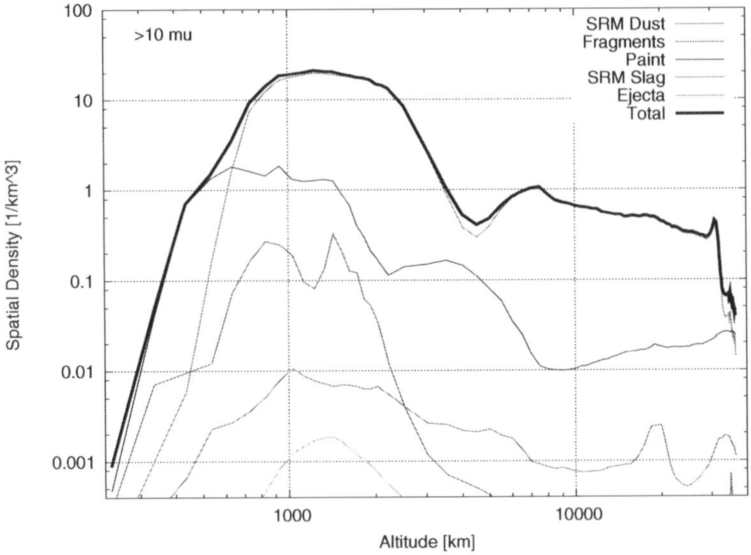

Fig. E.6. Spatial density versus altitude for objects of diameters $d > 10\,\mu\text{m}$ according to the MASTER-2001 model, for May 2001. Different sources are color coded. The thick line shows the envelope of total spatial density (see page 99).

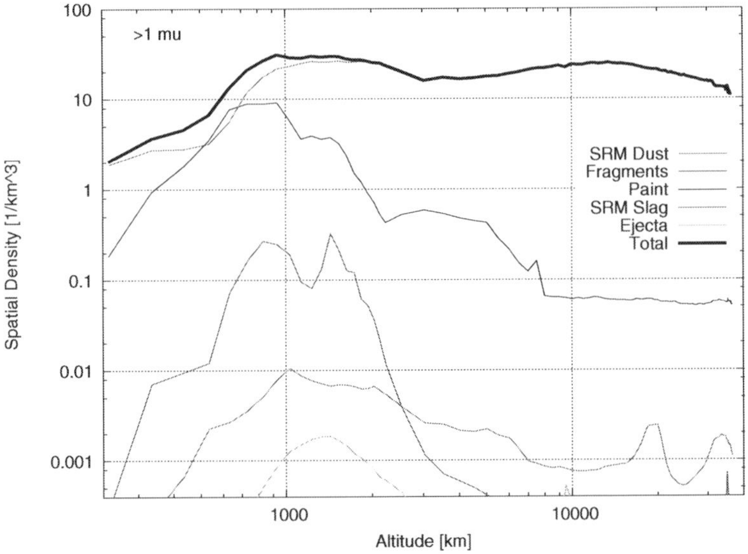

Fig. E.7. Spatial density versus altitude for objects of diameters $d > 1\,\mu\text{m}$ according to the MASTER-2001 model, for May 2001. Different sources are color coded. The thick line shows the envelope of total spatial density (see page 99).

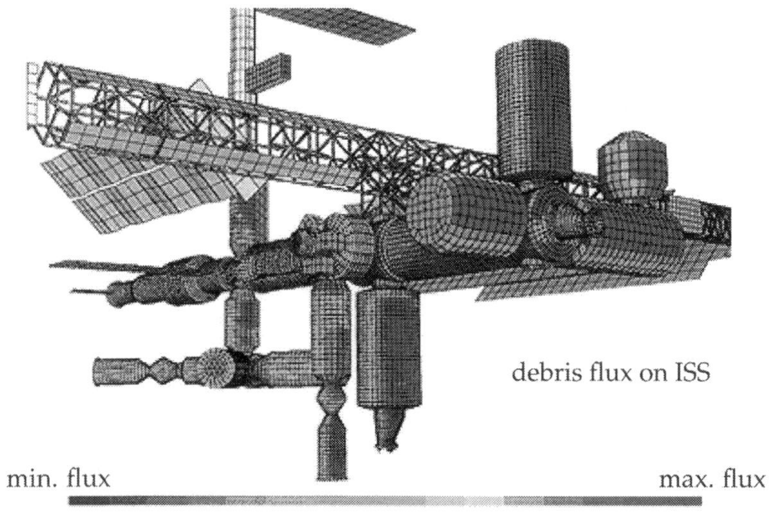

Fig. E.8. Impact flux distribution on the ISS geometry for space debris objects of $d >$ 1 cm (view from approach azimuth $A \approx +45^\circ$ and elevation $h \approx -10^\circ$; source: NASA; see page 118).

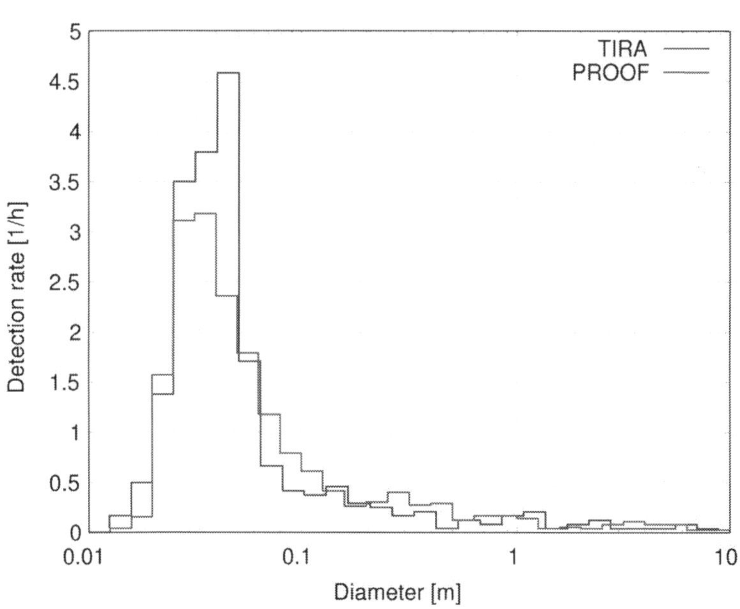

Fig. E.9. Number of detections versus object diameter for an FGAN/TIRA beam-park experiment in 2001, compared with PROOF simulations, using the MASTER space debris population (see page 107).

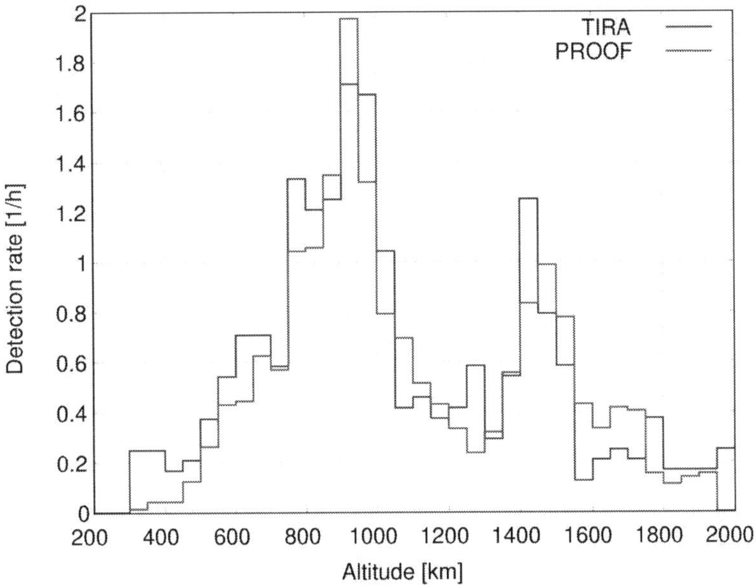

Fig. E.10. Number of detections versus altitude for an FGAN/TIRA beam-park experiment in 2001, compared with PROOF simulations, using the MASTER space debris population (see page 106).

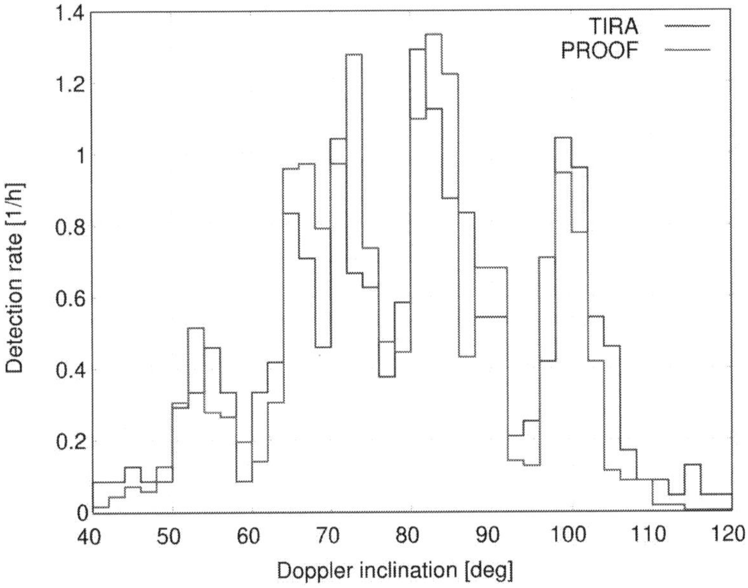

Fig. E.11. Number of detections versus Doppler-inclination for an FGAN/TIRA beam-park experiment in 2001, compared with PROOF simulations, using the MASTER space debris population (see page 106).

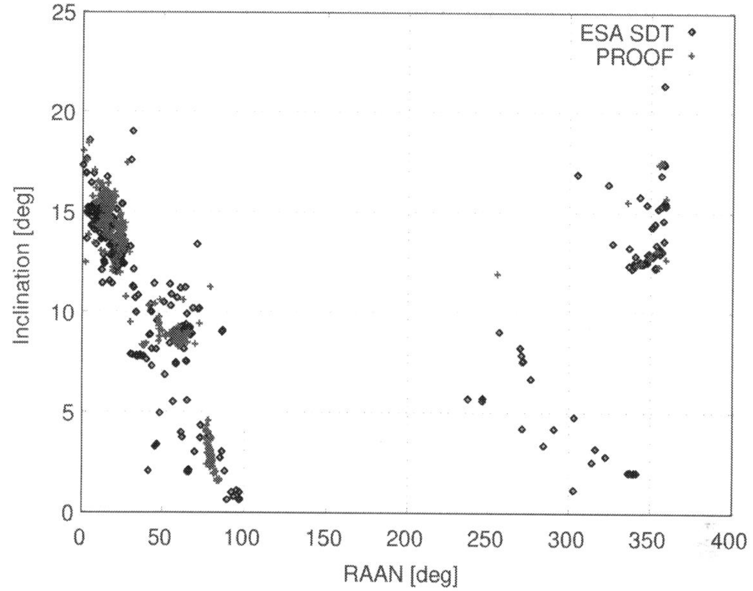

Fig. E.12. Right ascension and inclination of detectable near-GEO objects, as observed by the ESA telescope, and as simulated by PROOF for the MASTER space debris population of 2001 (see page 110).

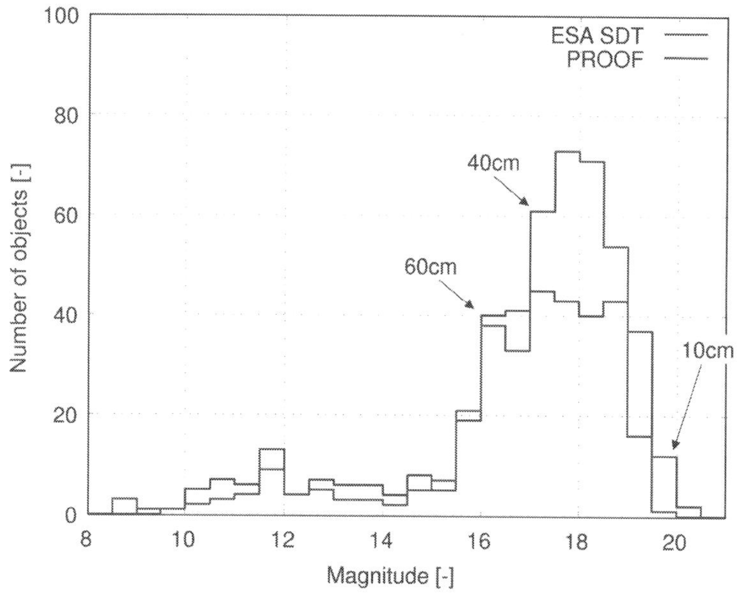

Fig. E.13. Visual magnitudes of detectable near-GEO objects, as observed by the ESA telescope, and as simulated by PROOF for the MASTER space debris population of 2001 (equivalent detection threshold diameters are marked; see page 110).

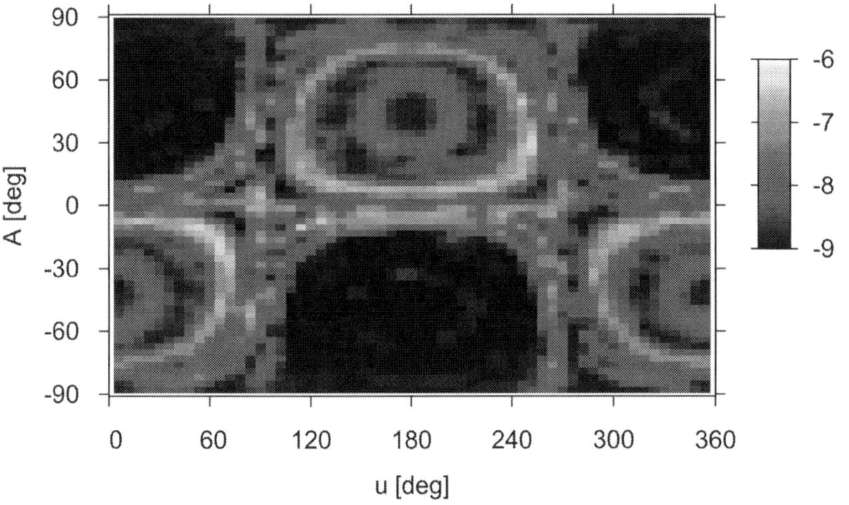

Fig. E.14. Debris flux distribution for $d > 1$ cm on an ERS orbit, as a function of impact azimuth A and orbit position $u = \omega + f$. Contour levels are given as $\log_{10}$ of the impact flux F [$\mathrm{m}^{-2}\,\mathrm{y}^{-1}$] (class widths: $\Delta u = 6°$, $\Delta A = 3°$; see page 128).

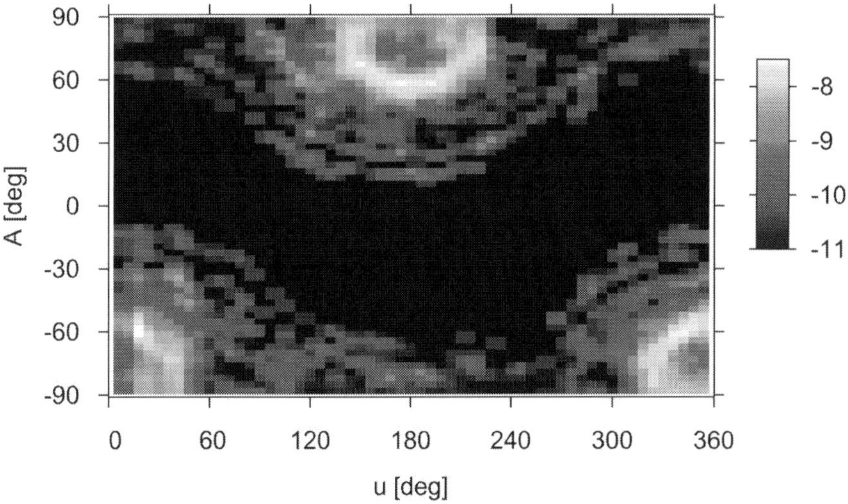

Fig. E.15. Debris flux distribution for $d > 1$ cm on an ISS orbit, as a function of impact azimuth A and orbit position $u = \omega + f$. Contour levels are given as $\log_{10}$ of the impact flux F [$\mathrm{m}^{-2}\,\mathrm{y}^{-1}$] (class widths: $\Delta u = 6°$, $\Delta A = 3°$; see page 129).

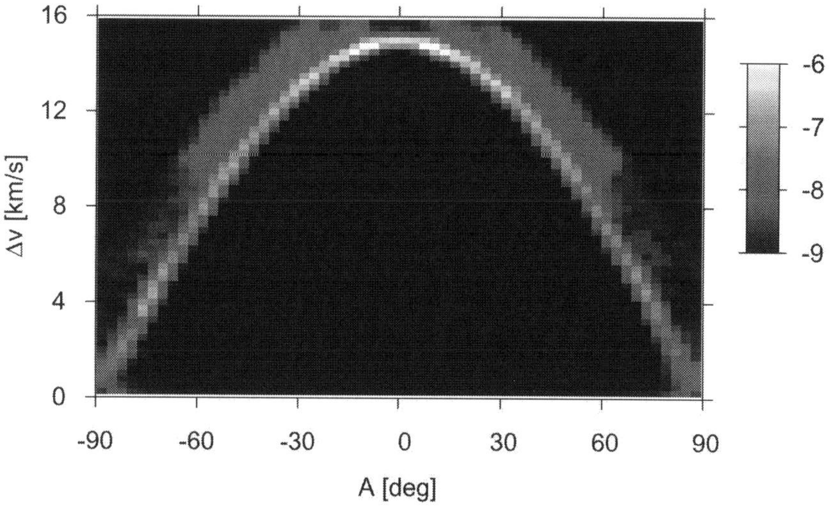

Fig. E.16. Debris flux distribution for $d > 1$ cm on an ERS orbit, as a function of impact azimuth A and impact velocity Δv. Contour levels are given as $\log_{10}$ of the impact flux $F\,[\mathrm{m}^{-2}\,\mathrm{y}^{-1}]$ (class widths: $\Delta A = 3°$, $\Delta(\Delta v) = 0.25$ km/s; see page 130).

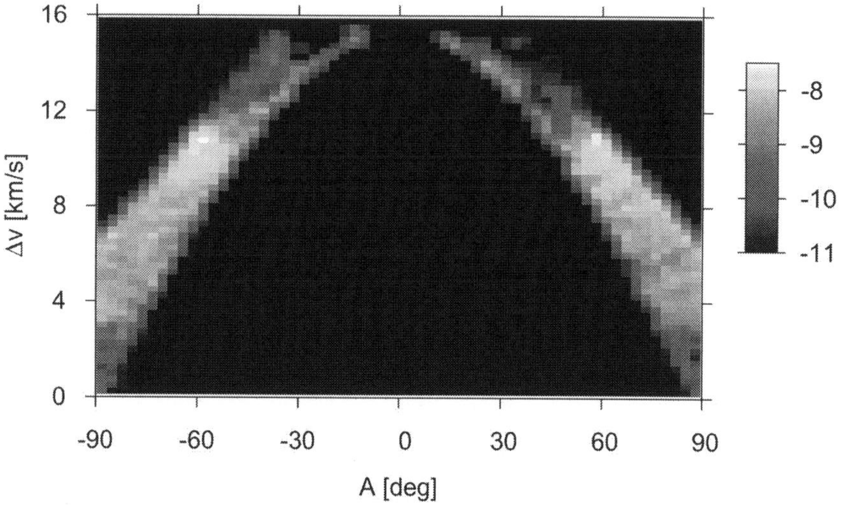

Fig. E.17. Debris flux distribution for $d > 1$ cm on an ISS orbit, as a function of impact azimuth A and impact velocity Δv. Contour levels are given as $\log_{10}$ of the impact flux $F\,[\mathrm{m}^{-2}\,\mathrm{y}^{-1}]$ (class widths: $\Delta A = 3°$, $\Delta(\Delta v) = 0.25$ km/s; see page 131).

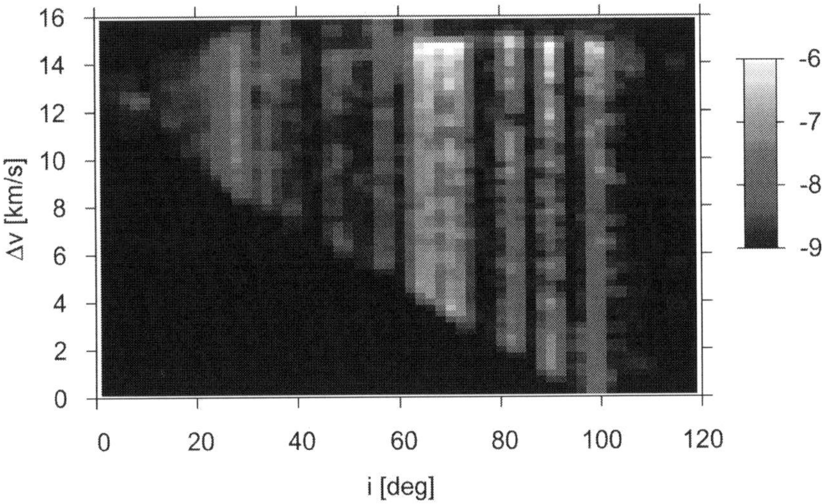

Fig. E.18. Debris flux distribution for $d > 1\,\text{cm}$ on an ERS orbit, as a function of the impactor orbit inclination i and the impact velocity Δv. Contour levels are given as $\log_{10}$ of the impact flux $F\,[\text{m}^{-2}\,\text{y}^{-1}]$ (class widths: $\Delta i = 2°$, $\Delta(\Delta v) = 0.25\,\text{km/s}$; see page 132).

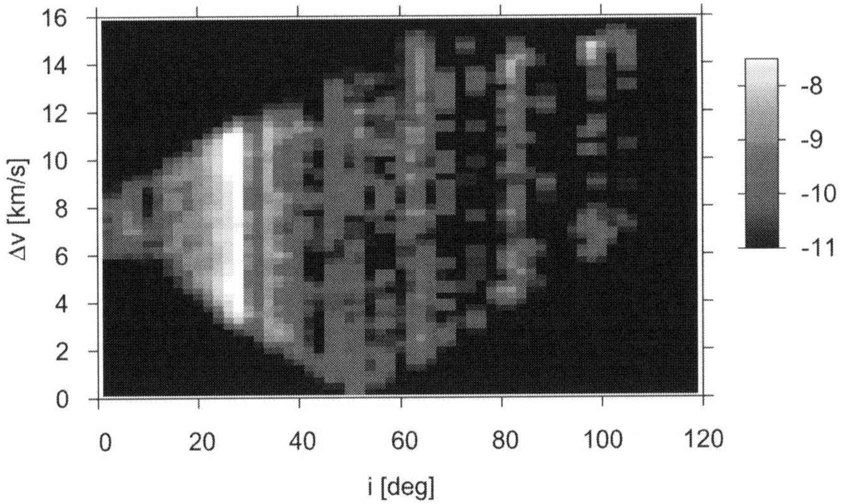

Fig. E.19. Debris flux distribution for $d > 1\,\text{cm}$ on an ISS orbit, as a function of the impactor orbit inclination i and the impact velocity Δv. Contour levels are given as $\log_{10}$ of the impact flux $F\,[\text{m}^{-2}\,\text{y}^{-1}]$ (class widths: $\Delta i = 2°$, $\Delta(\Delta v) = 0.25\,\text{km/s}$; see page 133).

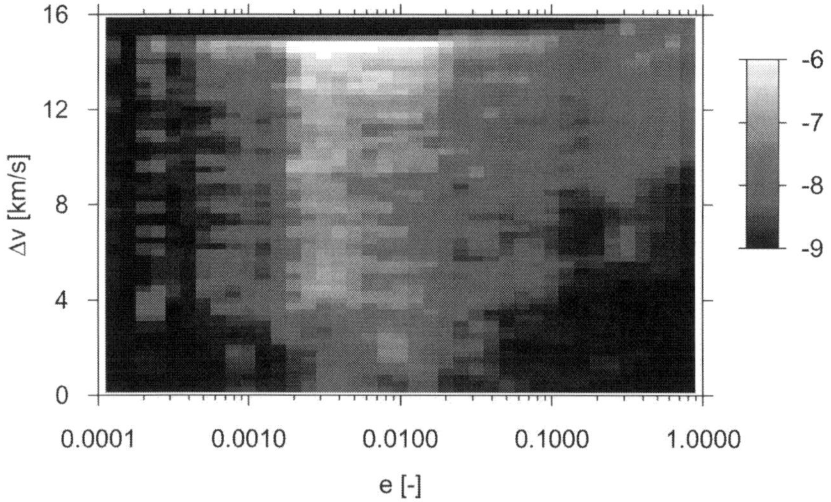

Fig. E.20. Debris flux distribution for $d > 1\,\mathrm{cm}$ on an ERS orbit, as a function of the impactor orbit eccentricity e and the impact velocity Δv. Contour levels are given as $\log_{10}$ of the impact flux $F\,[\mathrm{m}^{-2}\,\mathrm{y}^{-1}]$ (class widths: $\Delta(\Delta v) = 0.25\,\mathrm{km/s}$, 40 equal-size classes in $log_{10}\,e$; see page 134).

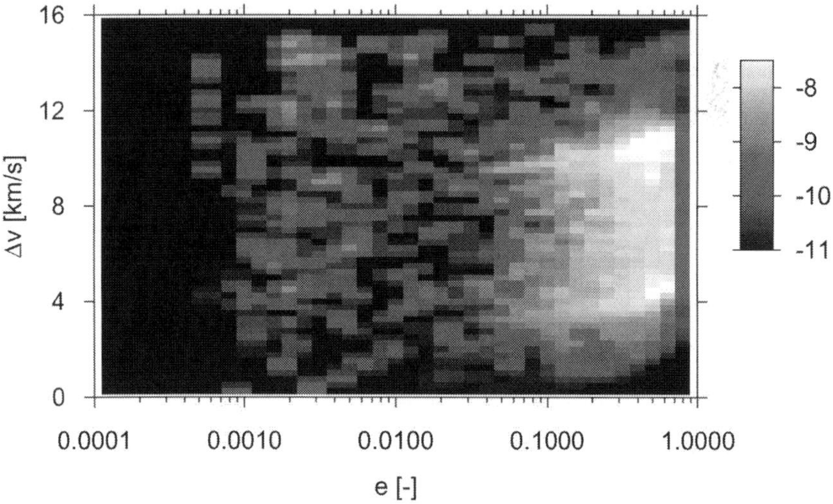

Fig. E.21. Debris flux distribution for $d > 1\,\mathrm{cm}$ on an ISS orbit, as a function of the impactor orbit eccentricity e and the impact velocity Δv. Contour levels are given as $\log_{10}$ of the impact flux $F\,[\mathrm{m}^{-2}\,\mathrm{y}^{-1}]$ (class widths: $\Delta(\Delta v) = 0.25\,\mathrm{km/s}$, 40 equal-size classes in $log_{10}\,e$; see page 134).

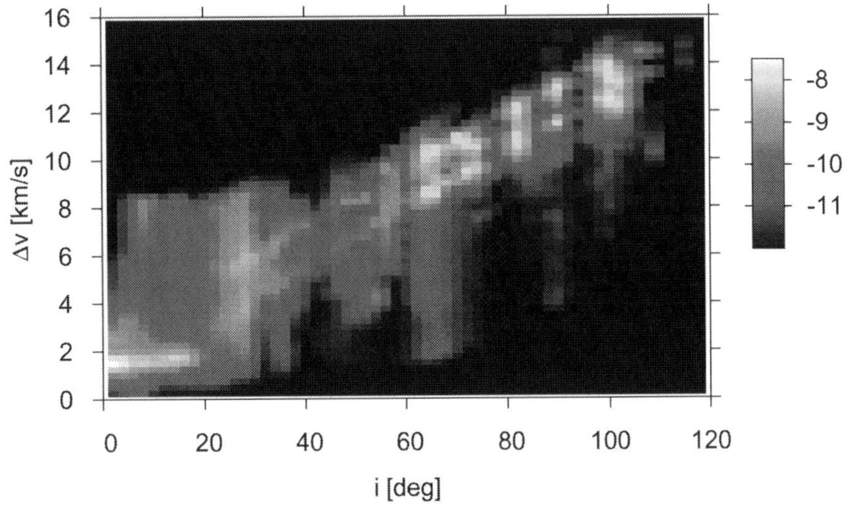

Fig. E.24. Debris flux distribution for $d > 1$ cm on a GTO orbit, as a function of impact velocity Δv and impactor orbit inclination i. Contour levels are given as $\log_{10}$ of the impact flux $F\,[\mathrm{m}^{-2}\,\mathrm{y}^{-1}]$ (class widths: $\Delta i = 2°$, $\Delta(\Delta v) = 0.25$ km/s; see page 137).

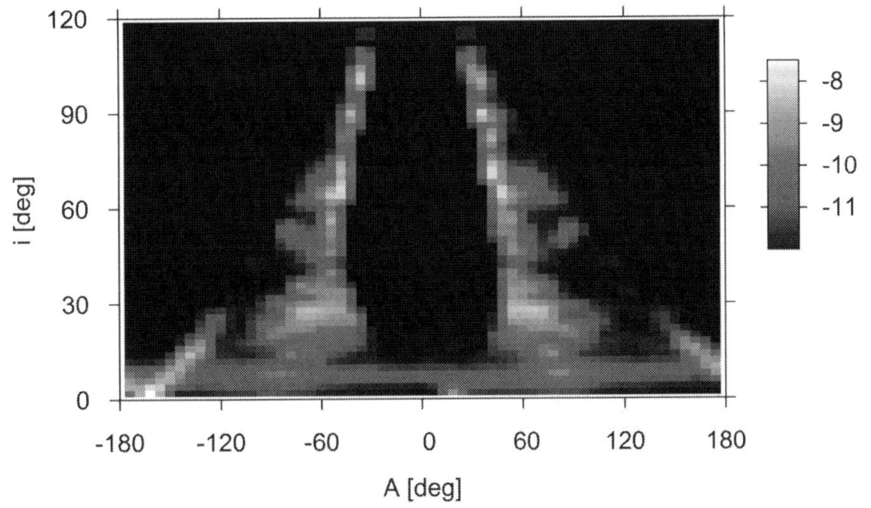

Fig. E.25. Debris flux distribution for $d > 1$ cm on a GTO orbit, as a function of the impact azimuth A and the impactor orbit inclination i. Contour levels are given as $\log_{10}$ of the impact flux $F\,[\mathrm{m}^{-2}\,\mathrm{y}^{-1}]$ (class widths: $\Delta A = 6°$, $\Delta i = 2°$; see page 138).

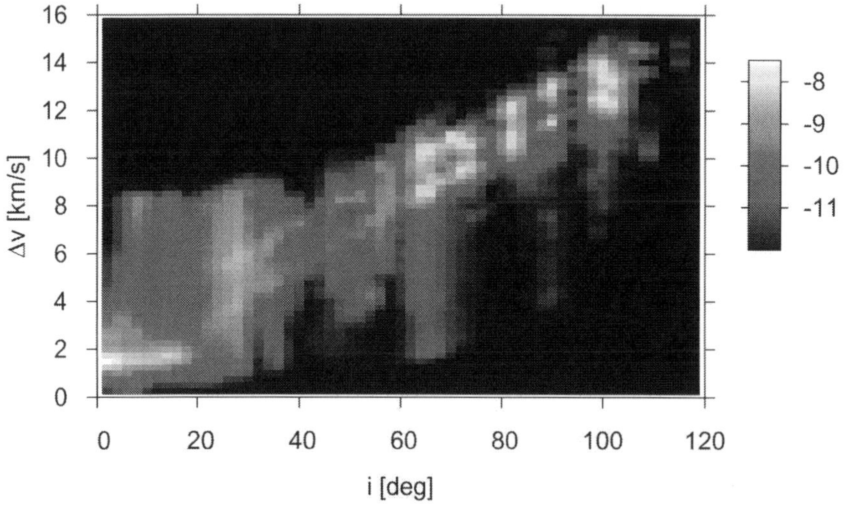

Fig. E.24. Debris flux distribution for $d > 1\,\mathrm{cm}$ on a GTO orbit, as a function of impact velocity Δv and impactor orbit inclination i. Contour levels are given as $\log_{10}$ of the impact flux $F\,[\mathrm{m}^{-2}\,\mathrm{y}^{-1}]$ (class widths: $\Delta i = 2°$, $\Delta(\Delta v) = 0.25\,\mathrm{km/s}$; see page 137).

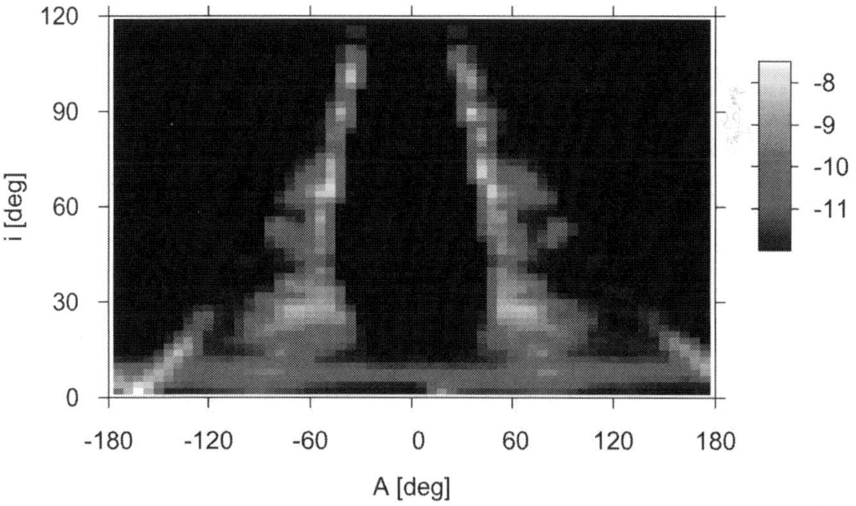

Fig. E.25. Debris flux distribution for $d > 1\,\mathrm{cm}$ on a GTO orbit, as a function of the impact azimuth A and the impactor orbit inclination i. Contour levels are given as $\log_{10}$ of the impact flux $F\,[\mathrm{m}^{-2}\,\mathrm{y}^{-1}]$ (class widths: $\Delta A = 6°$, $\Delta i = 2°$; see page 138).

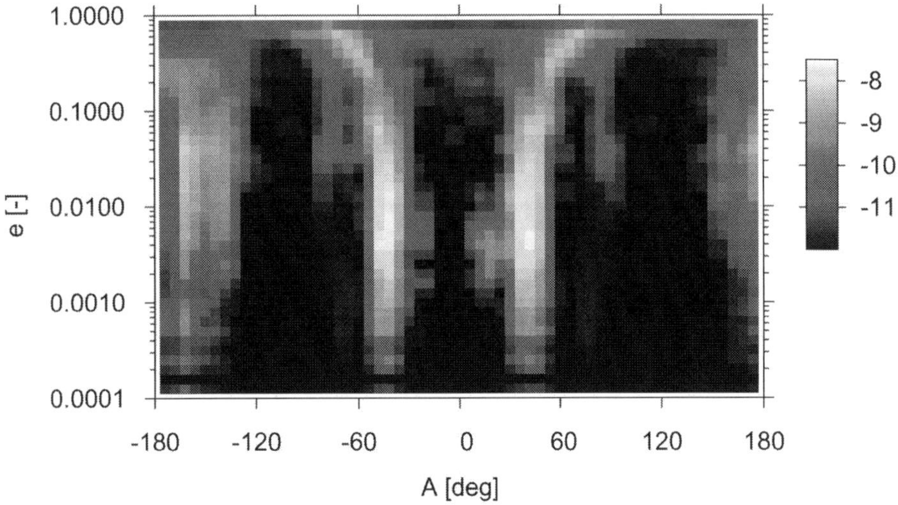

Fig. E.26. Debris flux distribution for $d > 1\,\mathrm{cm}$ on a GTO orbit, as a function of the impact azimuth A and the impactor orbit eccentricity e. Contour levels are given as $\log_{10}$ of the impact flux $F\,[\mathrm{m}^{-2}\,\mathrm{y}^{-1}]$ (class widths: $\Delta A = 6°$, 40 equal-size classes in $log_{10}\,e$; see page 138).

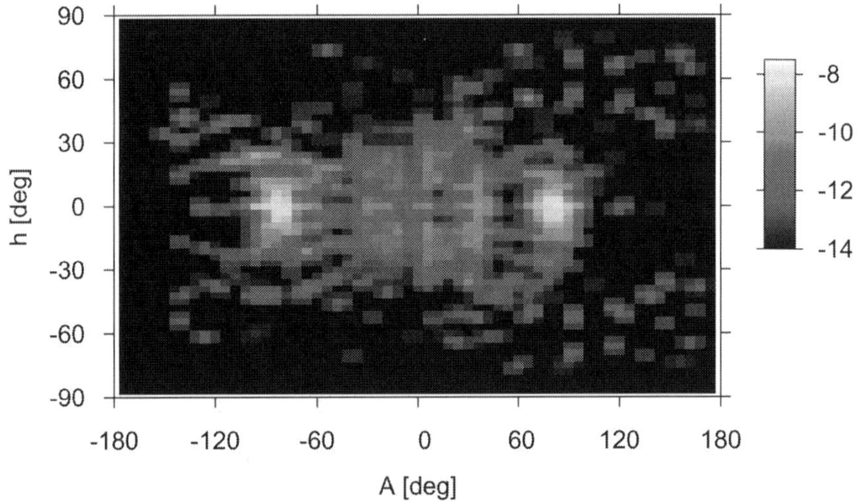

Fig. E.27. Debris flux distribution for $d > 1\,\mathrm{cm}$ on a GEO orbit, as a function of the impact elevation h and the impact azimuth A. Contour levels are given as $\log_{10}$ of the impact flux $F\,[\mathrm{m}^{-2}\,\mathrm{y}^{-1}]$ (class widths: $\Delta h = 3°$, $\Delta A = 6°$; see page 140).

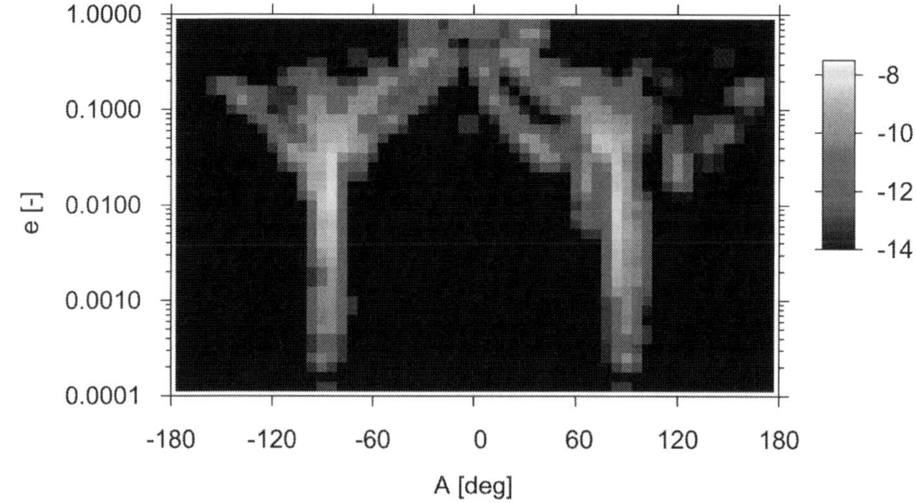

Fig. E.28. Debris flux distribution for $d > 1\,\mathrm{cm}$ on a GEO orbit, as a function of the impactor orbit eccentricity e and the impact azimuth A. Contour levels are given as $\log_{10}$ of the impact flux $F\,[\mathrm{m}^{-2}\,\mathrm{y}^{-1}]$ (class widths: $\Delta A = 6°$, 40 equal-size classes in $log_{10}\, e$; see page 140).

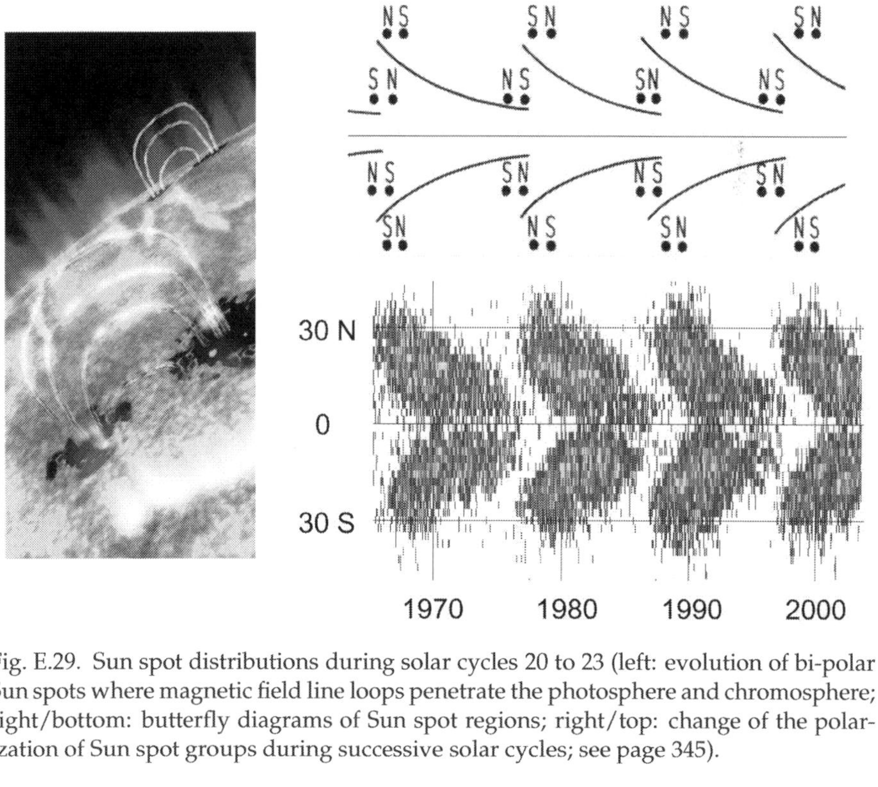

Fig. E.29. Sun spot distributions during solar cycles 20 to 23 (left: evolution of bi-polar Sun spots where magnetic field line loops penetrate the photosphere and chromosphere; right/bottom: butterfly diagrams of Sun spot regions; right/top: change of the polarization of Sun spot groups during successive solar cycles; see page 345).

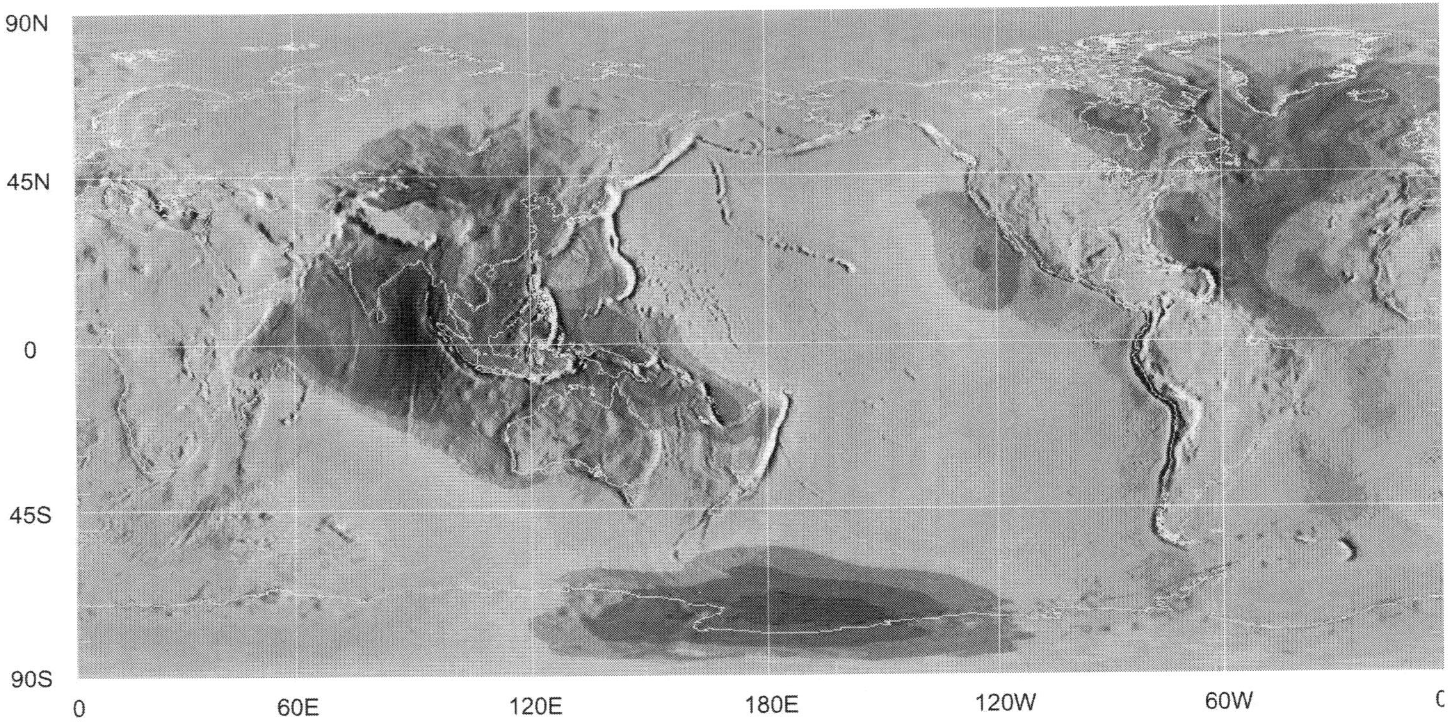

Fig. E.30. Geoid of the geopotential model EGM-96 (Lemoine et al., 1998). The geoid is an equipotential surface which approximates the mean sea level in the presence of an non-symmetric geopotential field. Level contours of geoid altitudes ΔH_g are provided relative to an Earth ellipsoid of $a_e \approx 6378.136$ km and $f_e = 298.257$. Bright (red) regions indicate mass concentrations leading to a global maximum of $\Delta H_{g,max} \approx +85$ m, North of New Guinea, while dark (blue) regions indicate mass deficits leading to a global minimum of $\Delta H_{g,min} \approx -105$ m, south of India (see page 353).

List of Symbols

Symbol	Dimension	Definition
a	[m]	semimajor axis, Fig. A.1;
	[m s^{-1}]	speed of sound, Eq. 7.2
a_e	[m]	Earth equatorial radius, $a_e = 6,378,135$ m
a_p	[–]	3-hourly planetary geomagnetic index
A	[m^2]	area; cross-section;
	[rad], [°]	azimuth angle, Fig. A.3
A_p	[–]	daily planetary geomagnetic index
A_B	[–]	conjunction parameter, Eq. 8.31
B	[Hz]	bandwidth;
	[m^2 kg^{-1}]	ballistic parameter, Eq. 6.1 & 6.2;
	[–]	slope parameter, Eq. 10.22
B_F	[–]	auxiliary quantity, Eq. 10.10
B_m	[V s m^{-2}]	magnetic induction, Eq. 6.11 to 6.13
B_S	[–]	auxiliary quantity, Eq. 10.12
c	[m s^{-1}]	speed of light, $c = 299,792,456.2$ m/s;
	[–]	collision flux parameter; constant
c_D	[–]	drag coefficient
c_h	[–]	heat transfer coefficient, Eq. 9.27
c_L	[–]	lift coefficient
c_p	[J kg^{-1} K^{-1}]	specific heat capacity at constant pressure
c_R	[–]	radiation coefficient
c_S	[–]	side force coefficient
C	[–]	covariance matrix, Eq. 8.3
$C_{n,m}$	[–]	geopotential harmonic coefficient, Eq. C.3
d	[m]	diameter
D	[m]	aperture; diameter;
	[m^{-3}]	spatial density

Symbol	Dimension	Definition
e	[–]	eccentricity, Fig. A.1;
	[–]	base of the natural logarithm, $e = \exp(1.0)$
e_χ	[–]	auxiliary variable, Eq. 10.4
E	[J]	energy;
	[rad], [°]	eccentric anomaly, Fig. A.1
$\underline{E}$	[–]	east unit vector, Fig. A.3
f	[rad], [°]	true anomaly;
	[m]	focal length;
	[–]	number of Sun spots
f_e	[–]	Earth oblateness, $f_e = 1/298.257$
F	[$\mathrm{m^{-2}\,s^{-1}}$]	object flux;
	[N]	thrust
$F_{10.7}$	[$10^{-22}\,\mathrm{W\,m^{-2}\,Hz^{-1}}$]	10.7 cm solar flux at $t-1$ day
$\overline{F}_{10.7}$	[$10^{-22}\,\mathrm{W\,m^{-2}\,Hz^{-1}}$]	mean 10.7 cm solar flux, averaged over the preceeding 81 days (3 solar rotations)
g	[$\mathrm{m\,s^{-2}}$]	gravitational acceleration;
	[–]	groups of Sun spots
h	[rad], [°]	elevation angle, Fig. A.3;
	[$\mathrm{J\,kg^{-1}}$]	specific enthalpy
$\underline{h}$	various	observation vector, Eq. 8.1
H	[km]	equatorial altitude, Eq. A.24;
	[–]	mass distribution parameter, Eq. 10.6
H	[–]	Jacobian matrix, Eq. 8.1
H_ρ	[km]	density scale height, Eq. B.9
H_{n_i}	[km]	number density scale height, Eq. B.2
i	[rad], [°]	inclination, Fig. A.2
I	[A]	plasma current
I_n	[–]	integer Bessel function of order n
I_{sp}	[s]	specific impulse
J	[–]	inertial tensor
$J_{n,m}$	[–]	geopotential harmonic coefficient, Eq. C.4
k	[$\mathrm{W\,K^{-1}\,Hz^{-1}}$]	Boltzmann constant, Eq. 2.9;
	[–]	scaling coefficient
k_p	[–]	3-hourly logarithmic geomagnetic index
K_p	[–]	daily mean logarithmic geomagnetic index
Kn	[–]	Knudsen number, Eq. 9.15
l	[m]	length
L	[–]	fraction of land masses
m	[kg]	mass;
	[$\mathrm{rad\,s^{-2}}$]	perturbing rotational acceleration
M	[$\mathrm{kg\,kmol^{-1}}$]	molecular mass
M_m	[$\mathrm{A\,m^2}$]	magnetic dipole moment, Eq. 6.11 to 6.13
n	[$\mathrm{rad\,s^{-1}}$]	mean orbit motion, Eq. A.4;

Symbol	Dimension	Definition
	$[\mathrm{m}^{-3}]$	number density, Eq. B.7
N	[–]	number of objects
$\underline{N}$	[–]	north unit vector, Fig. A.3
p	$[\mathrm{N\,m}^{-2}]$	pressure;
	various	probability density;
	$[\mathrm{m\,s}^{-2}]$	perturbing translational acceleration;
	[m]	parameter of an ellipse, Fig. A.1
$\underline{p}$	$[\mathrm{m\,s}^{-2}]$	perturbing acceleration vector
P	[–]	probability;
	$[\mathrm{m}^2\,\mathrm{s}^{-2}]$	perturbing potential;
	[W]	power
PDF_i	various	impact probability density function, Eq. 9.34ff
$\dot{q}$	$[\mathrm{W\,m}^{-2}\,\mathrm{s}^{-1}]$	heat flux
Q	[W]	heat energy;
	[C]	plasma charge
$\dot{Q}$	$[\mathrm{W\,s}^{-1}]$	heating rate, Eq. 9.16 & 9.17
r	[m]	radius
R	$[\mathrm{J\,kmol}^{-1}\,\mathrm{K}^{-1}]$	gas constant, $R = 8,314.41\,\mathrm{J\,kmol}^{-1}\,\mathrm{K}^{-1}$;
	[m]	radius;
	[–]	Zürich Sun spot number
R	[–]	rotation matrix
Re	[–]	Reynolds number
S	[m]	spacing between bumper and back wall
$S_{n,m}$	[–]	geopotential harmonic coefficient, Eq. C.3
SFU	$[10^{-22}\,\mathrm{W\,m}^{-2}\,\mathrm{Hz}^{-1}]$	solar flux unit = 10^4 Jy (Jansky)
St	[–]	Stanton number, Eq. 9.15
t	[s]	time;
	[m]	thickness
T	[s]	orbit period, Eq. A.5;
	[s]	libration period, Eq. 6.19;
	[K]	temperature
T_Ω	[s]	nodal orbit period, Eq. D.2
u	[rad], [°]	argument of true latitude, $u = \omega + f$
$\underline{u}$	[–]	unit vector
U	$[\mathrm{m}^2\,\mathrm{s}^{-2}]$	perturbing geopotential, Eq. C.3
$\underline{U}$	[–]	radial unit vector, Fig. A.2
v	$[\mathrm{m\,s}^{-1}]$	velocity
V	$[\mathrm{m}^3]$	volume
$\underline{V}$	[–]	transversal unit vector, Fig. A.2
w_e	$[\mathrm{m\,s}^{-1}]$	exhaust velocity
W	various	observation weighting matrix, Eq. 8.2
$\underline{W}$	[–]	out-of-plane unit vector, Fig. A.2
$\underline{x}$	various	state vector

Symbol	Dimension	Definition
$\underline{X}$	[–]	unit vector X, Fig. A.2
$\underline{Y}$	[–]	unit vector Y, Fig. A.2
z	[km]	geodetic altitude, Eq. A.25;
	[–]	drag parameter, Eq. 6.4
$\hat{z}$	[km]	geopotential altitude, Eq. B.3
Z	[–]	zenith hourly rate contribution, Eq. 10.22
$\underline{Z}$	[–]	unit vector Z, Fig. A.2
ZHR	[s^{-1}]	zenith hourly rate, Eq. 10.22
α	[rad], [°]	right ascension; incident angle;
	[–]	number density profile parameter, Eq. B.7
β	[–]	ratio of radii, Eq. A.21ff
γ	[rad], [°]	flight path angle, Fig. A.1;
	[–]	number density profile parameter, Eq. B.7
δ	[rad], [°]	declination
ΔS_{Ω}	[rad], [°]	equatorial ground track separation
ϵ	[rad], [°]	radiation incident angle;
	[–]	radiation emission coefficient, Eq. 9.16 & 9.17
ζ	[–]	fitting function, Eq. 3.52
η	[–]	efficiency parameter; performance index
θ	[rad], [°]	cone angle, Eq. 3.58;
	[rad], [°]	magnetic co-latitude, Eq. 6.11ff
ϑ	[–]	size distribution parameter, Eq. 3.34;
	[rad], [°]	tilt of the geomagnetic pole, Eq. 6.15
κ	[–]	ratio of specific heats, Eq. 7.2
λ	[rad], [°]	geodetic longitude;
	[m]	wavelength
λ_{Ω}	[rad], [°]	geodetic longitude at the ascending node
μ	various	mean value;
	[$m^3\ s^{-2}$]	central Earth attraction term, Eq. A.4;
	[$V\ s\ A^{-1}\ m^{-1}$]	magnetic field constant, Eq. 6.11 to 6.13
ν	[–]	velocity distribution parameter
ρ	[m]	range;
	[$kg\ m^{-3}$]	density
σ	various	standard deviation;
	[$W\ m^{-2}\ K^{-4}$]	Stefan–Boltzmann constant, Eq. 9.16 & 9.17;
	[$N\ m^{-2}$]	normal stress;
	[–]	temperature profile parameter, Eq. B.6
τ	[$N\ m^{-2}$]	shear stress; yield stress;
	[–]	temperature profile parameter, Eq. B.6
ϕ	[rad], [°]	geodetic latitude, Eq. A.26;
	[rad], [°]	thrust direction angle, Eq. 6.22
Φ	various	state transition matrix, Eq. 8.5

Symbol	Dimension	Definition
χ	[–]	area-to-mass distribution parameter, Eq. 3.35;
	[–]	auxiliary variable, Eq. 10.4
ψ	[J kg^{-1}]	specific energy, Eq. 5.6
$\underline{\psi}$	various	generalized state vector, Section A.8
ω	[rad], [°]	argument of pericenter, Fig. A.2;
	[rad s^{-1}]	angular velocity
ω_e	[rad s^{-1}]	Earth sidereal rotation rate, Section A.3
Ω	[rad], [°]	right ascension of the ascending node, Fig. A.2
$\square_a$		atmospheric; aerodynamic
$\square_{ap}$		apocenter
$\square_b$		meteoroid stream background
$\square_B$		B plane, Eq. 8.28
$\square_c$		collision; casualty; cut-off; continuum
$\square_{cat}$		cataloged
$\square_{ce}$		cone ejecta
$\square_{cond}$		conduction
$\square_{conv}$		convection
$\square_{cum}$		cumulated
$\square_d$		debris object; solid rocket motor dust; demise; day (of the year)
$\square_D$		airdrag
$\square_e$		eccentricity; emitter; Earth; equatorial; exhaust
$\square_{E/W}$		east–west
$\square_{\mathrm{f}}$		fatality; free molecular
$\square_{\mathrm{F}}$		gravitational focusing
$\square_{\mathrm{fa}}$		false alarm
$\square_g$		gravitational
$\square_i$		inclination; impact
$\square_{i,j,k}$		volume cell centered at r_i, δ_j, and α_k
$\square_{kin}$		kinetic
$\square_\ell$		ℓ-th object in a volume cell
$\square_{lam}$		laminar flow
$\square_L$		orbit lifetime; GEO libration; land
$\square_{LST}$		local solar time
$\square_m$		m-th cell passage; magnetic; melting; meteoroid
$\square_{man}$		maneuver
$\square_n$		sodium-potassium (NaK) reactor coolant; night
$\square_{N/S}$		north–south
$\square_p$		perihelion; projectile; propellant; population; meteoroid stream peak
$\square_{pe}$		pericenter
$\square_{pot}$		potential

Symbol	Dimension	Definition
$\square_{PNP}$		probability of no penetration
$\square_{r}$		receiver; radial
$\square_{rad}$		radiation
$\square_{rcs}$		radar cross-section
$\square_{red}$		reduced
$\square_{res}$		residual
$\square_{R}$		radiation
$\square_{s}$		shield; solid rocket motor slag; stream
$\square_{S}$		planetary shielding
$\square_{sd}$		surface degradation products
$\square_{sys}$		system
$\square_{t}$		target; third body
$\square_{tot}$		total
$\square_{T}$		transfer; tangential
$\square_{TLE}$		two-line elements
$\square_{U,E,N}$		in the U, E, N system, Fig. A.3
$\square_{U,V,W}$		in the U, V, W system, Fig. A.2
$\square_{v}$		voxel
$\square_{w}$		wind; back wall; Westford Needles
$\square_{X,Y,Z}$		in the X, Y, Z system, Fig. A.2
$\square_{\theta}$		magnetic co-latitude
$\square_{\rho}$		density
$\square_{\sigma}$		normal stress
$\square_{\tau}$		shear stress
$\square_{\infty}$		free stream conditions
$\square_{120}$		at 120 km altitude
$\square_{2w}$		two-way
$\square_{\circ}$		initial value; reference value
$\square_{\odot}$		with respect to the Sun
$\square_{\oplus}$		with respect to the Earth
$\square_{\parallel}$		parallel
$\square_{\perp}$		perpendicular
$\square_{\uparrow}$		ascending pass; outbound
$\square_{\downarrow}$		descending pass; inbound
$\square^{\star}$		threshold value
$\Delta\square$		difference
$\dot{\square}$		first time derivative
$\ddot{\square}$		second time derivative
$\hat{\square}$		normalized quantity
$\tilde{\square}$		intermediate quantity
$\overline{\square}$		mean value, singly averaged quantity
$\overline{\overline{\square}}$		doubly averaged quantity
$\underline{\square}$		vector quantity

List of Abbreviations

Abbreviation	Definition
AE	Atmospheric Explorer satellite
AIAA	American Institute of Aeronautics and Astronautics
AMOS	(US) Air Force Maui Optical and Supercomputing Site
ARIMA	Auto-Regressive Integrated Moving Average
ASAT	Anti Satellite
ASI	Agenzia Spaziale Italiana (Italian Space Agency)
AU	Astronomical Unit (1 AU $\approx 149.6 \times 10^{+6}$ km)
BAU	Business-As-Usual
BMEWS	(US) Ballistic Missile Early Warning System
BNSC	British National Space Centre
CAD	Computer Aided Design
CCD	Charge-Coupled Device
CDT	CCD Debris Telescope (NASA)
CFRP	Carbon Fiber Reinforced Plastic
CIO	Conventional International Origin
CIRA	COSPAR International Reference Atmosphere
CNDB	(US) Civil Needs Data Base
CNES	Centre National d'Etudes Spatiales (France)
CNSA	Chinese National Space Agency (PR of China)
CNUCE	Centro Nazionale Universitario di Calcolo Elettronico (Italy)
COIW	Center of Impact Window
CoM	Center of Mass
COSPAR	Committee on Space Research
CPE	Cell Passage Event
CPS	Center for Program Studies (ROSCOSMOS)
CPU	Central Processor Unit
CRASS	Collision Risk Assessment software (ESA)

Abbreviation	Definition
CSM	(Apollo) Command and Service Module
dB	decibel
dBsm	decibel square meter (see Eq. 2.7)
DCE	Centres d'Essais et d'Evaluation (France)
DEBIE	Debris In-Orbit Evaluator (ESA)
DELTA	Debris Environment Long-Term Analysis software (ESA)
DELTOP	DELTA Orbit Propagator
DGA	Délégation Générale pour l'Armement (France)
DISCOS	Database and Information System Characterizing Objects in Space (ESA)
DLR	Deutsches Zentrum für Luft- und Raumfahrt (Germany)
DOC	Department of Commerce (US)
DoD	Department of Defense (US)
DoF	Degrees of Freedom
DOT	Department of Transportation (US)
DPU	Data Processor Unit
DRAMA	Debris Risk Assessment and Mitigation Analysis software (ESA)
DTM	Density and Temperature Model
EDMS	European Debris Mitigation Standard
EGM	Earth Gravitational Model
EMI	Ernst Mach Institute (Germany)
ENVISAT	Environment Satellite (ESA)
EOIM	Evaluation of Oxygen Interaction with Materials
EOL	End of Life
EOS	Equations of State
EOVV	End On Velocity Vector
EPS	Etage Propergol Stockable (Ariane)
ERS	European Remote Sensing Satellite (ESA)
ESA	European Space Agency
ESOC	European Space Operations Center (ESA)
ESRO	European Space Research Center (ESA precursor)
EUMETSAT	European Meteorological Satellite Organization
EURECA	European Retrievable Carrier (ESA)
EUTELSAT	European Telecommunications Satellite Organization
EUV	Extreme Ultra-Violet
FAA	Federal Aviation Administration (US)
FCC	Federal Communications Commission (US)
FGAN	Forschungsgesellschaft für angewandte Naturwissenschaften (Germany)
FoV	Field of View
FSA	Federal Space Agency (Russia)
GEM	Goddard Earth Model
GEO	Geostationary Earth Orbit

Abbreviation	Definition
GEODSS	Ground-Based Electro-Optical Deep Space Surveillance (SSN)
GLONASS	Global Navigation Satellite System (Russia)
GORID	Geostationary Orbit Impact Detector (ESA)
GOST	Gosudarstvennyiu Standard (State Standard of Russia)
GPS	Global Positioning System (US)
GRAVES	Grande Réseau Adapté à la Veille Spatial (France)
GRO	Greenwich Royal Observatory (UK)
GSFC	Goddard Space Flight Center (US)
GTO	Geostationary Transfer Orbit
HAX	Haystack Auxiliary Radar (US)
HEO	Highly Eccentric Earth Orbit
HST	Hubble Space Telescope (US)
HVI	Hypervelocity Impact
HVIS	Hypervelocity Impact Symposium
HWM	Horizontal Wind Model
IAA	International Academy of Astronautics
IADC	Inter-Agency Space Debris Coordination Committee
IAC	International Astronautical Congress
IAF	International Astronautical Federation
IDES	Integrated Debris Evolution Suite (UK)
INMARSAT	International Maritime Satellite Organization
IR	Infra-Red
IRAS	Infra-Red Astronomy Satellite (The Netherlands and US)
ISAR	Inverse Synthetic Aperture Radar
ISO	International Organization for Standardization
ISRO	Indian Space Research Organization (India)
ISS	International Space Station
ISTI	Instituto di Scienza e Technologie dell' Informazione (Italy)
ISTS	International Symposium on Space Technology and Science (Japan)
ITU	International Telecommunication Union
IUS	Inert Upper Stage
JAXA	Japan Aerospace Exploration Agency (Japan)
JPL	Jet Propulsion Laboratory (US)
JSC	Johnson Space Center (US)
KSC	Kennedy Space Center (US)
LDEF	Long Duration Exposure Facility (US)
LEGEND	LEO-to-GEO Environment Debris Model
LEO	Low-Earth Orbit
LGG	Light-Gas Gun
LH	Liquid Hydrogen
LMRO	Launch- or Mission-Related Object

Abbreviation	Definition
LMT	Liquid Mirror Telescope (US)
LOX	Liquid Oxygen
LRIR	Long-Range Imaging Radar (US)
mag	magnitude (see Eq. 2.2)
MASTER	Meteoroid And Space debris Terrestrial Environment Reference model (ESA)
MCC	Mission Control Center (Russia)
MEO	Medium Earth Orbit
MIT	Massachusetts Institute of Technology (US)
MLI	Multi-Layer Insulation
MOD	Ministry of Defence (UK)
MOS	Metal Oxide Semiconductor
MOTIF	Maui Optical Tracking and Identification Facility (US)
MRO	Mission-Related Object
MSIS	Mass-Spectrometer and Incoherent Scatter (density model)
MSX	Mid-course Space Experiment (US)
NaK	sodium–potassium alloy
NASA	National Aeronautics and Space Administration (US)
NASDA	National Space Development Agency (Japan)
NATO	North Atlantic Treaty Organization
NAVSPASUR	Naval Space Surveillance (US)
NEO	Near-Earth Object
NOAA	National Oceanic and Atmospheric Administration (US)
NORAD	North American Aerospace Defense Command (US)
NPS	Nuclear Power Source
NRLMSIS	Naval Research Laboratory MSIS model (US)
NSAU	National Space Agency of the Ukraine
NSS	NASA Safety Standard
OCM	Orbital Conjunction Messages
ODCWG	Orbital Debris Coordination Working Group (ISO)
OIG	Orbit Information Group (NASA)
ORDEM	Orbital Debris Environment Model (NASA)
ORSAT	Orbital Re-entry Survival Analysis Tool (NASA)
PAM	Payload Assist Module
PAVE	program name of electronic systems used in PAWS
PAWS	Phased Array Warning System (US)
PIMS	Passive Imaging Metric Sensor (UK)
P/L	Payload
PNF	Probability of No Failure
PNP	Probability of No Penetration
PROBA	Project for On Board Autonomy (ESA)
PROOF	Program for Radar and Optical Observation Forecasting (ESA)
RAAN	Right Ascension of the Ascending Node

Abbreviation	Definition
RAL	Rutherford Appleton Laboratory (UK)
RASA	Russian Aeronautics and Space Administration
R/B	Rocket Body
RCS	Radar Cross-Section
RKA	Russian Space Agency (synonym for RSA)
RMS	Root Mean Squares
RORSAT	Radar Ocean Reconnaissance Satellite (USSR)
ROSACE	Restitution d'Orbite par Système Autonome CCD d'Ecartométrie
ROSCOSMOS	Russian Federal Space Agency
RSA	Russian Space Agency
RTG	Radio-isotope Thermo-electric Generator
SAR	Synthetic Aperture Radar
SBV	Space Based Visible
S/C	Spacecraft
SCARAB	Spacecraft Atmospheric Re-entry and Aerothermal Break-up prediction tool (ESA)
SDI	Strategic Defense Initiative (US)
SDM	Semi-Deterministic Model (ESA)
SDP	Simplified Deep-Space Perturbation theory (US)
SDPA	Space Debris Prediction and Analysis engineering model (Russia)
SDT	Space Debris Telescope (ESA)
SEM	Scanning Electron Microscope
SHO	Sufficiently High Orbit
SNAP	System for Nuclear Auxiliary Power (US)
SPASUR	Space Surveillance system (US)
SPH	Smoothed Particle Hydrodynamics
SPOC	Système Probatoire d'Observation du Ciel (France)
SPOT	Système Pour l'Observation de la Terre (France)
SRM	Solid Rocket Motor
SRON	Space Research Organization of the Netherlands
SSC	Space Surveillance Center (US)
SSN	Space Surveillance Network (US)
SSR	Satellite Situation Report (US)
SSS	Space Surveillance System (Russia)
STS	Space Transportation System (US)
TAROT	Télescope à Action Rapide pour les Objets Transitoires
TCA	Time of Closest Approach
TEA	Torque-Equilibrium Attitude
TEME	True Equator and Mean Equinox
TICCE	Time-Band Capture Cell Experiment
TIRA	Tracking and Imaging Radar (FGAN)
TLE	Two-Line Element set

Abbreviation	Definition
TNT	Trinitrotoluene
TUBS	Technical University of Braunschweig (Germany)
UHF	Ultra-High Frequency
UK	United Kingdom of Great Britain and Northern Ireland
UN	United Nations
UNCOPUOS	UN Committee on the Peaceful Uses of Outer Space
US	United States (of America)
USA	United States of America
USSPACECOM	US Space Command
USSR	Union of Socialist Soviet Republics
USSTRATCOM	US Strategic Command
UTC	Coordinated Universal Time
UV	Ultra-Violet
VHF	Very High Frequency
WARC	World Administrative Radio Conference
WGS	World Geodetic System
ZHR	Zenith Hourly Rate

List of Tables

List of Figures

Bibliography

Abramowitz, M. and Stegun, A. (1984). *Pocketbook of Mathematical Functions*. Verlag Harri Deutsch, Frankfurt a.M.

Agrawal, B. N. (1986). *Design of Geosynchronous Spacecraft*. Prentice-Hall, Inc., Englewood Cliffs, NJ.

Ailor, W., Hallman, W., Steckel, G., and Weaver, M. (2005). Analysis of Re-Entered Debris and Implications for Survivability Modeling. In *Proceedings of the Fourth European Conference on Space Debris*, ESA SP-587.

Akiba, R. et al. (1990). Behavior of Aluminum Particles Exhausted by Solid Rocket Motors. In *Orbital Debris Conference, Baltimore/MD*, AIAA/NASA/DoD.

Aksnes, K. (1976). Short-Period and Long-Period Perturbations of a Spherical Satellite Due to Direct Solar Radiation Pressure. *Celestial Mechanics*, vol.13:89–104.

Alarcón, J. (2002). Development of a Collision Risk Assessment Tool. Technical Report (final report) ESA contract 14801/00/D/HK, GMV S.A.

Alby, F. (2003). SPOT-1 End of Life Disposal Manoeuvres. *35th COSPAR Scientific Assembly*, PEDAS1/B1.6-0035-04.

Alby, F., Lansard, E., and Michal, T. (1997). Collision of Cerise with Space Debris. In *Proceedings of the Second European Conference on Space Debris*, ESA SP-393, pages 589–596.

Alfriend, K., Akella, M., Lee, D., Frisbee, J., and Foster, J. (1999). Probability of Collision Error Analysis. *Space Debris*, vol.1, no.1:21–35.

Allen, H. and Eggers, A. (1953). A Study of the Motion and Aerodynamic Heating of Missiles Entering the Earth's Atmosphere at High Supersonic Speeds. Technical Report NACA RM A53D28, National Advisory Committee for Aeronautics.

anon. (1976). U.S. Standard Atmosphere 1976. Technical Report NOAA-S/T 76-1562, NOAA/NASA/USAF, Washington, DC.

anon. (1985). Earth Upper Atmosphere – Density Model for Ballistic Support of Flights of Artificial Satellites. Technical Report GOST 25645.115-84, USSR State Standard, Moscow.

anon. (1988). Space Debris. Technical Report ESA SP-1109, ESA Space Debris Working Group, Darmstadt, Germany.

anon. (1989). Report on Orbital Debris. Technical report, Inter-Agency Group (Space) of the National Security Council, Washington, DC.

anon. (1995a). Hazard Analysis of Commercial Space Transportation. Technical Report DOT Office of Commercial Space Transportation.

anon. (1995b). *NASA Safety Standard – Guidelines and Assessment Procedures for Limiting Orbital Debris.* NSS 1740.14.

anon. (1995c). *Orbital Debris – A Technical Assessment.* National Academy Press, Washington, DC.

anon. (1996). *Space Debris Mitigation Standard.* NASDA-STD-18.

anon. (1997). *US Government Orbital Debris Mitigation Standard Practices.*

anon. (1999a). *CNES Exigence de Securité – Débris Spatiaux: Méthode et Procedure.* CNES, MPM-51-00-12, issue 1, rev. 0.

anon. (1999b). *Technical Report on Space Debris.* United Nations Publications, New York, USA.

anon. (2000a). Common Risk Criteria for National Test Ranges – Inert Debris. Technical Report MIL Standard 321-00.

anon. (2000b). Potentially Hazardous Near-Earth Objects. Technical Report by the NEO Task Force, BNSC, London, UK.

anon. (2000c). *Russian Aviation & Space Agency (RASA) Branch Standard – General Requirements for Space Debris Mitigation.* RASA.

anon. (2001a). *European Space Debris Safety and Mitigation Standard – Draft.* issue 1, rev. 3.

anon. (2001b). *Position Paper on Orbital Debris.* International Academy of Astronautics, Paris, France.

anon. (2002a). *IADC Space Debris Mitigation Guidelines.* issue 1, rev. 1.

anon. (2002b). *United Nations Treaties and Principles on Outer Space.* United Nations publications, Vienna, Austria.

anon. (2003). IADC Space Debris Mitigation Guidelines. Technical Report A/AC/105/C.1/L.2260, UNCOPUOS, Vienna, Austria.

anon. (2004a). Protection Manual. Technical Report IADC-WD-00-03, rel. 3.3, Inter-Agency Space Debris Coordination Committee (IADC).

anon. (2004b). Space Mission Priorities for Near-Earth Object Risk Assessment and Reduction. Technical Report by the NEOMAP Panel, ESA.

anon. (2005). Accidential Collisions of Cataloged Satellites Identified. *The Orbital Debris Quarterly News*, 9, Issue 2:1–2.

Barlier, F., Berger, C., Falin, L., Kockarts, G., and Tuillier, G. (1978). A Thermospheric Model Based on Satellite Drag Data. *Annales de Géophysique*, 34(1):9–24.

Bekey, I. (1997). Project Orion: Orbital Debris Removal Using Ground-Based Sensors and Lasers. In *Proceedings of the Second European Conference on Space Debris*, ESA SP-393, pages 699–701.

Bendisch, J., Bunte, K., Sdunnus, H., Wegener, P., Walker, R., and Wiedemann, C. (2002). Upgrade of ESA's MASTER Model. Technical Report (final report) ESA contract 14710/00/D/HK, ILR/TUBS, Braunschweig.

Bérend, N. (1997). Étude de la Probabilité de Collision Entre un Satellite et des Débris Spatiaux. Technical Report no. RT 38/3605 SY, ONERA.

Beutler, G. (2005a). *Methods of Celestial Mechanics – Volume I: Physical, Mathematical, and Numerical Principles*. Springer Verlag, Berlin, Heidelberg, New York.

Beutler, G. (2005b). *Methods of Celestial Mechanics – Volume II: Application to Planetary System, Geodynamics, and Satellite Geodesy*. Springer Verlag, Berlin, Heidelberg, New York.

Bleeker, J., Geiss, J., and Huber, M. (2001). *The Century of Space Science – Volume I*. Kluwer Academic Publishers, Dordrecht, Boston, London.

Bondarenko, S. and Lyagushin, S. (1997). Prospects of Using Military Lasers and Military Space Technology for Space Debris Removal. In *Proceedings of the Second European Conference on Space Debris*, ESA SP-393, pages 703–706.

Bonnal, C., Sanchez, M., and Naumann, W. (1997). Ariane Debris Mitigation Measures. In *Proceedings of the Second European Conference on Space Debris*, ESA SP-393, pages 681–688.

Bouslog, S., Ross, B., and Madden, C. (1994). Space Debris Re-Entry Risk Analysis. In *32nd Aerospace Sciences Meeting, Reno, USA*, AIAA Paper No.94-0591.

Brumberg, V. (1995). *Analytical Techniques of Celestial Mechanics*. Springer, Berlin, Heidelberg, New York.

Chao, C. (2001). MEO Disposal Orbit Stability and Direct Re-Entry Strategy. *11th AAS/AIAA Space Flight Mechanics Conference*, AAS-01-244.

Cook, G. (1962). Luni-Solar Perturbations of the Orbit of an Earth Satellite. *Geophysical Journal of the Royal Astronomical Society*, vol.6, no.3:271–291.

Cook, G. (1972). Basic Theory of PROD, a Program for Computing the Development of Satellite Orbits. *Celestial Mechanics*, Vol. 7:301–314.

Cour-Palais, B. (1969). Meteoroid Environment Model 1969 – Near Earth to Lunar Surface. Technical Report SP-8013, NASA.

Divine, N. (1993). Five Populations of Interplanetary Meteoroids. *Journal of Geophysical Research*, vol. 98:17,029–17,048.

Dreher, P., Little, R., and Wittenstein, G. (1980). Skylab Orbital Lifetime Prediction and Decay Analysis. Technical Report NASA TM 78308, NASA/MSFC.

Drolshagen, G. (1995). EURECA and HST Solar Array Post-Flight Impact Analysis. *Earth Space Review*, vol.4, no.2:15–18.

Drolshagen, G. (2001). Hypervelocity Impact Effects on Spacecraft. In *Proceedings of the Meteoroids Conference 2001, Kiruna, Sweden*, ESA SP-495, pages 533–541.

Foster, J. (1992). A Parametric Analysis of Orbital Debris Collision Probability and Maneuver Rate for Space Vehicles. Technical Report NASA JSC-25898, NASA.

Foster, J. (2001a). International Space Station Debris Avoidance Operations. *The Orbital Debris Quarterly News*, 6, Issue 2.

Foster, J. (2001b). The Analytical Basis for Debris Avoidance Operations of the International Space Station. In *Proceedings of the Third European Conference on Space Debris*, ESA SP-473, pages 441–445.

Fritsche, B., Klinkrad, H., Kashkovsky, A., and Grinberg, E. (1999). Spacecraft Disintegration During Uncontrolled Atmospheric Re-Entry. *50th IAC Congress*, IAA-99-IAA.6.7.02.

Fritsche, B., Koppenwallner, G., and Kashkovsky, A. (2000). Advanced Model for Spacecraft Disintegration During Atmospheric Re-Entry. Technical Report ESA Contract 12804/98/D/IM, ESA.

Goldstein, R., Goldstein, S., and Kessler, D. (1998). Radar Observations of Space Debris. *Planetary and Space Sciences*, vol.46, no.8:1007–1013.

Grün, E., Gustafson, B., Dermott, S., and Fechtig, H. (2001). *Interplanetary Dust*. Springer Verlag, Berlin/Heidelberg/New York.

Grün, E., Zook., H., Fechtig, H., and Giese, R. (1985). Collisional Balance in the Meteoritic Complex. *Icarus*, vol.62:244–272.

Hedin, A. (1987). MSIS-86 Thermospheric Model. *Journal of Geophysical Research*, vol.92, no.A5:4649–4662.

Hedin, A. (1991). Extension of the MSIS Thermosphere Model into the Lower Atmosphere. *Journal of Geophysical Research*, vol.96, no.A2:1159–1172.

Hedin, A., Bondi, M., Burnside, R., Hernandez, G., Johnson, R., Killeen, T., Mazaudier, C., Meriwether, J., Salah, J., Sica, R., Smith, R., Spencer, N., Wickwar, V., and Virdi, T. (1991). Revised Model of the Thermosphere Winds Using Satellite and Ground-Based Observations. *Journal of Geophysical Research*, vol.96, no.A5:7657–7688.

Hernández, C., Pina, F., Sánchez, N., Sdunnus, H., and Klinkrad, H. (2001). The DISCOS Database and Web Interface. In *Proceedings of the Third European Conference on Space Debris*, ESA SP-473, pages 803–807.

Hoots, F., Crawford, L., and Roehrich, R. (1984). An Analytical Method to Determine Future Close Approaches Between Satellites. *Celestial Mechanics*, vol.33.

Hoots, F. and Roehrich, R. (1980). Spacetrack Report No.3 – Models for the Propagation of NORAD Element Sets. Technical report, Office of Astrodynamics, Aerospace Defense Center, Peterson AFB, CO.

Hoyt, R. and Forward, R. (1999). Performance of the Termination Tether^{TM} for Autonomous De-Orbit of LEO Spacecraft. *50th IAF Congress*, IAF-99-IAA.6.6.09.

Jacchia, L. (1971). Revised Static Models of the Thermosphere and Exosphere with Empirical Temperature Profiles. Technical Report SAO Special Report 332, Smithsonian Astrophysical Observatory, Cambridge, MA.

Jacchia, L. (1977). Thermospheric Temperature, Density and Composition. Technical Report SAO Special Report 375, Smithsonian Astrophysical Observatory, Cambridge, MA.

Jehn, R., Hernández, C., Klinkrad, H., Flury, W., and Agapov, V. (2004). Space Traffic Management in the Geostationary Ring. *55th IAF Congress*, IAA-04-IAA.5.12.4.05.

Jenniskens, P. (1991). *Discoveries from Observations and Modeling of the 1998/99 Leonids*, pages 233–252. In (Grün et al., 2001).

Jenniskens, P. (1994). Meteor Stream Activity – I. The Annual Meteor Streams. *Journal of Astronomy and Astrophysics*, vol.287:990–1013.

Johnson, N., Krisko, P., Liou, J., and Anz-Meador, P. (2001a). NASA's New Break-Up Model of EVOLVE 4.0. *Advances in Space Research*, vol.28, no.9:1377–1384.

Johnson, N. and McKnight, D. (1991). *Artificial Space Debris*. Krieger Publishing Company, Malabar, FL, second edition.

Johnson, N. L., Anz-Meador, P., Cisek, E., and Portman, S. (2001b). History of On-Orbit Satellite Fragmentations, 12th edition. Technical Report JSC 29517, Orbital Debris Program Office, NASA/JSC, Houston, TX.

Kaula, W. (1966). *Theory of Satellite Geodesy*. Blaisdell Publishing Company, Waltam, MA/Toronto/London.

Kerridge, D., Carlaw, V., and Beamish, D. (1989). Development and Testing of Computer Algorithms for Solar and Geomagnetic Activity Forecasting. Technical Report WM/89/22C [ESA CR(P) 3039], British Geological Survey.

Kessler, D. (1981). Derivation of Collision Probability Between Orbiting Objects: The Lifetimes of Jupiter's Outer Moons. *Icarus*, vol.48:39–48.

Kessler, D. (1985). The Effects of Particulates from Solid Rocket Motors Fired in Space. *Advances in Space Research*, vol.5:77–86.

Kessler, D. (1991). Collisional Cascading: The Limits of Population Growth in Low Earth Orbit. *Advances in Space Research*, vol.11, no.12:63–66.

Khutorovsky, Z., Boikov, V., and Kamensky, S. (1993). Direct Method for the Analysis of Collision Probability of Artificial Space Objects in LEO – Techniques, Results, and Applications. In *Proceedings of the First European Conference on Space Debris*, ESA SD-01, pages 491–508.

King-Hele, D. (1987). *Satellite Orbits in an Atmosphere: Theory and Applications*. Blackie, Glasgow, UK.

Klinkrad, H. (1991). Salyut-7/Cosmos-1686 Re-Entry Prediction Activities at ESOC. In *Proceedings of an International Workshop on the Salyut-7/Cosmos-1686 Re-Entry*, ESA SP-345, pages 17–34.

Klinkrad, H. (1993). Collision Risk Analysis for Low Earth Orbits. *Advances in Space Research*, vol.13:551–557.

Klinkrad, H. (1997). One Year of Conjunction Events of ERS-1 and ERS-2 with Objects of the USSPACECOM Catalog. In *Proceedings of the Second European Conference on Space Debris*, ESA SP-393, pages 601–611.

Klinkrad, H. (1999). Evolution of the On-Ground Risk During Uncontrolled Re-Entries. *50th IAC Congress*, IAA-99-IAA.6.7.05.

Klinkrad, H. (2002). Monitoring Space – Efforts Made by European Countries. In *Proc. of an International Colloquium on "Europe and Space Debris"*, ANAE Proceedings (on CD ROM).

Klinkrad, H. (2004). A Standardized Method for Re-Entry Risk Evaluation. *55th IAC Congress*, IAC-04-IAA.5.12.2.07.

Klinkrad, H. et al. (2003). *ESA Space Debris Mitigation Handbook*. ESA/ESOC, second edition. issue 1.0.

Kompanietz, E. et al. (2000). Minimization of Space Debris Formation on Zenith-3SL ILV Launches Within the Sea Launch Program. *51st IAF Congress*, IAA-00-IAA.6.6.05.

Koppenwallner, G., Fritsche, B., and Lips, T. (2001). Survivability and Ground Risk Potential of Screws and Bolts of Disintegrating Spacecraft During Uncontrolled Re-Entry. In *Proceedings of the Third European Conference on Space Debris*, ESA SP-473, pages 533–539.

Krag, H., Bendisch, J., Rosebrock, J., Schildknecht, T., and Sdunnus, H. (2002). Sensor Simulation for Debris Detection. Technical Report (final report) ESA contract 14708/00/D/HK, ILR/TUBS, Braunschweig.

Lemoine, F., Kenyon, S., Factor, J., Trimmer, R., Pavlis, N., Chinn, D., Cox, C., Klosko, S., Luthcke, S., Torrence, M., Wang, Y., Williamson, R., Pavlis, E., Rapp, R., and Olson, T. (1998). The Development of the Joint NASA GSFC and NIMA Geopotential Model EGM96. Technical Report NASA TP-1998-206891, NASA/GSFC.

Liou, J., Matney, M., Anz-Meador, P., Kessler, D., Jansen, M., and Theall, J. (2001). The New NASA Orbital Debris Engineering Model ORDEM2000. In *Proceedings of the Third European Conference on Space Debris*, ESA SP-473, pages 309–313.

Liou, J.-C., Hall, D., Krisko, P., and Opiela, J. (2004). LEGEND – A Three-Dimensional LEO-to-GEO Debris Evolutionary Model. *Advances in Space Research*, vol.34, no.5:981–986.

Liu, J. (1974). Satellite Motion About an Oblate Earth. *AIAA Journal*, vol.12, no. 11:1511–1516.

Liu, J. and Alford, R. (1979). A Semi-Analytic Theory for the Motion of a Close-Earth Artificial Satellite with Drag. In *17th Aerospace Sciences Meeting, New Orleans*, AIAA Paper No. 79-0123.

Maclay, T. and McKnight, D. (1994). The Contribution of Debris Wakes from Resident Space Objects in the Orbital Debris Environment. *Safety & Rescue Science & Technology Series*, vol.88:215–228.

Marcos, F. (1993). Satellite Drag Models – Current Status and Prospects. *AAS/AIAA Conference*, paper AAS-93-621.

Matney, N., Anz-Meador, P., and Foster, J. (2002). Covariance Correlations in Collision Avoidance Probability Calculations. *COSPAR Congress*, PEDAS1-B1.4-0038-02.

McBride, N. (1997). The Importance of the Annual Meteoroid Streams to Spacecraft and their Detectors. *Advances in Space Research*, vol.20, no.8:1513–1516.

McDonnell, J. (1992). *Hypervelocity Impacts in Space.* University of Kent at Canterbury, Canterbury, UK.

McDonnell, J. et al. (1999). Meteoroid and Debris Flux and Ejecta Models. Technical Report (final report) ESA contract 1887/96/NL/JG, Unispace Kent, Canterbury, UK.

McDonnell, J., McBride, N., and Gardner, D. (1997). The Leonid Meteoroid Stream: Spacecraft Interactions and Effects. In *Proceedings of the Second European Conference on Space Debris*, ESA SP-393, pages 391–396.

McNish, A. and Lincoln, J. (1949). Prediction of Sunspot Numbers. *Transactions of the American Geophysical Union*, vol.30:673–685.

Merz, K. (2003). On the Stability of Near Semi-Synchronous, Near Circular Orbits. Technical Report no. 459, issue 2, rev. 0, ESA/ESOC/MAO.

Meyer, R. (1992). In-Flight Formation of Slag in Spinning Solid Propellant Rocket Motors. *Journal of Propulsion & Power*, vol.8, no.1:45–50.

Montenbruck, O. and Gill, E. (2000). *Satellite Orbits – Models, Methods, and Applications.* Springer, Berlin, Heidelberg, New York.

Montgomery, R. and Ward, J. (1995). Casualty Areas from Impacting Inert Debris for People in the Open. Technical Report RTI/5180/60-31F, Research Triangle Institute.

Morton, T. and Ferguson, D. (1993). Atomic Oxygen Exposure of Power System and Other Spacecraft Materials: Results of the EOIM-3 Experiment. Technical Report NASA TM 107427, NASA Lewis Research Center.

Nazarenko, A. and Menshikov, I. (2001). Engineering Model of the Space Debris Environment. In *Proceedings of the Third European Conference on Space Debris*, ESA SP-473, pages 293–298.

Ojakangas, G. et al. (1996). Solid-Rocket-Motor Contributions to the Large-Particle Orbital Debris Population. *Journal of Spacecraft & Rockets*, vol.33, no.4:513–518.

Opiela, J. and Krisko, P. (2001). Evaluation of Orbital Debris Mitigation Practices Using EVOLVE 4.1. *51st IAF Congress*, IAA-01-IAA.6.6.04.

Patera, R. (2001). General Method for Calculating Satellite Collision Probability. *Journal of Guidance, Control & Dynamics*, vol.24:716–722.

Peterson, G. (1999). *Dynamics of Meteor Outbursts and Satellite Mitigation Strategies.* The Aerospace Press, El Segundo, CA.

Portree, D. and Loftus, J. (1993). Orbital Debris and Near-Earth Environmental Management: A Chronology. Technical Report 1320, NASA Reference Publication.

Rawer, K., editor (1971). *Encyclopedia of Physics*, volume XLIX/3 of *Geophysics III*. Springer Verlag, Berlin, Heidelberg, New York.

Rochelle, W., Marichalar, J., and Johnson, N. (2004). Analysis of Re-Entry Survivability of UARS Spacecraft. *Advances in Space Research*, vol.34, no.5:1049–1054.

Rossi, A., Cordelli, A., Pardini, C., Anselmo, L., Farinella, P., and Jehn, R. (1995a). Long-Term Evolution of Earth Orbiting Objects. *47th IAF Congress*, IAA-95-IAA.6.4.06.

Rossi, A., Cordelli, A., Pardini, C., Anselmo, L., and Parinella, P. (1995b). Modeling of the Space Debris Evolution: Two New Computer Models. *Advances in the Astronautical Sciences (Space Flight Mechanics)*, pages 1217–1231.

Roth, E. (1996). Construction of a Consistent Semi-analytic Theory of a Planetary or Moon Orbiter Perturbed by a Third Body. *Celestial Mechanics*, vol.28:155–169.

Samson, P. (1999). Classification of Geostationary Objects – Users Guide. Technical Report no. 420, ESA/ESOC/MAO.

Sánchez-Ortiz, N., Belló-Mora, M., and Klinkrad, H. (2004). Collision Avoidance Maneuvers During Spacecraft Lifetime – Analysis of Operational Missions. *55th IAC Congress*, IAC-04-IAA.5.12.5.10.

Scala, K. (1998). A History of Nuclear Space Accidents. *Spaceflight*, vol.40.

Schildknecht, T., Ploner, M., and Hugentobler, U. (2001). The Search for Debris in GEO. *Advances in Space Research*, vol.28, no.9:1291–1299.

Seidelmann, P. et al., editors (1992). *Explanatory Supplement to the Astronomical Almanac*. University Science Books, Mill Valley, CA.

Skolnik, M. I. (1990). *Radar Handbook*. McGraw-Hill, Boston, MA, second edition.

Smith, P. (1999). Expected Casualty Calculations for Commercial Space Launch and Re-Entry Missions. Technical Report Advisory Circular No. 431.35-1, US Department of Transportation & Federal Aviation Administration.

Springmann, P. and de Weck, O. (2004). Parametric Scaling Model for Non-Geosynchronous Communication Satellites. *Journal of Spacecraft and Rockets*, vol.41, no.3:472–477.

Stansbery, G., Matney, M., Settecerri, T., and Bade, A. (1997). Debris Families Observed by the Haystack Orbital Debris Radar. *Acta Astronautica*, vol.41, no.1:53–56.

Staubach, P., Grün, E., and Matney, M. (2001). *Synthesis of Observations*, pages 345–384. In (Grün et al., 2001).

Steel, D. (2000). *Target Earth*. Quarto Publishing, London, UK.

Stickland, A., editor (1972). *COSPAR International Reference Atmosphere 1972*. Akademie Verlag, Berlin.

Terrillon, F., Warren, H., and Yelle, M. (1991). Orbital Debris Shielding Design of the RADARSAT Satellite. *42nd IAF Congress*, IAF-91-283.

Thomson, A., Clark, T., and Kerridge, D. (1992). Computer Algorithms and FORTRAN Programs for Forecasting Solar and Geomagnetic Activity in the Short Term. Technical Report WM/92/19C, British Geological Survey.

Tobler, W., Deichmann, U., Gottsegen, J., and Maloy, K. (1995). The Global Demography Project. Technical Report Technical Report TR-95-6, National Center for Geographic Information and Analysis, Santa Barbara, CA.

Turner, R., Taylor, E., McDonnell, J., Stokes, H., Marriot, P., Wilkinson, J., Catling, D., Vignjevic, R., Berthoud, L., and Lambert, M. (2002). Cost-Effective Debris Shields for Unmanned Spacecraft. In *International Journal of Impact Engineering*, volume vol.28, no.2, pages 365–378.

Vallado, D. (1997). *Fundamentals of Astrodynamics and Applications*. McGraw-Hill, New York, NY.

Verger, F., Ghiraradi, R., and Sourbès-Verger, I. (1997). *Atlas de la Géographie de l'Espace*. Éditions Belin, Paris, France.

Walker, J. (1971). Some Circular Orbit Patterns Providing Continuous Whole Earth Coverage. *Journal of the British Interplanetary Society*, vol.24:269–384.

Walker, R., Stokes, P., Wilkinson, J., and Swinerd, G. (1999). Enhancement and Validation of the IDES Orbital Debris Environment Model. *Space Debris Journal*, vol.1, issue 1:1–20.

Walker, R., Swinerd, G., Wilkinson, J., and Martin, C. (2000). Long Term Space Debris Environment Prediction. Technical Report (final report) ESA contract 12808/98/D/IM(SC), DERA, Farnborough, UK.

Wiedemann, C., Oswald, M., Stabroth, S., Vörsmann, P., and Klinkrad, H. (2004). NaK Droplet Size Distribution. Technical Report ILR-IB-2004-001, rel 0.9, Institute of Aerospace Systems, TU Braunschweig.

Zhang, W. (1997). Elimination of the Potential Hazard of the Long March 4 Launch Vehicle On-Orbit Break-Up. In *Proceedings of the Second European Conference on Space Debris*, ESA SP-393, pages 689–691.

Index

Printing: Mercedes-Druck, Berlin
Binding: Stein + Lehmann, Berlin

Printed in Great Britain
by Amazon

41578616R00251